Agricultural Land Use: Planning and Management

Agricultural Land Use: Planning and Management

Editor: Isabella Davis

RCALLISTO
REFERENCE

www.callistoreference.com

Callisto Reference,
118-35 Queens Blvd., Suite 400,
Forest Hills, NY 11375, USA

Visit us on the World Wide Web at:
www.callistoreference.com

ISBN: 978-1-63239-799-7 (Hardback)

Cataloging-in-publication Data

Agricultural land use : planning and management / edited by Isabella Davis.
   p. cm.
Includes bibliographical references and index.
ISBN 978-1-63239-799-7
1. Agriculture. 2. Land use, Rural--Planning. 3. Farm management. 4. Agricultural geography. I. Davis, Isabella.
S494.5.S95 A37 2017
630--dc23

Table of Contents

Preface

This book provides comprehensive insights into the field of agricultural land use. It strives to provide a fair idea about this discipline and to help develop a better understanding of the latest advances within this field. Agricultural land use refers to the optimum utilization of land for livestock management, food production and other agricultural activities. It comprises of planning methods for arable land, permanent cropland, permanent pastures, agriculturally-zoned area, etc. The planning and management of farmland is required to strike a balance between farming and environmental conservation. Most of the topics introduced in this book cover new techniques and the applications of agricultural land use. Those with an interest in this field would find it helpful. Coherent flow of topics, student-friendly language and extensive use of examples make this text an invaluable source of knowledge.

The information contained in this book is the result of intensive hard work done by researchers in this field. All due efforts have been made to make this book serve as a complete guiding source for students and researchers. The topics in this book have been comprehensively explained to help readers understand the growing trends in the field.

I would like to thank the entire group of writers who made sincere efforts in this book and my family who supported me in my efforts of working on this book. I take this opportunity to thank all those who have been a guiding force throughout my life.

Editor

Underperformance of African Protected Area Networks and the Case for New Conservation Models: Insights from Zambia

Peter A. Lindsey[1,2]*, **Vincent R. Nyirenda[3]**, **Jonathan I. Barnes[4]**, **Matthew S. Becker[5,6]**, **Rachel McRobb[7]**, **Craig J. Tambling[8]**, **W. Andrew Taylor[9]**, **Frederick G. Watson[6,10]**, **Michael t'Sas-Rolfes[11]**

1 Panthera, New York, New York, United States of America, 2 Mammal Research Institute, Department of Zoology and Entomology, University of Pretoria, Pretoria, Gauteng, South Africa, 3 Zambia Wildlife Authority, Chilanga, Lusaka, Zambia, 4 Design & Development Services, Windhoek, Namibia, 5 Department of Ecology, Montana State University, Bozeman, Montana, United States of America, 6 Zambian Carnivore Programme, Mfuwe, Zambia, 7 South Luangwa Conservation Society, Mfuwe, Zambia, 8 Centre for African Conservation Ecology, Department of Zoology, Nelson Mandela Metropolitan University, Port Elizabeth, South Africa, 9 Centre for Veterinary Wildlife Studies, Faculty of Veterinary Science, University of Pretoria, Pretoria, South Africa, 10 Division of Science and Environmental Policy, California State University Monterey Bay, Seaside, California, United States of America, 11 Cape Town, South Africa

Abstract

Many African protected areas (PAs) are not functioning effectively. We reviewed the performance of Zambia's PA network and provide insights into how their effectiveness might be improved. Zambia's PAs are under-performing in ecological, economic and social terms. Reasons include: a) rapidly expanding human populations, poverty and open-access systems in Game Management Areas (GMAs) resulting in widespread bushmeat poaching and habitat encroachment; b) underfunding of the Zambia Wildlife Authority (ZAWA) resulting in inadequate law enforcement; c) reliance of ZAWA on extracting revenues from GMAs to cover operational costs which has prevented proper devolution of user-rights over wildlife to communities; d) on-going marginalization of communities from legal benefits from wildlife; e) under-development of the photo-tourism industry with the effect that earnings are limited to a fraction of the PA network; f) unfavourable terms and corruption which discourage good practice and adequate investment by hunting operators in GMAs; g) blurred responsibilities regarding anti-poaching in GMAs resulting in under-investment by all stakeholders. The combined effect of these challenges has been a major reduction in wildlife densities in most PAs and the loss of habitat in GMAs. Wildlife fares better in areas with investment from the private and/or NGO sector and where human settlement is absent. There is a need for: elevated government funding for ZAWA; greater international donor investment in protected area management; a shift in the role of ZAWA such that they focus primarily on national parks while facilitating the development of wildlife-based land uses by other stakeholders elsewhere; and new models for the functioning of GMAs based on joint-ventures between communities and the private and/or NGO sector. Such joint-ventures should provide defined communities with ownership of land, user-rights over wildlife and aim to attract long-term private/donor investment. These recommendations are relevant for many of the under-funded PAs occurring in other African countries.

Editor: Danilo Russo, Università degli Studi di Napoli Federico II, Italy

Funding: The funding for this research was provided by the Wildlife Producers Association of Zambia. CJT was funded with a Claude Leon Fellowship. The funders had no role in study design, data collection and analysis, decision to publish, or preparation of the manuscript.

Competing Interests: JB is employed by Design and Development Services. PL was employed as a consultant during the research which provided the data used to compile this paper. There are no patents, products in development or marketed products to declare.

* E-mail: plindsey@panthera.org

Introduction

Many African countries have designated generous proportions of their land surface as protected areas. Such protected areas vary greatly in their makeup, from strictly protected areas with no human settlement to areas that have resident communities where multiple uses of wildlife are permitted. African governments find it difficult to fund protected area networks adequately and are facing severe threats from poaching and human encroachment [1,2]. These problems are pronounced where human settlement is permitted or tolerated inside protected areas, as occurs in parts of Ethiopia, Mozambique, Tanzania and Zambia, for example [3,4].

Zambia has a vast wildlife estate encompassing 20 national parks (~64,000 km²), 3 wildlife and bird sanctuaries (33.5 km²), 36 GMAs (167,000 km²) and several other protected area categories, comprising ~40% of the nation's land area [5] (Figure 1). Human settlement is not permitted in national parks, and land use is limited primarily to photo-tourism. National parks have generally not suffered from human encroachment, but are subject to widespread poaching, regular uncontrolled burning (which sometimes emanates from areas outside of the park boundaries) and in some cases, informal mining [6]. With the exception of Lusaka and Mosi-oa-tunya national parks, no protected areas in Zambia are fenced and most are simply demarcated with cut-lines or rivers, and in some cases, beacons.

GMAs were established as buffer-zones for national parks and have been used primarily for trophy hunting in recent years [7]. Unlike in the national parks, settlement is permitted in GMAs and there are large and expanding human populations in many of them, which is accompanied by widespread habitat loss. Habitat destruction is exacerbated by shifting agriculture, charcoal production and in some cases, mining [7,8,9]. In both national parks and GMAs, wildlife is under severe pressure from poaching, both for bushmeat and for trophies such as ivory [10].

In the 1980s, there was recognition of a need for greater community participation in wildlife-based land uses in GMAs [8]. In the early 1980s, subsidiary legislation was introduced to partially decentralize authority over wildlife to communities [8]. The Zambia Wildlife Act of 1998 provided for establishment of ZAWA as a parastatal responsible for managing protected areas [9]. The Wildlife Act identified Community Resource Boards as the institutions for communities to co-manage and benefit from

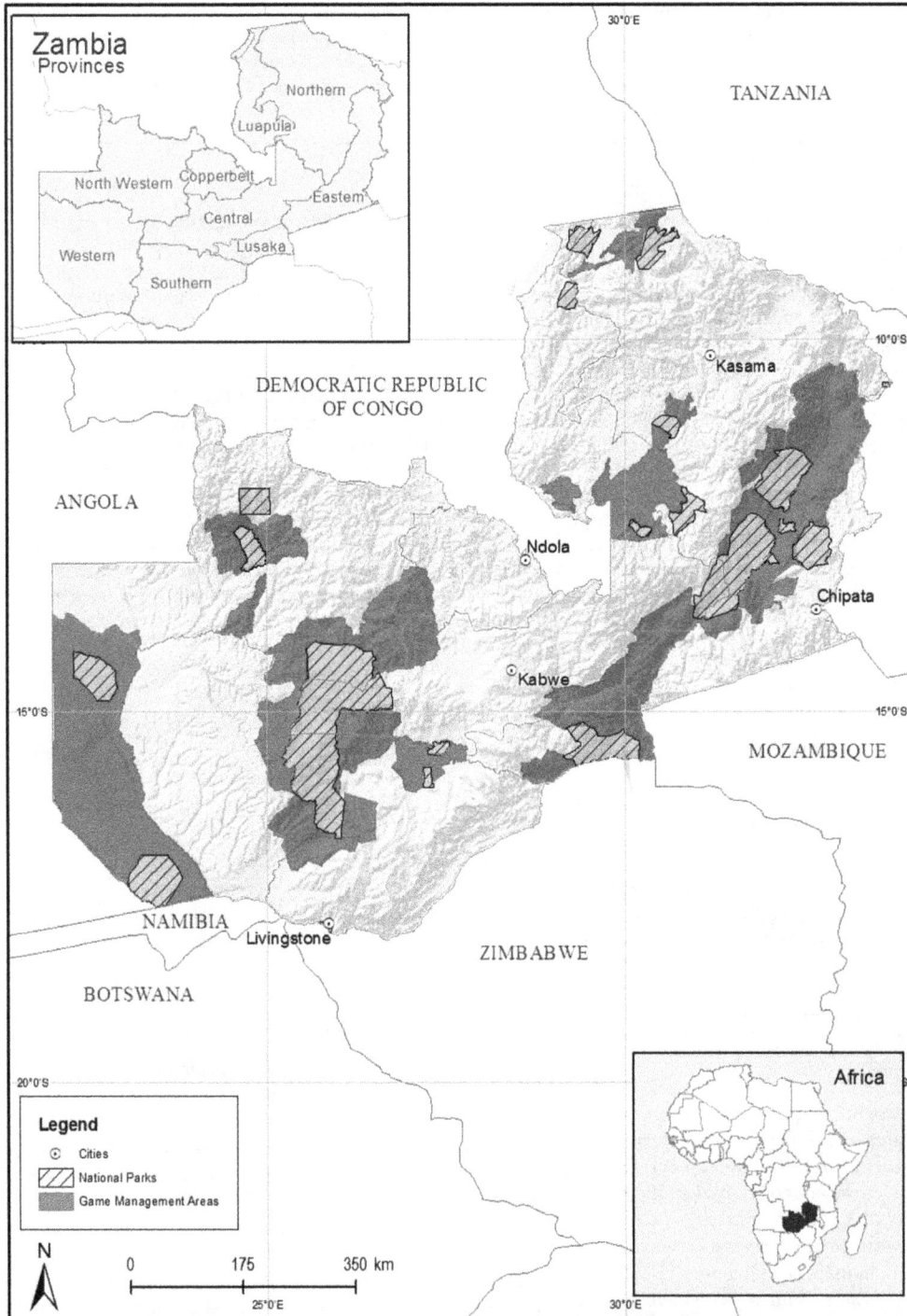

Figure 1. The Zambian protected area network.

wildlife in GMAs [9] though no mechanisms were created to enable communities to benefit from wildlife in national parks.

Wildlife-based land uses in the PA network have potential to improve livelihoods significantly for communities. People in GMAs are poorer and less educated than the national average, and GMAs have low agricultural potential and offer few alternative livelihood opportunities [11]. Trophy hunting in the GMAs has potential to generate significant incomes for communities if wildlife populations are allowed to recover and systems are put in place to ensure equitable benefit sharing and best-practices [12]. Similarly, national parks have potential to benefit rural communities through tourism-related employment and business opportunities. The PA network as a whole has enormous potential to contribute to rural and national economic growth by providing the basis for development of a major tourism industry [13]. However, wildlife populations are waning in many GMAs and national parks, and incomes from both trophy hunting and photo-tourism are limited to fractions of the GMAs and national parks [7]. In addition, mechanisms to enable communities to benefit legally from the PA area network are limited. Consequently the PA network is under-performing in ecological, economic and social terms.

There have been several attempts by the Zambian government to address the underperformance of the protected area network. For example, in 2006, the Zambian Government embarked upon a reclassification programme for protected areas [5] and in early 2013, a moratorium was imposed on hunting in GMAs. In addition, two protected areas have been added to the estate in recent years: the ~50 km^2 Lusaka National Park and the 5,104 km^2 Mukungule GMA. However, key challenges with regards to the functioning and effectiveness of Zambia's PA network remain.

In this paper we provide evidence of the under-performance of the Zambian protected area network, give reasons for that performance and suggest interventions needed to make the system more effective in ecological, economic and social terms. These recommendations have relevance for the PA networks of many other African countries.

Results and Discussion

Ecological Indicators of Protected Area Performance

Human encroachment of protected areas in Zambia is worse than in most other African countries [14], and ~2,500–3,000 km^2 of land are deforested annually [15]. Human population growth rates in GMAs (2.49±0.18%) are higher than elsewhere (2.31±0.24%, T-test 0.577, d.f. = 70, p = 0.566) [16]. Almost 40% of the total area of GMAs is now comprised of human-modified habitat (c.f. 71.2% outside of the protected area network) [6]. By contrast, habitat loss in national parks is limited (2.1%) (Table 1). The rate of habitat loss in GMAs (0.69% conversion per year) is faster than in national parks (0.05%) or outside protected areas (0.51%) (Table 1, Figure 2, [6]). Extrapolating from [6]'s sample area (Table 1, Figure 2), ~82 hectares of habitat are lost per daylight hour in GMAs on a national level. Human encroachment in GMAs is advancing from main roads towards national parks at a rate of up to 2 km per year ([6]). In some protected areas, and most notably Lukusuzi National Park [17], mining activity is evident and has potential to affect wildlife populations adversely through habitat degradation and bushmeat poaching [18]. In addition to habitat loss, human encroachment undermines the buffer zone role of GMAs for national parks, jeopardizes ecological connectivity among PAs and the concept of transfrontier conservation areas ([6]).

Table 1. Estimates of the extent and rate of habitat conversion in GMAs (from natural to human-modified habitat), national parks and land outside the protected area network in Zambia (taken from data extracted from [6]).

Land type	Total in study area (km²)	~1970 Area human (km²)	~1970 % human	~1985 Area human (km²)	~1985 % human	~2010 Area human (km²)	~2010 % human	~1970–~1985 Increase per year (km²)	~1970–~1985 % increase per year	~1980–~2010 Increase per year (km²)	~1980–~2010 % increase per year
Whole study area	159805	58926	36.9%	60935	38.1%	80157	50.2%	134	0.08%	769	0.48%
NPs	27098	252	0.9%	257	1.0%	571	2.1%	0	0.00%	13	0.05%
GMAs	47430	9468	20.0%	10616	22.4%	18744	39.5%	77	0.16%	325	0.69%
Non-NP, Non-GMA	85277	49206	57.7%	50061	58.7%	60841	71.3%	57	0.07%	431	0.51%
Luangwa Valley GMAs	26502	6815	25.7%	6190	23.4%	8878	33.5%	−42	−0.16%	108	0.41%
Southern Kafue GMAs	20928	2652	12.7%	4427	21.2%	9866	47.1%	118	0.57%	218	1.04%

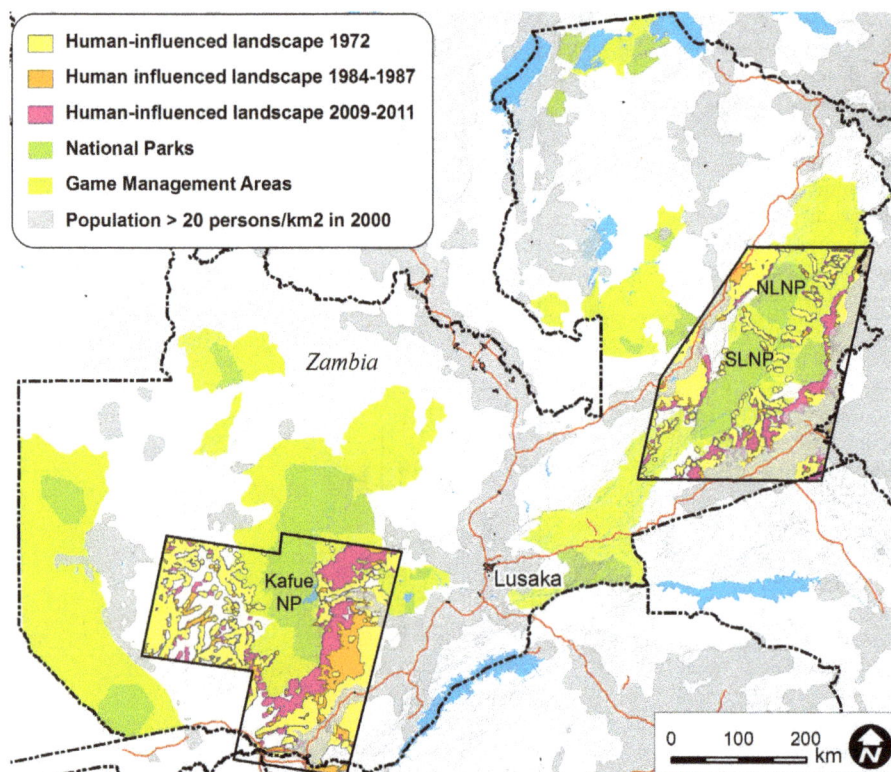

Figure 2. The extent of human encroachment of natural habitat in two focal areas in Zambia extracted from [6].

Data from aerial censuses indicate that wildlife populations in Zambian protected areas are relatively low (Table 2): ~169,000 wild ungulates occur in the ~61,000 km^2 of Zambian national parks for which data are available (excluding species of the size of a bushbuck or smaller and hippos) and ~143,000 in the ~160,000 km^2 of GMAs for which data were available. By contrast, ~63,000 ungulates occur on <6,000 km^2 of game ranches in Zambia [19]. Country-level population data for wildlife in other countries are scarce, but to provide a coarse comparisons, 1.8–2.8 million wild ungulates occur on 287,000 km^2 of Namibian wildlife ranches ~841,000 ungulates occurred on 27,000 km^2 of Zimbabwean game ranches prior to the land seizures [20] and ~215,000 ungulates occur in the ~20,000 km^2 Kruger National Park in South Africa [21,22].

The biomass of large wild ungulates is lower in GMAs (mean 212±59 kg/km^2) and national parks (791±240 kg/km^2) than in extensive game ranches (2,424±305 kg/km^2) (which are devoid of human settlement and rely primarily on trophy hunting for income) (Figure S1a) [19]. The diversity of wild ungulates is also lower in GMAs (4.7±0.58 species) and national parks (7.2±0.9 species) than on extensive unfenced game ranches 11.1±0.86 species) (Figure S1b) [19]. The higher biomass and diversity on private ranches is likely to be primarily due to the availability of greater resources for anti-poaching than in state protected areas. These findings reinforce the suggestion that trophy hunting need not have a negative impact on wildlife populations given appropriate land tenure arrangements [23].

Combining data from GMAs, national parks and extensive game ranches, wildlife ungulate biomass was negatively related to the presence of human settlement (in areas with settlement mean biomass was 268±70.8 kg/km^2 c.f. 1,755±281 kg/km^2), as was wild ungulate diversity (5.20±0.61 species c.f. 9.68±0.7) (F Ratio 26.2, d.f. = 2, p<0.001). Wild ungulate biomass was positively related to investment by the private sector/NGOs (in areas with such support, mean biomass was 1,592±222 kg/km^2 c.f. 233±113 kg/km^2 in areas without such investment, as was wild ungulate diversity (9.1±0.71 species c.f. 5.0±0.65) (F Ratio = 37.0, d.f. = 2, p<0.01) (Figures S2a, S2b).

Observed biomasses of large mammals in Zambian protected areas were lower than potential maximum carrying capacities by 93.7% in GMAs and 74.1% in national parks (Figures 3a, 3b). Wildlife densities are suppressed in some of Zambia's 'flagship' national parks and GMAs. For example, biomasses in Kafue, South Luangwa and Lower Zambezi national parks are 29%, 16% and 23% of potential carrying capacity, whereas that in Lupande GMA (which has in the recent past been categorised as a 'super-prime' concession), stands at 11% of carrying capacity. Depressed prey populations means that predator populations are almost certainly also occurring well below historic densities.

Population trend data are generally not available for Zambian protected areas. An exception is the Luangwa ecosystem, where the biomass of wildlife declined significantly in all five GMAs and four national parks during 2011–2012, and 80% of species showed declining trends during 2009–2012 [24]. By contrast, in Liuwa Plains (co-managed by African Parks/ZAWA since 2003), wildlife populations have recovered and large mammal biomass (excluding hippos and species of bushbuck size and smaller) increased from 966 kg/km^2 in 2003 to 1,921 kg/km^2 in 2013 [25].

Declining wildlife populations in GMAs have been reflected in changes to the way ZAWA classifies GMAs. In 2008, 24 GMAs were classified as depleted or secondary whereas in 1997, only 16 were categorized as such [7].

Table 2. Estimated wildlife populations in National Parks (data available for 61,462 km^2 of the ~64,000 km^2), Game Management Areas (GMAs, data available for 159,654 km^2 of the ~167,000 km^2) and game ranches (5,829 km^2) in Zambia (excluding species of bushbuck size and smaller, and hippopotamuses for which count data were not available) (data taken from [19]).

Species	National parks	GMAs	Game ranches	Total
Lechwe	9,737	75,808	1,513	87,058
Impala	27,820	13,507	27,998	69,325
Wildebeest	47,815	4,069	630	52,514
Buffalo	21,301	15,938	2,107	39,346
Puku	16,838	7,529	4,904	29,271
Elephant	10,830	8,094	1,710	20,634
Sable	8,172	4,895	3,682	16,749
Zebra, plains	8,375	1,050	2,060	11,485
Waterbuck	5,254	2,333	2,987	10,574
Kudu	1,908	1,976	6,287	10,171
Hartebeest	4,429	3,952	2,051	10,432
Roan	2,384	1,632	1,647	5,663
Reedbuck	1,137	852	2,735	4,724
Eland	1,069	237	1,558	2,864
Tsessebe	990	88	410	1,488
Giraffe	579	178	321	1,078
Sitatunga	40	369	328	737
Nyala	0	0	95	95
	168,678	142,507	63,023	374,208

Economic Indicators of Park Performance

Zambia attracts fewer tourists than other African countries known for wildlife: South Africa (~9.2 million), Botswana (~2.3 million), Zimbabwe (~2.2 million), Kenya (~1.8 million), Mozambique (~1.6 million), Tanzania (~1.25 Million), Namibia (~1.1 million), Zambia (~0.9 million) (www.wttc.org, accessed July 2013). In 2005 (when the latest data on tourist visitation to parks were available), just 59,350 tourists visited Zambia's national parks and 95% of those visited just five parks (Kafue, South Luangwa, Mosi-oa-Tunya, Lower Zambezi and Lochinvar): the remaining 15 parks comprising ~32,000 km^2 attracted just 1,384 tourists [26,27]. Most national parks thus fail to attract enough tourists to support viable photo-tourism [27]. Earnings from national parks and GMAs in Zambia are low. Photo-tourism generated USD9.1 million for ZAWA in 2011 (USD142/km^2) and UDD9.4 million in 2012 (USD146/km^2) (ZAWA unpublished data). The majority of those earnings (96.9%, USD154/km^2) were from four parks (South Luangwa – 44.4%, Mosi-o-tunya – 23.7%, Kafue – 14.7% and Lower Zambezi – 14.1%), which comprise 36,681 km^2 (57.3%) of the national parks network (ZAWA unpublished data). The remainder of the parks generated just USD182,229 (USD6.7/km^2) due to the relative (or actual) absence of tourism operators. Photo-tourism operations exist in only 10 of 36 GMAs [7] and in a marginal capacity given they currently are unable to bid for concessions in GMAs. Photo-tourism operations in GMAs are generally limited to lodges on national park boundaries with little investment or activity in the GMAs.

From trophy hunting income, ZAWA only earned concession fees from ~51% of the GMA estate in 2012 (ZAWA unpublished data). ZAWA earned a total of USD4.34 million from trophy hunting and resident hunting respectively in 2012 (equating to USD26/km^2) (ZAWA unpublished data). We estimate that an additional USD16.1 million was earned by operators from trophy hunting in GMAs, yielding total earnings of USD97/km^2 (averaged across all GMAs including those not hunted, Table 3). Earnings per km^2 from trophy hunting (when excluding hunting areas that do not generate any income to allow for comparison with other countries) (US$291±116/km^2) are lower than most other SADC countries: Zimbabwe – USD1,028/km^2; Tanzania – USD424/km^2; Namibia – USD378/km^2; Mozambique – USD130/km^2, [28] (F-Ratio 11.0, $d.f. = 1$, $p = 0.001$). Within Zambia, gross earnings per km^2 from trophy hunting in GMAs (USD97/km^2) are markedly lower than on extensive game ranches (USD878±226) (F Ratio 15.9, d.f. = 1, p<0.001) [19]. Contributing to the low earnings from hunting in Zambia in 2012 was the fact that 19 leases for hunting concessions ended prior to the end of the hunting season, and resident hunting was halted by the Zambian government (ZAWA unpublished data).

Social Indicators

The primary benefit to communities from national parks is employment and the development of wildlife-based economies around tourism hubs such as the growth point of Mfuwe, adjacent to South Luangwa National Park. Tourism generates an estimated 19,000 jobs in Zambia [27], though the proportion derived specifically from parks is not clear. In Mfuwe, ~900 jobs are created and such workers receive mean salaries of US$450/month, ~3x the minimum wage (A. Coley, pers. comm.). Tourism-related employment is significant because it can result in improved attitudes towards wildlife conservation [29]. However, most parks generate virtually no employment for communities due to the lack or small-scale of tourism operations.

In GMAs, benefits accruing to communities from trophy hunting include income generation for Community Resource

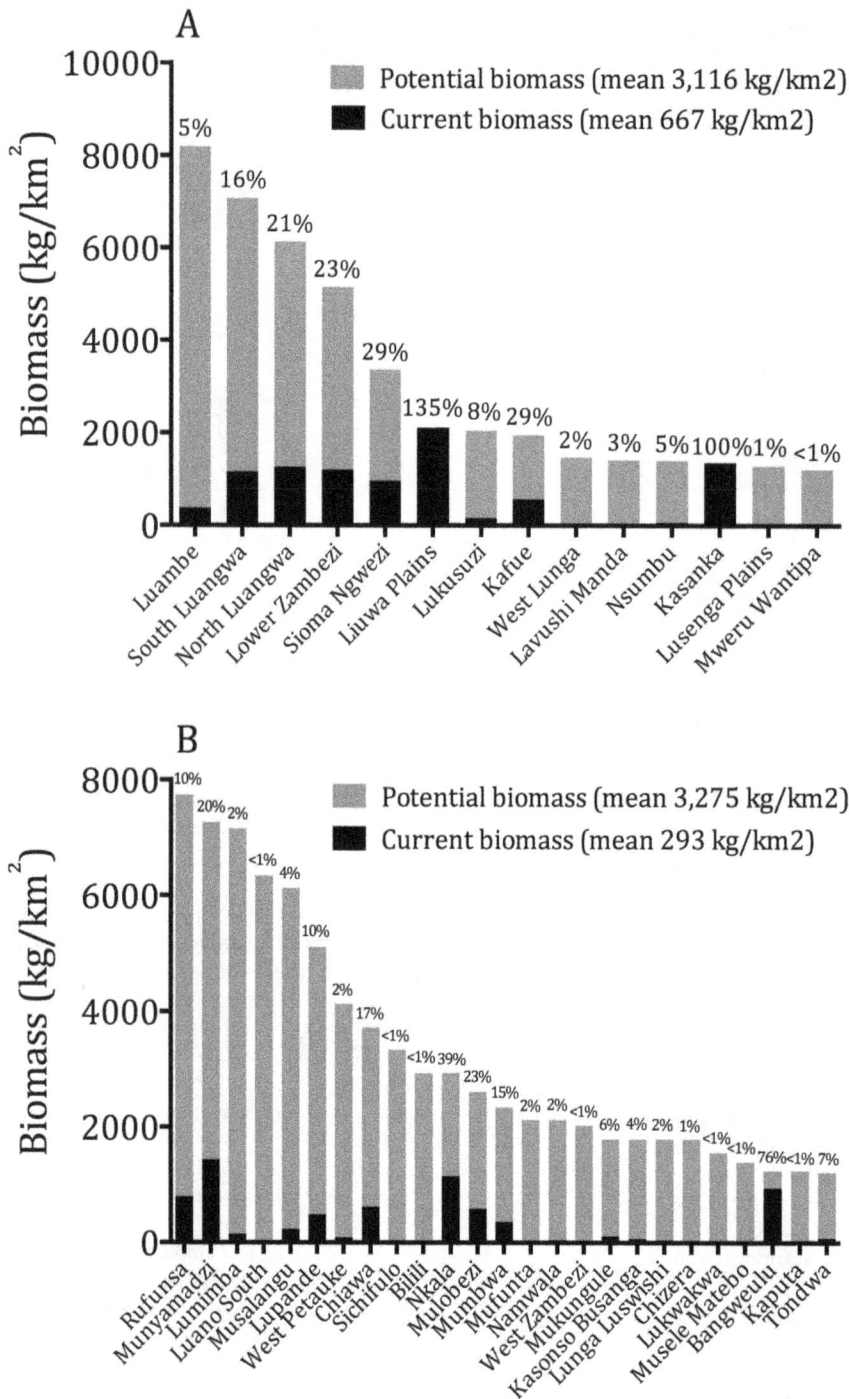

Figure 3. Observed large mammal biomass versus potential carrying capacity in Zambian (a) national parks and (b) GMAs.

Boards, employment and in some cases, various forms of development assistance from hunting operators. In some GMAs, such as those in the Luangwa Valley and Bangweulu system, livelihood improvements associated with income from trophy hunting are significant [30]: families in the most wildlife-rich GMAs are ~17% better off than those outside of the GMAs, and have a 7.8% higher chance of obtaining employment [8]. However, no earnings whatsoever are generated in half of the GMAs, and average earnings accruing to communities across all GMAs are low (USD11.9/km²) (ZAWA unpublished data,

Table 3). Concurrently, communities incur significant costs as a result of living with wildlife and ~50 people are killed annually by wild animals [31]. Overall, communities living in GMAs are 30% poorer than the national rural average [32].

Reasons for the Under-performance of Protected Areas

1. Community-related issues. There are no legal mechanisms to enable communities to benefit financially from photo-tourism in PAs. Furthermore, photo-tourism is under-developed and is practiced in a small fraction of the PA estate where the

Table 3. Gross earnings in USD from trophy hunting and non-resident hunting in Game Management Areas in 2012 (excluding land under 99 year lease within or near to GMAs) (NB that operators' net earnings are markedly lower than the gross income, due to the costs associated with paying concession and animal license fees to ZAWA, the costs of running and marketing safaris, and the costs of managing the concessions).

	Total	Operators*	ZAWA*	Community Resource Boards*	Chiefs*
Trophy hunting					
Concession fees	N/A	N/A	580,572	108,857	36,286
Animal license fees	N/A	N/A	1,697,878	1,528,090	169,788
Operators license fees	N/A	N/A	80,500	0	0
PH license fees	N/A	N/A	65,621	0	0
Daily rates	N/A	11,655,430	N/A	N/A	N/A
Trophy fees	N/A	4,506,987	N/A	N/A	N/A
Sub total	16,162,417	16,162,417	2,424,571	1,636,947	206,074
Resident hunting					
Total earnings	88,932	N/A	44,489	40,040	4,449
Non resident hunting earnings/km^2	0.6	N/A	0.3	0.3	0.0
Total earnings	16,251,349	16,162,417	2,469,060	1,676,988	210,522
Total earnings/km^2	97	N/A	14.8	10.0	1.3

*NB Operators were assumed to generate the total gross income from trophy hunting, from which the ZAWA and community income is derived. This income excludes that generated from extensive game ranches, which are on 99 year lease.

resource and existing conditions are suitable. For example, photo-tourism operations exist in only 10 of the 36 GMAs [7] and the existing legal framework does not require photo-tourism operators to pay Community Resource Boards [9]. Consequently, communities are only able to derive legal benefits from wildlife via trophy hunting in GMAs and through employment in national parks and GMAs.

The Wildlife Act fails to recognize communities as the rightful owners of the land or wildlife in GMAs (in contradiction to the Lands Act of 1995, [33]). There are no mechanisms for specific communities to obtain exclusive rights over land in GMAs or over the wildlife resources therein. Consequently, mechanisms to prevent in-migration of external communities into GMAs are weak. Immigration is likely fuelled by the availability of bushmeat, firewood and other natural resources, and in some cases, due to potential employment opportunities from tourism in adjacent national parks. Clearing of lands for settlement and shifting agriculture and tree-cutting for charcoal production all contribute to habitat loss in GMAs [9,15]. Uncontrolled immigration means that land use planning is difficult to enforce [6] and creates an open-access system whereby it pays communities to occupy land or kill wildlife before someone else does.

ZAWA collects revenues from trophy hunting and remits 20% of the concession fees to communities and 50% of animal license fees for wildlife shot as trophies in GMA [8]. ZAWA thus retains most of the income from hunting in GMAs even though the land is under customary tenure and belongs to the community. Wildlife-based land uses are effectively heavily taxed, whereas livestock production, small-scale agriculture and others are not (Table 3). Furthermore, income from wildlife is often paid late and as hand-outs that do not create a clear link between conservation and earnings [9,30]. Earnings for communities from trophy hunting are lower than estimated earnings from illegal bushmeat hunting and create weak incentives for conservation [34].

Communities living in GMAs are also marginalized from the decision-making relating to wildlife management. The concession agreements that are signed between ZAWA, hunting operators and Community Resource Boards on the leasing of GMAs favour ZAWA. ZAWA retain most of the income, delegate numerous responsibilities to other stakeholders, allocate minimal rights, are able to dismiss Community Resource Boards or to cancel the leases of hunting operators, and are not obliged to make their financial records available [11]. There is widespread resentment towards ZAWA, hunting and tourism operators (anon survey respondent, pers. comm.).

The make-up and functioning of Community Resource Boards is also problematic. More affluent community members benefit disproportionately from trophy hunting [30]. There is a lack of appropriately educated or skilled community members to form Community Resource Boards, which limits their ability to negotiate effectively with ZAWA or operators [9]. Skills shortages are exacerbated by the fact that the Community Resource Boards are re-formed each 2–3 years [35]. Furthermore, local chiefs can dismiss board members and 'village scouts' (community members employed by the community resource boards for anti-poaching) at their discretion, making Community Resource Boards unstable. There are frequent financial irregularities associated with CRB income from wildlife and funding allocated to resource protection is inadequate [9]. The 850 village scouts employed throughout Zambia are poorly and irregularly paid, insufficiently trained or equipped and are too few in number to effectively patrol the vast GMA estate [11].

2. ZAWA-related issues. When ZAWA was formed in 2000, a key objective was to increase efficiency and financial self-sufficiency [35]. ZAWA was meant to retain government support, but in practise funding was reduced to just 15% of operational budget. Including the ~USD4.6 million generated from trophy hunting, ZAWA's resources equate to USD20–60/km^2/year, which compares poorly with the USD358–USD455/km^2 required to manage protected areas effectively [36] ZAWA has a field staff complement of 1,179 to protect 231,000 km^2, which equates to 1/196 km^2 and compares poorly with the southern African regional average of 1/40 km^2 [1]. As little as 8% of the ZAWA budget is spent on GMAs even though they generate >50% of ZAWA earnings and comprise >70% of land under their jurisdiction [7]. Furthermore, ZAWA is encumbered by large numbers of sick and poorly trained staff, and an increasing proportion of ZAWA funds have been accruing to head office (anonymous survey respondent pers. comm.). Consequently, field capacity is low when the threat to wildlife from the bushmeat trade and is unprecedented and that from ivory poachers resurgent [2,37]. Funding shortages mean that ZAWA's mandate of protecting the vast wildlife estate is impossible to achieve.

To rectify this situation, ZAWA have allocated partial responsibility for resource protection to Community Resource Boards and hunting operators with the effect that roles are blurred and none of the stakeholders contribute sufficiently. This is despite the fact that responsibility for anti-poaching falls on ZAWA according to the Wildlife Act [11]. Technical support to Community Resource Boards from ZAWA is inadequate and communication between Community Resource Boards and ZAWA is limited [9].

Forced to generate their own funding, ZAWA rely on safari hunting in GMAs for ~45–67% of their revenue [11,35]. This reliance means that ZAWA are sometimes forced to make decisions to achieve financial survival at the expense of the wildlife they are mandated to conserve. For example, in 2003, ZAWA increased quotas and reduced the size of hunting blocks [7]. In addition, ZAWA have imposed high 'fixed-quotas' (of 60–100%) whereby operators are forced to pay for animal license fees before commencement of hunting [7,11]. Such quotas create a perverse incentive, forcing operators to harvest wildlife regardless of sustainability. Due to lack of funds, there is a lack of monitoring of wildlife populations or of trophies. Trophy quotas are established arbitrarily, quota utilization is low (averaging 40%) and prior to the ban trophy quality was falling, implying that quotas were not sustainable [7]. Quotas of lions have been particularly excessive [38,39].

The requirement for ZAWA to generate their funds means that it is not in their best interests to devolve user-rights over wildlife to communities in GMAs or to private landowners on extensive wildlife ranches (who are forced to pay license fees for animals hunted to ZAWA), as that would reduce income in the short-term and create perceived competition [19]. A similar conflict arose when the Zimbabwe Parks and Wildlife Management Authority became a parastatal responsible for generating their own revenue [40]. In Zimbabwe, that shift resulted in a gradual shift towards centralized authority over wildlife and gradual reversal of the devolution that made the wildlife ranching industry in that country such a success [40].

Wildlife in GMAs is affected by several other forms of legal harvest including resident hunting conducted by Zambian citizens and residents. Resident hunting licenses cost ~1/3 the meat value of the animals hunted, and consequently resident hunters often shoot wildlife specifically to obtain meat to sell, and in many cases, quotas are exceeded (Table 4). Such abuses are made possible by inadequate supervision of resident hunts and corruption. Prior to the hunting ban, a varying amount of wildlife was also killed under 'special licenses' allocated on a discretionary basis by the Minister of Tourism and Arts [35]. Special licenses and non-resident hunting licenses strip the value of wildlife and create minimal incentives for conservation by communities. The multiple forms of legal off-take compound the effects of habitat loss, predation and poaching and confer heavy depletion of wildlife in most GMAs (Figure 3b).

3. Operator-related issues. In national parks, tourism operators are not required to conduct anti-poaching and input is generally limited to sporadic provision of support for NGOs involved in resource protection [41]). In GMAs, the concession allocation system created disincentives for investment and good practice by hunting operators. Leases are granted for 10 years (or 15 for depleted blocks), which is not sufficient to encourage adequate investment in the area or to a sense of ownership of the areas [42]. Where wildlife populations are depressed, 15 years does not allow for sufficient time for operators to recoup the investments needed to allow wildlife populations to recover [43].

The hunting concession agreements (which outline commitments to anti-poaching and community outreach) are not effectively enforced [9]. Furthermore, operators typically vacate the hunting blocks during the rainy season, leaving their areas vulnerable to poachers. Some of the hunting blocks in GMAs are extremely large, and several operators complained that their size renders effective enforcement impossible. Some operators appear to invest a significant amount in anti-poaching and others virtually nothing. However, there is no system to link past performance of hunting operators to the prospects of them obtaining an extension of a lease or a new area. Consequently, responsible operators are not adequately rewarded, and unscrupulous operators not adequately punished, reducing incentives for good practice and allowing abuses (including alleged over-shooting of quotas by some operators) to continue [44]. In general, declining wildlife populations have resulted in falling incomes [7] and thus declining resources available protect the resource.

In early 2013, ZAWA imposed a moratorium on hunting in GMAs in response to alleged corruption in the tender process and due to concern over wildlife population trends. Consequently, hunting operators have vacated the GMAs, resulting in loss of their contribution to anti-poaching and creating a vacuum in which illegal activities are more likely to proceed unhindered. Evidence from Kafue National Park suggests that the simple presence of operators has a significant deterrent effect for poachers [41]. Furthermore, there have been extended periods in 2013 when village scouts went without pay, and many likely relied on poaching for income (Anon survey respondent, pers. comm.). While government has subsequently stepped in to pay the salaries of village scouts, it is not clear as to how such payments will be sustained in the absence of hunting income. There have been few proposals from the photo-tourism industry to take over GMAs in the wake of the hunting ban. The hunting moratorium is thus likely to fuel wildlife declines by reducing: anti-poaching effort and presence in GMAs; working capital for ZAWA; and incentives for conservation by communities.

4. Other factors. There are multiple authorities in GMAs with jurisdiction over the management of different resources [35]. In addition, conflicting legislation precludes effective land use planning. For example, the wildlife act states that ZAWA is responsible for wildlife resources in the area, and by implication habitat, whereas the Local Government Act says that local councils are responsible for planning and development [35]. Consequently development-related decisions are sometimes made in GMAs with little consideration of their impacts on wildlife or the potential for wildlife-based land uses. Furthermore, chiefs are able to allocate land to private investors in the middle of GMAs without consulting Community Resource Boards [9].

Changes Needed to Improve the Functioning of the Protected Area Network

In this section we outline a number of key steps that we consider to be necessary to improve the functioning of the Zambian PA

Table 4. The price of citizen licenses for hunting in GMAs, the meat value of those species and the value of the meat relative to the price of citizen licenses.

	Citizens	Meat value*	Meat value relative to license fee
Buffalo	493	1398	3
Bushbuck	40	142	4
Bush pig	16	146	9
Duiker	32	40	1
Eland	592	1425	2
Hartebeest	158	355	2
Impala	40	150	4
Kudu	493	551	1
Oribi	59	33	1
Puku	69	159	2
Reedbuck	79	159	2
Warthog	79	185	2
Waterbuck	158	615	4
Wildebeest	158	593	4
Zebra	296	757	3
Average ± S.E.			3±0.75

*Assuming male animals are hunted, a meat price of USD4.3/kg (the price paid by Lusaka butchers for dressed meat) and assuming the mean mass of dressed carcasses presented by [72].

network, many of which are likely to apply to PAs in various other parts of Africa.

1. Empowering communities to benefit more from the PA network. There is a need to develop systems to enable communities to benefit more from PAs so as to enhance their social and economic value. Potential models to for such community engagement are discussed in the next section.

2. Re-defining the role of ZAWA. The role of ZAWA should be re-defined so that the organisation focuses primarily on national parks, and elsewhere plays more of a hands-off coordinating role to facilitate the development of wildlife-based land uses by community, private, and NGO partners. There is need for recognition from government that unless earnings are vastly higher than they have been in the past, the GMAs cannot be used to fund other parts of the PA network or to support ZAWA headquarters. If ZAWA continue to extract revenues from GMAs, then a commensurate reinvestment in law enforcement and wildlife management in those areas is required.

3. Greatly increasing funding from government for ZAWA. Drastically increasing government funding for ZAWA. Such funding should be seen as an investment in the development of the tourism industry rather than as a cost (as discussed further below). Even at current rates of funding for ZAWA, there is evidence of impressive returns on investment: in 2005, the government of Zambia received USD8 million from wildlife-related tourism from an investment of just USD1 million [32]. The tourism industry in Zambia comprises just 5% of GDP compared to a global mean of 14%, and creates just 3.6% of national employment compared to a regional mean of 7.2% [13,45,46]. In addition, there has been relatively minimal capital investment in the tourism industry in Zambia (1.7% of GDP c.f. a regional average of 6.9%) [13]. There is thus major scope for growth in the Zambian tourism industry, particularly given projections of rapid increases in visitor arrivals to the SADC region [13]. To achieve that growth, however, there is a need for much greater investment in protecting the wildlife product.

4. Harnessing international willingness to pay for conservation. There is a need for greater international support for PA management in Zambia. The Zambian PA network constitutes ~40% of the country's land area, compared to a global average of 12% [47]. A developing country such as Zambia cannot reasonably be expected to pay for all of the costs of the maintenance of that global asset in addition to bearing the associated opportunity costs. Furthermore, tourism and trophy hunting (in the case of GMAs) may never generate enough income to cover the costs needed for effective protected area management. There are numerous potential avenues for harnessing international willingness to pay for wildlife conservation, including *inter alia*:

a) Encouraging co-management agreements for PA management. In national parks, ZAWA could pursue the development of co-management agreements with NGOs and the private sector to share the burden of PA management and attract additional technical capacity. Encouragingly, there has been an increasing trend towards development of such agreements in Zambia in recent years [5,48]. There are official co-management agreements in place for five national parks and two GMAs (ZAWA unpublished data). In addition, there is significant NGO involvement in resource protection in several other national parks (e.g. Kafue, South Luangwa, Lower Zambezi, Sioma Ngwezi, and Nyika) and 11 other GMAs. If ZAWA could focus primarily on coordinating and regulating the management of PAs by partner organisations, the effectiveness and cost-effectiveness of parks could be greatly enhanced. NGOs and the private sector could be encouraged to take over the management of entire PAs or of concessions within PAs. Such a set up also would provide NGOs that are opposed to trophy hunting with the opportunity to reduce the prevalence of the practice by making payments in lieu of hunting revenues in the GMAs.

b) Attracting funding for the development of a national CBNRM programme. In addition to funding ZAWA and the PA network, there is a need for funding to allow for a national community-based natural resource management (CBNRM) programme to facilitate the capture of a greater proportion of benefits from the PA network by communities. Such funding could build on the progress made by the Administrative Management Design for Game Management Areas (ADMADE) during the 1990s. The successes of the Namibian community conservancy programme and the Zimbabwean CAMPFIRE programme have been dependent on long-term and substantial injections of technical capacity and funding (USD173 million and USD35 million respectively) ([49], C. Weaver pers. comm.). Likewise, the community conservancies in the northern Kenyan rangelands are supported by a coordinating NGO with funding of ~USD1.2 million annually [50]. A similarly well funded, supported and coordinated national CBNRM programme is needed in Zambia.

c) Encouraging allocation of a greater portion of overseas development aid towards PAs. There is a strong case for donors to direct a portion of international development aid towards PA management, capitalizing ZAWA and/or co-management agreements between ZAWA and NGOs, and/or developing a national CBNRM programme. Recent estimates suggest that for every 1% increase in tourism-related investment in the SADC region, a 0.3% increase in GDP per capita accrues [13]. An allocation of just 2–3% of the ~USD1 billion of overseas development aid that Zambia receives annually [51] would cover the costs of managing the national parks effectively and of protecting the main tourism asset [32]. However, there is a need for checks and balances to ensure efficient use of donor funds by ZAWA.

A potentially cost-effective way of using donor funds to achieve both conservation and development objectives (ideally in the context of a coordinated national CBNRM programme) is through schemes that channel payments to communities living in or near PAs for the provision of ecosystem services. For example, communities could be paid an annual fee for desisting from converting habitat or for protecting wildlife from poaching. In the Maasai steppe in Tanzania for example, such an approach has achieved notable conservation gains for a cost of just USD48/km^2 [52], which compares favourably with the costs of traditional PA management. A key potential value of PES approaches is that they can help correct 'market failures' whereby wildlife that is valuable to the nation as a whole is not valuable to the people living with it, who thus over-exploit the resource or invest little in protecting it [50]. PES approaches could be combined with efforts to provide communities with stable markets and fair prices for livestock and crops, as is being conducted in the Kenyan community conservancies, and as part of the community markets for conservation approach in parts of Zambia [50,53]. However, the latter approaches are only likely to be successful if combined with efforts to actively protect wildlife populations via anti-poaching.

d) Other options. Capturing the willingness of international philanthropists to pay for conservation represents another potential means of funding PAs [54]. Several precedents for such investment exist, such as that provided in Gorongosa National Park in Mozambique and the Grumeti Game Reserve in Tanzania. Additionally, attracting carbon-related investment via projects such as REDD+ could potentially generate funds for the protection of woodlands and associated biodiversity, notwithstanding the constraints currently associated with that programme [55]. Further, government could potentially generate additional revenue by taxing stakeholders who benefit substantively from ecological services provided by PAs, such as commercial farmers and power-generating or mining companies.

5. Attracting significant private investment. To develop tourism and hunting businesses in national parks and GMAs, and to protect and manage wildlife in GMAs (assuming that ZAWA focuses their efforts on national parks) would require substantial private and/or donor investment. Rehabilitating a single depleted GMA is predicted to cost millions of dollars and such investments would likely take many years to recoup [43]. Attracting such investment is most likely under the following circumstances: an enabling policy environment; simple safe and standardized processes for investing; long leases (of at least 40 years for depleted areas [43]); attractive terms (e.g. as would be conferred if ZAWA desisted from taxing wildlife-based land uses in GMAs); minimal red-tape or interference from ZAWA, a functioning national programme for the development of community wildlife conservancies (see below) and knowledge that communities are supportive of wildlife investments on their land. In GMAs, investors and community partners should be able to choose any forms of wildlife-based land uses and the combinations that will yield the best returns for their particular spatial setting. Consequently, hunting bans or bans on the hunting of high-value species are to be avoided so long as hunting can be managed in a manner that ensures sustainability [12]. Finally, resident hunting should not be allowed to undermine private investments in GMAs and should only be permitted if desired by the community and investor partners and if priced appropriately.

6. Addressing key conservation threats decisively. For the PA networks to function better, there is a need to decisively address key threats such as human encroachment, bushmeat and other forms of poaching. Human encroachment could be addressed through linking the allocation of leases to communities with agreed land use plans (see below). Alternatively, portions of GMAs close to national parks could be re-gazetted as 'buffer-zones' where human settlement is not permitted (though the leases for such areas could still be leased to communities to enable them to benefit from legal wildlife-based land uses, see below). In such instances and at the edge of community conservancies (see below), fencing external boundaries (if supported by communities) may play a significant role in reducing edge-effects, reducing human-wildlife conflict, demarcating boundaries and helping to prevent further encroachment [58,59]. Where human settlement has reached right up to the boundaries of national parks, the fencing of such sections may be justified if appropriate materials are used and if adequate funding for maintenance exists.

Addressing poaching requires elevated opportunities for communities to benefit legally from wildlife, stiffer legal frameworks relating to poaching (with penalties that reflect the value of wildlife and the threat posed by poachers to the life of PA staff) and improved anti-poaching are required [2]. To achieve professional, well-funded anti-poaching requires either a much greater investment from ZAWA, and/or significant investments from the private and/or NGO sector.

7. Prioritising conservation efforts. Zambia's PA network is vast and funding limited. Furthermore, some of the GMAs are probably damaged beyond repair due to heavy human settlement and habitat modification. Given these factors plus the high human population growth rates, conserving the entire PA network in the long term is unlikely. There is a case for a scientific priority setting exercise to identify the PAs that should be the priority for investment of available funding.

Potential Models for Achieving Elevated Community Participation in the PA Network

The degree of involvement of communities in national parks versus GMAs should arguably differ as people reside the latter but generally not the former. For national parks, one option would be to allocate ownership of PAs, shareholdings of PAs, or tourism concessions within parks to neighbouring communities for them to lease out to tourism operators. Precedents for such arrangements have been established in South Africa, for example, through the creation of contractual parks [56]. Such changes would require clear definition on who comprises 'the community' as at present, membership is poorly defined, compromising effective plan implementations and revenue sharing.

In the GMAs, we recommend changes that empower communities and enable them to participate in and benefit from wildlife-based land uses through the formation of Community Wildlife Conservancies (CWCs). Precedents for such conservancies have been developed in both Namibia and the northern rangelands of Kenya, both of which have achieved significant conservation and livelihood gains [50,57]. While there are many potential variants of such models that could applied to the GMAs, there are a few general principles that we believe should be adhered to: a) communities should be allocated the maximum permissible degree of ownership over land and wildlife; b) that ownership should be structured such that it is exclusive for specific communities to avoid perpetuation of the tragedy of the commons; c) communities should accrue benefits from wildlife directly, and not via remittances from ZAWA; d) communities must actively participate in wildlife management decisions and not be passive recipients of hand-outs; e) community structures that are used to administer finances relating to wildlife must be democratic, transparent and regularly audited to ensure equitable distribution of benefits and avoid elite capture, and f) there must be mechanisms to ensure funding for high-quality anti-poaching security given the level of threat in GMAs. Two examples of potential models for the establishment and functioning of CWCs in GMAs are as follows:

1. Complete devolution model. Here CWCs in the GMAs would be established as community conservancies based on joint ventures between communities and the private sector. ZAWA would not extract income from the CWC and would play a purely regulatory, facilitating and over-seeing role. ZAWA currently earn nothing from half of the GMAs as it is and so in such areas this kind of arrangement would not cause loss of revenue for ZAWA. The community would create a democratic, accountable and transparent body or trust to administer the area as a conservancy. A long lease would then be allocated by government to that community conservancy trust, the validity of which should be contingent on a land use plan that ensures that a particular area is set aside for wildlife only. Communities would then sub-lease the land to or engage in business partnerships with private or NGO investors, ideally for long periods to attract significant investment. That leasing process could follow either a public auction or an open tender process. The communities and successful bidding investors would then form a second body or trust with the mandate of managing wildlife in the GMA, ensuring professional anti-poaching and effective communication and cooperation between the community and investors. Alternatively, the investors could gain representation on the community conservancy trust after signing a partnership agreement, and then that body would coordinate wildlife management. Investors would then pay: a) an annual land rental to communities (which means they would derive some income without waiting years for wildlife populations to recover); b) an annual resource use fee (e.g. bed night levies or licence fees for animals hunted) (which means that communities would receive income proportional to their conservation 'performance'); and, c) an annual levy to capitalize the body with the responsibility for managing wildlife. Investors could generate income either by acting as their own hunting or tourism operators,

by auctioning hunting packages to the highest bidding operator, or sub-leasing tourism concessions.

2. Partial devolution model. In this model, GMAs would be administered as community conservancies based on tripartite public-private-community-partnerships involving communities, ZAWA and private investors. As in the previous model, the communities would form a body that obtains a lease for the land, and would sub-lease the land to investors (or engage in a long term business partnership). A largely independent not-for-profit body would be established with representation from ZAWA, investors/ participating NGOs and communities with the mandate of managing the wildlife in the area. The fees that have traditionally been paid to ZAWA by operators would be paid into that not-for-profit body to ensure that they are reinvested in the area.

Re-designating GMAs as CWCs would confer multiple benefits: CWCs would secure land rights for communities and protect against the loss of land and natural resources that would arise from the current open access system; CWCs could provide an effective buffer role for national parks; communities would generate significant and sustainable incomes, meat supplies and employment; if wildlife populations were successfully rehabilitated, CWCs could generate economic outputs at least 20x greater than currently being earned in GMAs [43]; CWCs would attract external investment by a wide-spectrum of donors; CWCs would create scope for communities to sell carbon and biodiversity credits by securing land rights [15]; and ZAWA would be relieved of the burden and costs of protecting wildlife in GMAs.

Legislative Changes Needed to Allow CWCs to Happen

Ownership of wildlife in Zambia is vested in the President on behalf of the country [19]. On private land, user-rights over wildlife can be conferred to landowners via certificates of ownership, but there is no such provision for communities. Similarly, there is scope for investors, but not communities, to alienate land [9]. New legislation is required to enable communities to obtain 99-year leases for their land in GMAs following formation of a CWC, and to enable them to obtain full-user rights over wildlife.

At present, fencing is a pre-requisite for obtaining ownership over wildlife on wildlife ranches in Zambia [19]. In some contexts, as discussed, fencing can confer clear benefits [58,59]. However, fencing can also reduce ecological connectivity and provide massive supplies of snare-material if inappropriate wires are used [58,60]. Fencing should not be a pre-requisite for communities to obtain user-rights over wildlife and should not be permitted between adjacent CWCs or between CWCs and national parks. If needed, fencing should also be composed of kinked wire mesh to prevent snare construction, and should be accompanied by an environmental impact assessment and a clear long-term maintenance plan and budget.

Conclusions

Wildlife populations are faring poorly in many African protected areas [61] and many of the challenges and solutions highlighted in this paper occur in other African countries. The under-funding of protected area networks is a widespread problem and parks agencies are often required to generate their own income, which creates the kinds of conflicts of interest outlined in this paper. There is a need for vastly elevated funding for PA management from both African and international governments and institutions. There is the need for improved mechanisms to enable communities to participate in and benefit more from wildlife in many African countries. In addition, creating frameworks for safe and secure private and NGO investment in PA management is an intervention with widespread applicability. Strong measures to address unplanned human encroachment in PA networks are also needed in many areas, as are efforts to tackle high levels of ivory and bushmeat poaching. The net result of these interventions is likely to be significant improvements in the effectiveness of parks networks, substantial job creation and economic gains due to growth in tourism industries. In the absence of such changes, wildlife populations in protected areas in Zambia and many other countries are likely to continue to wane due to on-going poaching and human encroachment.

Methods

Insights into the Performance of Protected Areas

We provided insights into the performance of PAs using: a) a literature review; b) data obtained from ZAWA and other sources; c) and, semi-structured interviews with key stakeholders, including the highest-ranking ZAWA officials (n = 7); representatives from relevant NGOs (n = 14); wildlife industry experts/photo-tourism operators (n = 11); and trophy hunting operators (n = 13). The literature review was conducted using key words such as 'Zambia', 'GMAs', 'wildlife policy', 'CBNRM', 'ADMADE', 'trophy hunting', 'wildlife ranching', 'co-management', 'bushmeat', 'and encroachment', etc. We included both published papers and unpublished consultancy reports. We searched for references using Google and Google Scholar. Selection of survey respondents was conducted by contacting and meeting as many individuals from each group that we could during our fieldwork period (September–November, 2012). Refusal rate was zero.

1. Ecological performance of protected areas. We assessed ecological performance of protected areas by looking at the degree of human encroachment and the size and diversity of wildlife populations. Data on human encroachment were obtained from [6]. The 2010 Zambian census was used to obtain district-level estimates of human population growth rates [16].

Data on wildlife abundance in protected areas were derived from aerial census reports [17,24,25,62–67]. The most recent reports were used, though some abundance estimates were made using census reports as old as 2003, and for some PAs no census data were available at all. Census data were available for 39 Zambian PAs (14 National Parks comprising 61,812 km^2 and 25 GMAs comprising 152,122 km^2) (or ~93% of the national park and GMA estate). Estimates of mammalian biomass were made by removing species of bushbuck *Tragelaphus scriptus* size or smaller, hippopotamuses *Hippopotamus amphibius* and predators as most reports did not provide estimates for those species. The typical mass of an individual in a population for each species following [68] was multiplied by population sizes to estimate biomass.

We used rainfall, soil nutrient status and large herbivore biomass for 28 wildlife areas in eastern and southern Africa [69] to create five regression curves for predicting herbivore biomass for: 1) medium soil nutrient areas for moist-adapted species; 2) medium soil nutrient areas for arid-adapted species; 3) low soil nutrient areas for moist-adapted species; 4) low soil nutrient areas for arid-adapted species with annual rainfall <700 mm; 5) low soil nutrient areas for arid-adapted species with annual rainfall > 700 mm. In each case, herbivore biomass was plotted against rainfall using the software programme GraphPad Prism. These five regression curves were then used to predict *potential* herbivore standing crop biomasses (kg/km^2) for protected areas in which annual rainfall and soil nutrient data were available. Estimates for annual rainfall were determined from literature and internet sources, while soil nutrient status was determined using a combination of two sources: 1) soil maps [70] and 2) vegetation types identified from the literature and vegetation maps [71].

In protected areas where there was more than one soil or vegetation type, we estimated the proportion of each type within the area, and used these to calculate an average soil nutrient status. In many cases, the soil nutrient status estimated from soil and vegetation types corresponded well, but in cases when they differed, the lower estimate was selected for the sake of conservatism.

2. Economic performance of protected areas. Data on earnings from photo-tourism in national parks, and from trophy and resident hunting in GMAs were obtained from ZAWA to assess economic performance of protected areas. Earnings of safari operators in GMAs were estimated using trophy off-takes for 2012 obtained from ZAWA, following [28] and using the mean 2013 pricing for Zambian trophy hunts (from a survey of n = 10 websites). Current per km^2 earnings from trophy hunting in Zambia were compared with regional estimates derived from [28]. Further insights into the performance of the hunting and tourism industries were obtained from the literature.

3. Social performance. Insights into the social performance of protected areas were obtained from the literature and through surveys. Comprehensive community surveys in GMAs have been completed by other authors recently (e.g. [8,9,30] and their findings provided insights into community-related issues.

Ethics Statement

The University of Pretoria Ethics Committee approved this research, and approved the procedure for obtaining consent for the surveys conducted during the research. We were issued with written consent for this study from the Wildlife Producers' Association of Zambia. In addition Zambia Wildlife Authority provided verbal approval and participated in the research. From respondents we obtained verbal consent prior to conducting the surveys. Written consent from individual respondents was not considered practical or necessary and the requirement for written consent was waived by the University of Pretoria. We documented any cases where respondents did not wish to participate in order to calculate refusal rates.

Acknowledgments

Thanks to ZAWA for permission to conduct the study, for participating in surveys and for provision of information, thanks to the Wildlife Producers Association of Zambia, Don Stacey and to the ranchers and other respondents.

Author Contributions

Conceived and designed the experiments: PL WAT VN JB MT CT. Performed the experiments: PL WAT VN JB MT CT. Analyzed the data: PL WAT VN JB MT MB RM FW CT. Contributed reagents/materials/analysis tools: PL WAT VN JB MT MB RM FW CT. Wrote the paper: PL WAT VN JB MT MB RM FW CT.

References

1. Cumming D (2004) Performance of parks in a century of change. In: Child B, editor. Parks in Transition: Biodiversity, Rural Development, and the Bottom Line. UK: Earthscan.
2. Lindsey PA, Balme G, Becker M, Begg C, Bento C, et al. (2013) The bushmeat trade in African savannas: Impacts, drivers, and possible solutions. Biol Conserv 160: 80–96.
3. Caro TM, Pelkey N, Borner M, Campbell K, Woodworth B, et al. (1998) Consequences of different forms of conservation for large mammals in Tanzania: Preliminary analyses. Afr J Ecol 36: 303–320.
4. Nelson F, Lindsey P, Balme G (2013) Trophy hunting and lion conservation: a question of governance? Oryx 47: 501–509.
5. Government of Zambia (2010) Reclassification and effective management of the national protected areas system. Lusaka, Zambia: Ministry of Tourism, Environment and Natural Resources.
6. Watson FGR, Becker MS, Nyirenda MA (In press) Human encroachment into protected area networks in Zambia: Implications for large carnivore conservation. Environmental Change.
7. Simasiku P, Simwanza H, Tembo G, Bandyopadhyay S, Pavy J (2008) The impact of wildlife management policies on communities and conservation in game management areas in Zambia. Zambia: Natural Resources Consultative Forum.
8. Fernandez A, Richardson RB, Tschirley DL, Tembo G (2010) Wildlife conservation in Zambia: Impacts on rural household welfare. Michigan State University: Food Security Collaborative Working Papers.
9. Chemonics International Inc (2011) Situational and livelihoods analysis study in nine Game Management Areas, surrounding the Kafue National Park, Zambia. Washington DC, USA: Millenium Challenge Corporation.
10. Becker M, McRobb R, Watson F, Droge E, Kanyembo B, et al. (2013) Evaluating wire-snare poaching trends and the impacts of by-catch on elephants and large carnivores. Biol Conserv 158: 26–36.
11. Manning I (2011) Wildlife conservation in Zambia and the Landsafe customary commons. Natural Resources Journal 52: 195–214.
12. Lindsey P, Balme G, Funston P, Henschel P, Madzikanda H, et al. (2013) The trophy hunting of African lions: Scale, current management practices and factors undermining sustainability. Plos One 8(9): e73808.
13. Makochekanwa A (2013) An analysis of tourism contribution to economic growth in SADC countries. Botswana Journal of Economics 11(5): 42–56.
14. Pfeifer M, Burgess ND, Swetnam RD, Platts PJ, Willcock S, et al. (2012) Protected areas: Mixed success in conserving east Africa's evergreen forests. Plos One 7: e39337.
15. Vinya R, Syampungani S, Kasumu EC, Monde C, Kasubika R (2011) Preliminary study on the drivers of deforestation and potential for REDD+ in Zambia. Lusaka, Zambia: FAO/Zambian Ministry of Lands and Natural Resources.
16. Zambia Central Statistical Office (2011) 2010 census of population and housing: Preliminary population figures. Lusaka, Zambia: Government of Zambia.
17. Simukonda C (2011) Wet season survey of the African elephant and other large herbivores in selected areas of the Luangwa Valley. Chilanga, Zambia: Zambia Wildlife Authority.
18. Poulsen JR, Clark CJ, Mavah G, Elkan PW (2009) Bushmeat supply and consumption in a tropical logging concession in northern Congo. Conserv Biol 23: 1597–1608.
19. Lindsey P, Barnes J, Nyirenda V, Pumfrett B, Taylor A, et al. (2013) The Zambian wildlife ranching industry: Scale, associated benefits, and limitations affecting its development. Plos One 8(12): e81761.
20. Bond I (2009) CBNRM as a mechanism for addressing global environmental challenges. In: Roe D, Nelson F, Sandbrook C, editors. Community management of natural resources in Africa: impacts, experiences and future directions. London, UK.: International Institute for Environment and Development. 95–104.
21. Owen-Smith N, Ogutu J (2003) Interactions between biotic components. In: Du Toit J, Rogers K, Biggs H, editors. The Kruger experience: ecology and management of savanna heterogeneity. Washington DC: Island Press. 310–331.
22. Whyte I, van Aarde R, Pimm S (2003) Kruger's elephant population: Its size and consequences for ecosystem heterogeneity. In: Du Toit J, Rogers K, Biggs H, editors. The Kruger experience: ecology and management of savanna heterogeneity. Washington DC: Island Press. 332–348.
23. Lindsey PA, Roulet PA, Romañach SS (2007) Economic and conservation significance of the trophy hunting industry in sub-Saharan Africa. Biol Conserv 134: 455–469.
24. Frederick H (2013) Aerial survey report: Luangwa valley, 2012. Chilanga, Zambia: Zambia Wildlife Authority.
25. ZAWA (2013) Liuwa Plain National Park aerial wildlife survey results 2013. Lusaka, Zambia: ZAWA.
26. ZAWA (2013) Kafue National Park general management plan. Chilanga, Zambia: Zambia Wildlife Authority.
27. Hamilton K, Tembo G, Sinyenga G, Bandyopadhyay S, Pope A, et al. (2007) The real economic impact of nature tourism in Zambia. Lusaka, Zambia: Natural Resources Consultative Forum.
28. Lindsey P, Balme G, Booth V, Midlane N (2012) The significance of African lions for the financial viability of trophy hunting and the maintenance of wild land. Plos One : e29332.
29. Snyman SL (2012) The role of tourism employment in poverty reduction and community perceptions of conservation and tourism in southern Africa. Journal of Sustainable Tourism 20: 395–416.
30. Bandyopadhyay S, Tembo G (2010) Household consumption and natural resource management around national parks in Zambia. Journal of Natural Resources Policy Research 2: 39–55.
31. Chomba C, Senzota R, Chabwela H, Mwitwa J, Nyirenda V (2012) Patterns of human–wildlife conflicts in Zambia, causes, consequences and management responses. Journal of Ecology and the Natural Environment 4: 303–313.
32. World Bank (2007) Economic and poverty impact of nature-based tourism. Lusaka, Zambia: World Bank.
33. Manning I (2011) Wildlife conservation in Zambia and the landsafe customary commons. Pretoria: University of Pretoria, PhD thesis.
34. Lindsey P, Taylor A, Nyirenda V, Barnes J (2014) Through severe impacts on wildlife populations and net negative economic and food security impacts, bushmeat poaching is a market failure. Harare, Zimbabwe: FAO/Panthera/ IUCN SULi/ZSL report.
35. Sichilongo M, Mulozi P, Mbewe B, Machala C, Pavy J (2013) Zambian wildlife sector policy: Impact analysis and recommendations for the future policy. Lusaka, Zambia: Africa Technical Environment and Natural Resources Unit, World Bank.
36. Cumming DH (2008) Large scale conservation planning and priorities for the Kavango-Zambezi Transfrontier Conservation Area. Unpublished report commissioned by Conservation International.
37. Douglas-Hamilton I (2009) The current elephant poaching trend. Pachyderm 45: 154–157.
38. Yamazaki K (1996) Social variation of lions in a male-depopulated area in Zambia. The Journal of Wildlife Management 60: 490–497.
39. Becker M, Watson F, Droge E, Leigh K, Carlson R, et al. (2012) Estimating past and future male loss in three Zambian lion populations. Journal of Wildlife Management DOI:10.1002/jwmg.446.
40. Wels H (2003) Private wildlife conservation in Zimbabwe: Joint ventures and reciprocity. Leiden, Netherlands: Brill.
41. Lindsey P, Midlane N, Sayer E, van der Westhuizen H (2013) Kafue National Park resource protection strategy review. New York: Panthera/Frankfurt Zoological Society.
42. Lindsey P, Balme G, Funston P, Henschel P, Hunter L, et al. (2013) The trophy hunting of African lions: Scale, current management practices and factors undermining sustainability. Plos One 8(9): e73808.
43. Lindsey P, Nyirenda V, Barnes J, Becker M, Taylor A, et al. (2013) The reasons why Zambian Game management areas are not functioning as ecologically or economically productive buffer zones and what needs to change for them to fulfill that role. Lusaka, Zambia: Wildlife Producers Association of Zambia.
44. Leader-Williams N, Baldus RD, Smith RJ (2009) The influence of corruption on the conduct of recreational hunting. In: Dickson B, Hutton J, Adams W, editors. Recreational Hunting, Conservation and Rural Livelihoods. Oxford, UK: Wiley-Blackwell. 296–316.
45. WTTC (2013) Economic impact of travel and tourism: Mid-year update, October 2013. London, UK: World Travel and Tourism Council.
46. WTTC (2012) Travel and tourism, economic impact 2012 Zambia. London, UK: World Travel and Tourism Council.
47. Secretariat of the Convention on Biological Diversity (2008) Protected areas in today's world: Their values and benefits for the welfare of the planet. Montreal, Canada: CBD.
48. Nyirenda VR, Nkhata BA (2013) Collaborative governance and benefit sharing in Liuwa Plain National Park, western Zambia.
49. Taylor RD (2009) Community based natural resource management in Zimbabwe: The experience of CAMPFIRE. Biodiversity and Conservation 18(10): 2563–2583.
50. Pye-Smith C (2013) The story of the northern rangelands trust. Isiolo, Kenya: Northern Rangelands Trust.
51. OECD (2013) Development aid at a glance. Paris, France: Organisation for Economic Co-operation and Development.
52. Ingram J, Wilkie D., Clements T, McNab R, Nelson F, et al. (In press) Evidence of payments for ecosystem services as a mechanism for supporting biodiversity conservation and rural livelihoods. Ecosystem Services.
53. Lewis D, Bell SD, Fay J, Bothi KL, Gatere L, et al. (2011) Community markets for conservation (COMACO) links biodiversity conservation with sustainable improvements in livelihoods and food production. Proceedings of the National Academy of Sciences 108: 13957–13962.

54. Spierenburg M, Wels H (2010) Conservative philanthropists, royalty and business elites in nature conservation in southern Africa. Antipode 42: 647–670.

55. Phelps J, Webb EL, Koh LP (2011) Risky business: An uncertain future for biodiversity conservation finance through REDD+. Conservation Letters 4: 88–94.

56. Reid H (2001) Contractual national parks and the Makuleke community. Hum Ecol 29: 135–155.

57. Jones B, Weaver C (2008) CBNRM in Namibia: Growth, trends, lessons and constraints. In: Suich H, Child B, Spenceley A, editors. Evolution and innovation in wildlife conservation in southern Africa. London, UK: Earthscan. 223–242.

58. Lindsey PA, Masterson CL, Beck AL, Romañach S (2012) Ecological, social and financial issues related to fencing as a conservation tool in Africa. In: Somers MJ, Hayward M, editors. Fencing for Conservation.: Springer New York. 215–234.

59. Packer C, Canney S, Loveridge A, Garnett S, Zander K, et al. (2013) Effective conservation of large carnivores. Ecological Letters 16: 635–641.

60. Lindsey PA, Romañach SS, Tambling CJ, Chartier K, Groom R (2011) Ecological and financial impacts of illegal bushmeat trade in Zimbabwe. Oryx 45: 96–111.

61. Craigie ID, Baillie JEM, Balmford A, Carbone C, Collen B, et al. (2010) Large mammal population declines in Africa's protected areas. Biol Conserv 143: 2221–2228.

62. Simwanza H (2004) Aerial survey of large herbivores in West Lunga National Park, Chizera, Musele Matebo, Lukwakwa and Chibkwika Ntambo Game Management Areas. Chilanga, Lusaka: Zambia Wildlife Authority.

63. Simwanza H (2004) Aerial survey to establish the status of large herbivores in the Upper Luano hunting block. Chilanga, Lusaka: Zambia Wildlife Authority.

64. Simwanza H (2005) Aerial survey of large herbivores in the Zambezi heartland, Zambia: October 2005. Kariba, Zimbabwe: African Wildlife Foundation.

65. Simukonda C (2008) A country-wide survey of large mammals in Zambia. Chilanga, Zambia: Zambia Wildlife Authority.

66. Frederick H (2011) Aerial survey of Kafue ecosystem 2008. Lusaka, Zambia: Zambia Wildlife Authority.

67. Viljoen P (2013) Liuwa plain national park and adjacent Game Management Area, Zambia: Aerial wildlife survey April 2013. Johannesburg: African parks network report.

68. Coe M, Cumming D, Phillipson J (1976) Biomass and production of large African herbivores in relation to rainfall and primary production. Oecologia 22(4): 341–354.

69. East R (1984) Rainfall, soil nutrients status and biomass of large African mammals. African Journal of Ecology 22: 245–270.

70. Jones A, Breuning-Madsen H, Brossard M, Dampha A, Deckers J, et al. (2013) Soil atlas of Africa. Luxembourg: European Commission.

71. Wild H, Fernandes A (1967). Vegetation map of the Flora Zambesiaca area. Salisbury, Rhodesia: Collins.

72. du P Bothma J, du Toit JG (2010) Game ranch management. Pretoria, South Africa: Van Schaik.

2

Methods for Estimating Population Density in Data-Limited Areas: Evaluating Regression and Tree-Based Models in Peru

Weston Anderson[1,3]*, Seth Guikema[1], Ben Zaitchik[2], William Pan[4]

1 Department of Geography and Environmental Engineering, The Johns Hopkins University, Baltimore, Maryland, United States of America, **2** Department of Earth and Planetary Sciences, The Johns Hopkins University, Baltimore, Maryland, United States of America, **3** International Food Policy Research Institute, Washington, D.C., United States of America, **4** Nicholas School of Environment and Duke Global Health Institute, Duke University, Durham, North Carolina, United States of America

Abstract

Obtaining accurate small area estimates of population is essential for policy and health planning but is often difficult in countries with limited data. In lieu of available population data, small area estimate models draw information from previous time periods or from similar areas. This study focuses on model-based methods for estimating population when no direct samples are available in the area of interest. To explore the efficacy of tree-based models for estimating population density, we compare six different model structures including Random Forest and Bayesian Additive Regression Trees. Results demonstrate that without information from prior time periods, non-parametric tree-based models produced more accurate predictions than did conventional regression methods. Improving estimates of population density in non-sampled areas is important for regions with incomplete census data and has implications for economic, health and development policies.

Editor: Hiroshi Nishiura, The University of Tokyo, Japan

Funding: This research was supported in part by NASA Applied Sciences award NNX11AH53G. The funders had no role in study design, data collection and analysis, decision to publish, or preparation of the manuscript. No additional funding was received for this study.

Competing Interests: The authors have declared that no competing interests exist.

* Email: Weston.B.Anderson@gmail.com

Introduction

Estimates of the distribution and growth of human population are invaluable. They are used as input to research-focused and operational applications, including emergency response, infectious disease early warning systems, resource allocation projections and food security analysis, to list only a few examples. However, obtaining reliable population estimates at the spatial resolutions required for many of these applications is a significant challenge. Census data, the primary source of population size, is often incomplete or unreliable - particularly in less-developed countries - which causes considerable problems for policy planning and decision makers. For this reason, models that can refine existing estimates of human populations or that can estimate populations in areas that lack population data altogether are of considerable importance.

Small area estimation (SAE) refers to methods for estimating small-scale characteristics of populations when there is little data, or in some cases no data at all. SAE methods are often used to produce estimates of population counts for small geographical areas, and to assess the accuracy of these estimates. Two distinct components make up SAE: design-based methods and model-based methods, which may be further divided into area-level models and unit-level models that employ either frequentist or Bayesian frameworks [1–2]. Unit-level models correspond to models for which information on the covariate and response variables are available for individuals, whereas area-level models only require area-averaged data for the covariates and the response. Design-based methods calculate the bias and variance of estimates from their randomization distribution induced by repeated application of the sample, while model-based methods produce inferences that are with respect to the underlying model. A significant limitation of design-based methods is that they have no means of producing predictions for areas in which no samples exist. SAE models, however, will draw information either from previous time periods or from similar areas in lieu of accurate population information [2]. This paper explores the extent to which the availability of data from previous time periods affects the choice of optimal model structure for area-level SAE models. The paper will focus on model-based methods for estimating population when no direct samples are available in the administrative unit of interest. The analysis and models discussed are limited in scope to spatial population estimation and should not be considered interchangeable with problems of temporal population prediction.

A variety of model-based methods are applicable to SAE problems, including microsimulation, areal interpolation and statistical modeling. While microsimulation and areal interpolation are valid approaches to SAE problems, this paper focuses on an assortment of statistical modeling methods. The brief review of relevant model-based estimation methods offered below is intended only to position the current research in relation to previous work on the same problem, and should not be considered a complete review of SAE as a whole.

Microsimulation produces SAE population estimates by modeling specific individuals or households and, in the case of dynamic

Table 1. Summary of model structures, strengths and weaknesses.

	Model Description	Advantages	Disadvantages
Linear Model (LM)	Linear model	Simple to implement, transparent model structure	Unable to capture nonlinear relationships
Linear Mixed Model (LMM)	Linear model incorporating spatial correlation	Explicitly accounts for spatial correlation, transparent model structure	Unable to capture nonlinear relationships
Generalized Additive Model (GAM)	Non-linear extension of a LM using a smoothing function	Able to represent nonlinear relationships	Vulnerable to model over fit, which degrades predictive accuracy
Multiple Adaptive Regression Splines (MARS)	Penalized spline, extension of a LM using multiple basis functions	Able to represent nonlinear relationships	Vulnerable to model over fit, which degrades predictive accuracy
Random Forest (RF)	Bagged classification and regression tree (CART) method	Nonparametric, designed to reduce variance and improve predictive accuracy of CART methods	Complex model structure, more difficult to succinctly measure variable importance
Bayesian Additive Regression Tree (BART)	Sum-of-trees method	Nonparametric, provides a flexible inference of the relationship between response variables and covariates	Complex model structure, difficult to interpret variable importance, computationally intensive

microsimulation, life events of those individuals [3–4]. Spatial microsimulation builds upon static or dynamic microsimulation by explicitly representing the spatial dimension inherent in population modeling. Because the method is so computationally intensive, the simulation is often limited in its scope of application but has the unique advantage of being able to model the impact of detailed alternative policy scenarios.

In contrast to the bottom-up approach of microsimulation, areal interpolation entails distributing administrative level census data across a finer scale to produce a detailed population surface. Areal interpolation is of interest for SAE problems when coarse scale population information is available, but small-scale measurements in areas of interest are not. The most commonly used technique for producing heterogeneous population density surfaces from homogeneous zones is dasymetric mapping, which uses ancillary information to divide each zone of the source data into subzones [5]. Each subzone is assigned a population density such that the sum of population over all subzones equals the population of the original source zone [6]. In recent years areal interpolation has been used to produce gridded estimates that are more readily compatible with external modeling frameworks.

Statistical modeling in the context of SAE refers to model-based methods of producing estimates, often described as synthetic estimates, which may be used directly or blended with design-based measurements to produce a final estimate [1–2]. The phrase "synthetic estimate" alludes to the fact that these estimates are inferred using a model of relationships drawn from a larger domain. Synthetic estimates may be produced using either indirect implicit methods, meaning that the model assumes a homogenous relationship between dependent and independent variables across the entire small area, or indirect explicit methods, meaning that the model takes into account the spatial heterogeneity present within the small area domain. This paper analyzes both indirect implicit and indirect explicit methods, as discussed further in the following section. We emphasize that the analysis is not a complete assessment of SAE methods, but is specifically focused on methods of producing estimates when no direct estimates are available for a particular administrative unit of interest.

Many statistical models employed for SAE are regression based. One of the fundamental models is the area level model, originally employed by Fay and Herriot (1979), which takes the form

$$y_i = \theta_i + \epsilon; \ \theta_i = x_i\beta + u_1,$$

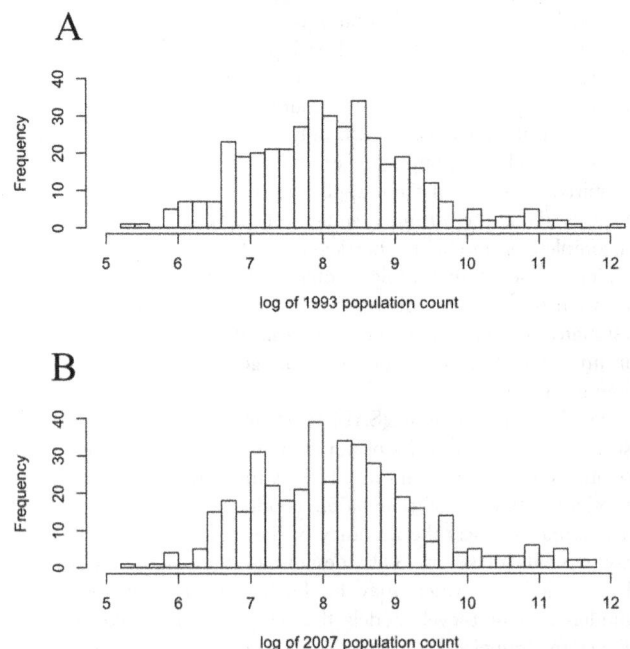

A

log of 1993 population count

B

log of 2007 population count

Figure 1. Population count by district for all regions included in the analysis for A) 1993 and B) 2007.

Table 2. Population counts for the five regions included in the analysis.

	1993 Population	2007 Population
Apurimac	381,997	404,190
Arequipa	916,806	1,112,858
Ayacucho	483,341	584,959
Cusco	1,001,898	1,154,969
Madre de Dios	67,008	102,577
Total:	2,851,050	3,359,553

Districts excluded due to irresolvable redistricting issues (see Data Consistency) are not included in the table.

$$u_i \sim N(0, \sigma_u^2),\ \epsilon \sim N(0, \sigma_\epsilon^2) \qquad (1)$$

where y_i is the direct estimator of θ_i, x_i is the associated covariate for area i and β is a coefficient for fixed effects [7]. u_i and ϵ are mutually independent error terms where u_i represents the random effects of area characteristics not accounted for in the covariates. The best linear unbiased predictor (BLUP), $\hat{\theta}_i$, under this model is defined as

$$\hat{\theta}_i = \gamma_i y_i + (1 - \gamma_i) x_i \hat{\beta} \qquad (2)$$

where γ_i is a tuning coefficient defined using the variances of u_i and ϵ as $\gamma_i = \sigma_u^2 / (\sigma_u^2 + \sigma_\epsilon^2)$. Note that Eq. 2 reduces to $\hat{\theta}_i = x_i \hat{\beta}$ for areas without any direct samples. This model may be adapted to the unit level as proposed by Battese, Harter and Fuller (1988); however, in the present application the covariate and response variables are available at the area-level only [8].

The BLUP, $\hat{\theta}_i$, in Eq. 2 is also the Bayesian predictor under normality of the error terms when using a diffuse prior for β. When the variance σ_u^2 is unknown – as is often the case – it is common to replace it with a sample estimate, yielding the empirical BLUP under the frequentist approach, or the Empirical Bayes predictor under the normality assumption (the prediction then being the mean of the posterior). For area-level data, the error variance (σ_ϵ^2) must be specified from external sources. The posterior distribution of θ_i may alternatively be calculated by specifying prior distributions for σ_u^2 and β, a technique known as the Hierarchical Bayes method [1].

The linear model outlined in equation (1) may be expanded to include spatial autocorrelation to address problems that are inherently spatially dependent. These models are often used, for example, in problems of disease mapping [9]. One such model used to account for spatial autocorrelation is a linear mixed model [10]. Our analysis includes versions of both linear models and linear mixed models as established linear modeling structures against which the performance of nonlinear methods may be evaluated.

Recent studies have explored semi-parametric variations of the area-level model (equation 2), including penalized splines [11] and the M-quantile method [12–13], in an effort to produce a more robust inference. Models using penalized splines allow for a more flexible representation of the relationship between covariate and response variables, while an M-quantile method uses area-specific models for the regression M-quantiles of the response for estimation. A variation of the penalized splines method is included in this study, as discussed further in the following section.

The work described above demonstrates the potential of using parametric and non-parametric regression-based models to improve small area estimates. Despite this potential, the application of SAE methods in resource-poor areas is somewhat limited. In Peru, for example, although the Center for International Earth Science Information Network (CIESIN) has produced gridded estimates of population density following from statistical data, the only attempt to utilize SAE models has been in the area of poverty estimation and follows from the methods outlined in Ghosh & Rao (1994) and Rao (1999) [14–19]. One potential barrier to using models outlined in the SAE literature may be a lack of expertise. In our analysis we propose a number of model structures that provide accessible alternatives to more complicated methods, which often require precise user specification to produce accurate estimates. In fact, two of the models explored in our analysis – the tree-based models – require no tuning or parameterization at all, as do regression models. This makes them particularly accessible tools for providing robust model estimates.

In order to systematically identify and understand alternative model structures for population prediction in data-limited regions, we compare the predictive accuracy of six different model techniques–including both regression and tree-based methods. Each model predicts population density for districts in Peru with no available direct samples in two separate circumstances: once in which population from a previous time period is available and once in which it is not. The models use information on transportation corridors, satellite-derived land surface conditions, economic indicators and – when included – population information from the 1993 census to predict population density in 2007. Sporadic population data collection such as the gap between the 1993 and 2007 census in Peru is common in low- and middle-income countries. In the following sections we will detail the structure of the models included (Section 2), describe data sources and the required data processing (Section 3), present and discuss results (Section 4), and offer general conclusions (Section 5). The analysis is relevant for regions with limited reliable census data. In the circumstance that policy or planning scenarios require population counts as opposed to density, the area of each district can be used to transform model estimated population density back to count estimates.

Materials and Methods

Understanding and selecting the appropriate model structure is perhaps the most important decision in the process of population modeling. The fundamental act of choosing a model structure will significantly affect the population estimate and the understanding of covariate influence. The most appropriate model structure often depends on the data available. In this analysis, six regression and

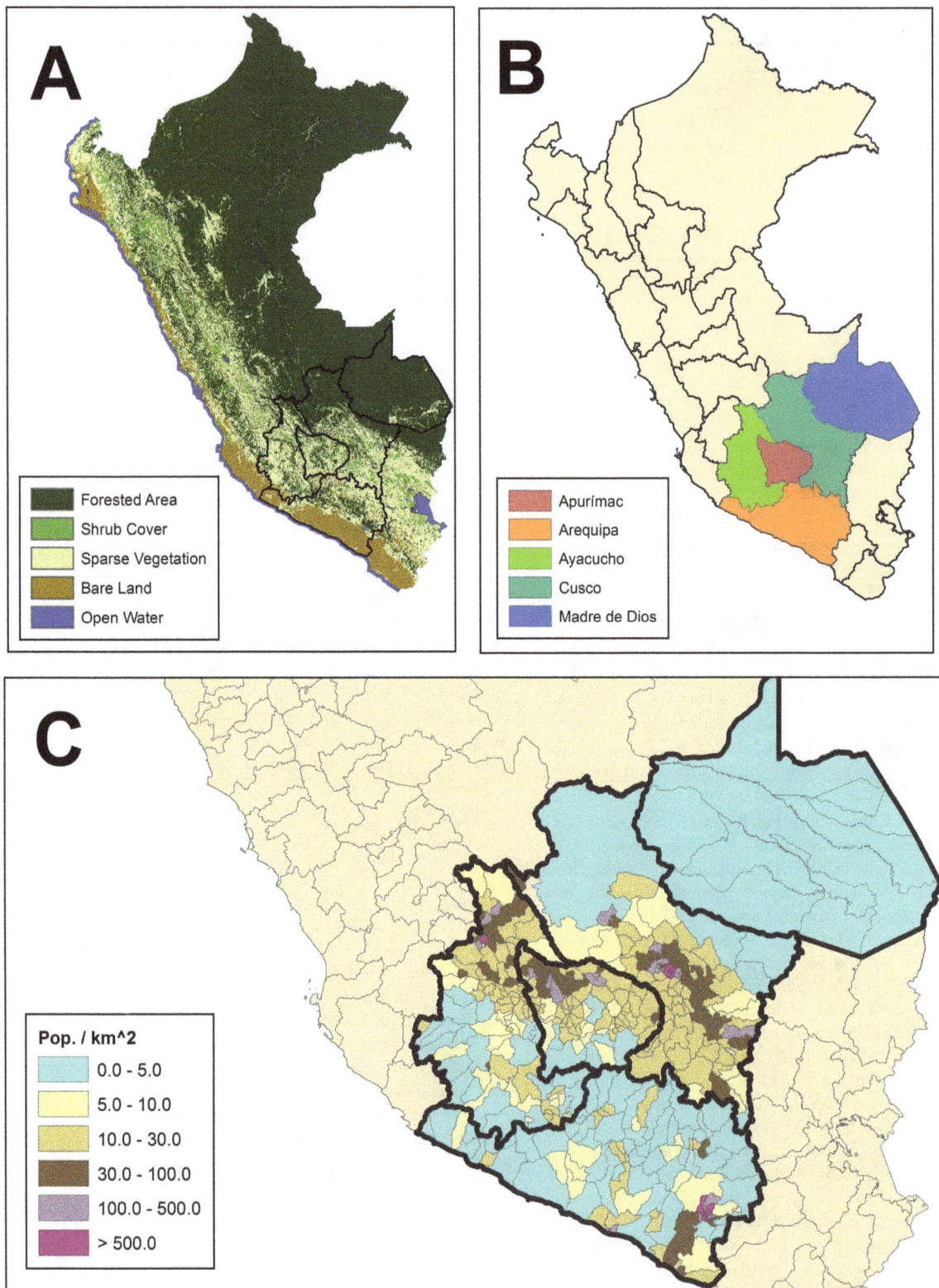

Figure 2. Political boundaries and topographical features of Peru. A) Classified land cover. B) Regions included in the analysis (Apurimac, Arequipa, Ayacucho, Cusco and Madre de Dios). C) Population density at the district level derived from the 2007 census.

tree-based models were chosen to explore how predictive accuracy and variable importance changes in the presence or absence of population information. The regression-based model structures include a linear model (LM), linear mixed model (LMM), a generalized additive model (GAM), and a multivariate adaptive regression spline (MARS) structure. The tree-based models

Table 3. Model errors and p-values assessed using density, population data included in the covariates.

	Average MAE	MAE Standard Error	LM	GAM	RF	MARS	BART	LMM
LM	0.051	0.033	-					
GAM	0.089	0.095	**	-				
RF	0.108	0.091	**	-	-			
MARS	0.051	0.036	-	**	**	-		
BART	0.085	0.060	**	-	*	**	-	
LMM	0.053	0.032	-	**	**	-	**	-
Mean	0.296	0.256	**	**	**	**	**	**

Columns 3–8 display p-values corresponding to the t-test between the MAE distributions of each row-column pair. Stars indicate statistical significance at a level of 0.01 (**) or 0.05 (*), while dashes indicate not significant.

Table 4. Model errors and p-values assessed using population count error as a proportion of actual district population: |Predicted−Actual|/Actual.

	Average MAE	MAE Standard Error	LM	GAM	RF	MARS	BART	LMM
LM	0.105	0.036	-					
GAM	0.108	0.042	-	-				
RF	0.150	0.054	**	**	-			
MARS	0.109	0.036	-	-	**	-		
BART	0.133	0.049	**	**	**	**	-	
LMM	0.104	0.036	-	-	**	-	**	-
Mean	0.491	0.207	**	**	**	**	**	**

Population data included in the covariates. Stars indicate statistical significance at a level of 0.01 (**) or 0.05 (*), while dashes indicate not significant.

Table 5. Model errors and p-values assessed using density, population data not included in the covariates.

	Average MAE	MAE Standard Error	LM	GAM	RF	MARS	BART	LMM
LM	0.371	0.123	-					
GAM	0.412	0.121	**	-				
RF	0.207	0.121	**	**	-			
MARS	0.371	0.174	-	**	**	-		
BART	0.298	0.157	**	**	**	**	-	
LMM	0.302	0.124	**	**	**	**	-	-
Mean	0.296	0.256	*	**	**	-	-	-

Columns 3–8 display p-values corresponding to the t-test between the MAE distributions of each row-column pair. Stars indicate statistical significance at a level of 0.01 (**) or 0.05 (*), while dashes indicate not significant.

Table 6. Model errors and p-values assessed using population count error as a proportion of actual district population: $|$Predicted $-$ Actual$|$/Actual.

	Average MAE	MAE Standard Error	LM	GAM	RF	MARS	BART	LMM
LM	0.515	0.118	-					
GAM	0.496	0.116	-	-				
RF	0.408	0.106	**	**	-			
MARS	0.479	0.129	-	-	**	-		
BART	0.436	0.110	**	**	-	**	-	
LMM	0.453	0.125	**	**	**	-	-	-
Mean	0.491	0.207	-	-	**	-	**	-

Population data not included in the covariates. Stars indicate statistical significance at a level of 0.01 (**) or 0.05 (*), while dashes indicate not significant.

Table 7. Measures of variable importance, population data included in the covariates.

	Previous Popdensity	Roads	River Water	X coordinate	Y coordinate	NDVI	LST Day	GDP	Perm Water
LM Beta Values*	0.979	-	-0.0199	-	-	-	-	-	-
LMM Beta Values*	0.979	0.0198	-	NA	NA	-	-	-	-
GAM Percent reduction in MSE	505.22	-0.92	-0.87	-0.07	1.6	1.78	-	-	-
MARS GCV*	100	5	-	-	-	-	-	-	-
RF Percent reduction in MSE	22.19	1	2.26	1.35	3.9	3.33	3.74	2.15	1.21
RF Inc. Node Purity	204.94	81.85	22.21	16.25	27.27	19.65	18.34	3.11	0
BART Mean number of splits	68.28	19.89	13.84	11.95	10.76	11.71	31.21	17.88	27.29

Stars indicate the models producing the most accurate estimates. Dashes indicate variables that were discarded by the model during variable selection.

include Random Forest (RF) and Bayesian additive regression tree (BART). A no model alternative was also included in the suite of models for reference. The strengths and weaknesses of each model structure are described in the following sections and summarized in Table 1.

The models were run twice: once with population density from 1993 included as a covariate, and once with it excluded, leaving only socioeconomic and environmental covariates. The two groups are intended to contrast the efficacy of model structures in the presence or absence of consistent census data. Previous population information, when included, was not modeled as a lagged effect, but rather as part of the covariate matrix. These two analyses are hereafter referred to as being with or without population data, although neither uses current period population information to estimate population density.

Linear Models (LM)

A LM is a linear function of the form

$$E(Y) = X\beta + \epsilon; \; \epsilon \sim N(0, \sigma_\epsilon^2) \qquad (3)$$

where Y is the vectorized form of the response variable, X is the covariate matrix, β is a vector of coefficients and ϵ is a vector of the normally distributed errors [20]. In this case β may be interpreted as the relative influence of each variable. The LM model structure provides a point of comparison with the LMM to determine the marginal benefit of adding spatial correlation, and provides perspective on the performance of each of the more complex models.

Linear Mixed Models (LMM)

A LMM is a LM in which the linear predictor may contain random effects with correlated errors [21], and takes the form:

$$E(Y) = \eta + \epsilon; \; \eta = X\beta + \upsilon;$$

$$\epsilon \sim N(0, \sigma_\epsilon^2) \; \upsilon \sim N(0, \sigma_\upsilon^2); \qquad (4)$$

where η is the linear predictor, β is a coefficient for fixed effects, X denotes the explanatory variables associated with these fixed effects and υ is a set of district-specific and possibly spatially correlated random effects that model between district variability in the response. For the purpose of population estimation in this study, errors between districts are modeled using exponential spatial autocorrelation according to the centroid of each district.

Generalized Additive Models (GAM)

A GAM is an extension of the LM, in which the assumption of linear relationships between covariates and response variables is relaxed by replacing the linear relationship with a nonparametric smoothing function, $f(X)$, such that the form of the function becomes

$$Y = f_1(X_1) + f_2(X_2) + ... + f_n(X_n) + \epsilon; \; \epsilon \sim N(0, \sigma_\epsilon^2) \qquad (5)$$

In this case a cubic spline was chosen for the nonparametric smoother with restricted degrees of freedom. In this way the GAM allows for non-linear relationships between the covariates and response variables [22]. The GAM model structure was included in this study as a relaxation of the linear features of LMs and

Table 8. Measures of variable importance, population data not included in the covariates.

	Previous PopDensity	Roads	River Water	X coordinate	Y coordinate	NDVI	LST Day	GDP	Perm Water
LM Beta Values	NA	0.154	-0.184	0.117	-0.102	-	-	-	-
LMM Beta Values	NA	0.166	-	NA	NA	-	-	-	-
GAM Percent reduction in MSE*	NA	-4.37	-1.3	-	-2.33	-2.59	-	-	-
MARS GCV	NA	100	45.9	15.8	26	22.2	16.8	-	-
RF Percent reduction in MSE*	NA	3.19	-0.91	2.45	4.7	4.88	5.61	4.33	-0.63
RF Inc. Node Purity*	NA	116.79	45.36	35.23	47.67	32.61	41.14	7.78	0.18
BART Mean number of splits	NA	55.03	28.43	18.09	33.01	30.33	39.62	37.19	26.61

Stars indicate the models producing the most accurate estimates. Dashes indicate variables that were discarded by the model during variable selection.

LMMs, and as an additional non-parametric alternative to the MARS model.

Multivariate Adaptive Regression Splines (MARS)

MARS is an extension of the linear class of models that allows for nonlinearity in the relationship between covariates and response variable by way of multiple basis functions that take the form $(x-t)_+$ or $(t-x)_+$ where t is a "knot point" determined in the model training process and x is the covariate. The model first enumerates basis functions to fit the data and then prunes back these functions, as would a tree-based model [23]. This gives the model the form

$$Y = \beta_0 + \sum_{m=1}^{M} \beta_m h_m(X) \qquad (6)$$

where $h_m(X)$ is a basis function, or product of basis functions, and β_m are coefficients estimated by minimizing the sum-of-squares. The MARS model was included in this study as a variation of a penalized spline, which has been previously explored as a means of providing more robust inferences [12].

Random Forest (RF)

Tree-based methods are often most useful for models that are highly non-linear. The most basic tree-based structure is the Classification and Regression Tree (CART), which recursively partitions the data into i subspaces and applies a very simple model to each subspace. If the loss measure used is the sum of squares, the model takes the form $\mu_i = \text{mean}(Y_i \mid x_i \in A_i)$ where μ_i is the parameter to be predicted in subspace A_i, Y_i is the set of values of the response variable on which the model is trained in that subspace and x_i is the matrix of the associated covariates.

One downside of CART is that the hierarchical nature of the model means relatively small changes in the data set can result in drastically different partitions within the data space, which makes drawing insight from the model structure difficult. One approach to reduce the variability inherent in predictions from CART models is to use model averaging based on bootstrapping, a method known as bagging [22]. The RF model structure is similar to a bagged CART method, except that a random subset of variables less than the total number of variables are chosen to use at the splitting point for each tree. The method originally proposed by Breiman [24] to grow B trees, each denoted by T_b is summarized below for a training dataset X containing M classifier variables:

1. Form bootstrap datasets x_b by sampling with replacement from X.
2. Select m < M variables at each node of tree T_b. Calculate the best split for the bootstrapped dataset based on the m selected variables. Repeat this step until the specified minimum node size is reached.
3. Repeat steps 1 and 2 for each of the B trees.

Each tree is thus grown to its maximal depth. This process may be represented as:

$$f_{RF}^{(B)}(X) = \frac{1}{B} \sum_{i=1}^{B} f_{RF}^{(b)}(x_b) \qquad (7)$$

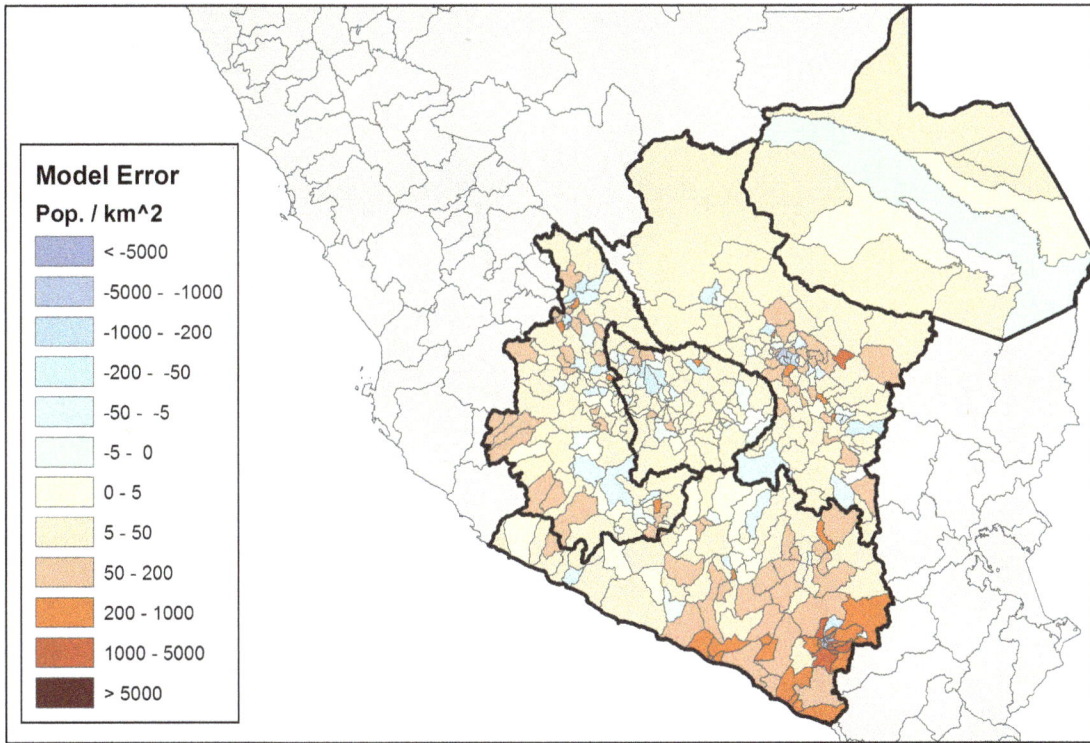

Figure 3. Random Forest model error by district.

The randomization process is intended to produce uncorrelated trees (although in reality some correlation may remain) such that the aggregate result is a reduction in the variance [24]. If σ^2 is taken to be the variance of an individual tree, and ρ is the correlation between tree predictions, then the variance may be represented as

$$Var\left(f_{RF}^{(B)}(X)\right) = \rho\sigma^2 + \frac{(1-\rho)}{B}\sigma^2 \tag{8}$$

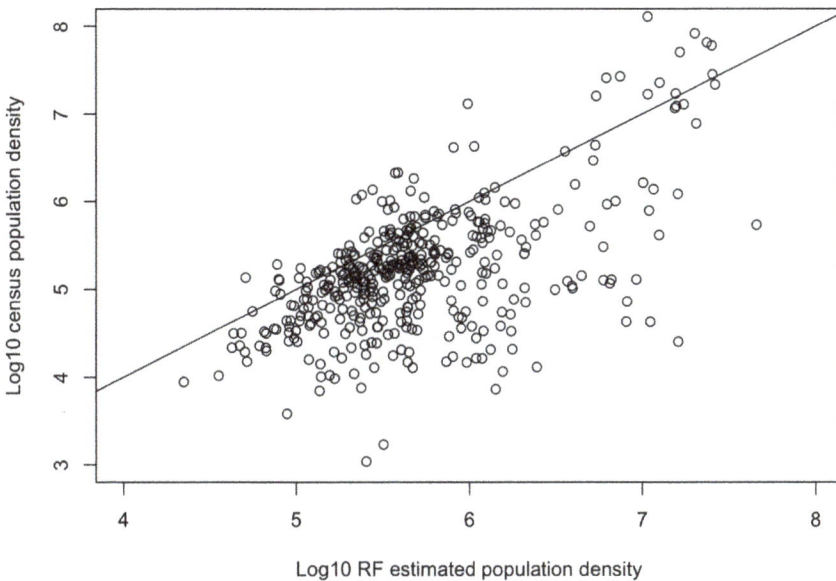

Figure 4. Census population density vs. Random Forest estimated density. 1:1 line plotted for reference.

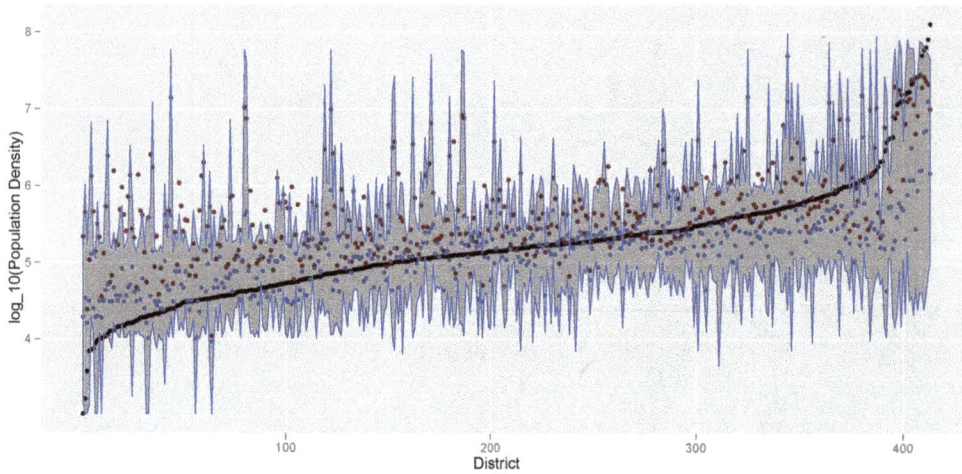

Figure 5. Random Forest uncertainty analysis. Observations of district population density (black points) are ordered from lowest to highest density. The Random Forest mean (red point), median (blue point) and interval between the 5th/95th percentiles (blue lines) illustrate the uncertainty in each corresponding model estimate.

Bayesian Additive Regression Tree (BART)

The BART model builds on regression and classification trees as a "sum-of-trees" method. The model places a prior probability on the nodes of each tree such that the tree is constrained to be a "weak learner," biasing the tree towards a shallower, simpler structure [25]. This constraint ensures that each tree contributes only minimally to the overall fit. The model is designed to produce a flexible inference of the relationship between the sum of trees and the response variable.

Following the notation of Chipman et al. (2010), let T represent a single binary tree containing a set of interior decision nodes, terminal nodes and $M = \{\mu_1, \mu_2, \dots \mu_n\}$ parameter values associated with each of the n terminal nodes [25]. Each decision rule is a binary split of the form $\{x \in A\}$ vs $\{x \notin A\}$, where A is a subset of the range of x. Each value of x, by means of binary decision nodes, is assigned to a single terminal node and therefore to a value μ_i. for a given T and M, $g(x; T, M)$ denotes the function that assigns $\mu_i \in M$ to x. A single tree model is therefore represented as:

$$E(Y|x) = g(x; Y, M) + \hat{\epsilon}; \ \hat{\epsilon} \sim N(0, \sigma^2) \qquad (9)$$

Where $\hat{\epsilon}$ represents the normally distributed residual term centered on 0 with variance σ^2. In the single tree model represented by Eq. 9, $E(Y|x)$ equals the parameter μ_i assigned by $g(x; T, M)$. Using the same notation, the sum of m trees model may therefore be expressed as:

$$E(Y|x) = \sum_{j=1}^{m} g(x; T_j, M_j) + \hat{\epsilon}; \ \hat{\epsilon} \sim N(0, \sigma^2) \qquad (10)$$

Under the sum of trees model (Eq. 10), $E(Y|x)$ equals the sum of all μ_{ij}s assigned to x by the $g(x; T_j, M_j)$s. Each μ_{ij} therefore only represents part of $E(Y|x)$ under the sum of trees model. Because each $g(x; T_j, M_j)$ may be based on one or more x's, each tree has the ability to represent either a "main effect" (single component of x, single variable tree) or an "interaction effect" (multiple components of x, multi-variable tree).

The final specification of the BART model is a prior that is imposed on all parameters in the sum of trees model (i.e. $(T_1, M_1)\dots(T_m, M_m)$ and σ). The prior is designed to regularize individual tree influence such that the effect from no one tree dominates the model. The prior on T puts a larger weight on small trees and the prior on μ shrinks the fit of each terminal node proportional to the number of trees such that the contribution of any one tree decreases as number of trees increases. The full specifications of this prior may be found in Chipman et al. [25].

Mean Model

Each of the previously described models was compared against the no-model mean alternative. The no-model mean estimate was simply calculated as the mean of available response data in the holdout dataset.

Data

Peru is divided administratively into regions then provinces followed by districts. The variables used, discussed below, were calculated annually at the district level for five regions (Ayacucho, Cusco, Madre de Dios, Arequipa and Apurimac). Table 2 and Figure 1 provide descriptive statistics of the population in the five chosen regions. The total population count of all five regions increased from 1993 to 2007 (see Table 2) with a slight skew towards growth in more populous districts (see Figure 1), which broadly reflects the overall population dynamics of the country as a whole [16].

Combined, these regions contain 42 provinces, 417 districts and span a reasonable cross section of Peruvian land cover (see Fig. 2). The country exhibits a broad range of climatic variability with land cover including rainforest, mountains and coastal areas. The five regions included in the study were chosen to be representative of Peruvian topography and to minimize redistricting within the study domain. Eight districts were created due to redistricting between 1993 and 2007, seven of which resulted from splitting an existing administrative area into two separate districts. Affected districts were recombined to pre-2007 boundaries, and all relevant variables were re-calculated.

Response variable

Population density in 2007 was used as the response variable (Y) and was calculated using the 2007 census data downloaded from the Peruvian Institute of Statistics and Information (El Instituto Nacional de Estadística e Informática; INEI). Information on the area of each district was obtained from the GADM Global Administrative Database [26].

Covariates

While obtaining past measures of population is a priority for improving the accuracy of population estimates, previous modeling efforts have also used a variety of ancillary variables to improve model accuracy, including information on roads, slope, nighttime lights, measures of urban areas, land use, socioeconomic characteristics and dwelling counts [27–28]. The scale of our study precluded the use of direct estimates of density such as counting dwellings, but measures of the land surface conditions were incorporated in the form of NDVI and daytime LST, as described in detail below. We chose to expand upon past studies that demonstrated the utility of including information on roads by selecting measures of potential transportation corridors such as rivers and inland open water. In addition to the physical characteristics of the study region, we used an index of GDP as an economic indicator for each region. In the following sections we describe the model covariates in detail and provide a brief explanation of why each was chosen for the analysis.

Population Density in 1993. Population density obtained from the 1993 census was incorporated as a measure of population in a previous time period. Peru conducted a census in 2005, but the methodology and results were considered flawed resulting in the 2007 census When data are available, previous population metrics have obvious value for predicting current population. The population density for 1993 was calculated similarly to that for 2007.

Connectivity Variables. Proximity to transportation corridors has previously proven useful in mapping population, as demonstrated by the LandScan Global Population Project, which uses the Digital Chart of the World to incorporate distance to major roads as part of a population estimation model [27]. In this study, we have extended the idea of incorporating information on transportation corridors to include navigable bodies of water. A map detailing inland roads, rivers and permanent bodies of water in Peru was downloaded from the Digital Chart of the World and aggregated to the third administrative level (districts) to obtain the density of roads, rivers and permanent bodies of water in each district. All data that originates from the Digital Chart of the World represents infrastructure circa 1992, and is static across analyses.

Density of inland roads is an indicator of urbanization and accessibility for each district. Proximity to transportation corridors–both roads and navigable rivers–may be especially important in the Amazon where large tracts of forest make transportation difficult.

Information on inland water was included because urban areas have a history of developing around bodies of water. Rivers are a valuable natural resource and provide a method of transportation. The predictive accuracy of river density may be understated in the Amazon basin because the river network is so dense that almost all land areas are accessible by river. Data for density of permanent bodies of water was categorized as a variable as well, separate from density of rivers.

Satellite Derived Land Surface Conditions. Remotely sensed data derived from aircraft or satellite has a fairly long tradition in population modeling. Remote sensing provides a method for detailing landscape characteristics for each district, which may in turn be linked to the population density [27,29–31]. Early uses of remote sensing for population estimation were logical extensions of aerial photography, which has been used to count dwellings since the mid 1950s in areas without reliable population information [32–33]. Although counting dwellings becomes prohibitively time-consuming as the size of the study region increases, high-resolution remote sensing has still been used to disaggregate population counts in urban spaces under the assumption that areas with similar land cover will have similar population densities. In recent years, remote sensing has become a prominent source of environmental information, including land use and transportation patterns, which can provide valuable input to population models.

In addition to being used to disaggregate population densities, remote sensing is perhaps even more valuable for statistical population models in which it is used to estimate population density. The most common way that this is done is to relate the remotely sensed data to land use and to include that information in a regression-based model that is identified and trained using one dataset and evaluated using a separate dataset from a culturally and demographically similar area [30,34]. While remotely sensed data are often used to derive social or economic information relevant to population density, satellite observations may also be included directly in a population model. Liu and Clarke (2002) demonstrated this by using high-resolution satellite-derived reflectance and landscape texture information to estimate population distribution within a single city [35]. The inclusion of these remotely sensed data-either directly or indirectly-allows modelers to draw insight into the underlying drivers of local population processes. For this study, remote sensing data collected by the MODIS Terra sensor was included directly as an indicator of land surface conditions. The two chosen characteristics were the Normalized Difference Vegetation Index (NDVI) and the daytime land surface temperature (LST).

NDVI measures the difference between reflectance in the near infrared and the visible spectrum. The chlorophyll in healthy vegetation strongly absorbs visible radiation while the plant cell structure reflects it. NDVI may therefore be used both as a measure of vegetative distribution and as an indicator of vegetative health. The difference in vegetative distribution may also provide information on patterns of topographical features in the landscape, as vegetative differences are often indicative of topography. NDVI was available as a 1 km resolution gridded product, but was aggregated to district averages using the administrative boundaries to match the resolution of the other covariates. Data were available as monthly composites. For this study a consistent month during the dry season was chosen (July).

Daytime LST may act to differentiate between the diverse land cover of Peru, which includes open water, bare soil, forested areas, rock and urban areas. The diurnal thermal signal of each category of land cover may provide insight into the potential habitability of that area. Daytime LST can also give an indication of the heat island effect of cities for some of the smaller districts in which impervious cover is an appreciable portion of total surface area. LST was available at 1 km resolution in 8-day composites. For this study the same composite was chosen from each year (mid July to match the NDVI). LST was aggregated to the district level using the GADM Global Administrative Areas dataset.

NDVI and LST are not perfect or comprehensive indicators of land surface conditions or specific land use. However, they are well-understood and physically meaningful variables that are indicative of differing land cover patterns and physiographic conditions that are likely to be relevant to the distribution of population. As satellite-based thematic land use classification in humid tropical regions is still a topic of active research [36–40], we choose to use NDVI and LST instead of relying on an error-prone land use dataset.

Economic Variables. While transportation networks and land cover and conditions may be important predictors of population density, they are unlikely to provide useful information for areas that have already been urbanized. A GDP index derived from the 2007 and 1993 censuses was downloaded from the INEI as an economic indicator for the analysis. Although the available GDP index is on a coarse spatial scale (provincial), it may provide significant information on interannual variability not present in other predictors.

Data Consistency

Not all of the data from the Digital Chart of the World matched the INEI districting, although discrepancies between datasets were minor. After standardizing the data, out of 417 districts present in each year (according to the most recent INEI report) 412 districts mapped to those in the Digital Chart of the World. Districts that were omitted from the study include Jesus Nazareno, Llochegua, Huepetuhe, Majes and Kimbiri. The missing districts were due to redistricting between 1992–the year that the Digital Chart of the World was created–and the 1993 census.

Analysis Structure

The analysis may be broadly categorized into two sections: one in which population density from 1993 is included as a covariate and one in which it is excluded, leaving only socioeconomic and environmental covariates to predict population density in 2007. The predictive accuracy of each model, both with and without population information, was evaluated using 300 repetitions of a random 10% holdout analysis. This process involves withholding 10% of the data at random (response variable and associated covariates), fitting the parameters of each model using the remaining 90% of data, and then producing predictions for the withheld 10%. The absolute difference between the prediction from each model k for each district i and the actual withheld value was calculated as mean absolute error (MAE), displayed below:

$$MAE_{kj} = \frac{\sum_{i}^{m} |\tilde{Y}_i - Y_i|}{m} \text{ and } AMAE_k = \frac{\sum_{j=1}^{N} (MAE_{kj})}{N} \quad (11)$$

Where $\tilde{Y}_i$ is the actual population density of district i, Y_i is the model prediction for district i, m is the total number of districts in the holdout dataset, MAE_{kj} is the mean absolute error of all predictions for model k in repetition j and N is the total number of repetitions. $AMAE_k$ is the average mean absolute error for model k over all repetitions j. Only one year of census data was available for the response variable dataset, meaning that every prediction made in the holdout analysis was an out-of-sample prediction. Models are therefore drawing strength from the surrounding districts in that they formulate inferences based on the 90% of districts not in the holdout dataset and make predictions for the 10% of districts in the holdout sample. Such an approach is useful in many contexts but still has important limitations. Our holdout analysis cannot be generalized to countries or large regions that completely lack population data and does not test the ability of models to predict current population following calibration using old data from the same region. While calibrating each model using past data is a feasible alternative to using the information from

surrounding administrative units, we were unable to test the relative strength of this methodology due to the limited temporal availability of our covariates.

The suite of models is assessed using average mean absolute error (AMAE) as a measure of general model accuracy as well as average root mean square error (ARMSE). The ARMSE penalizes large model errors more heavily than does AMAE, meaning that the difference between AMAE and ARMSE is used to assess the skill of each model in providing population density estimates for the outlier districts in the dataset. In evaluating the results of each analysis, reported levels of statistical significance are measured to a significance level of 0.05 following a Bonferroni correction for multiple pair-wise comparisons and are based on AMAE (See Tables 3 through 6). Tables 3 and 5 provide measures of model accuracy when measured using population density, while Tables 4 and 6 measure model accuracy using population count error as a proportion of actual district population to provide a more intuitive measure of model performance.

The diversity of model structures included in the analysis required the use of multiple measures of variable influence in the analysis of the results. The relative importance of each variable in the LM and LMM was measured using the β coefficient from the final fitted model. In this case the β coefficient indicates the linear relationship between covariate and response variable. Variable influence in the MARS model was based on the contribution of a variable towards reducing the model's generalized cross-validation (GCV) score. GCV is an approximation of the leave-one-out cross-validation using a squared error loss measure [22]. The measure of variable importance used for a GAM was the increase in average MSE that results from removing a specific variable. Variable importance in the RF model was evaluated using two separate indices. The first is based on perturbing each variable and recording the effect on the out-of-bag accuracy as measured by average MSE, while the second measures the decrease in node impurities- measured by the residual sum of squares- that results from splitting on a variable. Variable importance in the BART model was evaluated by the number of times a variable was used as a splitting decision in a tree, averaged over all trees. Due to the discrepancy between measures of variable influence, direct comparisons between models cannot be made. Instead, the shift in relative variable influence between analyses is explored within each model to understand how each model is affected by the presence of population density information in the covariates.

The analysis was conducted using R, with the following packages and functions: the linear model was fit using the glm() function from the stats package with a Gaussian link function and variable selection conducted using the step() command; the generalized additive model was fit using the gam() function from the mgcv package with added penalty terms for each new variable using the select command; the linear mixed model was fit using the glmmPQL() function from the MASS package with exponential spatial correlation and a Gaussian link function; the random forest model was fit using the function randomForest() from the package of the same name; the multivariate adaptive regression splines model was fit using the earth() function from the package of the same name; and the Bayesian Additive Regression Tree was fit using the bart() function from the BayesTree library. No additional specifications were required for the RF, MARS or BART models.

Results and Discussion

Despite the fact that the regression based models (LM, LMM, GAM and MARS) provided the most skilled predictions of population density when 1993 population density was included in the covariates, these models- with the exception of the LMM-provided among the worst predictions when no population information was included (see Tables 3 and 5). The inclusion of spatial correlation (represented by the LMM) produced the best predictive accuracy among regression models, but the model still performed no better than the no-model alternative (Table 5). Although there are minor differences in model performance when assessed using population counts, notably the improved performance of GAM, the relative model accuracy remains unchanged (see Tables 4 and 6). This result indicates that when population information was not available regression-based models (both parametric and non-parametric) were unable to capture the relationship between indicator variables and current population density.

In contrast to the regression models, the RF model – a non-parametric tree-based model - provided among the most skilled estimates when 1993 population density was not included in the covariates, but among the least skilled estimates when it was (see Tables 3 through 6). Notably, when population information was not included the RF model was the only model to significantly outperform the no-model alternative as assessed by population density. When measured using population counts instead of density, both tree-based models (RF and BART) significantly outperform the no-model alternative. The shift in relative model performance indicates that the relationship between previous population and current population at a district scale can be modeled effectively using regression methods, but the relationships between ancillary variables and population require a more flexible model structure better able to handle nonlinearity and high variance. This discrepancy in predictive accuracy demonstrates the potential of tree-based models for estimating population in data-limited areas.

The differences in model accuracy as evaluated by ARMSE as opposed to AMAE are minimal in terms of ordinal rank but entail consistently larger mean error estimates with increased standard errors. The systematic difference in model accuracy as measured by AMAE and ARMSE (results not shown) implies that the model estimates contained disproportionately large errors in a small number of predicted districts. Although an expanded dataset containing a greater number of districts may help to reduce the standard error of the model estimates across models, in data-scarce regions it's very likely that the data available to train models is limited.

The covariate influence of all models was explored to understand the differences in variable importance between the two analyses. Although the most direct measure of variable importance is model dependent, which precludes direct comparisons between measures of variable importance, relative comparisons between analyses are instructive. When population density from 1993 was included in the covariates, LM, LMM and MARS – the three models that provided the best population density estimates for the analysis– all indicated that previous population density was the most significant variable as assessed by their respective measures of variable importance (Table 7). This relative variable importance is not surprising, but is an important point of comparison for evaluating the models that do not include 1993 population density information in the covariates.

When 1993 population information was not included in the analysis, nearly all of the models indicated a greater number of covariates were significant, many of which the models had previously excluded (see Tables 7 and 8). Random Forest – the model that provided the best population density estimates for the analysis - indicated that the majority of remaining covariates had comparable variable importance (Table 8). The flexibility of the RF model structure compensated for a lack of previous population density information more effectively than regression-based models by incorporating information from the available covariates. The second tree-based method, BART, similarly produced estimates that were more accurate than all regression-based models (both parametric and non-parametric), although not statistically distinct from the mean model.

Random Forest population density estimates and model errors are explored spatially and in their relation to actual district population density to better understand the performance of the model. Figure 3 demonstrates some spatial dependencies in the model errors, particularly in the southeast, although performance was mixed across much of the study domain. Isolated districts across the domain display large model errors, which demonstrates the limitations of using model-based population estimates alone. The population dynamics in these districts are likely controlled by variables not captured in our analysis or exhibit an anomalous relationship between the covariates and response variable. This serves as a point of caution when interpreting any single prediction from a model optimized for predictive skill across a large and diverse study region. Figure 4 shows that RF tended to overestimate population density in general but slightly underestimated the population density of the most dense ~10% of districts. The overestimation bias for lower-density districts is not surprising given the relatively small margin available for underestimation in such districts. The inability of the RF model to produce accurate population density estimates for the most population dense districts implies that the resolution of the analysis – which in this case is the district level – may have been insufficient to capture the upper extreme of population-density due to heterogeneity of the response variable within each district. Dense urban areas may account for the majority of a district's population but a relatively minimal proportion of its land area, on which many covariates were based.

To explore the uncertainty in estimates made using the Random Forest model, a Quantile Regression Forest (QRF) was used to characterize the distribution around each model estimate [41]. Just as the RF model is used to estimate the conditional mean of a response variable, $E(Y | X = x)$, the QRF model is used to estimate quantiles in the conditional distribution of a response variable as $E(1\{Y \leq y\} | X = x)$. Figure 5 depicts the actual district population densities, ordered from smallest to largest, the RF mean estimate for each district and the QRF estimate of the 5th, 50th and 95th percentiles. The RF mean is shown to overestimate density in all districts with the exception of the ~10% of districts with the highest population density. Although nearly all of the population density observations fall within the bounds defined by the 5% and 95% QRFs, these bounds are often quite wide, demonstrating the uncertainty inherent in predicting population density without prior information on population.

Figure 5 demonstrates that the RF mean is significantly skewed towards the high end of the QRF distribution, at times falling outside of the 95th percentile. Such behavior is an indicator that the RF model is producing some trees in the ensemble with significantly higher population density estimates than the rest of the ensemble. This is likely another indicator that, as previously discussed, the data is highly nonlinear and may not be well represented in some of the area-level covariates due to heterogeneity within individual districts. The QRF median (50th percentile) was included in Figure 5 as an illustration of a potential alternative to the RF mean for problems that demonstrate such nonlinearities. While the QRF median provides a more stable estimate that is more accurate for all but the highest density districts, it also significantly reduces the range of the estimates. Therefore, while the median may be preferable in this particular model, it is still logical to begin with the RF mean as an estimate in most situations.

Conclusions

For regions in which data limitations preclude the use of reliable demographic information, it is important that model structures effectively incorporate all available data. Producing reliable small area estimates of population density for areas that lack direct samples is a problem of interest for resource allocation, disease early warning, and food security analysis. Such estimates are vital for decision makers operating in regions limited by incomplete or unreliable census data. It is often these same data-limited regions that lack information from previous time periods for use in training and testing models.

The improvement in predictive accuracy demonstrated by the RF model represents practical value for decision makers and policy makers. The average MAE of MARS, GAM and LM are twice as large as for RF, and that of LMM is one and a half times as large. Even when compared to the average model in which population information is available, the RF model produces errors that are only two to four times greater and therefore still useful in an applications context. While the improvements to predictive accuracy are limited to the case in which no previous population

information is available, it is in just such cases that model estimates must be relied upon most heavily.

The results of this study illustrate that for non-sampled areas a regression-based model may not be the most effective model structure depending on the continuity and consistency of available census data. Without information from prior time periods, the flexibility provided by the non-parametric tree-based models produced more accurate predictions than did conventional parametric and non-parametric regression methods. The predictive accuracy of tree-based non-parametric models in population modeling is an area that has been largely unexplored, but which warrants further study as a flexible alternative to conventional regression based methods.

Acknowledgments

The authors would like to thank Dr. Laurie Brown and Dr. Filipe Batista e Silva for providing constructive and insightful comments in their reviews, as well as the helpful comments of three anonymous reviewers.

Author Contributions

Conceived and designed the experiments: WA SG. Performed the experiments: WA. Analyzed the data: WA SG BZ WP. Wrote the paper: WA SG BZ WP.

References

1. Pfeffermann D (2002) Small Area Estimation – New Developments and Directions. International Statistical Review, 70(1): 125–143 doi:10.1111/j.1751-5823.2002.tb00352.x.
2. Pfeffermann D (2013) New Important Developments in Small Area Estimation. Statistical Science, 28 40–58. doi:10.1214/12-STS395.
3. O'Neill B, Balk D, Brickman M, Ezra M (2001) A Guide to Global Population Projections. Demographic Research, 4(8) p203–288. doi:10.4054/DemRes.2001.4.8.
4. Ballas D, Clarke G, Dorling D, Eyre H, Thomas B, et al. (2005) SimBritain: a spatial microsimulation approach to population dynamics. Population, Space and Place 11, 13–34.
5. Eicher C, Brewer C (2001) Dasymetric mapping and areal interpolation: Implementation and evaluation. Cartography and Geographic Information Science, 28, 125–138.
6. Langford M, Maguire DJ, Unwin DJ (1991) "The Aerial Interpolation Problem: Estimating Population Using Remote Sensing in a GIS Freamework" in Handling Geographical Information: Methodology and Potential Applications, Masser, I. and Blakemore, M. (Eds), New York, NY: Wiley, 55–77.
7. Fay RE, Herriot R (1979) Estimates of income for small places: An application of James-Stein procedures to census data. J. Am. Statist. Ass., 74, 269–277.
8. Battese GE, Harter RM, Fuller WA (1988) An error components model for prediction of county crop area using survey and satellite data. J. Amer. Statist. Assoc. 83 28–36.
9. Clayton D, Kaldor J (1987) Empirical Bayes estimates of age-standardized relative risks for use in disease mapping. Biometrics, 43: 671–681.
10. Wakefield J (2007) Disease Mapping and Spatial Regression with Count Data. Biostatistics, 8: 158–183.
11. Opsomer JD, Claeskens G, Ranalli MG, Kauermann G, Breidt FJ (2008) Nonparametric small area estimation using penalized spline regression. J. R. Stat. Soc. Ser. B Stat. Methodol. 70 265–286. MR2412642.
12. Chambers R, Tzavidis N (2006) M-quantile models for small area estimation. Biometrika 93 255–268. MR2278081.
13. Tzavidis N, Ranalli MG, Salvati N, Dreassi E, Chambers R (2014) Robust small area prediction for counts. *Statistical Methods in Medical Research.* To appear. doi:10.1177/0962280214520731.
14. Escobal J, Ponce C (2011) Spatial patterns of growth and poverty changes in Peru (1993–2005). Documento de Trabajo N° 78. Programa Dinámicas Territoriales Rurales. Rimisp, Santiago, Chile, 1–43.
15. Ghosh M, Rao JNK (1994) Small Area Estimation?: An Appraisal. Statistical Science, 9(1), 55–76.
16. Rao JNK (1999) Some recent advances in model-based small area estimation. Survey Methodology, Vol. 25, No. 2, 175–186. Statistics Canada, Catalogue No. 12-001.
17. Tobler W, Deichmann U, Gottsegan J, Maloy K (1997) World population in a grid of spherical quadrilaterals. International Journal of Population Geography, V3: 203–225.
18. Deichmann U, Balk D, Yetman G (2001) Transforming Population Data for Interdisciplinary Usages: From Census to Grid, Palisades, NY: CEISEN,

Columbia University. Working paper online: Available: http://sedac.ciesin.columbia.edu/plue/gpw/GPWdocumentation.pdf.
19. Center for International Earth Science Information Network - CIESIN - Columbia University (2005) Poverty Mapping Project: Small Area Estimates of Poverty and Inequality. Palisades, NY: NASA Socioeconomic Data and Applications Center (SEDAC). Available: http://sedac.ciesin.columbia.edu/data/set/povmap-small-area-estimates-poverty-inequality.
20. Cameron AC, Trivedi PK (2005) Microeconometrics: Methods and Applications. Cambridge: Cambridge University Press.
21. Breslow NE, Clayton DG (1993) Approximate inference in generalized linear mixed models. J. Am. Stat. Assoc. 88: 9–25.
22. Hastie T, Tibshirani R, Friedman J (2009) Additive Models, Trees and Related Methods. Chapter 9 of The Elements of Statistical Learning: Data Mining, Inference, and Prediction. Springer-Verlag.
23. Friedman JH (1991) Multivariate Adaptive Regression Splines. Annals of Statistics 19 (1): 1–67. doi:10.1214/aos/1176347963.
24. Breiman L (2001) Random forests, Machine Learning. Vol 45(1), No 5–32.
25. Chipman HA, George EI, McCulloch RE (2010) BART: Bayesian Additive Regression Trees. Ann. Appl. Stat. Volume 4, Number 1, 266–298.
26. Hijmans R, Garcia N, Rala A, Maunahan A, Wieczork J, et al. (2012) GADM Global Administrative Areas Version 2, www.gadm.org.
27. Dobson JE, Bright EA, Coleman PR, Durfee RC, Worley BA (2000) LandScan: A Clogal Population Database for Estimating Populations at Risk, Photogrammetric Engineering & Remote Sensing. Vol 66, No 7.
28. Wu S, Qiu X, Wange L (2005) Population Estimation Methods in GIS and Remote Sensing: A Review. GIScience and Remote Sensing, 42 p58–74.
29. Monmonier MS, Schnell GA (1984) Land-Use and Land-Cover Data and the Mapping of Population Density. The International Yearbook of Cartography, 24: 115–121.
30. Harvey JT (2002) Population Estimation Models Based on Individual TM Pixels. Photogrammetric Engineering and Remote Sensing, 68(11): 1181–1192.
31. Langford M, Harvey J (2001) "The Use of Remotely Sensed Data for Spatial Disaggregation of Published Census Population Counts" IEEE/ISPRS Joint Workshop on Remote Sensing and Data Fusion over Urban Areas, 260–264.
32. Boudot Y (1993) Application of remote sensing to urban population estimation: a case study of Marrakech, Morocco. EARSeL Advances in Remote Sensing, Vol 3, No 3.
33. Puissant A (2010) Estimating population using remote sensing imagery. Panel contribution to the Population-Environment Research Network Cyberseminar, "What are the remote sensing data needs of the population-environment research community?".
34. Lo CP (2003) Zone Based Estimation of Population and Housing Units from Satellite-Generated Land Use/Land Cover Maps. In V. Mesev (Ed.), Remotely Sensed Cities (157–180). London, UK/New York, NY: Yaylor & Francis.
35. Liu X, Clarke KC (2002) Estimation of Residential Population Using High Resolution Satellite Imagery, Proceedings of the 3rd Symposium in Remote Sensing of Urban Areas, Istanbul, Turkey. June 11–13.
36. Hansen MC, Stehman SV, Potapov PV (2010) Quantification of global gross forest cover loss. PNAS 107 (19) 8650–8655; doi:10.1073/pnas.0912668107.

37. Walker WS, Stickler CM, Kellndorfer JM, Kirsch KM, Nepstad DC (2010) Large-Area Classification and Mapping of Forest and Land Cover in the Brazilian Amazon: A Comparative Analysis of ALOS/PALSAR and Landsat Data Sources, IEEE Journal of Selected Topics in Applied Earth Observations and Remote Sensing 3(4) 594–604: doi:10.1109/JSTARS.2010.2076398.

38. Asner GP, Rudel TK, Aide TM, Defries R, Emerson RE (2009) A Contemporary Assessment of Change in Humid Tropical Forests. Conservation Biology 23: 1386–1395.

39. Fritz S, See L, McCallum I, Schill C, Obersteiner M, et al. (2011) Highlighting continued uncertainty in global land cover maps for the user community. Environmental Research Letters, 6(4), p.044005.

40. Herold M, Mayaux P, Woodcock CE, Baccini A, Schmullius C (2008) Some challenges in global land cover mapping: An assessment of agreement and accuracy in existing 1 km datasets. Remote Sensing of Environment, 112(5), 2538–2556.

41. Meinshausen N (2006) Quantile Regression Forests. Journal of Machine Learning Research 7, 983–999.

3

A Century of the Evolution of the Urban Area in Shenyang, China

Miao Liu[1], Yanyan Xu[1,2], Yuanman Hu[1]*, Chunlin Li[1,2], Fengyun Sun[1,2], Tan Chen[1,2]

1 State Key Laboratory of Forest and Soil Ecology, Institute of Applied Ecology, Chinese Academy of Sciences, Shenyang, China, **2** University of Chinese Academy of Sciences, Beijing, China

Abstract

Analyzing spatiotemporal characteristics of the historical urbanization process is essential in understanding the dynamics of urbanization and scientifically planned urban development. Based on historical urban area maps and remote sensing images, this study examined the urban expansion of Shenyang from 1910 to 2010 using area statistics, typology identification, and landscape metrics approaches. The population and gross domestic product were analyzed as driving factors. The results showed that the urban area of Shenyang increased 43.39-fold during the study period and that the growth rate has accelerated since the 1980s. Three urban growth types were distinguished: infilling, edge-expansion, and spontaneous growth. Edge-expansion was the primary growth type. Infilling growth became the main growth type in the periods 1946–70, 1988–97, and 2004–10. Spontaneous growth was concentrated in the period of 1997 to 2000. The results of landscape metrics indicate that the urban landscape of Shenyang originally was highly aggregated, but has become increasingly fragmented. The urban fringe area was the traditional hot zone of urbanization. Shenyang was mainly located north of the Hun River before 1980; however, the south side of the river has been the hot zone of urbanization since the 1980s. The increase of urban area strongly correlated with the growth of GDP and population. Over a long time scale, the urbanization process has been affected by major historical events.

Editor: Guy J-P. Schumann, NASA Jet Propulsion Laboratory, United States of America

Funding: This project was supported by the National Natural Science Foundation of China (No. 41171155 and 40801069) and national science and technology major project: water pollution control and governance (No. 2012ZX07505-003). The funders had no role in study design, data collection and analysis, decision to publish, or preparation of the manuscript.

Competing Interests: The authors have declared that no competing interests exist.

* E-mail: lium79@163.com

Introduction

Nearly 50% of the human population (3.3 billion) lived in urban areas in 2008, and this number is expected to reach 5 billion by 2030 [1]. The world is currently undergoing an unprecedented process of urbanization [2]. The urban area plays an important role in the regional economy as the spatial unit where most economic activities occur. Both the scale and rate of this urban expansion are extraordinary. The first urban transition took place in Europe and North America from 1750 to 1950, when the urban population of these places increased from 15 million to 423 million. The second urban transition (1950–2030) is happening largely in Africa and Asia, and it will increase their urban population from 309 million to 3.9 billion in only 80 years [3,4].

The conversion of rural lands to urban or other built-up uses is the most drastic form of land-use change [5–7], and it is a key research topic in landscape ecology [8,9]. Although urbanized areas cover only about 3% of the earth's land surface, they account for more than 78% of carbon emissions, 60% of residential water use, and 76% of the wood used for industrial purposes [10]. The influence of urbanized areas on biodiversity and ecosystem functioning and services extends far beyond the limits of cities [11,12]. To understand the process of urbanization as well as its ecological consequences, it is necessary to quantify the spatiotemporal patterns of urbanization [13].

Urban patterns and dynamics have been extensively studied over the past century. Many classical urban theories have been developed, such as the concentric zone theory [14], sector theory [15], multiple nuclei theory [16], catastrophe theory [17]. In the past several decades, some theories were applied in urban studies, such as chaos theory [18], dissipative structure theory [19], percolation theory [20], self-organization theory [21], and fractal theory [22]. With the development of computers, geographic information systems (GIS), and remote sensing (RS), new technologies and methods were used in the spatial analysis and simulation of urbanization, such as with non-equilibrium and non-linear system perspectives [23], models and forecasting patterns of urban systems, cellular automata [24,25], agent-based simulation [26], and the entropy method [27]. These methods and models have provided a deeper understanding of urban structure and dynamics.

RS is a significant data source for urban analysis, offering high spatial and temporal accuracy and consistency [28]. With development of commercial satellites since the 1970s, a number of remote sensing data types can be accessed, such as Landsat, Spot, Quickbird and so on. These data present useful sources for studying urban dynamics and improving the modeling of urban systems [29]. Over recent decades, various approaches for urban land-use classification and change detection have been developed to facilitate urban analysis based on RS data [30–32]. Historical geographic data and maps, such as aerial photographs and

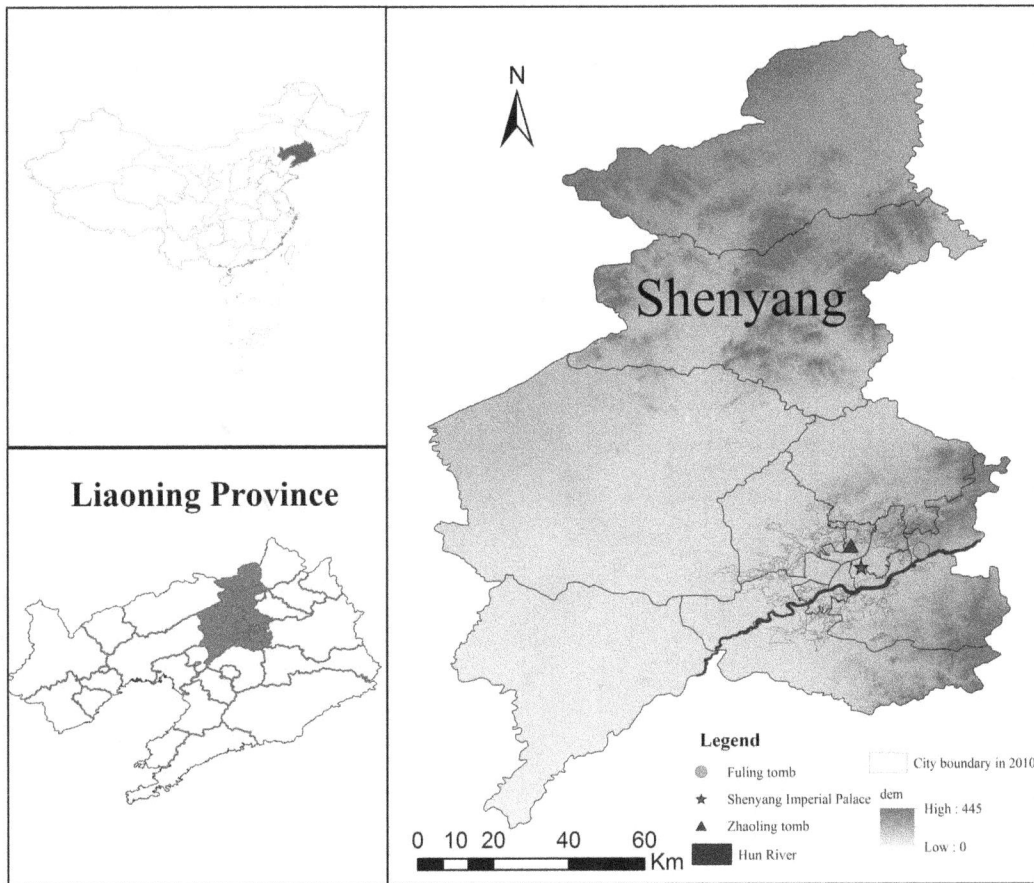

Figure 1. Location of the study area.

cartographic maps, enable the evaluation of land use change before RS application. The old maps were usually processed and analyzed through digitization with GIS. Numerous authors have studied ways of using these maps [33–35].

The urbanization process is a consequence of the interaction of various kinds of driving forces, including natural and socioeconomic factors [36]. The spatial heterogeneity of these factors causes different typologies of urban sprawl. The study of driving factors of land-use change is one of the main research topics of landscape ecology [7], and is also relevant for ecology [37].

Some studies have focused on the urban structure and fractal distribution, especially with Fractal geometry [38–41]. Camagni et al. (2002) distinguished five types of urban growth: infilling, expansion, linear development, sprawl, and large-scale projects. Wilson et al. (2002) identified five types of urban growth: infill, expansion, isolated, linear branch, and clustered branch. Basically, three main types of urban growth have been documented: infilling, edge-expansion, and spontaneous growth [12]. Infilling signifies a non-urban area surrounded by an urban area being converted to urban land; edge-expansion, also called urban fringe development, refers to a newly developed urban area spreading out from the fringe of an existing urban area; and spontaneous growth means that a new urban area is formed without direct spatial connection to an existing urban area.

In the last few decades, landscape metrics were developed with landscape ecology and GIS technology. Landscape pattern metrics are useful methods for quantifying spatial patterns, which have been wildly employed in landscape pattern analysis and urban landscape studies [32,42,43]. Many studies have shown the availability of landscape metrics for describing the temporal dynamics of urban landscapes [13,23,44]. Large numbers of landscape metrics were developed in the last few decades [45]. The validity of landscape metrics, response to scaling change and relevance among metrics were analyzed [8,46], providing the basis of choosing landscape metrics for different purposes.

Shenyang is the central and largest city in northeast China. Some cases studied the urban expansion [47,48] and land use change [49] in the last several decades based on remote sensing images. Wu et al. (2009) predicted the future urban expansion from 2005 to 2030 with the SLEUTH model [50]. Most data in these studies were derived from RS images. The temporal scale mainly began after RS images became available (Landsat MSS and TM images since 1972).The main purpose of the present study was to quantify historical urban land use changes and driving factors during the last century in Shenyang, using typology of urban growth and landscape metrics methods. Through this long-time urbanization process study, we aim to explore the characters of urban spatial expansion in China.

Methods

Study area

The city of Shenyang ($41°11'$–$42°02'$N; $122°25'$–$123°48'$E) lies in the transition zone between a branch of the Changbai Mountains and the flood plain of the Liao River in Northeast China (Fig. 1). The major topography is characterized by an alluvial plain in the west, with low hilly lands in the northeast and southeast. Shenyang is the capital of Liaoning Province, and it is the communication, commercial, scientific, and cultural centre of Northeast China. Its annual average temperature is 7.8°C and its annual precipitation is 707 mm (1906–2002). As a key investment and industrial base designated by the Chinese government since 1948, Shenyang developed into the centre of Chinese heavy industrial development before the 1980s. With the growth of its population and industry, Shenyang has continuously expanded its borders.

Data collection

In this study, many spatial data sets were collected, including multi-temporal historical city maps, and a time series of Landsat Thematic Mapper data (Table 1). All the historic city maps include the information of build-up area and main roads. The city boundary from 1910 to 1970was interpreted from the downtown area, which had clear boundary. The image-to-image method was used for the geo-referenced registration of other images with a total root mean square error of less than 0.5 pixels (cell size 30 m). The historical maps did not have a coordinate system before 1980. The relief map for 1980 was geometrically calibrated with the surveying maps, which were produced by the Chinese Bureau of Land and Resources in 1979. All maps in other years were calibrated based on the map in 1980. Three landmark buildings with a long history and unchanged locations in Shenyang were employed as reference points: Zhaoling tomb, the Qing dynasty imperial mausoleum, located in the north side was built in 1651; Fuling tomb, another Qing dynasty imperial mausoleum which was built in 1629; and the Shenyang Imperial Palace, which was completed in 1636 (Fig.1). Main roads around these reference points have not changed since the Shenyang city was built. Historical maps before 1980 were geometrically corrected based on the three landmark buildings and the main roads around them.

Visual interpretation (using local knowledge) from Landsat TM images and aerial images was carried out to form a binary map of urban/non-urban classes. In interpretation, we classified the urban area based on built-up areas surrounding the downtown region. To determine the accuracy of the image classification, the stratified random sampling method (Jensen 1996) was used to generate 96 reference points for each of the classified images. 140 reference points were located in the field with the help of a global positioning system (GPS) with ±5 m error for ground-truthing in 2010. The attributes of each point from 1981 to 2010 were collected through field survey with local residents. The kappa accuracy index (Congalton 1991) was 85.5% in 1981, 87.6% in 1988, 88.3% in 1992, 89.1% in 1997, 90.2% in 2000, 95.8% in 2004 and 92.2% in 2010. The accuracy of urban maps compared to historical maps was not measured owing to the absence before 1980 of coordinate systems. All the maps were converted to raster data in ARCGIS with spatial resolution 30 m, which is the pixel size of Landsat images.

Except for the spatial data, the statistical books of Shenyang in 1979, 1985, 1990, 2005 and 2010 were used to collect social-economic data.

The images from 1910 to 1970 were urban area maps, which have a clean city boundary. The maps of the urban area from 1988 to 2010 were interpreted based on remote sensing images. The boundary of the urban area was delineated based on the emergence of new urban patches at the edge of the urban area in each previous period.

Typology of urban growth

A simple quantitative method was used to distinguish the three growth types (Fig.2) using the following equation:

$$S = \frac{L_C}{P},$$

Where Lc is the length of the common boundary between a newly grown urban area and pre-growth urban patches, and P is the perimeter of this newly grown area. The urban growth type was identified as follows: infilling when $S \geq 0.5$; edge-expansion when $0 < S < 0.5$; and spontaneous growth when $S = 0$, which indicates no common boundary. Xu et al. (2005) proved the effectiveness of this approach.

Landscape metrics

To characterize urban landscape pattern effectively and succinctly, four metrics were chosen to reflect urban patch characters and spatial distribution based on previous studies of the meaning and representativeness of metrics [12,13,46,51], including number of patches (NP), Patch density (PD), landscape

Table 1. List of spatial dataset used in the study.

No	Dataset	Time (Year)	Source	Producer	Original resolution or scale
1	Elevation	1981	Chinese second-generation 1:100,000 relief maps	Liaoning surveying and mapping bureau	25 m
2	TM and ETM images	1988, 1992, 1997, 2000, 2010	USGS (United States Geological Survey)	USGS (United States Geological Survey)	30 m
3	Aerial images	2004	Liaoning surveying and mapping bureau	Liaoning surveying and mapping bureau	10 m
4	Shenyang city map	1910,	Shenyang Urban Construction Archives	Surveying and mapping agency of Qing dynasty	1:100,000
5		1920,	Shenyang Urban Construction Archives	Surveying and mapping agency of local government	1:100,000
6		1931, 1939,	Shenyang Urban Construction Archives	Surveying and mapping agency of puppet Manchurian regime	1:100,000
7		1946, 1961, 1970	Shenyang Urban Construction Archives	Liaoning surveying and mapping bureau	1:50,000

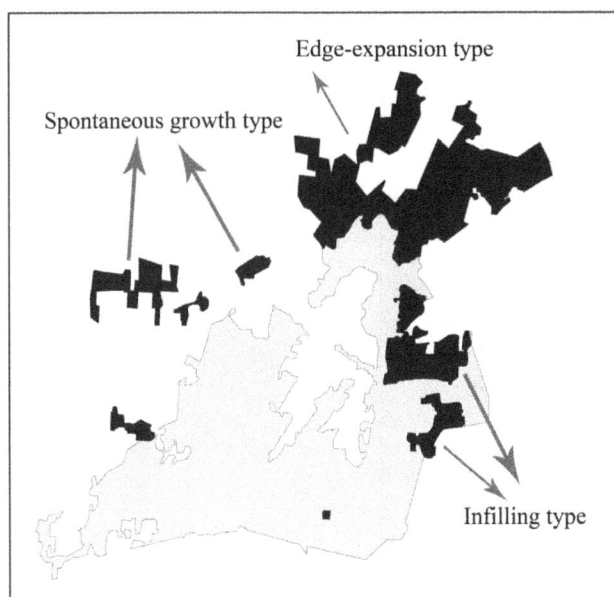

Figure 2. Typology of urban growth. The grey area represents the pre-growth urban patches and the dark area represents the newly grown urban patches.

shape index (LSI), and aggregation index (AI) (Table 2). These four metrics can reflect the urban expansion landscape pattern comprehensively from the patch class level. The landscape metrics were calculated using FRAGSTATS 4.1 software [52].

Results

As the Fig.3 show, the Hun River flows through Shenyang, and the old urban area was located north of the river in 1910. Expanded urban patches were also concentrated north of the Hun River before 1980. Since the 1980s, large urban area patches appeared south of the river. Between 1980 and 2004, more than half of the new urban patches still were located north of the Hun River. Since 2004, more than half of the new patches have been located south of the river.

Over the last century (1910–2010), the urban area of Shenyang has shown a continuous increase (Fig.4). The urbanized area was 15.07 km^2 in 1910 and 653.91 km^2 in 2010—a 43.39-fold increase. The mean annual growth rate of urban area was 1.61 km^2/year from 1910 to 1980, 7.29 km^2/year from 1980 to 1997, and 17.54 km^2/year from 1997 to 2010. The urban area growth rate and urban population has significantly increased since the 1980s, which indicates that the urbanization process of Shenyang has been accelerating.

The contribution of the three urban area growth types to the increased area is presented in Fig.5. Over the 100 years of the study period, the area of edge-expansion, infilling, and spontaneous growth was 775.40, 406.83, and 117.77 ha respectively. Edge-expansion was the primary growth type, accounting for 59.65%. Before 1946, the growth areas were mostly of the edge-expansion type. In the periods 1970–88 and 2000–04, edge-expansion accounted for over 70% of the growth. However, infilling became the main growth type during the periods 1946–70, 1988–97, and 2004–10. In contrast, spontaneous growth accounted only for 9.06% of the growth over the past century, and it was concentrated in the period of 1997 to 2000.

The trajectories of landscape metrics are shown in Fig.6. NP increased with the emergence of the spontaneous growth type (Fig.6(a)). NP showed an increasing trend with the urban expansion. However, NP decreased in the 1990s because the open space in the urban patches showed infilling growth. PD quantifies the density of patches, which kept decreasing in the study period (Fig.6(b)). The decrease of PD indicates that the mean scale of patches kept increasing and single urban patches became larger. LSI qualifies the shape complexity of patches (Fig.6(c)). LSI maintained an increasing trend, which indicates that the patch shapes became more and more complicated. LSI had a dramatic increase from 2000 to 2004, which was associated with fast development. The value of AI was nearly 100 from 1910 to 2010, which indicates that urban patches show a very high degree of concentrated spatial distribution (Fig.6(c)).

Driving factor analysis

Driving factors are the forces that cause observed land use changes, i.e., they are influential processes in the evolutionary trajectory of the land use. Urban expansion and land use change are forced by geographical and social-economical factors. Urban expansion and related land use change is decided by a combination of these factors. Shenyang city locates in the flood plain of the Liao River. The relief of study area is smooth (elevation ranges from 58 to 124 m), so the geographical factors, such as elevation, slope, aspect, do not influence or limit urban expansion. The population of Shenyang city from 1910 to 2010 was available (Fig.7). However, due to the absence of data before 1949, gross domestic product (GDP) of Shenyang city only was available from 1949 to 2010 The GDP data were converted to the comparative GDP in 1952.

The increase of urban area strongly correlates with the growth of GDP and population in a linear form ($r^2 = 0.93$ and 0.86 correspondingly). The GDP correlation demonstrates the econo-

Table 2. Landscape metrics used to quantify spatial pattern.

Acronym	Scale	Index name	Description
NP	Class or landscape	number of patches	Number of patches for each class
PD	Class or landscape	patch density	Quantifies the number of patches of the corresponding patch type divided by total landscape area (m^2).
LSI	Class or landscape	landscape shape index	A modified perimeter-area ratio of the form that measures the shape complexity of the whole landscape or a specific patch type.
AI	Class or landscape	aggregation index	Shows the frequency with which different pairs of patch types appear side-by-side on the map

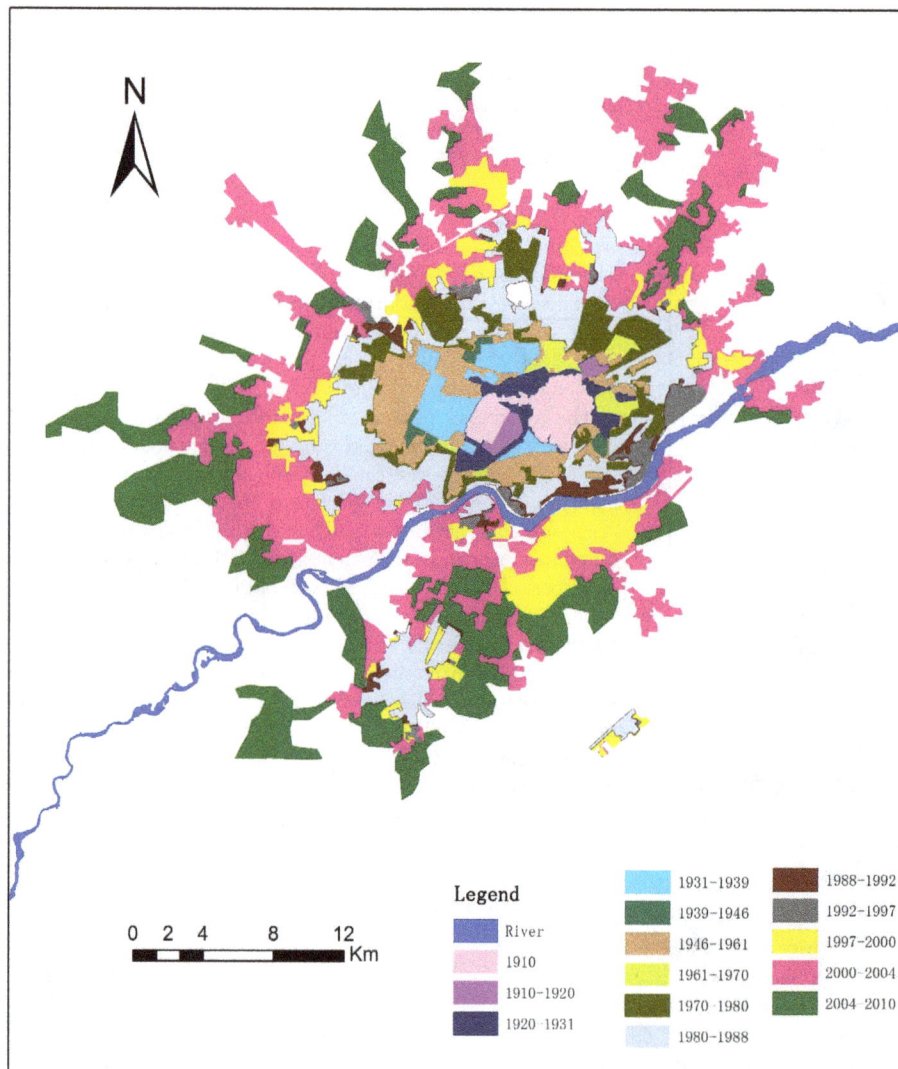

Figure 3. Urban spatial expansion of Shenyang from 1910 to 2010. (Buffer distance from the Shenyang Imperial Palace: 0–15 kilometers step 1 kilometer; 15–25 kilometers step 5 kilometers.).

my was the more dominant factor for urban expansion than population in Shenyang from 1910 to 2010.

Discussion

Shift in the growth hot zone

Urbanization is driven by natural and cultural factors. Analyzing the shift in the urban growth hot zone has some useful implications for urban planning and modeling [24]. All the new-growth urban areas before 1980 in Shenyang occurred in the zones around the pre-growth urban patches before 1980, which are located north of the Hun River. Over 90% of the total growth took place in zones within a distance of 3 km from the edge of the pre-growth urban patches. The buffer zone of 0–3 km of built-up urban area, in which the Shenyang Imperial Palace north of the Hun River was considered the centre of the city, was the growth hot zone before 1980. Since the 1980s, major spontaneous growth patches have appeared around the urban area. In 1980 to 1988, new urban patches developed in many locations, and new urban patches started appearing south of the Hun River. Many cities in the world developed on both side of a main river, but Shenyang

was largely located north of the Hun River before 1980. In traditional Chinese culture, north of the river represents "yang", which signifies that it is better for living. The name "Shenyang" derives from the city location north of the Hun River, which before 1675 was called the Shen River. With economic development and urban expansion, the urban area of Shenyang has extended south of the Hun River. The buffer zone of 0–4 km of built-up urban area located on either side of the river became a growth hot zone from 1980 to 1988. In 1988 to 2000, urban expansion was concentrated in an area south of the Hun River; thus, south of the Hun River became a hot zone. In 2000 to 2010, the urban area showed explosive growth in all directions, with extensive spontaneous patches in the south, north, and west. The explosive urban growth was caused not only by China's economic development but also the local government's land financing in the 2000s. Land financing refers to some local governments relying on the income produced from land-use rights to finance local expenditure. The buffer zone from 0 to 7 km of built-up urban area extending to the east was the growth hot zone from 2000 to 2010. The distance of the hot zone to the built-up urban area has

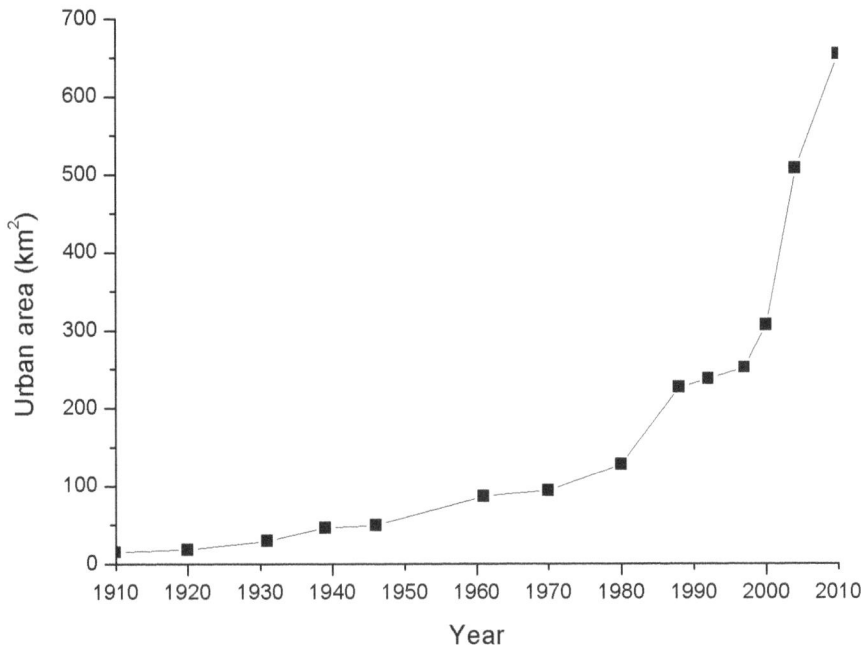

Figure 4. Urban area change of Shenyang from 1910 to 2010.

progressively increased, and the new-growth patches have become larger. The speed of urban growth has also become more rapid.

Major historical events and urbanization

Various kinds of driving forces, including natural and socioeconomic factors, influence the urbanization process [7]. As many studies showed that economic and demographic factors are the main driving forces [53–55], these are also the main driving forces in Shenyang urban expansion. However, the effect of national and local historic events may affect in city's long historic expansion. The major events in the long history of Shenyang are listed in Table 3. Shenyang was a small town before the building of its Imperial Palace in 1625. Northeast China was the home region of the Qing dynasty (1636–1911) emperors. The Qing dynasty

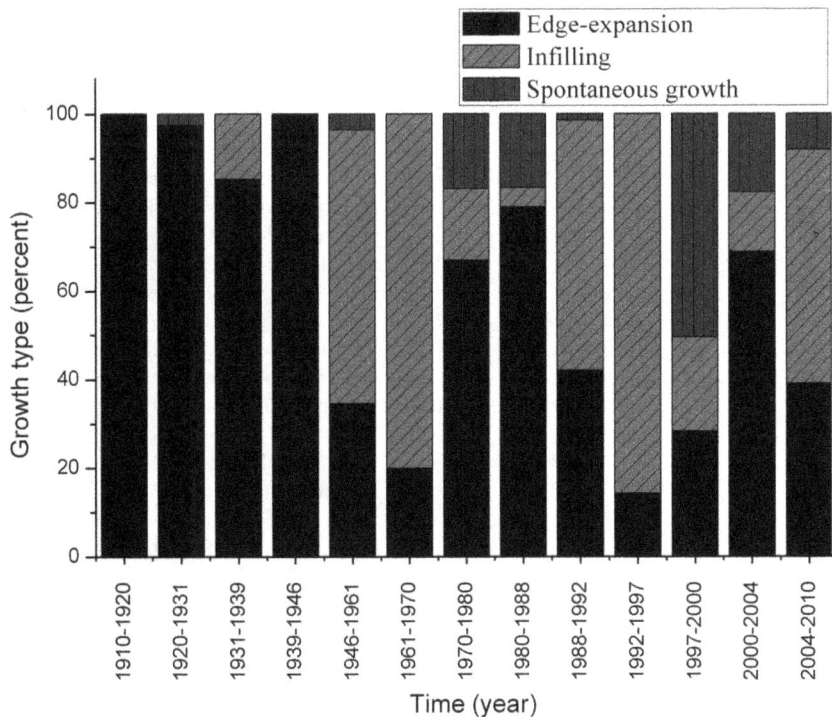

Figure 5. Proportion of the three growth types in different periods.
doi:10.1371/journal.pone.0098847.g005

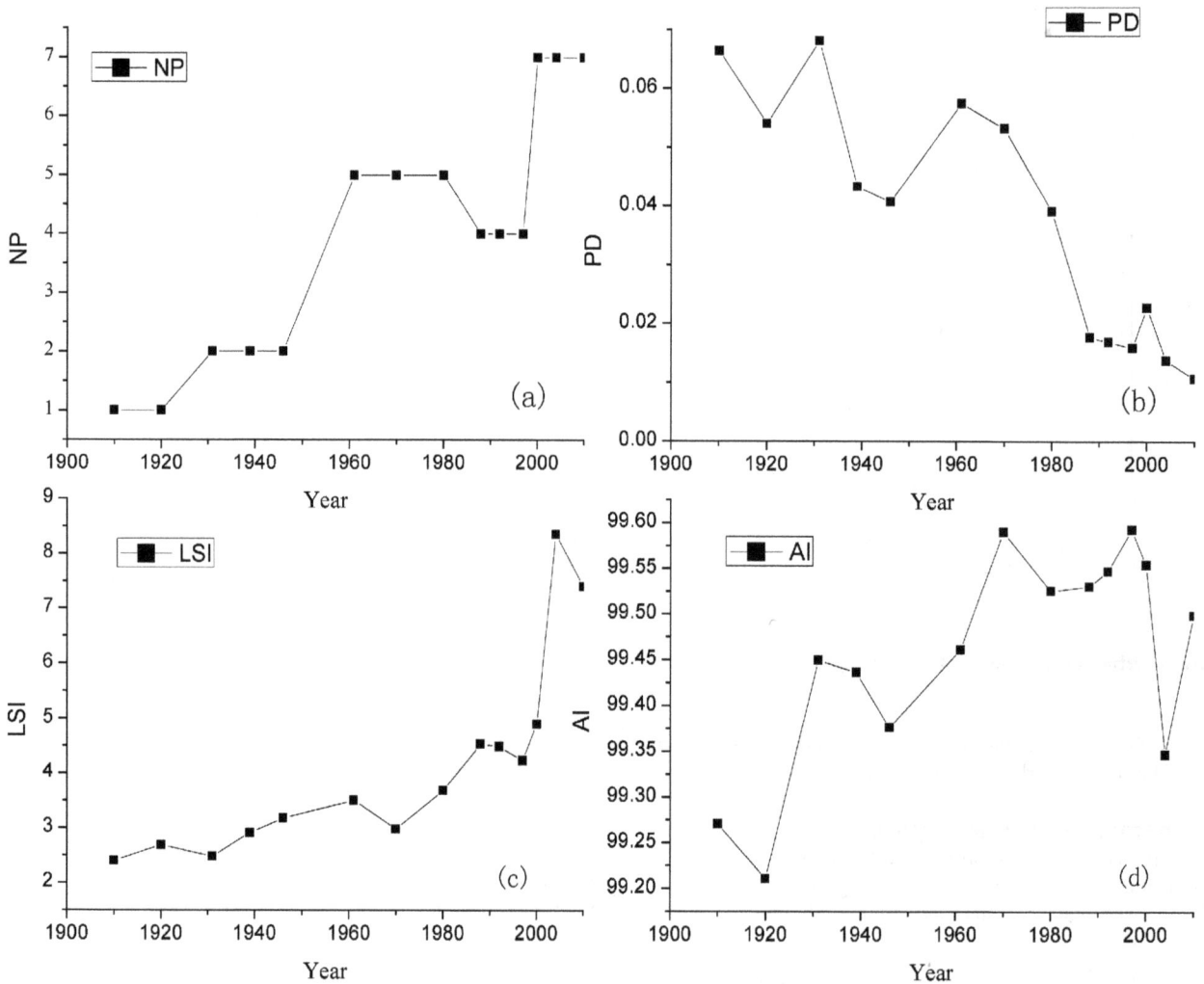

Figure 6. Change in the landscape indices during the period 1910–2010: (a) number of patches (NP), (b)Patch density (PD), (c) landscape shape index (LSI), and (d) aggregation index (AI).

capital moved to Beijing from Shenyang in 1644. Shenyang was subjected to little development from 1644 to 1911 because as the home of the Qing dynasty, such development was forbidden. After the Xinhai Revolution of 1911, Shenyang was controlled by the

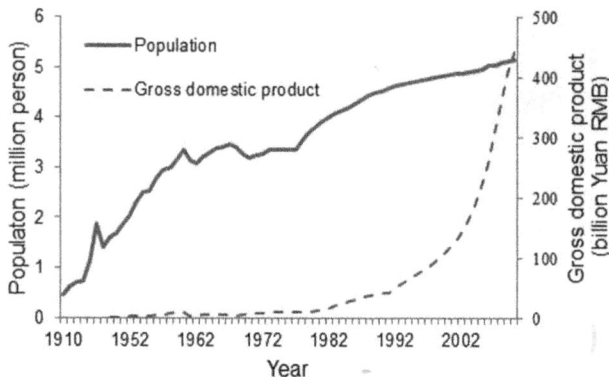

Figure 7. The gross domestic product and population of Shenyang from 1919 to 2010.

Fengtian clique of warlords, and the city began to expand slowly. During the period of the city's occupation by the Japanese (1931–45), railways and factories were built, and new edge-expansion patches appeared in western and northern areas as industrial districts. After the establishment of the People's Republic of China in 1949, Shenyang was positioned as a heavy industrial city. Urban expansion, which was mainly concentrated in the western part, was related to industrial development. The urban area of Shenyang showed a moderate increase from 1949 to 1979. In the first 10 years (1980–90) of reform and opening up, the urban area of Shenyang underwent accelerated growth in line with its economic development. However, numerous factories in Shenyang closed down in and after 1990, which led to substantial job losses and an economic downturn. In 1990–97, urbanization almost came to a halt. Since 1999, the construction of private residential buildings has not been the responsibility of the government, though policies for the construction of commercial residential buildings have been in place. The new accelerated growth in the urban area has mainly been driven by residential construction. To help the heavy industrial cities in Northeast China, policies for reviving previous key industrial bases in that region have been implemented by the central government.

Table 3. Major historical events related to Shenyang.

Year	Historic events
1644	The city was renamed 'Fengtian', which was the second capital of Qing Dynasty
1911	The city was capital of the Fengtian clique of warlords after Revolution of 1911
1926	The city was renamed 'Shenyang'
1931	The city was occupied by Japanese, and renamed 'Fengtian'
1945	The city was renamed 'Shenyang' after the victory of Anti-Japanese War
1949	The People's Republic of China was built
1979	The reform and opening-up police was carried out in China
2003	The police of reviving previous key industrial bases in North-eastern China was carried out
2004	The government of Shenyang plans moving to south of Hun River

Through the integrated effects of these policies and the local government's land financing, the urban area of Shenyang showed an explosive increase. The government of Shenyang city planned to move to south of Hun River. The urban growth hotspot has moved to the area before publishing the plan due to this information of 'government moving'. The major historical events in the city may not have been the direct driving forces of urbanization, but the economic and political changes brought about by those events have determined the city character of Shenyang and its urbanization. Major historic events have brought a corresponding increase in spontaneous growth (Fig.5). However, the landscape metrics are not sensitive to the major historic events. The reason may be landscape metrics reflect the whole landscape status, but the major historic events mainly affect the new-growth patches.

Historical data availability

Data quality is an important and difficult problem in land-use and land cover change analysis [30,56], and this is especially true when using historical information that dates back well before remotely sensed data were available. The historical data, especially historical maps, generally carry a higher degree of positional inaccuracy and uncertainty through the use of different sources and lack of geographic coordinates [57]. These "inaccuracy" and "uncertainty" can affect the results of land use change and landscape metrics calculation. Several standardizing historical maps methods have been proposed [58–60]. Different from land use change studies with historic maps [13,34,60], this research focused on urban landscape change. So, we chose old urban administrative maps, which were acquirable and relatively accurate. The remote sensing data was first geographically corrected, and the historical data were geographically corrected using remote sensing data based on unchanged significant landmarks, such as Shenyang Imperial Palace, Fuling tomb, Zhaoling tomb and unchanged roads around them. Although the historical data contain errors, the data are for the most part meaningful in terms of the urbanization process and the city's historical trajectory.

Because of data limitations, we were unable to study the internal changes in urban development. Major internal changes have occurred in Shenyang through urban renewal and relocation of houses, which have taken place since 2000. The Tiexi district of Shenyang, which was a heavy industrial district before 2000, has been completely transformed into a residential district. Internal change and the building height increase are important parts of urbanization studies, and it will be emphasized in our future research.

Social and environmental problems

The rapid growth of urban area brought a series of social and environmental problems both in local and regional regions. The area of urban expansion mainly transformed farmlands, which caused the loss of farmland. Population increase and industry development caused the increase in energy and water consumption, which caused the environmental problems including: groundwater funnel area expansion, surface and ground water pollution and air pollution, etc. The Liao River watershed, which includes the city of Shenyang, has become one of the most severely polluted area in China in the last several decades. Concentration of population in urban area had caused social problems, such as the traffic jam and the shortage of public resources, common issues in huge cities.

Acknowledgments

We also thank Prof. Chang Yu for his pre-submission review.

Author Contributions

Conceived and designed the experiments: ML YH. Wrote the paper: ML YH YX. Data collection: ML YX TC. The calibration and interpretation of spatial data: ML YX FS. The data analysis: ML YX. Figures preparation: ML CL.

References

1. United Nations (2010) World Urbanization Prospects. The 2009 Revision. Highlights, United Nations.

2. Seto KC, Sánchez-Rodríguez R, Fragkias M (2010) The New Geography of Contemporary Urbanization and the Environment. Annual Review of Environment and Resources 35: 167–194.

3. United Nations Population Fund (2007) State of the World Population 2007. Unleashing the Potential of Urban Growth, United Nations Population Fund.

4. Montgomery MR (2008) The Urban Transformation of the Developing World. Science 319: 761–764.

5. Antrop M (2000) Changing patterns in the urbanized countryside of Western Europe. Landscape Ecology 15: 257–270.

6. Pickett STA, Cadenasso ML, Grove JM, Nilon CH, Pouyat RV, et al. (2008) Urban Ecological Systems: Linking Terrestrial Ecological, Physical, and Socioeconomic Components of Metropolitan Areas. Urban Ecology. In: Marzluff JM, Shulenberger E, Endlicher W, Alberti M, et al, editors: Springer US. pp. 99–122.

7. Bürgi M, Hersperger A, Schneeberger N (2004) Driving forces of landscape change - current and new directions. Landscape Ecology 19: 857–868.

8. Wu J, Hobbs R (2002) Key issues and research priorities in landscape ecology: An idiosyncratic synthesis. Landscape Ecology 17: 355–365.

9. Dietzel C, Herold M, Hemphill JJ, Clarke KC (2005) Spatio-temporal dynamics in California's Central Valley: Empirical links to urban theory. International Journal of Geographical Information Science 19: 175–195.

10. Brown LR (2001) Eco-Economy: Building an Economy for the Earth. W W Norton, New York: 330–334.

11. Grimm NB, Grove JM, Pickett STA, Redman CL (2008) Integrated Approaches to Long-Term Studies of Urban Ecological Systems. Urban Ecology. In: Marzluff JM, Shulenberger E, Endlicher W, Alberti M, Bradley G, et al, editors: Springer US. pp. 123–141.

12. Berling-Wolff S, Wu J (2004) Modeling urban landscape dynamics: A case study in Phoenix, USA. Urban Ecosystems 7: 215–240.

13. Wu J, Jenerette GD, Buyantuyev A, Redman CL (2011) Quantifying spatiotemporal patterns of urbanization: The case of the two fastest growing metropolitan regions in the United States. Ecological Complexity 8: 1–8.

14. Burgess EW (1925) In: Park RE, Burgess EW, McKenzie RD (eds) The city. The Chicago University Press, Chicago, USA: 47–62.

15. Hoyt H (1939) The structure and growth of residential neighborhoods in American cities. Federal Housing Administration, Washington DC, USA.

16. Ullman CDHEL (1945) The nature of cities. Annals of the American Academy of Political and Social Science.

17. Wilson AG (1976) Catastrophe theory and urban modelling: an application to modal choice. Environment and Planning A 8: 351–356.

18. Wong D, Fotheringham AS (1990) Urban systems as examples of bounded chaos: exploring the relationship between fractal dimension, rank-size, and rural to urban migration. Geografiska Annaler B, 72: 89–99.

19. Allen PM, Sanglier M (1979) A dynamic model of urban growth: II. Journal of Social and Biological Structures 2: 269–278.

20. Franceschetti G, Marano S, Pasquino N, Pinto IM (2000) Model for urban and indoor cellular propagation using percolation theory. Physical Review E 61: R2228–R2231.

21. Portugali J (2000) Self-organization and the city. Springer, Berlin, Germany.

22. Parker DC, Manson SM, Janssen MA, Hoffmann MJ, Deadman P (2003) Multi-Agent Systems for the Simulation of Land-Use and Land-Cover Change: A Review. Annals of the Association of American Geographers 93: 314–337.

23. Luck M, Wu J (2002) A gradient analysis of urban landscape pattern: a case study from the Phoenix metropolitan region, Arizona, USA. Landscape Ecology 17: 327–339.

24. Batty M (1997) Cellular Automata and Urban Form: A Primer. Journal of the American Planning Association 63: 266–274.

25. Engelen G, White R, Uljee I, Drazan P (1995) Using cellular automata for integrated modelling of socio-environmental systems. Environmental Monitoring and Assessment 34: 203–214.

26. Batty M (2005) Agents, cells, and cities: new representational models for simulating multiscale urban dynamics. Environment and Planning A 37: 1373–1394.

27. Yeh AGO, Li X (2001) Measurement and monitoring of urban sprawl in a rapidly growing region using entropy. Bethesda, MD, ETATS-UNIS: American Society for Photogrammetry and Remote Sensing.

28. Longley PA (2002) Geographical Information Systems: will developments in urban remote sensing and GIS lead to 'better' urban geography? Progress in Human Geography 26: 231–239.

29. Herold M, Goldstein NC, Clarke KC (2003) The spatiotemporal form of urban growth: measurement, analysis and modeling. Remote Sensing of Environment 86: 286–302.

30. Masek JG, Lindsay FE, Goward SN (2000) Dynamics of urban growth in the Washington DC metropolitan area, 1973–1996, from Landsat observations. International Journal of Remote Sensing 21: 3473–3486.

31. Seto KC, Liu WG (2003) Comparing ARTMAP neural network with the maximum-likelihood classifier for detecting urban change. Photogrammetric Engineering and Remote Sensing 69: 981–990.

32. Zhu M, Xu J, Jiang N, Li J, Fan Y (2006) Impacts of road corridors on urban landscape pattern: a gradient analysis with changing grain size in Shanghai, China. Landscape Ecology 21: 723–734.

33. Cousins SO, Ohlson H, Eriksson O (2007) Effects of historical and present fragmentation on plant species diversity in semi-natural grasslands in Swedish rural landscapes. Landscape Ecology 22: 723–730.

34. Pärtel M, Mändla R, Zobel M (1999) Landscape history of a calcareous (alvar) grassland in Hanila, western Estonia, during the last three hundred years. Landscape Ecology 14: 187–196.

35. Skaloš J, Weber M, Lipský Z, Trpáková I, Šantrůčková M, et al. (2011) Using old military survey maps and orthophotograph maps to analyse long-term land cover changes – Case study (Czech Republic). Applied Geography 31: 426–438.

36. Foster DR, Motzkin G, Slater B (1998) Land-Use History as Long-Term Broad-Scale Disturbance: Regional Forest Dynamics in Central New England. Ecosystems 1: 96–119.

37. Dale VH, Brown S, Haeuber RA, Hobbs NT, Huntly N, et al. (2000) Ecological Principles and Guidelines for Managing the use of LAND1. Ecological Applications 10: 639–670.

38. Batty ML, Longley PA (1994) Fractal Cities: A Geometry of Form and Function. London: Academic Press.

39. Frankhauser P (1998) The Fractal Approach: A New Tool for the Spatial Analysis of Urban Agglomerations. Population: An English Selection 10(1): 205–240.

40. Makse HA, Andrade JS, Batty M, Havlin S, Stanley HE (1998) Modeling urban growth patterns with correlated percolation. Physical Review E 58: 7054–7062.

41. Chen Y (2013) Fractal analytical approach of urban form based on spatial correlation function. Chaos, Solitons & Fractals 49: 47–60.

42. Wu J, Jelinski DE, Luck M, Tueller PT (2000) Multiscale Analysis of Landscape Heterogeneity: Scale Variance and Pattern Metrics. Annals of GIS 6: 6–19.

43. Jenerette GD, Wu J (2001) Analysis and simulation of land-use change in the central Arizona – Phoenix region, USA. Landscape Ecology 16: 611–626.

44. Seto KC, Fragkias M (2005) Quantifying Spatiotemporal Patterns of Urban Land-use Change in Four Cities of China with Time Series Landscape Metrics. Landscape Ecology 20: 871–888.

45. Uuemaa E, Mander Ü, Marja R (2013) Trends in the use of landscape spatial metrics as landscape indicators: A review. Ecological Indicators 28: 100–106.

46. Li X, He HS, Bu R, Wen Q, Chang Y, et al. (2005) The adequacy of different landscape metrics for various landscape patterns. Pattern Recognition 38: 2626–2638.

47. Wang HJ, Li XY, Zhang ZL, He XY, Chen W, et al. (2008) Analysis on urban spatial expansion process in Shenyang city in 1979–2006. Chinese Jourban of Applied Ecology 19: 2673–2679.(in Chinese)

48. Zhou R, Li YH, Hu YM, Wu XQ, He HS, et al. (2009) Characterization of spatial expansion of urban land in Shenyang city based on GIS. Resources Science 31: 1947–1956. (in Chinese)

49. Chang Y, Su WG, Gao RP (1997) Changes of land use pattern in eastern Shenyang. Chinese Journal of Applied Ecology 8: 421–425. (in Chinese)

50. Wu XQ, Hu HY, He HS, Bu RC, Xi FM (2009) Research for scenarios simulation of future urban growth and land use change in Shenyang city. Geographical Reesearch 28: 1264–1274. (in Chinese)

51. Xu C, Liu M, Zhang C, An S, Yu W, et al. (2007) The spatiotemporal dynamics of rapid urban growth in the Nanjing metropolitan region of China. Landscape Ecology 22: 925–937.

52. McGarigal KM, Barbara J (1995) FRAGSTATS: Spatial Pattern Analysis Program for Quantifying Landscape Structure. U.S. Department of Agriculture, Forest Service, Pacific Northwest Research Station, Portland, OR, USA, pp122.

53. Krausmann F, Haberl H, Schulz NB, Erb KH, Darge E, et al. (2003) Land-use change and socio-economic metabolism in Austria—Part I: driving forces of land-use change: 1950–1995. Land Use Policy 20: 1–20.

54. Long H, Tang G, Li X, Heilig GK (2007) Socio-economic driving forces of land-use change in Kunshan, the Yangtze River Delta economic area of China. Journal of Environmental Management 83: 351–364.

55. Stephenne N, Lambin EF (2004) Scenarios of land-use change in Sudano-sahelian countries of Africa to better understand driving forces. GeoJournal 61: 365–379.

56. Iverson LR (2007) Adequate data of known accuracy are critical to advancing the field of landscape ecology. In: Wu, J., Hobbs, R. (Eds.), Key Topics in Landscape Ecology. Cambridge University Press, Cambridge, UK, pp. 11–38.

57. Rhemtulla J, Mladenoff D (2007) Why history matters in landscape ecology. Landscape Ecology 22: 1–3.

58. Haase D, Walz U, Neubert M, Rosenberg M (2007) Changes to Central European landscapes: Analysing historical maps to approach current environmental issues, examples from Saxony, Central Germany. Land Use Policy 24: 248–263.

59. Käyhkö N, Skånes H (2006) Change trajectories and key biotopes—Assessing landscape dynamics and sustainability. Landscape and Urban Planning 75: 300–321.

60. Petit CC, Lambin EF (2002) Impact of data integration technique on historical land-use/land-cover change: Comparing historical maps with remote sensing data in the Belgian Ardennes. Landscape Ecology 17: 117–132.

4

Much beyond Mantel: Bringing Procrustes Association Metric to the Plant and Soil Ecologist's Toolbox

Francy Junio Gonçalves Lisboa[1,4]*, **Pedro R. Peres-Neto**[2], **Guilherme Montandon Chaer**[3], **Ederson da Conceição Jesus**[3], **Ruth Joy Mitchell**[4], **Stephen James Chapman**[4], **Ricardo Luis Louro Berbara**[1]

1 Soil Science Department, Agronomy Institute, Federal Rural University of Rio de Janeiro, Seropédica-RJ, Brazil, 2 Canada Research Chair in Spatial Modelling and Biodiversity; Université du Québec à Montréal, Département des sciences biologiques, Québec, Canada, 3 Embrapa Agrobiologia, Seropédica-RJ, Brazil, 4 The James Hutton Institute, Craigiebuckler, Aberdeen, United Kingdom

Abstract

The correlation of multivariate data is a common task in investigations of soil biology and in ecology in general. Procrustes analysis and the Mantel test are two approaches that often meet this objective and are considered analogous in many situations especially when used as a statistical test to assess the statistical significance between multivariate data tables. Here we call the attention of ecologists to the advantages of a less familiar application of the Procrustean framework, namely the Procrustean association metric (a vector of Procrustean residuals). These residuals represent differences in fit between multivariate data tables regarding homologous observations (e.g., sampling sites) that can be used to estimate local levels of association (e.g., some groups of sites are more similar in their association between biotic and environmental features than other groups of sites). Given that in the Mantel framework, multivariate information is translated into a pairwise distance matrix, we lose the ability to contrast homologous data points across dimensions and data matrices after their fit. In this paper, we attempt to familiarize ecologists with the benefits of using these Procrustean residual differences to further gain insights about the processes underlying the association among multivariate data tables using real and hypothetical examples.

Editor: Andrew R. Dalby, University of Westminster, United Kingdom

Funding: This work was carried out with the aid of a grant from Conselho Nacional de Desenvolvimento Científico e Tecnológico (CNPq - 563304/2010-3 and 562955/2010-0), Fundação de Amparo à Pesquisa de Minas Gerais (FAPEMIG CRA - APQ-00001-11), and the Inter-American Institute for Global Change Research (IAI-CRN II-021). Francy Lisboa greatly acknowledges a research scholarship from CAPES/EMBRAPA (Carbioma) and Dr. Beata Madari. The funders had no role in study design, data collection and analysis, decision to publish, or preparation of the manuscript.

Competing Interests: The authors have declared that no competing interests exist.

* Email: agrolisboa@gmail.com

Introduction

In multidimensional data analysis, ecologists often encounter situations where they need to choose between two or more numerical approaches that are able to tackle the same question of interest. The preference between approaches is based, among other factors, on the familiarity of the user with the method, which in turn depends on the time a particular method has been available in statistical packages and the ease in implementing and interpreting its results. Another relevant factor to consider is "literature–induced use" in which renowned research groups involved in the development, improvement and generation of statistical ecological approaches have a strong influence on the types of statistical approaches other ecologists use.

Determining the strength of the relationships between multivariate datasets is a routine analysis when trying to understand the environmental factors driving the composition and structure of ecological communities. Two approaches, the Mantel test [1] and Procrustes analysis [2], though considered analogous by the literature in the questions they can tackle [3], have not been used to the same extent. Despite the advantages of Procrustes analysis over the Mantel test [3] regarding greater statistical power in detecting significant relationships (i.e., lower type II errors) and the possibility of analyzing further the patterns of association between

multivariate matrices (visually and by further statistical analyses), the Procrustean approach remains relatively unused in tackling questions regarding the relationships between data matrices involving plant and soil information or between soil matrices (Fig. 1).

The Mantel test and the Procrustes approach can be both used in many similar situations where the aim is to assess how multivariate data matrices are associated (correlated), though for unknown reasons they have been used in quite different ways in the ecological literature. For example, while the Mantel test has often been applied when testing the relationship between above and below ground data matrices [4], [5], [6], [7], [8], [9], [10], [11], [12], Procrustes analysis has predominantly been used to contrast the results of different ecological ordinations on the same data [13], [14], [15], [16], to compare fingerprinting tools for assessing microbial communities [17], [18], [19] and for deciding between methodological choices [20], [21]. Indeed the Procrustean framework has been rarely used to make inferences about plant and soil relationships [22], [23], [24], [25], [26] and other types of ecological associations between data sets. However there are instances in which the Procrustean and Mantel tests cannot be used interchangeably. Unlike Mantel, the Procrustean approach can be used to compare multiple data matrices. However, when ecologists are interested in correlating distance (or similarity)

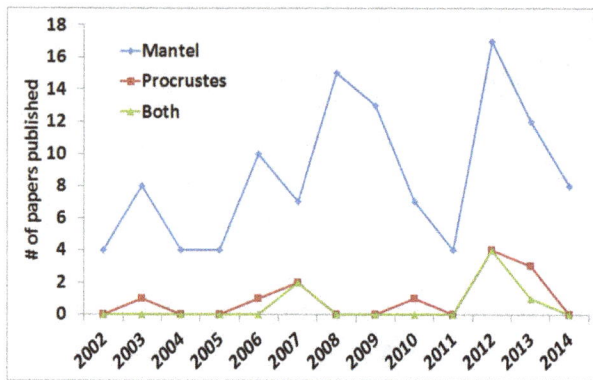

Figure 1. Papers published using Mantel and Procrustes for relating data matrices from soil or plant studies in the ten years since [3] **stated the advantages of Procrustes over the Mantel approach.** Data obtained using Thompson Reuters database (May, 12, 2014). We searched for papers using uniquely the Mantel approach, uniquely the Procrustes approach and papers using both approaches. The search was based on Procrust* (Procrustean or Procrustes) and PROTEST.

matrices, rather than testing the association among data matrices in their raw form (i.e., not transformed by the property of distance measures), Mantel, rather than Procrustes, is more appropriate. One particular case is the distance-decay of similarity in ecological communities [27] in which one is interested in testing the hypothesis that the similarity in community composition decreases in relation to linear (or log transformed) geographic distance between communities. The differences between raw-based and distance-based approaches have been discussed extensively elsewhere [28], [29].

Despite the relative merits of the Procrustean framework over the Mantel test shown by the relatively well-cited paper by Peres-Neto and Jackson [3], its potential has not yet been tapped. Perhaps the reason for Procrustes analysis not being as popular as the Mantel test among ecologists is the lack of a paper showing that in many situations traditionally investigated by Mantel, the Procrustean analysis can be equally well used. Here, we attempt to familiarize ecologists with the use of Procrustes analysis by using real and hypothetical examples where the Mantel test tends to be preferred. Most importantly, we highlight little explored limits of Procrustes by using its residual vector of association between data tables, hereafter referred as to PAM, in three common statistical approaches: multivariate ordination, variation partitioning and ANOVA.

Procrustes analysis: a foundation for soil and plant ecologists

In ancient Greek mythology there was a character named Procrustes who was a resident of Eleusis Mountain, a known travelers' route. As a "good" host, Procrustes always invited travelers to spend the night at his home; more specifically, he invited them to lie down on his iron bed, which was tailored to fit Procrustes' own body. The guests who did not fit the dimensions of his bed either had their limbs cut off or were stretched until their dimensions approached those of Procrustes's bed. Ironically, none of the guests ever fitted the iron bed because Procrustes secretly had two beds of different sizes [30]. One can easily make a parallel here with ecological data in which data from different sources will almost never easily compare or fit to one another.

Procrustes analysis is based on the search for the best fit between two data tables, hereafter referred to as matrices, where one is kept fixed ("Procrustes' bed" or target matrix), while the other ("Procrustes' guest" or rotated matrix) undergoes a series of transformations (translation, mirror reflection and rotation; [2]) to fit the fixed matrix. Although in this paper we concentrate on fitting two matrices, the extension of Procrustes analysis to multiple matrices is straightforward [3] in which the reference matrix can be either one of the original matrices or their averages (or medians). Hereafter, the target matrix (target) will be referred as to $\mathbf{X}$, and the data matrix to be fitted as $\mathbf{Y}$. $\mathbf{X}$ and $\mathbf{Y}$ are both $n \times p$ matrices, where n is the number of rows and p is the number of columns. The goal of the transformations in $\mathbf{Y}$ is to minimize the residual sum of squared differences between the corresponding n dimensions between $\mathbf{X}$ and $\mathbf{Y}$; the sum of the squares of these residual differences is termed m^2 (Gower's statistic), representing the optimal fit between the two data matrices, such that the higher the value of m^2, the weaker the relationship between the two data tables is. The significance of m^2 can be estimated through a permutation test (termed PROTEST after [31]; see [3] for further details).

Procrustean association metric (PAM)

The least squares superimposition between the corresponding n observations of $\mathbf{X}$ and $\mathbf{Y}$ is one of the main advantages (in addition to the increased statistical power) of the Procrustean framework in contrast to the Mantel test. The Procrustes superimposition generates a $(n \times p)$ matrix of residuals that can be further used to contrast the differences between homologous observations (rows) across matrices in the form of a vector (PAM). Given that within the Mantel approach differences between observations across all dimensions are packed down into a single distance, it cannot be used to assess differences across observations across dimensions. Consistent small and large differences across homologous observations across matrices in regard to other factors of interest can further assist in understanding how $\mathbf{X}$ and $\mathbf{Y}$ are related. For example, we could use PAM to assess the degree of observation matching between a plant function trait matrix and a composition matrix and assess whether smaller or greater residual values are a function of the time elapsed since some disturbance event.

PAM is simply a vector of residual differences between the corresponding n observations. For example, assuming that an ecologist wants to correlate two matrices of data $\mathbf{X}$ and $\mathbf{Y}$, both of which are formed by four rows (i.e. sites, plots, observational units), Procrustes analysis will generate four residual differences between the $\mathbf{X}$ and $\mathbf{Y}$ configurations. The compilation of these residual differences between homologous rows (observations) across dimensions in the form of a vector – PAM – represents a useful way to represent information on the relationship between two matrices and make it available for further statistical analysis, both parametric and non-parametric; this feature is not offered by the Mantel approach.

The use of the residual vector from Procrustes (PAM) has been quite restricted in the plant and soil ecological literature. To our knowledge the first study was by [32] who assessed the plant-pollinator interaction during three consecutive summers in the southeastern portion of California, USA. These authors employed the PAM to identify which pollinating species exhibited the greatest deviation between two consecutive years. Singh et al. [22] used the PAM in a study on soil microbiology to verify the effect of soil pH on the relationship between arbuscular mycorrhizal fungi (AMF) and plant assemblages. These authors employed the following strategy: 1) Procrustes analysis was applied between the matrices representing the AMF community and that representing the plant community; 2) after detecting a significant relationship ($m_{12} = 0.28$; P<0.001), these authors extracted the PAM and used it as a response in a simple regression analysis with the soil pH. No effect of pH on the association between the AMF and plant communities was detected, suggesting that neither the pH nor the identity of the plant species that composed the community affected the AMF community. Other applications can be certainly found (e.g., [24], [25], [26][33]) but its flexibility and general usage remains largely unexplored.

Constructing a practical roadmap for applying PAM

There are few studies in the ecological literature that have used PAM for analyzing relationships between plant and soil datasets. The lack of examples partially explains the low popularity of Procrustes analysis among plant and soil ecologists and ecologists in general as an alternative tool to the more traditional Mantel test. In order to make the possible uses of Procrustean residuals more familiar, we will introduce a number of examples in the form of schematic roadmaps for applying PAM in association with three common statistical approaches: ordination, regression analysis and ANOVA.

Plant and soil ecologists must keep in mind that Procrustes analysis requires that the **X** and **Y** have the same number of rows and columns, though the last dimension is less restricting (see below). Given that the data for both matrices usually originate from the same sites, it is most common in ecology that only the number of columns (descriptors or variables) varies between the two matrices. Therefore, the question arises of how to make the number of columns equal across the two matrices, i.e., how to reduce them to the same dimensionality. Although Procrustes analysis can be performed between matrices having different number of dimensions (i.e., the fit is based on a singular value decomposition (svd) of $X^T Y$, where X and Y are scaled prior to svd and T stands for matrix transpose), traditionally the matrix with the fewer number of columns ("missing columns") is made equal in dimension to the larger matrix by adding columns of zeros in order to keep (Fig. 2a; [2]). Although there are some criticisms related to this practice and alternatives have been suggested [34], the addition of zero columns does not affect the distances between columns among observations and is a convenient device rather than a hurdle [35].

Another convenient way to make **X** and **Y** have the same number of columns is to represent most of the variation in their raw data by matrices formed by the same number of orthogonal axes (Fig. 2b; [3],[35], [36]), i.e., matrices formed by axes derived through ordination methods such as Principal Components Analysis (PCA), Non-metric dimensional scaling, Correspondence Analysis (CA), Principal Coordinate Analysis (PCoA), the choice being dependent on the nature of the data (continuous, presence-absence data, abundance data). Moreover, raw data matrices can be transformed prior to ordination (see [37] for different transformations and their characteristics) or alternatively have pairwise distance matrices calculated from the data matrices that are then orthogonolized via PCoA to extract ordination axes based on the chosen distance measure (e.g., Bray-Curtis, Jaccard, Sorensen, Gower).

Here, for simplicity, we use a PCA in all applications. In cases, where species data (presence/absence or abundance) was used, the data was Hellinger-transformed and PCAs were extracted on species correlation matrix calculated from the transformed data. The Hellinger transformation alleviates the issue of double-zeros

in species data matrix transformed into correlation or Euclidean-distance pairwise matrices prior to PCA in which sites sharing no species in common can be found to be more similar than sites sharing a reduced number of species in common (e.g., the horse shoe effect in ordination plots).

The general strategy is as follows:

1) Subject the raw data matrices to an ordination method (here PCA but see above for other strategies);

2) After ordinating **X** and **Y**, use the same number of ordination axes for both matrices (Fig. 2b).

Given that the higher the number of ordination axes used, the higher is the amount of variation explained in **X** and **Y**, it would be interesting run the Procrustean analysis sequentially using matrices made up of an increasing number of ordination axes. It could help ecologists check the consistency of the relationship between **X** and **Y** based on different numbers of ordination axes, which will give more reliability to the results.

The use of PAM in ecological ordination

The first form of PAM shown here is based on ordination methods. Ordination is the graphical representation of the variation of objects (sites), descriptors (species/environmental parameters) or both, in a reduced space formed by orthogonal axes [38].

To illustrate the use of Procrustes analysis associated with ordination we use data derived from Mitchell et al. [39]. This study aimed to compare the plant communities and soil chemistry in their ability to predict changes in the structure of the soil microbial community in three moorland areas established in Northern Scotland called Craggan, Kerrow and Tulchan. The plant community matrices from each area were based on the percent cover. Three matrices for the soil microbial community were obtained for each site: one based on the fatty acid profile of the soil (PLFA analysis), and the other two on the T-RFLP analysis of the communities of fungi and bacteria, respectively. The matrix representing the soil chemistry was based on the concentrations of Na, K, Ca, Mg, Fe, Al, P, total C, total N in addition to pH, loss on ignition and moisture.

There is some consensus that the variation in vegetation can act as a proxy for changes in the soil microbial community, either directly in the case of symbionts, for example, or indirectly via changes in soil chemistry itself. We use Procrustes analysis associated with ordination techniques to verify potential drivers of the soil microbial community and to determine if plant community and soil chemistry are equally related to the microbiological variation. The sequence of analysis was as follows:

1) Ordination analysis: All data matrices (community plant, soil chemistry and soil microbial communities) containing the three chronosequences were subjected to separate PCAs based on correlation matrix. The community plant was Hellinger-transformed prior to PCA. Then, the first six PCA ordination axes from each matrix were retained in order to assemble four PCA matrices representing the variation summarized in the first 3, 4, 5 and 6 PCA axes. Thus, four PCA matrices were obtained from each dataset: plant community, soil chemistry and soil microbial community (PLFA, bacterial and fungal T-RFLP) (Fig. 3a).

2) Procrustes analysis: The PCA matrices of plant community and soil chemistry were used to run Procrustean analyses with the PCA matrices of soil microbial community based on PLFA, and fungal and bacterial T-RFLP datasets.

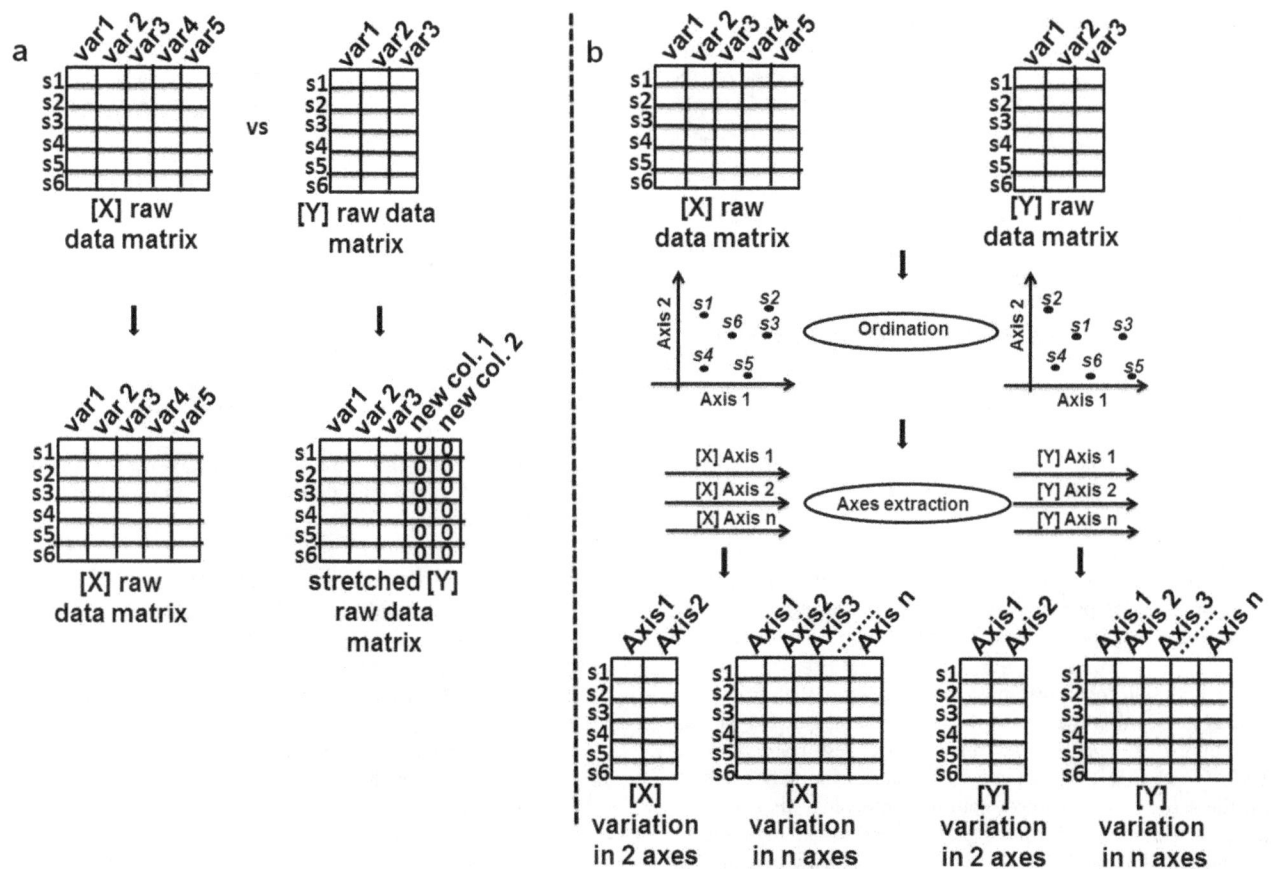

Figure 2. Roadmap for two alternative ways to reach the same dimensionality between matrices, and so relating it by Procrustes analysis. a) Addition of columns containing zeros to the **Y** raw data matrix for matching the **X** raw data matrix dimension; **b)** Application of ordination to raw data matrices to make matrices have equal dimensionality prior to Procrustes analysis.

3) PAM extraction: Since all Procrustean relationships based on PCA matrices with n axes were significant, for simplicity, only the PAM obtained from relationships of PCA matrices with 6 axes were used for subsequent analyses. Six PAMs were generated: PAM1 (soil chemistry on PLFA), PAM2 (soil chemistry on bacteria), PAM3 (soil chemistry on fungi), PAM4 (plant on PLFA), PAM5 (plant on bacteria), and PAM6 (plant on fungi) (Fig. 3c).

4) PAM ordination: The PAMs were assembled in a single matrix ("effect matrix") with one PAM per row (Fig. 3c). Therefore, the effect matrix compiled the effects of plant community and soil chemistry on soil microbial community structure derived from the three methods. This effect matrix was submitted to PCA ordination to verify whether the plant community effect on soil microbial community structure differed from the effect of soil chemistry (Fig. 3c).

The results showed that for all chronosequences the plant effect on microbial structure was divergent in relation to the soil chemistry effect, as suggested by the separation along the axis of greatest variation (Fig. 4). Although we cannot apply a proper statistical significance test in one-table based ordination methods (PCA, NMDS, PCoA, etc), visual inferences can be made. For example the Craggan area exhibited a clear distortion between plant community and soil chemistry variation in terms of their effects on the soil microbial community structure depicted by PLFA, bacterial T-RFLP and fungal T-RFLP (Fig. 4a). Also in this

area, the response of the microbial community based on PLFA was distant from the response based on molecular data (T-RFLP) (Fig. 4a).

We expected that the effects of soil chemistry on microbial structure were closer to the effect of plant community once the plant community is considered to be a direct and indirect driver for the biotic component of soil [39]. However, these results suggest that plant communities and soil chemistry are acting differently on the soil microbial community structure [24], [40]. They also suggest that the effects of soil chemical properties on the microbial communities may be weakly mediated by above ground alterations [24]. This example shows the usefulness of Procrustes analysis to raise additional evidence in plant and soil ecology studies. (See Text S1 containing the R code used for this example).

The PAM and regression analysis

In regression analysis, 'response' and 'predictor' are common terms. In ecology, predictors can have different natures. Space, time, organic matter and moisture, among other factors, are some examples of predictors. On the other hand the microbial communities are often used as a response variable because they are considered better indicators of a given ecosystem.

Some authors familiar with soil microbial ecology have been using the Mantel test to assess the individual contribution of deterministic and stochastic processes on the soil microbial structure variation [41], [42]. As an example of the utility of the

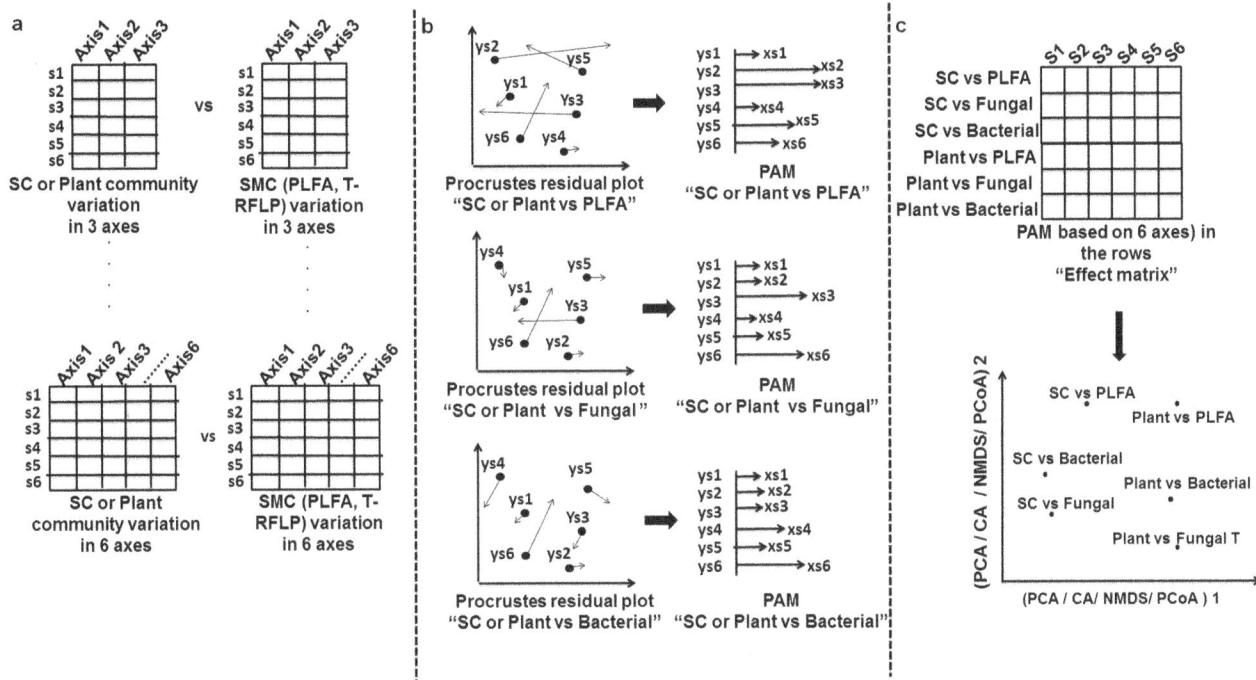

Figure 3. Roadmap for applying the Procrustes association metric (PAM) in the multivariate ordination context using data of [39]. **a)** Assembling matrices with different ordination axes, through Procrustes analysis, soil chemistry (SC) and plant community with soil microbial community (PLFA, and bacterial and fungal T-RFLP); **b)** Extraction of PAM from Procrustean relationships based on matrices with 6 ordination axes; **c)** Assembling of PAM based PCA matrices with 6 axes as rows in a single matrix ("effect matrix"), and using it in an ordination technique (e.g., PCA, PCoA, NMDS) to verify if the different effects diverge.

Procrustes analysis in the context of variation partitioning we can take a hypothetical scenario with four datasets from a given area, corresponding to soil microbial community structure (PLFA), soil microbial functioning (enzyme activities), soil properties and spatial variation. Spatial variation can be represented, for example, by 100 sampling points generated from a 10 m×10 m transects. The matrix of geographical coordinates of the sampling points can be submitted to PCNM (principal coordinates neighbour matrix) analysis generating a matrix of spatial eigenfunctions termed PCNMs [34]. In this scenario, we can assume that the ecologist aims to assess the relative contributions of individual soil properties (deterministic processes) and spatial variation (stochastic event) on the relationship between microbial community structure and soil microbial functioning rather than on

these components individually. To use the Procrustean association metric (PAM) in this context, one can use the following steps:

1) Ordinate the two matrices (i.e., the soil microbial community and soil microbial functioning) via PCA (the soil microbial community matrix was Hellinger-transformed) and select a similar number of ordination axes. The multivariate scores of the two matrices across the selected number of axes are subjected to a Procrustes analysis and a PAM was then calculated.

2) Use individual PAMs (based on 2, 3 or more PCA axes) as response variable and soil properties and spatial variation as independent (predictor) variables in a multiple regression framework (Fig. 5b).

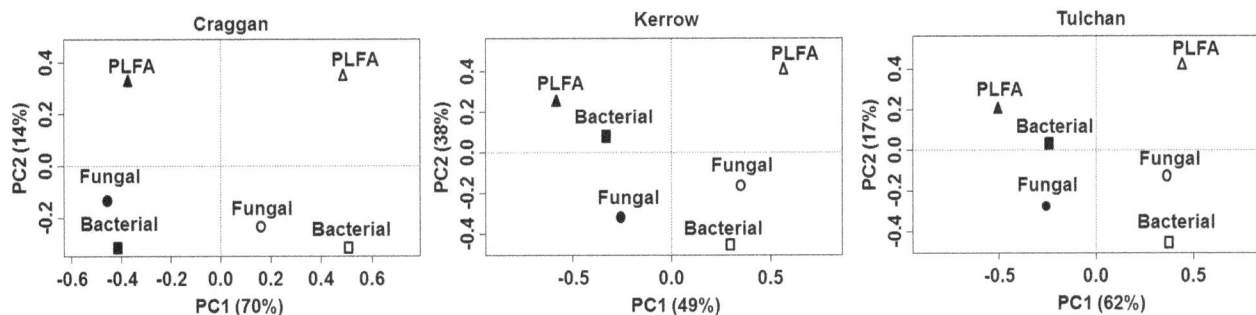

Figure 4. Results from PCA ordination of the Procrustes association metric matrix ("effect matrix") gathering the interactions of soil chemistry and plant community with soil microbial matrices (PLFA, and bacterial and fungal T-RFLP). The filled symbols are the Procrustes relationships between soil chemistry and soil microbial matrices, and the open symbols between plant community and soil microbial matrices. Data from three chronosequences (Craggan, Kerrow and Tulchan) obtained by [39].

3) Finally, the independent contributions of soil properties (independent of space) and unmeasured spatial process and/or factors (spatial variation independent of soil properties) to the microbial structure can be estimated via variation partitioning [43] and represented by a Venn diagram (Fig. 5c). (See Text S2 containing the R code for this example).

The PAM and Analysis of Variance

Although regression and analysis of variance are ultimately the same analysis in which the response is either continuous (regression) or ascribed to factors (ANOVA), we provide examples for each of them in different sections given that often they are seen as distinct forms of analyses. Evaluation of the effects of land use on soil microbial communities has been a common case-study issue in soil ecology. Some of these studies have been carried out using the Mantel approach to assess how land use type effects soil microbial structure and functioning [44], [45]. However, Mantel does not yield a vector of structure – functioning relationship, that is, a continuous variable, able to be partitioned by categorical variables like land use types. In the following example we show how to use PAM to evaluate the effect of land use type on the relationship between microbial community structure and microbial function in the form of PAM.

In a hypothetical scenario, a researcher is interested in studying whether four different land use types within the Amazon biome are affecting the relationship between microbial structure and microbial functioning. In each of the land uses (original forest fragment, silvipastoral system, improved pasture and unimproved pasture) six plots (10 m×10 m) were established and one composite soil sample (0–10 cm) collected per plot (Fig. 6a). The X dataset (soil microbial structure) was represented by PLFA data,

and the Y dataset (microbial functioning) by the abundance of genes associated with microorganisms involved in greenhouse gas emission processes, such as nitrifiers, denitrifiers and methanotrophic organisms. The researcher's hypothesis is that in the original forest (non-altered environment) there is a better matching between microbial structure and microbial function. Thus, in anthropogenically disturbed environments (silvipastoral system, improved pasture, and unimproved pasture) the change in microbial structure relative to the original (forest) is not followed by a change in the microbial functioning to the same magnitude. This hypothesis can be tested using an integration of Procrustes analysis and ANOVA through the following steps:

1) Reduce the datasets **X** (soil microbial structure) and **Y** (soil microbial functioning) to similar dimensions using PCA. Then, run the Procrustean analysis between the PCA matrix of the soil microbial community structure and the PCA matrix of soil microbial functioning and extract the PAM (Fig. 6a).

2) Run an ANOVA with land use type as fixed factor and the PAM as the response variable (Fig. 6b).

3) If the F value of ANOVA is significant, a means test can be performed to compare the mean PAMs of the land use types (Fig. 6c). (See Text S3 containing the R code for this example).

Discussion

In this paper we have attempted to show the advantages of the Procrustean analysis over the Mantel test, in which the former can be used for gaining further information on underlying drivers of data table associations. In particular we have shown how the Procrustean association metric (PAM) constructed of the residuals

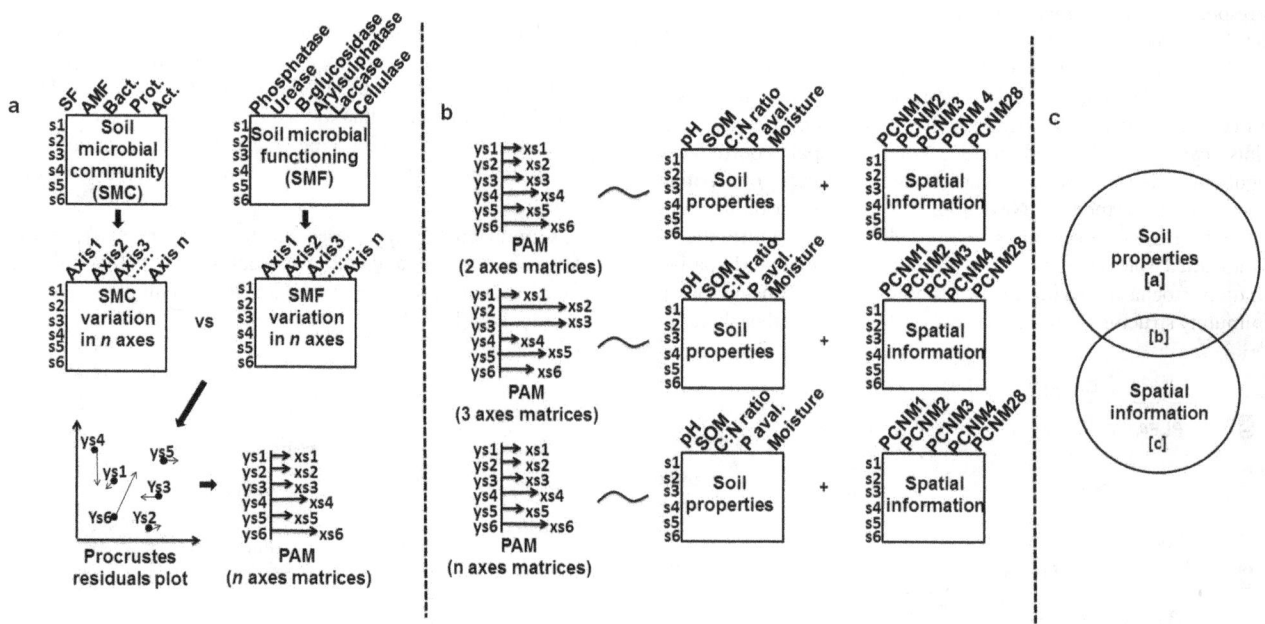

Figure 5. Roadmap for using Procrustes Association Metric (PAM) in a multiple regression analysis framework (variation partitioning). a) Soil microbial community (SMC) and soil microbial functioning (SMF) matrices are submitted to an ordination to reach the same dimensionality, and SMC and SMF matrices formed by 2, 3 and n axes related through Procrustes analysis in order to generate PAMs; **b)** PAMs generated were used as response variables in a variation partitioning to verify the individual contribution of soil properties and spatial information (PCNM eigenfunctions) on the SMC-SMF relationship; **c)** Venn diagram depicting the relative contribution of soil properties (niche processes **[a]**) and unmeasured spatial factors (neutral processes **[c]**).

of the vectors after the Procrustes analysis. We concentrated on showing how patterns of concordance between data matrices can be displayed and individual observations contrasted separately using the Procrustean framework, allowing further examination of the common and different association patterns among multiple data matrices. Given that in the Mantel framework, multivariate information is translated into a pairwise distance matrix, we lose the ability to contrast homologous data points across dimensions and data matrices. It is important to notice that it was not our goal to show the statistical advantages of Procrustes over Mantel as done by previous work [3]. Instead, we concentrated on generating different analytical schemes, especially for plant and soil ecologists, to incorporate Procrustes into their statistical toolbox.

What is unique about Procrustean framework? There are at least four characteristics of the approach not shared by others. First, because the approach is correlative rather than regressive, the number of observations (e.g., sites) in the matrices does not have to be greater than the number of columns as in common regression approaches such as RDA and CCA. Second, we can fit as many matrices as we have available; this latter issue is particularly restrictive under a regression approach given the limitation of number of rows versus number of columns. Moreover, all matrices are treated in equal footing as no matrix is treated as response or predictor. Third, the relationships within (only across) matrix columns do not affect the analysis. Fourth, residual values across observations and dimensions can be calculated and explored as shown here. These characteristics should not be necessarily seen as advantages per se over other methods but rather features that are unique and may be useful in many situations. There are certainly other tools that can be used to look at the associations between data sets. RDA and CCA are well-established tools in ecology and are based on regression (asymmetric) methods. Traditionally these approaches may have been thought to be more appropriate for analysis of the examples given in this paper, since they establish relations of cause and effect. However, because these analyses include a regression step, they are limited to situations where the number of rows (sites) in the environmental matrix **X** is higher than the number of columns (variables) [36], [46]. This is not a limitation in Procrustes analysis and moreover, it is not clear how residual variation among homologous observations across dimensions should be explored in the case of RDA and CCA.

At least two other symmetric approaches are similar to the Procrustean approach, namely Co-inertia analysis [47] and symmetric Co-correspondence analysis [48], a form of Co-inertia analysis in which a correspondence analysis is applied to two species matrices prior to the analysis. The main difference resides in the fact that fit is influenced by all variables pairs in Co-inertia analysis (within and between matrices), whereas fit is influenced only by variation between matrices in Procrustean. Co-inertia is always based on ordination within data matrices, whereas in Procrustes either the raw data or their ordination axes can be used. Co-inertia can also take into account row (e.g., sites) and column (e.g., species) weights in the analysis, though the standardization and fit processes in Procrustean analysis could also take these into account [36]. Co-inertia and Procrustean analysis are certainly related in the sense that they both treat matrices as symmetrical during the fitting process, though more studies are necessary to assess in which conditions (e.g., correlation within and across matrices, differences in dimensionality between matrices, outliers within and across matrices) they differ. Finally Dray et al. [36] showed the advantages of merging Co-inertia and Procrustean analysis, where the latter is used as a precursor of the former. In reality, future studies are required to contrast Co-inertia and Procrustean analysis, but in either form of analyses we can produce residual vectors (PAM) that can be further analyzed.

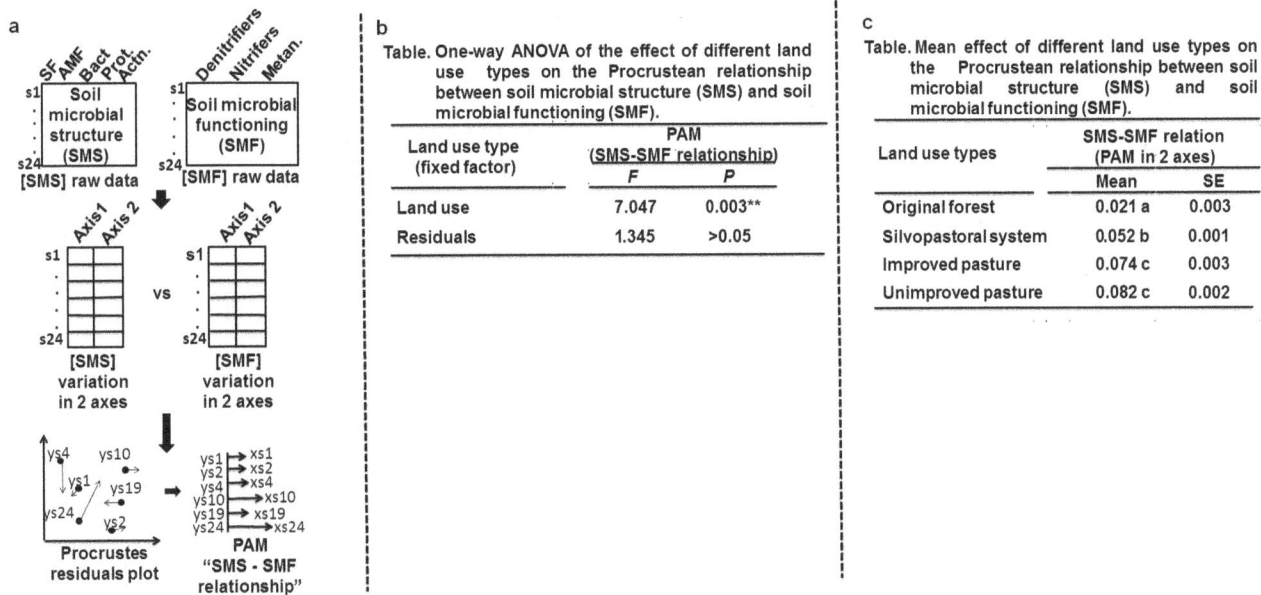

Figure 6. Roadmap for using Procrustes association metric (PAM) in an ANOVA context. a) PCA ordination of each SMC and SMF raw data matrices, and then Procrustes correlation from 2 axes-based PCA matrices in order to generate the PAM depicting the SMC-SMF relationship. **b)** Table showing results of a one-way ANOVA for using PAM as response and land use type as fixed factor. **c)** Multiple comparisons test (Tukey, 95%) for means of the Procrustean relationship between soil microbial structure and functioning (PAM in 2 axes) across land use types.

Procrustes can be perhaps best justified when the number of predictors is greater than the number of observations or when **X** and **Y** matrices are equally applicable as explanatory and response variables. In plant-soil ecology, for example, above- and below-ground data matrices can be interchanged as explanatory and response variables. Plant community variation has been shown to be related to variation in below-ground compartments [24]. In addition, soil components such as fertility and the microbial community have been proven to influence aspects of vegetation [49]. Thus, with the literature showing that both types of datasets under analysis can structure each other, the use of Procrustes analysis, as a symmetric canonical analysis method, should be encouraged among plant and soil ecologists and ecologists in general. We hope that this paper has provided enough examples of the potential for using the Procrustes framework as a precursor to further explore ecological data.

References

1. Mantel N (1967) The detection of disease clustering and a generalized regression approach. Cancer Res 27: 209–220.
2. Gower JC (1971) Statistical methods of comparing different multivariate analyses on the same data. In: Hodson FR, Kendall DG, Tautu P, editors. Mathematics in the archeological and historical sciences. Edinburgh University Press, Edinburgh. 138–149.
3. Peres-Neto PR, Jackson DA (2001) How well do multivariate data sets match? The advantages of a Procrustean superimposition approach over the Mantel test. Oecologia 129: 169–178.
4. Tuomisto H, Poulsen AD, Ruokolainen K, Moran RC, Quintana C, et al. (2003a) Linking floristic patterns with soil heterogeneity and satellite imagery in Ecuadorian Amazonia. Ecol App 2: 352–371.
5. Tuomisto H, Ruokolainen K, Yli-Halla M (2003b) Dispersal, environment, and the floristic variation of western Amazonian forests. Science 299: 241–244.
6. Tuomisto H, Ruokolainen K, Aguilar M, Sarmiento A (2003c) Floristic patterns along 43-km long transect in an Amazonian rain forest. J Ecol 91: 743–756.
7. Kang S, Mills AL (2004) Soil bacterial community structure changes following disturbance of the overlying plant community. Soil Sci 169: 55–65.
8. Poulsen AD, Tuomisto H, Balsev H (2006) Edaphic and florist variation within a 1-ha plot of lowland Amazonian rain forest. Biotropica 38: 468–478.
9. Fitzsimons MS, Miller RM, Jastrow JD (2008) Scale-dependent niche axes of arbuscular mycorrhizal fungi. Oecologia 158: 117–127.
10. Powers JS, Becknell JM, Irving J, Pères-Aviles D (2009) Diversity and structure of regenerating tropical dry forests in Costa Rica: Geographic patterns and environmental drivers. Forest Ecol Manag 258: 959–970.
11. Castilho-Monroy AP, Bowker MA, Maestre FT, Rodriguez-Echeverria S, Martinez I, et al. (2011) Relationships between biological soil crusts, bacterial diversity and abundance, and ecosystem functioning: insights from a semi - arid Mediterranean environment. J. Veg. Sci 22: 165–174.
12. Pomara LY, Ruokolainen K, Tuomisto H, Young K (2012) Avian composition co-varies with floristic composition and soil nutrient concentration in Amazonian upland forests. Biotropica 44: 545–553.
13. Artz RRE, Chapman SJ, Campbell CD (2006) Substrate utilization profiles of microbial communities in peat are depth dependent and correlate with whole soil FTIR profiles. Soil Biol Biochem, 38: 2958–2962.
14. Trivedi MR, Marecroft MD, Berry PM, Dowson TP (2008) Potential effects of climate change on plant communities in three montane nature reserves in Scotland, UK. Bio Conserv 141: 1665–1675.
15. Jesus EC, Marsh TL, Tiedje JM, Moreira FMS (2009) Changes in land use alter the structure of bacterial communities in Western Amazon soils. ISME J 3: 1004–1011.
16. Merilä P, Lämsa- MM, Stark S, Spetz P, Vierikko K, et al. (2010) Soil organic matter quality as a link between microbial community structure and vegetation composition along a successional gradient in a boreal forest. App Soil Ecol 46: 259–267.
17. Grayston SJ, Campbell CD, Bardgett RD, Mawdsley JL, Clegg CD, et al. (2004) Assessing shifts in microbial community structure across a range of grasslands of differing management intensity using CLPP, PLFA, and community DNA techniques. Appl Soil Ecol 25: 63–84.
18. Singh BK, Munro S, Reid E, Ord B, Potts M, et al. (2006) Investigating microbial community structure in soils by physiological, biochemical and molecular fingerprinting methods. Eur J Soil Sci 57: 72–82.
19. Vinten AJA, Artz RRE, Thomas N, Potts JM, Avery L, et al. (2011) Comparison of microbial community assays for the assessment of stream biofilm ecology. J Microbiol Methods 85: 190–198.
20. Hirst CN, Jackson DA (2007) Reconstructing community relationships: the impact of sampling error, ordination approach, and gradient length. Divers. Distrib 13: 361–371.
21. Poos MS, Jackson DA (2012) Addressing the removal rare species in multivariate bioassessments: The impact of methodological choices. Ecol Indic 18: 82–90.
22. Singh BK, Nunan N, Ridgway KP, Mcnicol J, Young JPW, et al. (2008) Relationship between assemblages of mycorrhizal fungi and bacteria on grass roots. Environ Microbiol 10: 534–542.
23. Burke L, Irwin R (2009) The importance of interannual variation and bottom up nitrogen enrichment for plant – pollinator networks. Oikos 118: 1816–1829.
24. Lisboa FJG, Chaer GM, Jesus EC, Gonçalves FS, Santos FM, et al. (2012) The influence of litter quality on the relationship between vegetation and below-ground compartments: a Procrustean approach. Plant Soil 367 551–562.
25. Landeiro VL, Bini LM, Costa FRC, Franklin E, Nogueira A, et al. (2012) How far can we go in simplifying biomonitoring assessments? An integrated analysis of taxonomy surrogacy, taxonomic sufficiency and numerical resolution in a mega diverse region. Ecol Indic 23: 366–373.
26. Siqueira T, Bini LM, Roque FO, Cottiene K (2012) A metacommunity framework for enhancing the effectiveness of biological monitoring strategies. PLoS One 7: e43626.
27. Nekola JC, White PS (1999) The distance decay of similarity in biogeography and ecology. J Bio 26: 867–878.
28. Legendre P, Borcard D, Peres-Neto PR (2005) Analysing beta diversity: partitioning the spatial variation of community composition data. Ecol Mon 75: 435–450.
29. Tuomisto H, Ruokolainen K (2006) Analysing or explaining beta diversity? Understanding the targets of different methods of analysis. Ecology 87: 2697–2708.
30. Kuehnelt-Leddihn ER (2007) The menace of the herd or Procrustes at large. The Bruce Publishing Company Milwaukee, Auburn. 385 p.
31. Jackson DA (1995) PROTEST: a Procrustean randomization test of community environment concordance. Ecoscience 2: 297–303.
32. Alárcon R, Waser NM, Orlleton J (2008) Year-to-year variation in the topology of a plant - pollinator interaction network. Oikos 117: 1796–1807.
33. Burke L, Irwin R (2009) The importance of interannual variation and bottim up nitrogen enrichment for plant – pollinator networks. Oikos 118: 1816–1829.
34. ten Berge JMF, Kiers HAL, Commandeur JJF (1993) Orthogonal Procrustes rotation for matrices with missing values. B J Math Stat Psyc 46: 119–134.
35. Dijksterhuis GB, Gower JC (1992) The interpretation of Generalized Procrustes Analysis and allied methods. Food Qual Prefer 3: 67–87.
36. Dray S, Chessel D, Thioulouse J (2003) Procrustean Co-inertia analysis for the linking of multivariate datasets. Ecoscience 10: 110–119.
37. Legendre P, Gallagher ED (2001) Ecologically meaningful transformations for ordination of species data. Oecologia 129: 271–280.
38. Legendre P, Legendre L (2012) *Numerical Ecology*, 3rd English edn. Elsevier Science BV, 516 Amsterdam.
39. Mitchell RJ, Hester AJ, Campbell CD, Chapman SJ, Cameron CM (2010) Is vegetation composition or soil chemistry the best predictor of soil microbial community? Plant Soil 333: 417–430.
40. Mitchell RJ, Hester AJ, Campbell CD, Chapman SJ, Cameron CM, et al. (2012) Explaining the variation in the soil microbial community: do vegetation composition and soil chemistry explain the same or different parts of the microbial variation? Plant Soil 351: 355–362.
41. Dumbrell AJ, Nelson M, Helgason T, Dytham C, Fitter AH (2009) Relative roles of niche and neutral processes in structuring a soil microbial community. ISME J doi:10.1038/ismej.2009.122.

Supporting Information

Text S1 R code showing how to use PAM associated to ordination methods (Fig. 4 in the main text). For this example we used data from Mitchell et al. [39].

Text S2 R code showing how to use the PAM in a Regression framework (Fig. 5 in the main text).

Text S3 R code showing how to use the PAM in an ANOVA framework (Fig. 6 in the main text).

Author Contributions

Conceived and designed the experiments: FJGL. Analyzed the data: FJGL. Wrote the paper: FJGL PRPN GMC ECJ SJC RJM RLLB. Contributed with real datasets: SJC RJM.

42. Zheng YM, Cao P, Fu B, Hughes JM, He JZ (2013) Ecological Drivers of Biogeographic Patterns of Soil Archaeal Community. PloS One 8: e63375.

43. Peres-Neto PR, Legendre P, Dray S, Borcard D (2006) Variation partitioning of species data matrices: estimation and comparison of fractions. Ecology 87: 2603–2613.

44. Chaer GM, Fernandes MF, Myrold DD, Bottomley PJ (2009) Shifts in microbial community composition and physiological profiles across a gradient of induced soil degradation. Soil Sci Soc Am J 73: 1327–1334.

45. Peixoto RS, Chaer GM, Franco N, Reis Junior FB, Mendes IC, et al. (2010) A decade of land use contributes to changes in the chemistry, biochemistry and bacterial community structures of soils in the Cerrado. A Van Leeuw 98: 403–413.

46. Thiouleuse J, Simier M, Chessel D (2004) Simultaneous analysis of a sequence of paired ecological tables. Ecology 85: 272–283.

47. Dolédec S, Chessel D (1994) Co-inertia analysis: an alternative method for studying species–environment relationships. Freshwater Biol 31: 277–294.

48. Ter Braak, CJF, Schaffers AP (2004) Co-correspondence analysis: a new ordination method to relate two community compositions. Ecology 85: 834–846.

49. van der Heijden MGA, Klironomos J, Ursic M, Moutoglis P, Streitwolf-Engel R, et al. (1998) Mycorrhizal fungi determines plant biodiversity, ecosystem variability and productivity. Nature 396: 69–72.

5

Functional Changes in Littoral Macroinvertebrate Communities in Response to Watershed-Level Anthropogenic Stress

Katya E. Kovalenko[1,2]*, Valerie J. Brady[2], Jan J. H. Ciborowski[1], Sergey Ilyushkin[3], Lucinda B. Johnson[2]

1 University of Windsor, Windsor, Ontario, Canada, 2 Natural Resources Research Institute, University of Minnesota Duluth, Duluth, Minnesota, United States of America, 3 Colorado School of Mines, Golden, Colorado, United States of America

Abstract

Watershed-scale anthropogenic stressors have profound effects on aquatic communities. Although several functional traits of stream macroinvertebrates change predictably in response to land development and urbanization, little is known about macroinvertebrate functional responses in lakes. We assessed functional community structure, functional diversity (Rao's quadratic entropy) and voltinism in macroinvertebrate communities sampled across the full gradient of anthropogenic stress in Laurentian Great Lakes coastal wetlands. Functional diversity and voltinism significantly decreased with increasing development, whereas agriculture had smaller or non-significant effects. Functional community structure was affected by watershed-scale development, as demonstrated by an ordination analysis followed by regression. Because functional community structure affects energy flow and ecosystem function, and functional diversity is known to have important implications for ecosystem resilience to further environmental change, these results highlight the necessity of finding ways to remediate or at least ameliorate these effects.

Editor: Christopher J. Salice, Texas Tech University, United States of America

Funding: Data collection for this project was supported by grants from the U.S. Environmental Protection Agency Science to Achieve Results Estuarine and Great Lakes Program through the Great Lakes Environmental Indicators Project (R-8286750) to G.L. Niemi et al. and the Protocols for Selection of Classification Systems and Reference Conditions project (R-828777) to L.B. Johnson et al. Other parts of this project were supported by the second stage GLEI-II Indicator Testing and Refinement project funded by a Great Lakes Restoration Initiative grant from the U.S. Environmental Protection Agency Great Lakes National Program Office to L.B. Johnson et al. (GL-00E00623-0). Although the research described in this work has been partly funded by the US EPA, it has not been subjected to the agency's required review and, therefore, does not necessarily reflect the views of the agency and no official endorsement should be inferred. The funders had no role in study design, data collection and analysis, decision to publish, or preparation of the manuscript.

Competing Interests: The authors have declared that no competing interests exist.

* Email: philarctus@gmail.com

Introduction

Widespread anthropogenic modification of the landscape is jeopardizing freshwater ecosystems and their services. Watershed-level anthropogenic stress has negative effects on macroinvertebrate communities in lakes, streams and wetlands [1,2,3]. Development in the watershed is characterized by reduction/removal of natural vegetation, increased road density and increased proportion of other impervious surfaces, as well as higher human population density. The effects of development-associated stressors and those associated with agricultural land use on macroinvertebrate communities are mediated through increased nutrient loading, point and non-point source pollution, sediment loads, altered hydrologic and temperature regimes, and habitat destruction and fragmentation in riparian zones and littoral areas [1].

Several functional characteristics (*e.g.*, flow, drag or silt adaptations, respiration and locomotion techniques, feeding habits, voltinism, *reviewed in* [3]) are considered to be important indicators of the state of an ecosystem and its potential resilience to further anthropogenic modification. In particular, functional diversity (FD) is a critical property of a group of organisms at any scale because increased trait space breadth is likely to be associated with a greater diversity of ecosystem processes and

nutrient pathways, which in turn increase resistance and resilience to perturbations [4,5]. FD can be related to reticulation of the food web, which has important implications for the resilience of food webs as demonstrated in a theoretical study [6]. In terrestrial plant assemblages, greater functional diversity has been shown to maximize resource use in heterogeneous environments, and affect energy flow and ecosystem function (*reviewed in* [4,7,8]), and similar patterns were observed in theoretical studies [9]. However, little is known about the effects of reduced FD in littoral systems. Another important functional trait, voltinism, may have important implications for temporal redistribution of nutrient processing and ecosystem stability, and proportion of univoltine and other longer-lived organisms was shown to decrease with increasing land use stress [10]. Furthermore, changes in the relative abundance of different invertebrate functional groups can alter nutrient processing characteristics [11,12], potentially triggering cascading effects in higher trophic levels, which may also affect littoral-pelagic and aquatic-terrestrial habitat coupling, by virtue of this group's central position in aquatic food webs.

Although invertebrate functional trait responses to increasing anthropogenic stress have been described in streams-particularly changes in voltinism and respiration type [3] - those findings may not apply to lake and wetland systems because the effects of watershed-scale stressors are no longer mediated primarily through hydrological alteration and canopy clearing [1]. In addition, some of the more sensitive functional attributes, such as the proportion of semi- and merovoltine taxa and anoxia-intolerant taxa, are already under-represented in lentic systems because these systems routinely experience greater variability in parameters such as dissolved oxygen, temperature, and other factors [13–15]. It is important to understand whether functional attributes of littoral macroinvertebrate communities change in response to watershed-scale anthropogenic stress in order to design management and conservation strategies specific to freshwater littoral systems.

To assess the functional responses of lentic macroinvertebrates, we used wetland macroinvertebrate community composition data from the Great Lakes Environmental Indicators project [16] collected across a full gradient of watershed development stress (basin-wide minimum to maximum) to 1) test whether FD and the relative abundance of longer-lived (uni-, semi- and merovoltine) organisms are affected by watershed-level development and agriculture, and 2) investigate the functional changes in invertebrate communities. The *a priori* prediction was that FD and proportion of longer-lived taxa would decline with increasing development and agriculture due to an overall decrease in taxonomic diversity and reduced habitat availability resulting from the combined effects of direct habitat degradation and pollution commonly associated with these anthropogenic stressors [1]. Due to the large number of taxa and their wide range of physiologic tolerances and habitat requirements, macroinvertebrates can serve as an indicator of the changes impacting other assemblages, as well as a warning signal of changes in littoral food webs.

Materials and Methods

Macroinvertebrates were sampled in 101 coastal wetlands of the U.S. coastline of the Laurentian Great Lakes in the summer of 2002 and 2003 (Fig. 1). This dataset was previously collected for a different purpose, as part of a multidisciplinary effort to identify indicators of anthropogenic stress in the Great Lakes coastal zone (the Great Lakes Environmental Indicators project). No permits were required for macroinvertebrate sampling by the U.S. at the time of the original study and no recognized endangered or threatened invertebrate taxa occur in those coastal wetlands. All appropriate protected area sampling permits were secured by the original study.

The Laurentian Great Lakes include five glacial till lakes: Superior, Michigan, Huron, Erie and Ontario, located in the temperate part of North America. Lakes range from oligotrophic to eutrophic, with the surrounding land use spanning a gradient from completely unimpacted to highly impacted by anthropogenic activities including development and agriculture. Wetlands were equally distributed among lakes and across a full gradient of anthropogenic stress, previously defined as a composite of five major classes of anthropogenic pressure: agriculture, atmospheric deposition, land cover, human development, and point source pollution. The site selection procedure ensured representation of four geomorphic wetland types, including riverine, barrier-beach protected, and lacustrine coastal wetlands and embayments [17]. All wetlands were hydrologically connected to a Great Lake, and most wetlands had well-developed submerged and emergent macrophyte communities.

Proportion of human development and agriculture (by area) in a wetland's watershed was derived from the USGS National Land Cover Dataset (2001) for each catchment delineated using ArcHydro with 10-m Digital Elevation Models (see [18] for details). Development was defined to include all residential, commercial and industrial areas, but did not include road density.

Invertebrate sampling and functional metrics

To ensure the most complete sampling of the different habitats present within a wetland, macroinvertebrates were collected from all representative near shore land-use and shoreline (*e.g.*, sand beach, cobble beach, lawn, etc.) zones in each wetland. In each land-use-shoreline zone, two transects were extended perpendicular to shore, and macroinvertebrates were collected using 30-second D-frame dipnet sweeps at 0.25 and 0.75-m water depths along each transect. Sweeps were done through the water column from the bottom to the surface, in a forward direction parallel to the shore, regardless of vegetation type or presence. Samples were rinsed in a 250 μm sieve net or bucket to remove fine particles, and preserved in Kahle's solution for laboratory processing. In the laboratory, macroinvertebrates were identified to the highest possible resolution (genus for most insects) using the most current keys available at that time [19,20]. Data from the two transects in each zone were first averaged by depth, then averaged across zones to achieve site-level data. After taxonomic identification and sample averaging were completed, traits were assigned to each taxon using the latest reviews [14,19,21–27] and expert judgment (see Table S1 for details).

FD was measured using Rao's Quadratic entropy (Q), which accounts for the relative abundances of species and for the functional differences between species by measuring differences between two randomly selected individuals with replacement (*see* [28,29] for formulas and performance evaluation). It is closely correlated with the index of functional dispersion based on the centroid-distance approach [30], although it has also been demonstrated that this metric can be conservative under some scenarios due to negative covariation with species richness [29]. This analysis was performed with trait variables including trophic status, feeding mechanism, locomotion and primary and secondary functional feeding groups (Table S2), because those were the traits for which information was most consistently available across all encountered taxa. All traits were combined into a single Q-space.

The voltinism measure was expressed as the proportion of a sample comprised of taxa with long-lived aquatic phases (i.e. the proportion of individuals in a sample that were uni-, semi or merovoltine) to all other taxa. Voltinism is defined only for insect taxa; this life-history information was available for 77 taxa and unavailable for the remaining 85 insect taxa (of the total of 222 insect and non-insect taxa). This was mostly due to incomplete knowledge, as voltinism is one of the most difficult traits to describe, and to a lesser extent due to the level of resolution (variable life history at the species level, but identification to genus level) and differences in life history across a large geographic range of a taxon. Although all studies looking at voltinism unavoidably have this limitation, we chose to analyze this trait because it is a very important indicator of the state of macroinvertebrate assemblages and there is no *a priori* reason to suspect that the effect of this missing information is directional and not conservative on our ability to understand how longer-lived insects are affected by land use. Fifty-eight taxa in our samples were uni-, semi- or merovoltine, primarily belonging to Odonata and several genera of Trichoptera and Ephemeroptera, along with a few rarely occurring groups.

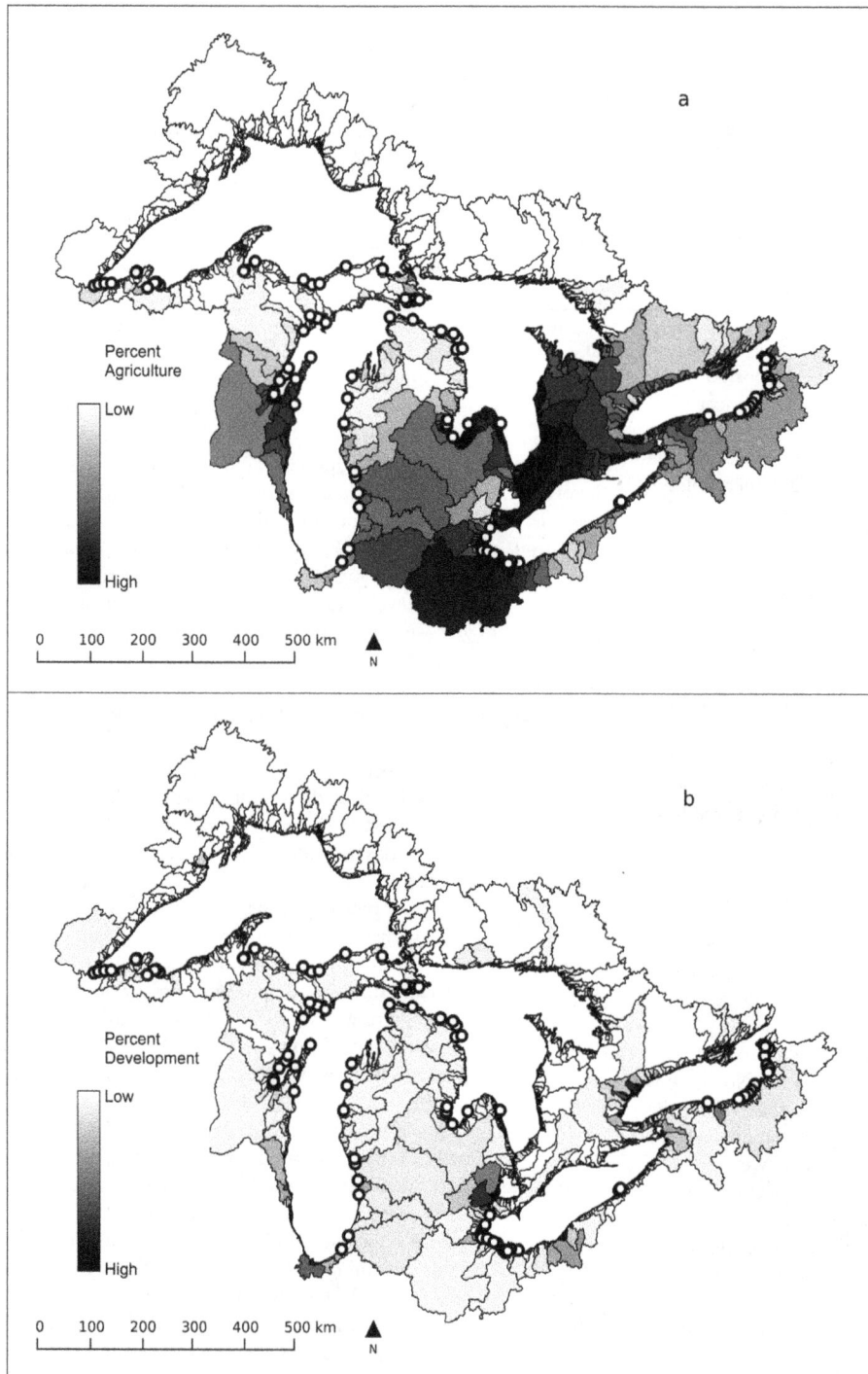

Figure 1. A map of the study sites in the coastal wetlands of the Laurentian Great Lakes, overlaid with the land-use stressor gradient for a) agriculture and b) development.

Statistical Analyses

Multiple regression (MR) analyses were used to test for relationships between watershed stressor variables and voltinism and FD. Percent development was log-transformed, whereas percent agriculture was not transformed to satisfy the assumptions of regression residuals distribution, as tested using Shapiro-Wilk tests. To address the possibility of the confounding effect of sampling across large geographic scales, we conducted additional multiple regression analyses using latitude as a third predictor for each of the response variables. Latitude was significantly, but weakly correlated with agriculture (linear regression permutation $p = 0.010$, $R^2 = 0.06$), but not with development ($p = 0.26$). Functional community structure was summarized using Principal Components Analysis (PCA) conducted on the log-transformed relative abundances of 12 functional traits. The resulting site Principal Component scores for each factor were regressed against the predictor variables as described above. Analyses were done in R 2.12.2 (R Development Core Team, Vienna, Austria), using packages *vegan* and *lattice*. Package *relaimpo* was used to calculate the "*lmg* relative importance metric" [31] for predictors in the multiple regression models, which produces averages of sequential sums of squares over all orderings of regressors (*see* [31] and package documentation for details). Although computationally intensive, the *lmg* procedure has been recommended because it decomposes R^2 into non-negative contributions and accounts for direct effects as well as adjustments for other regressors in the model [32]. Natgrid, a two-dimensional random data interpolation package, was used to resample data to a regular grid and produce contour plots. Natgrid, based on nngridr package [33], implements a natural neighbour interpolation method and uses a weighted average method. The package was implemented through *matplotlib* [34], plotting library for the Python programming language.

Results

Watershed development exerted stronger effects on all functional response variables than did watershed agriculture (Fig. 2). Rao's functional diversity was significantly reduced by both development and agriculture (Fig. 3, MR $F_{2, 97} = 9.60$, adjusted $R^2 = 0.16$, p = 0.003 and 0.004 for development and agriculture, respectively; see Table S3 for details and all regression coefficients). Development was the more important driver of this relationship (Fig. 2, Fig. S1a). To illustrate the extent of this effect, sites with more than 10% development had a 25% reduction in functional diversity compared to the less developed sites. The latitudinal predictor did not contribute significantly to the MR (p = 0.58), and its presence had almost no effect on the relative importance of the two stressors, confirming the lack of confounding effects (Table S3).

Relative abundance of those longer-lived taxa declined with increasing development in the watershed (MR $F_{2, 97} = 11.17$, adjusted $R^2 = 0.17$; Fig. 4, p<0.001 for development, also Fig. S1b). The weak (Fig. 2) but significant negative effect of agriculture observed in the two-factor model (p = 0.017) was no longer significant (p = 0.062) when latitude was included as a predictor (see Table S3 for model details and regression coefficients).

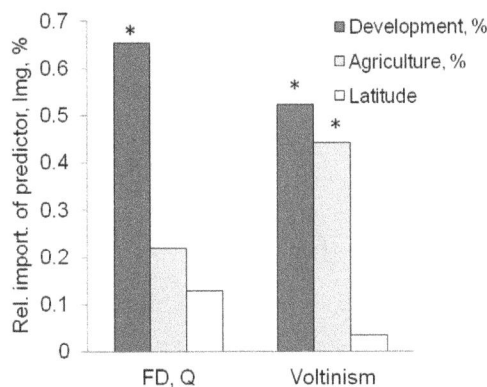

Figure 2. Relative importance of the two watershed stressors and latitude in explaining variation in the functional attributes of macroinvertebrate communities, as proportional *lmg* contribution. Note the overriding importance of development over agricultural stress and the lack of significant geographic confounding. Asterisks indicate predictors that were significant in the multiple regression (p<0.05).

The overall functional community structure was weakly but significantly affected by the extent of development in the watershed. The first PC axis, explaining 32% of the variation in functional community structure (eigenvalue = 3.74), was driven primarily by burrowers and filterers on one end vs. scrapers and clingers on the other end (*see* Table S4 for loadings). This axis was significantly positively correlated with percent development (MR $F_{2, 97} = 5.82$, adjusted $R^2 = 0.11$; p = 0.004) but not with agriculture. In other words, the proportions of positively-loading groups, including burrowers and filterers, were positively correlated with greater amounts of watershed development, whereas scrapers and clingers were negatively correlated. This axis score tended to be negatively correlated with agriculture, although not significantly (p = 0.053), with this trend disappearing when we added latitude as a predictor, similar to the pattern observed for other response variables. The second PCA axis (eigenvalue = 3.10, percent variation explained = 19%) was mostly driven by omnivores (Table S4). This axis was weakly negatively correlated with percent agriculture (p = 0.011, $R^2 = 0.06$).

Discussion

This study demonstrates that macroinvertebrate functional diversity and abundance of longer-lived taxa in Laurentian Great Lakes coastal wetlands decreased significantly with greater watershed-level stress. Although macroinvertebrate communities are influenced by factors acting at many spatial scales, as has been demonstrated in a number of stream studies (e.g. [3,35,36] but see [37] for lake margins), our results indicate that the negative effects of watershed-level development were sufficiently robust to be detected against the background of a strong geographic gradient, the presence of additional stressors and a wide range of local habitat features. Compin and Céréghino [38] used self-organizing maps to show that, in human-modified landscapes, effects of stream watershed-scale stressors over-rode the influence of natural physical factors. It also means that large-scale stressors can produce detectable changes in ecosystem function, at least with respect to macroinvertebrate functional traits. The effect of watershed development was significant, but not highly predictive, reflecting the hierarchical complexity of the underlying factors affecting macroinvertebrate assemblages and leading to the high variability seen in many such datasets [15,35].

A reduction in macroinvertebrate FD examined in this study translates into reduced community-level variation in foraging mechanisms and locomotion/substrate relations, which is likely to lead to significant alterations in food web structure and energy flow in those coastal wetland systems. Reduced FD is also likely to translate into decreased asynchrony of taxa responses to environmental perturbations (because more functionally similar taxa would have more synchronized responses to changes in resource abundance), which in turn has been proposed to be one of the key mechanisms reducing ecosystem stability in a theoretical modeling study [39].

The greater relative importance of development vs. agricultural stress is not surprising considering previously reported macroinvertebrate community thresholds at very low levels of development [40–44]. Similarly, a previous study found that development in the watershed was the best predictor of several macroinvertebrate metrics, including *Aeshna* abundance [37]. However, further study is needed to identify the more proximal factors driving this relationship, specifically in littoral systems. For instance, decline in longer-lived organisms could be mediated through the greater cumulative effects of development-associated stressors such as point-source pollution, in particular organic pollution, on these

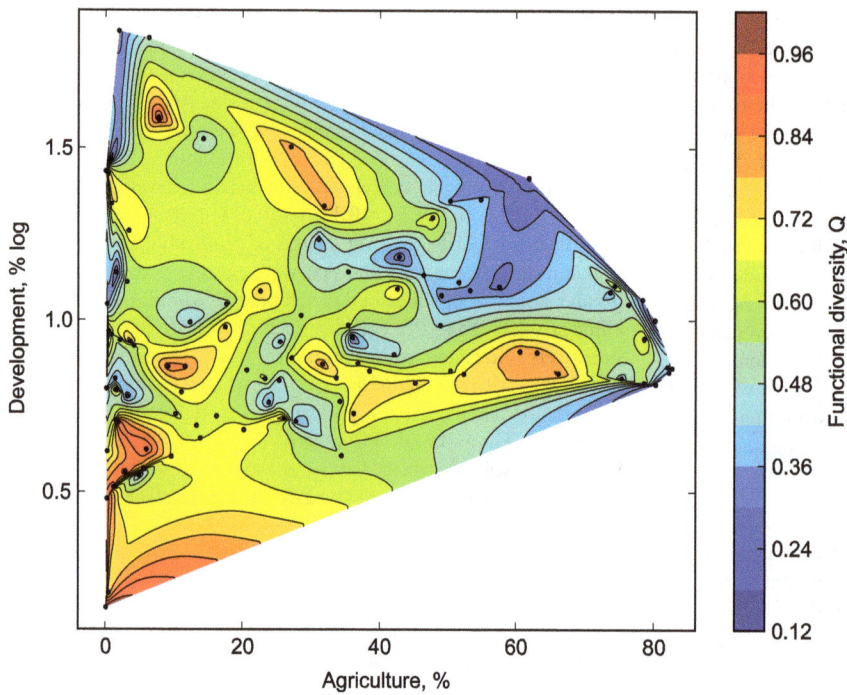

Figure 3. Macroinvertebrate functional diversity (Rao's Q) as a function of development and agricultural stress in the watershed. Note that maximum values of the functional trait are observed mostly below a certain proportion of development (around mid-point of the y-axis, corresponding to 10% untransformed % development), indicating the over-riding contribution of that stressor; and if high values are observed at the higher levels of development, those occur only at sites with minimal agriculture. *[figure footnote] Contour lines divide the figure into a region where values are higher than they are on the contour line itself and a region where they are lower; values change across the line but not along the contour line, and the gradient is larger where contour lines are packed closer together.

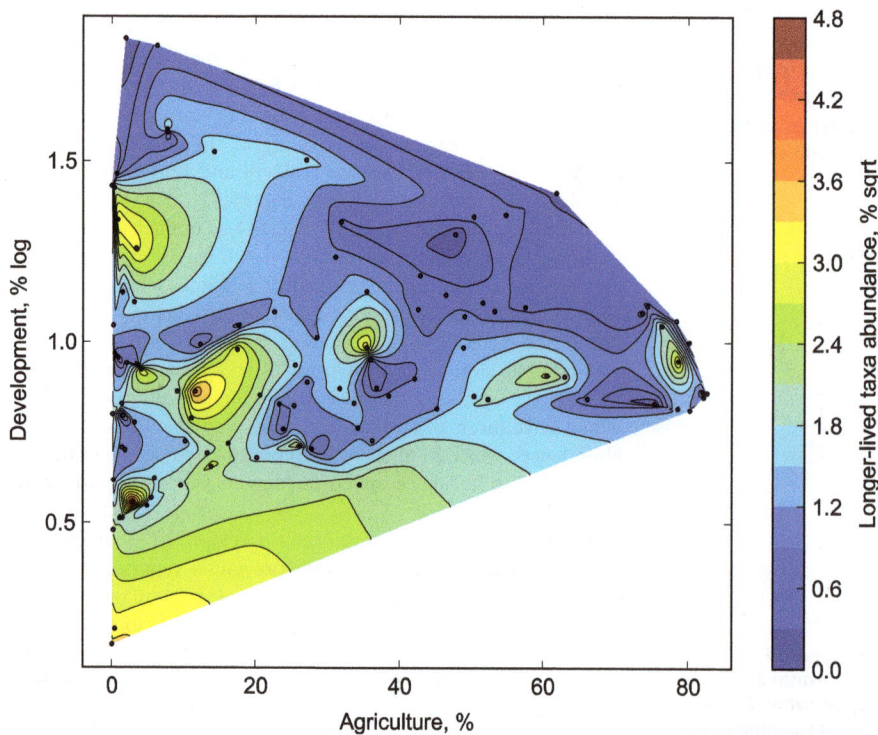

Figure 4. Relative abundance of long-lived taxa (uni-, semi- and merovoltine) as a function of development pland agricultural stress in the watershed.

organisms, or through the destruction of biogenic (macrophyte) complexity, which has been shown to support greater abundance of several types of predators [45] and likely sustains more stable, and thereby more diverse, predator-prey populations [46]. It has proven difficult to elucidate repeatable patterns in these coastal wetlands, possibly due to their highly variable hydrology [35,47], e.g. half a meter or more changes in water level on diel and annual basis due to fetch and climate change. Despite this, there is considerable interest in finding community metrics responsive to anthropogenic stress, and several such metrics have been found including Ephemeroptera-Trichoptera-Odonata richness-based metrics [48].

The observation of shift in functional community composition towards greater relative abundance of burrowers and filterers is consistent with other studies and has been previously related to increased sedimentation in streams [49]. Yet flashier hydrology and increased siltation, which are cited as the most common causes of anthropogenic changes in stream assemblages [1], are less likely to be the same factors responsible for observed changes in littoral assemblages, and lake-specific environmental variables responsible for these effects need to be investigated. Several functional groups (clingers, burrowers and insect filter-gatherers) were previously demonstrated to be affected by the extent of anthropogenic development in this system; however, responses were complex and dependent on several predictor variables as well as land use predictors' buffer sizes [37].

Trait responses are generally less frequently reported in aquatic studies than trends in diversity and abundance [50]. Effects of watershed land use on FD have garnered even less attention, particularly in non-stream ecosystems. For lotic macroinvertebrates, it has been shown that trait type changed [51] and trait diversity decreased [52] with increasing agricultural land use. The latter study demonstrated that this effect was detectable both at the watershed scale and local patch scale, and was related to increasing sedimentation [52]. Significant effects on these important functional variables (*e.g.*, trait diversity) emphasize the need to further investigate functional responses to anthropogenic stress in lentic macroinvertebrates, the mechanisms potentially underlying these responses and the local factors that may mitigate those effects. Considering additional local factors as well as integrating traits with explicit consideration of trait linkages [53] may increase the predictive capability of future littoral trait models. However, what is of greater interest for future studies is that this uncertainty potentially reflects trait responses to smaller-scale habitat factors,

which, if amendable to manipulation or restoration, could be used to ameliorate effects of watershed-scale land use.

In summary, we observed statistically and biologically significant reduction in macroinvertebrate FD and abundance of longer-lived taxa, and noticeable differences in functional community structure associated with increasing proportion of development in the watershed. These findings, along with the previously-observed threshold changes in macroinvertebrate community composition [44], show that lentic fauna exhibit significant functional changes associated with greater levels of watershed-scale anthropogenic stress.

Supporting Information

Figure S1   Univariate relationships with the development stressor.

Table S1   List of observed taxa with taxonomic information and trait values.

Table S2   List of possible states for each trait category used in calculating Rao's Q.

Table S3   Details of multiple regression analyses.

Table S4   Trait loadings for the Principal Component Analysis.

Acknowledgments

We thank field and lab personnel from NRRI-University of Minnesota Duluth, and University of Windsor for their assistance in collecting and processing samples and data used in this manuscript, and Terry Brown for creating the maps. We are grateful to Arthur Compin and an anonymous reviewer for their constructive comments. This is contribution number 572 of the Center for Water and the Environment, Natural Resources Research Institute, University of Minnesota Duluth.

Author Contributions

Conceived and designed the experiments: KK JC LJ. Performed the experiments: KK VB JC LJ. Analyzed the data: KK. Contributed reagents/materials/analysis tools: SI. Wrote the paper: KK. Provided comments on the manuscript: JC LJ VB SI.

References

1. Allan JD (2004) Landscapes and riverscapes: the influence of land use on stream ecosystems. Annu Rev Ecol Evol Syst 35: 257–284.

2. Heino J (2010) Are indicator groups and cross-taxon congruence useful for predicting biodiversity in aquatic ecosystems? Ecol Indic 10: 112–117.

3. Statzner B, Bêche L (2010) Can biological invertebrate traits resolve effects of multiple stressors on running water ecosystems? Freshw Biol 55: 80–119.

4. Díaz S, Cabido M (2001) Vive la différence: plant functional diversity matters to ecosystem processes. Trends Ecol Evol 16: 644–655.

5. Folke C, Carpenter SR, Walker BH, Scheffer M, Elmqvist T, et al. (2004) Regime shifts, resilience and biodiversity in ecosystem management. Annu Rev Ecol Evol Syst 35: 557–581.

6. Dunne JA, Brose U, Williams RJ, Martinez ND (2005) Modeling food web dynamics: complexity-stability implications. In: Belgrano A, Scharler UM, Dunne J, Ulanowicz RE, editors. Aquatic food webs: an ecosystem approach. Oxford University Press, New York. 117–129.

7. Hooper DU, Solan M, Symstad A, Diaz S, Gessner MO, et al. (2002) Species diversity, functional diversity and ecosystem functioning. Chapter 17 in Loreau M, Naeem S, Inchausti P, editors. Biodiversity and Ecosystem Functioning: a Current Synthesis. Oxford University Press. 147–154.

8. McGill BJ, Enquist BJ, Weiher E, Westoby M (2006) Rebuilding community ecology from functional traits. Trends Ecol Evol 21: 178–185.

9. Loreau M (1998) Biodiversity and ecosystem functioning: A mechanistic model. Proc Natl Acad Sci USA 95: 5632–5636.

10. Dolédec S, Phillips N, Scarsbrook M, Riley RH, Townsend CR (2006) Comparison of structural and functional approaches to determining landuse effects on grassland stream invertebrate communities. J North Am Benthol Soc 25: 44–60.

11. Covich AP, Palmer MA, Crowl TA (1999) The role of benthic invertebrate species in freshwater ecosystems. BioScience 49: 119–127.

12. Cardinale BJ, Palmer MA, Collins SL (2002) Species diversity increases ecosystem functioning through interspecific facilitation. Nature 415: 426–429.

13. Merritt RW, Cummins KW, Berg MB, editors (2008) An Introduction to the Aquatic Insects of North America: Fourth Edition. Kendall/Hunt Publishing Company. 1214 p.

14. Huryn AD, Wallace JB, Anderson NH (2008) Habitat, life history, secondary production and behavioral adaptations of aquatic insects. Chapter 5 in: Merritt RW, Cummins KW, Berg MB. An Introduction to the Aquatic Insects of North America. Kendall/Hunt Publishing Company. 55–104.

15. Batzer DP (2013) The seemingly intractable ecological responses of invertebrates in North American wetlands: a review. Wetlands 33: 1–15.

16. Niemi GJ, Kelly JR, Danz NP (2007) Environmental indicators for the coastal region of the North American Great Lakes: Introduction and prospectus. J Great Lakes Res 33: 1–12.

17. Danz NP, Regal RR, Niemi GJ, Brady VJ, Hollenhorst TP, et al. (2005) Environmentally stratified sampling design for the development of Great Lakes environmental indicators. Environ Monit Assess 102: 41–65.

18. Hollenhorst TP, Johnson LB, Ciborowski J (2011) Monitoring land cover change in the Lake Superior basin. Aquat Ecosyst Health Manag 14: 433–442.

19. Merritt RW, Cummins KW, editors (1995) An introduction to the aquatic insects of North America. Third edition. Kendall/Hunt Publishing Company. 862 p.

20. Thorp JH, Covich AP, editors (2001) Ecology and classification of North American freshwater invertebrates. Second edition. Academic Press. 1056 p.

21. Courtney GW, Merritt RW (2008) Aquatic Diptera Part One. Larvae of aquatic Diptera. Chapter 22 in: Merritt RW, Cummins KW, Berg MB. An Introduction to the Aquatic Insects of North America. Kendall/Hunt Publishing Company. 687–722.

22. Morse JC, Holzenthal RW (2008) Trichoptera Genera. Chapter 18 in: Merritt RW, Cummins KW, Berg MB. An Introduction to the Aquatic Insects of North America. Kendall/Hunt Publishing Company. 481–552.

23. Polhemus JT (2008) Aquatic and semiaquatic Hemiptera. Chapter 15 in: Merritt RW, Cummins KW, Berg MB. An Introduction to the Aquatic Insects of North America. Kendall/Hunt Publishing Company. 385–423.

24. Solis MA (2008) Aquatic and semiaquatic Lepidoptera. Chapter 19 in: Merritt RW, Cummins KW, Berg MB. An Introduction to the Aquatic Insects of North America. Kendall/Hunt Publishing Company. 553–569.

25. Tennessen KJ (2008) Odonata. Chapter 12 in: Merritt RW, Cummins KW, Berg MB. An Introduction to the Aquatic Insects of North America. Kendall/Hunt Publishing Company. 237–294.

26. Waltz RD, Burian SK (2008) Ephemeroptera. Chapter 11 in: Merritt RW, Cummins KW, Berg MB. An Introduction to the Aquatic Insects of North America. Kendall/Hunt Publishing Company. 181–236.

27. White DS, Roughly RE (2008) Aquatic Coleoptera. Chapter 20 in: Merritt RW, Cummins KW, Berg MB. An Introduction to the Aquatic Insects of North America. Kendall/Hunt Publishing Company. 571–671.

28. Rao CR (1982) Diversity and dissimilarity coefficients – a unified approach. Theor Popul Biol 21: 24–43.

29. Botta-Dukát Z (2005) Rao's quadratic entropy as a measure of functional diversity based on multiple traits. J Veg Sci 16: 533–540.

30. Laliberté E, Legendre P (2010) A distance-based framework for measuring functional diversity from multiple traits. Ecology 91: 299–305.

31. Lindeman RH, Merenda PF, Gold RZ (1980) Introduction to Bivariate and Multivariate Analysis, Glenview, IL: Scott, Foresman. 444 p.

32. Grömping U (2006) Relative importance for linear regression in R: the package relaimpo. J Stat Softw 17: 1–27.

33. Watson D (1994) Nngridr-An Implementation of Natural Neighbor Interpolation. D. Watson. 170 p.

34. Hunter JD (2007) Matplotlib: A 2D graphics environment. Comput Sci Eng 9: 90–95.

35. Poff NL (1997) Landscape filters and species traits: towards mechanistic understanding and prediction in stream ecology. J North Am Benthol Soc 16: 391–409.

36. Weigel BM, Wang L, Rasmussen PW, Butcher JT, Stewart PM, et al. (2003) Relative influence of variables at multiple spatial scales on stream macroinvertebrates in the Northern Lakes and Forest ecoregion, U.S.A. Freshw Biol 48: 1440–1461.

37. Brazner JC, Danz NP, Niemi GJ, Regal RR, Hollenhorst T, et al. (2007) Responsiveness of Great Lakes wetland indicators to human disturbances at multiple spatial scales: a multi-assemblage assessment. J Great Lakes Res 33: 42–66.

38. Compin A, Céréghino R (2007) Spatial patterns of macroinvertebrate functional feeding groups in streams in relation to physical variables and land-cover in Southwestern France. Landsc Ecol 22: 1215–1225.

39. Loreau M, de Mazancourt C (2013) Biodiversity and ecosystem stability: a synthesis of underlying mechanisms. Ecol Lett 16: 106–115.

40. King RS, Baker ME (2010) Considerations for analyzing ecological community thresholds in response to anthropogenic environmental gradients. J North Am Benthol Soc 29: 998–1008.

41. Hilderbrand RH, Utz RM, Stranko SA, Raesly RL (2010) Applying thresholds to forecast potential biodiversity loss from human development. J North Am Benthol Soc 29: 1009–1016.

42. Utz RM, Hilderbrand RH (2011) Interregional variation in urbanization-induced geomorphic change and macroinvertebrate habitat colonization in headwater streams. J North Am Benthol Soc 30: 25–37.

43. Kail J, Arle J, Jähnig SC (2012) Limiting factors and thresholds for macroinvertebrate assemblages in European rivers: Empirical evidence from three datasets on water quality, catchment urbanization, and river restoration. Ecol Indic 18: 63–72.

44. Kovalenko KE, Brady VJ, Brown TN, Ciborowski JJH, Danz NP, et al. (2014) Congruence of community thresholds in response to anthropogenic stressors in Great Lakes coastal wetlands. Freshw Sci (In press).

45. Heino J (2008) Patterns of functional biodiversity and function–environment relationships in lake littoral macroinvertebrates. Limnol Oceanogr 53: 1446–1455.

46. Kovalenko KE, Thomaz SM, Warfe DM (2012) Habitat complexity: approaches and future directions. Editorial review. Hydrobiologia 685: 1–17.

47. Gathman JP, Burton TM (2011) A Great Lakes coastal wetland invertebrate community gradient: relative influence of flooding regime and vegetation zonation. Wetlands 31: 329–341.

48. Uzarski DG, Burton TM, Genet JA (2004) Validation and performance of an invertebrate index of biotic integrity for Lakes Huron and Michigan fringing wetlands during a period of lake level decline. Aquat Ecosyst Health Manag 7: 269–288.

49. Larsen S, Pace G, Ormerod SJ (2011) Experimental effects of sediment deposition on the structure and function of macroinvertebrate assemblages in temperate streams. River Res Appl 27: 257–267.

50. Stendera S, Adrian R, Bonada N, Cañedo-Argüelles M, Hugueny B, et al. (2012) Drivers and stressors of freshwater biodiversity patterns across different ecosystems and scales: a review. Hydrobiologia 696: 1–28.

51. Richards C, Haro RJ, Johnson LB, Host GE (1997) Catchment and reach-scale properties as indicators of macroinvertebrate species traits. Freshw Biol 37: 219–230.

52. Larsen S, Ormerod SJ (2010) Combined effects of habitat modification on trait composition and species nestedness in river invertebrates. Biol Conserv 143: 2638–2646.

53. Verberk WCEP, van Noordwijk CGE, Hildrew AG (2013) Delivering on a promise: integrating species traits to transform descriptive community ecology into a predictive science. Freshw Sci 32: 531–547.

6

Pollinator Interactions with Yellow Starthistle (*Centaurea solstitialis*) across Urban, Agricultural, and Natural Landscapes

Misha Leong*, Claire Kremen, George K. Roderick

Department of Environmental Science Policy and Management, University of California, Berkeley, California, United States of America

Abstract

Pollinator-plant relationships are found to be particularly vulnerable to land use change. Yet despite extensive research in agricultural and natural systems, less attention has focused on these interactions in neighboring urban areas and its impact on pollination services. We investigated pollinator-plant interactions in a peri-urban landscape on the outskirts of the San Francisco Bay Area, California, where urban, agricultural, and natural land use types interface. We made standardized observations of floral visitation and measured seed set of yellow starthistle (*Centaurea solstitialis*), a common grassland invasive, to test the hypotheses that increasing urbanization decreases 1) rates of bee visitation, 2) viable seed set, and 3) the efficiency of pollination (relationship between bee visitation and seed set). We unexpectedly found that bee visitation was highest in urban and agricultural land use contexts, but in contrast, seed set rates in these human-altered landscapes were lower than in natural sites. An explanation for the discrepancy between floral visitation and seed set is that higher plant diversity in urban and agricultural areas, as a result of more introduced species, decreases pollinator efficiency. If these patterns are consistent across other plant species, the novel plant communities created in these managed landscapes and the generalist bee species that are favored by human-altered environments will reduce pollination services.

Editor: Steven M. Vamosi, University of Calgary, Canada

Funding: This work was supported by The Margaret C. Walker Fund, Department of Environmental Science Policy and Management, Essig Museum of Entomology, the Berkeley Institute for Global Change Biology, the Gordon and Betty Moore Foundation, and UC Berkeley's Undergraduate Research Apprentice Program. The funders had no role in study design, data collection and analysis, decision to publish, or preparation of the manuscript.

Competing Interests: The authors have declared that no competing interests exist.

* E-mail: mishaleong@berkeley.edu

Introduction

Human-altered landscapes are expanding globally and are often associated with declining natural habitat, non-native species, fragmentation, and transformations in structure, inputs, climate, and connectivity [1,2,3,4]. These changes collectively have resulted in shifts in both spatial distributions and species diversity across many taxa including birds, mammals, reptiles, amphibians, invertebrates, and plants [5,6]. One common driver of global change is urbanization, which in the extreme is associated with a reduction in biodiversity compared to habitats in their more natural state [7]. However, in moderately urbanized areas, the effects of urban impacts on species distribution and diversity can vary greatly and depends on region, type of change, and taxonomic group, among other factors [8,9].

Documenting the effects of urbanization compared to natural communities has proven problematic, making predictions of community change associated with urbanization difficult. Human-altered landscapes are often associated with many non-native species which add to species diversity [6,10,11] but also can obscure changes in community dynamics. Thus, to assess accurately the complex impacts of land use change on ecological communities, one must look beyond species richness to investigate ecological processes themselves. Ecological processes are the links between organisms in a functioning ecosystem, and are critical in understanding how altered biodiversity can lead to changes in ecosystem functioning [12].

Global environmental change has been found to have a wide variety of impacts on ecological processes in different systems [13]. Pollinator-plant relationships in particular are found to be particularly vulnerable to land use change, resulting in decreases in interaction strength and frequency [14]. Pollination services are crucial ecosystem processes in natural systems, but also in agricultural and urban areas [15]. Bees provide the majority of animal-mediated pollination services on which it is estimated 87.5% of flowering plants depend [16]. The value of pollination in agriculture is estimated at $200 billion worldwide [17], largely due to many foods that are essential for food security and a healthy human diet, including numerous fruits, vegetables, and nuts that require bee pollination. As urban areas expand, there has been increasing interest in urban agriculture to ensure food security and access to healthy foods for growing populations, and these systems also depend on pollination. For example, Kollin [18] estimated that the economic value of urban fruit trees (many of which require pollination) to be worth $10 million annually in San Jose, California.

Despite the important role of pollinators and concerns about bee declines [19,20], there remain many uncertainties regarding the impact of land use change on pollinators [21]. Urbanization has resulted in more interfaces with both natural and agricultural

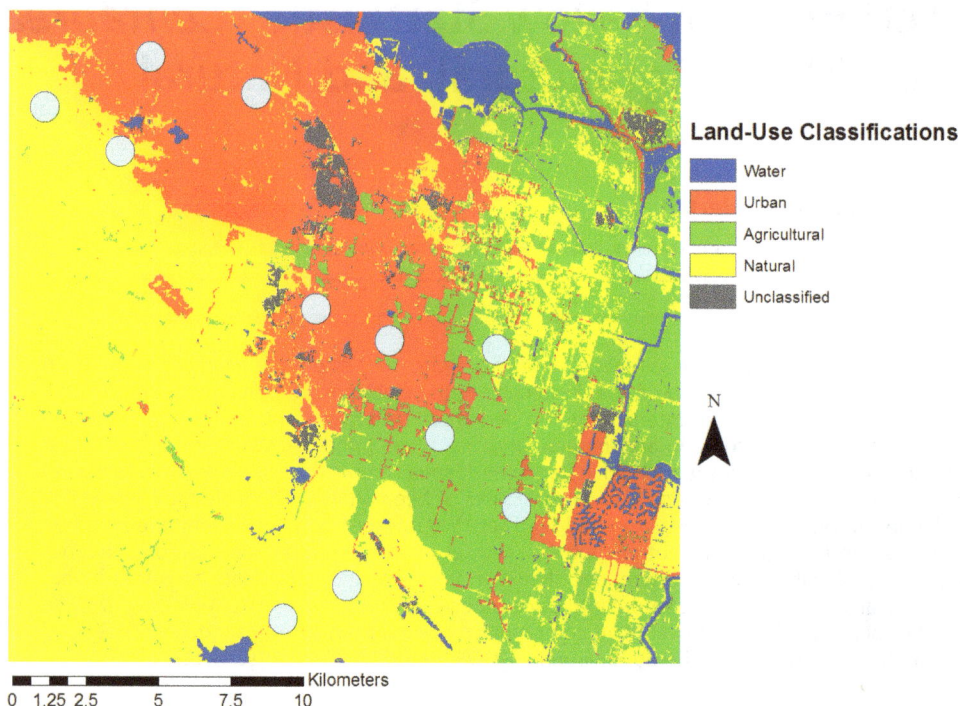

Figure 1. Map of study area and locations of plots in East Contra Costa County, California. Light blue dots represent a 500 m radius around the center point of each of the 12 sites. The sites were chosen to be located in agricultural (green), urban (red), and natural (yellow) land use types.

landscapes, creating new transitional zones of peri-urbanization [22]. While there has been extensive pollinator research in agricultural and natural systems [23,24,25,26,27], less attention has focused on pollination in neighboring urban areas and how the changing landscape has impacted pollination [9,28]. In addition, very few studies of urban areas have looked beyond changes in bee diversity to understand explicitly the effect of urbanization on pollinator-plant interactions [10,29,30].

Here, we investigate the effect of land use change on pollinator-plant ecosystem processes. We make use of a "natural experimental design" in which urban, agricultural, and natural areas intersect. Bees visit flowers for both pollen and nectar resources, and floral visitation is a commonly used as an index of pollination services. However, depending on the flower, certain bee groups are much more effective pollinators than others [9,21,31]. Thus, while visitation is important, it alone does not definitively indicate whether pollination services were received by the plant [32]. When pollen is limited by other factors, consequences for plant fitness can include failure to set seed, production of smaller fruits, and even complete lack of reproduction [33,34]. By looking at rates of bee visitation and comparing this with other measures of plant fitness, such as seed set, we can develop a more complete understanding of how shifts in bee distributions between areas that differ in land use are impacting pollination services.

To study the impact of changing land use on pollinator-plant interactions, we focus on bee pollination of a widespread plant, yellow starthistle (*Centaurea solstitialis*), a common weed found in natural, agricultural, and urban habitats. Using standardized observations of floral visitation and seed set measurements of yellow starthistle, we test the hypotheses that increasing urbanization decreases 1) rates of bee visitation, 2) viable seed set, and 3) the efficiency of pollination (relationship between bee visitation and seed set). In addition to contributing to a better understanding

of how change in landscape use, particularly urbanization, affects pollination-plant interactions, the study illustrates the importance of use of neighboring lands for pollination services.

Methods

Ethics Statement

No protected species were sampled in this field study. Permits and approval were obtained for field observations on public land from the East Bay Regional Park District, Contra Costa County Flood Control and Water Conservation District, and the Los Vaqueros Reservoir.

Study System

Our study system was located around Brentwood, in east Contra Costa County, California, where natural, agricultural, and urban areas intersect with each other within a 20×20 km region (Figure 1). A county water district (Los Vaqueros Watershed), regional park district (East Bay Regional Parks: Black Diamond Mines, Round Valley, and Contra Loma), and California state park (Mount Diablo) all fall within the region, leaving large areas of land protected from development. This protected (hereafter referred to as "natural") land consists mainly of grasslands and oak woodlands, some portions of which are managed for grazing. East Contra Costa County has had a farming community presence since the late 19th century. The agricultural areas of Brentwood, Knightsen, and Byron mostly consist of orchards (cherries, stone fruit, grapes and walnuts), corn, alfalfa, and tomatoes [35]. A housing boom in the 1990s led to massive residential growth in the area. The city of Brentwood has grown from less than 2500 people in the 1970s to over 50,000 today (2010 U.S. Census), and nearby Antioch has now over 100,000 residents (2010 U.S. Census).

We selected 12 sites dominated by yellow star thistle in a stratified design to span the different land use types (Figure 1). Yellow starthistle (*Centaurea solstitialis*) is a common weedy plant that forms homogenous flowering patches in grassy areas throughout this region. Many different bee taxa in a range of functional groups and size classes have been observed to visit yellow starthistle [36], in part because it flowers late in the season relative to other floral resources [37]. Despite being considered a serious introduced weed, yellow starthistle is unusual as an invasive species in that it depends on animal pollinator visits in order to set seed [38].

Within each site we selected a 50 m×50 m plot such that each plot was at least 2 km away from all others, a distance larger than the maximum assumed typical bee foraging ranges [39]. Although certain bee species have been recorded foraging as far as 1400 m [40], most bees in this type of habitat have nesting and foraging habitat within a few hundred meters of each other [39,40,41]. Within each plot we estimated number of flowering yellow starthistle blooms by randomly placing 10, 1 m×1 m quadrats and counting the number of flowering blooms in each. We also measured the spatial area of yellow starthistle patches within each 50 m×50 m plot to obtain an estimate of total flowering blooms within each plot. We categorized total blooms/plot on a log scale: $<10^3$ (Category 1), $10^3–10^4$ (Category 2), and $>10^4$ (Category 3).

Using NOAA's 2006 Pacific Coast Land Cover dataset (developed using 30 meter resolution Landsat Thematic Mapper and Landsat Enhanced thematic Mapper satellite imagery, USGS Products), a 500 m buffer (representing estimated bee foraging ranges [39,40,41]) was created around each plot, and the number of pixels classified as agricultural, urban, natural, water, or bare land was extracted. These categories were obtained by lumping finer categories in NOAA's classification scheme using the following definitions: Urban–"High Intensity Developed", "Medium Intensity Developed", "Low Intensity Developed", and "Developed Open Space"; Agricultural–"Cultivated", "Pasture/Hay"; Natural– "Grassland", Deciduous Forest", "Mixed Forest", "Scrub/Shrub". Each plot was classified as a proportion of each of the 3 different land use categories, as well as for the category that was dominant. By this latter measure, of our 12 sites, 4 of each were classified as "urban", "agricultural", and "natural".

Bee Visitation

We observed visits by all bee species to yellow starthistle at all sites 3 times (AM, mid-day, and PM) for a 30 min period for a total of 90 min of total observation time per site within the same 2 wk period in August 2011. AM was defined as being between the hours of 9:30–11:30, Mid-Day as between 11:30–13:30, and PM as between 13:30–15:30. All observations were conducted by the same individual (ML) to avoid sampling biases. Also recorded at each observation period were approximate number of blooms, and wind and temperature simultaneously (using a Kestrel 3000 Pocket Weather Meter). Bees were not netted for later identification as we did not want to interfere with visitation to starthistle during this study. Instead, we used a modified protocol of citizen scientist observation surveys [42] with 15 expected bee morphotypes (Table 1) that correspond to 30 possible genera known to occur in the region (Leong, unpublished data). The observer slowly walked through the yellow starthistle patch, and upon reaching patch edge, returned on a path at least 3 m away from the previous, and recorded the morphotype classification of all bee visitors within 1.5 m on either side of the transect.

Seed Set

Yellow starthistle (Asteraceae) has composite flowers, which are aggregations of anywhere from 20–80 florets [43]. At each site, 12 yellow star thistle buds were randomly selected from different plants and covered with a mesh bag. Yellow starthistle blooming cycles have been described in detail in other publications [43]. We selected buds at stage BU-4 [43], when buds had no yellow petals exposed, but had well-developed straw-colored spines. When in full flowering, 10 bags were opened for a 4 hour period from 10 am to 2 pm, while 2 were kept closed as controls to verify that self-pollination was not occurring. At the opening and re-closing of the bags, the number of florets that had their stigmas extended (and thus, available for pollination) were counted. Later, when flowers were fully mature (dry and straw-colored), seed heads were collected, and later dissected in the lab. Viable and non-viable seeds in yellow starthistle seed heads are easily distinguishable based on color and shape [38]. Because yellow starthistle requires pollination to produce viable seeds (also confirmed by our controls), non-viable seeds represent pollen limitation occurring during the 4-hour period that the flowers were exposed to pollinators. All seeds were counted to compare ratios of viable to non-viable seeds. Any seed predation was noted, and when possible, the seed predator was identified.

Analyses

All analyses were done in R 2.15.1 (R Development Core Team, 2011). Because each site had an AM, Mid-Day, and PM observation event, there were a total of 36 observation events, each with unique wind and temperature recordings, and visit observations of the 15 bee morphotypes. From these, we calculated the total number of bee visitors, total number of bee morphotypes, Shannon diversity of morphotypes, and morphotype evenness. Shannon diversity and evenness were calculated using the R package *vegan* [44]. The spatial autocorrelation of all bee visitor response variables (each morphotype abundance, total abundance, morphotype richness, diversity, and evenness) was assessed by Mantel tests in R package *ade4* [45], using the average values for each time of day at each site. Spatial autocorrelation was not detected ($p\geq0.14$).

To test for the effect of land use type on each of the response variables we used a generalized linear mixed model using the R package *lme4* [46]. We designated land use type, bloom category of flowering patch, observation time period, wind, and temperature as fixed effects and site as a random effect. Natural land use and AM observation time period were the model baselines for the categorical variables of land use type and observation time. Shannon diversity and evenness were fit with Gaussian distributions while all other variables were fit with Poisson distributions.

In comparing the ratios of viable seeds to total seeds vs. the ratio of viable seeds to counted stigmas, we found that there was a strong correlation between these metrics. To look at the effect of land use type on seed-set, we therefore decided to utilize the ratio of viable seeds to total seeds in each seed head that did not experience seed predation, because of error in counting the number of stigmas (in some cases, we had slightly more viable seeds than counted number of stigmas, suggesting errors in this measurement). We then used a generalized linear mixed model fit with a Binomial distribution, with land use type as a fixed effect and site as a random effect.

Finally, we tested for an effect of floral visitor observations on yellow starthistle seed set at each site. We averaged the number of visits from each morphotype across temporal observation events at the same site. Morphotypes that averaged at least one visit per 30 minute observation window were included as fixed effects in a

Table 1. The 15 bee morphotypes observed and their associated genera and species in East Contra Costa County, California.

Morphotype	Possible Species
Honey bee	Apis mellifera
Bumblebee	Bombus spp.
Carpenter bee	Xylocopa spp.
Hairy leg bee, medium	Melissodes spp., Anthophora spp., Eucera spp., Peponapis spp., Exomalopsis spp., Diadasia spp.
Hairy leg bee, large	Svastra spp.
Green sweat bee	Agapostemon texanus
Striped sweat bee, medium	Halictus ligatus, Halictus spp., >0.5 cm
Striped sweat bee, small	Halictus tripartitus, Halictus spp., <0.5 cm
Small dark bee, rounded tip	Lasioglossum spp.
Small dark bee, shield tip	Ceratina spp.
Striped hairy belly bee, small	Ashmeadiella spp., Megachile spp., <0.5 cm
Striped hairy belly bee, medium	Megachile, >0.5 cm, <1.5 cm
Striped hairy belly bee, large	Megachile, >1.5 cm)
Wasp-like hairy belly bee	Dianthidium app., Anthidium spp.
Cuckoo bee	Sphecodes spp., Nomada spp., Nomia spp., Calliopsis spp.

linear mixed model fit with a binomial distribution, with site as a random effect and the ratio of viable to total seeds as the response variable. We also modeled the effects of total bee visitation, morphotype richness, and morphotype diversity on seed set ratios.

Data on bee visitation rates for all observation events and seed set ratios for each plant are available from the Dryad Digital Repository: http://dx.doi.org/10.5061/dryad.b5np1 [47].

Results

Bee Visitation and Land Use

A total of 2816 total bee visits were recorded, representing 15 bee morphotypes. Total bee visitation was significantly higher in urban and agricultural areas with respective effect sizes (± standard errors) of 0.885 ± 0.26 (p = 0.0007) and 0.813 ± 0.22 (p = 0.0002) (Figure 2, Tables 2 & 3). The effect of land use type on visitation rates when analyzed separately for each bee morphotype, was a significant variable for 6 of 15 morphotypes. Bloom category, time of observation, wind, and temperature were only occasionally significant in some of the models.

Agricultural sites (Table 2) had the highest total bee visitation; 62% of total bee observations were honey bees (*Apis mellifera*), which were observed significantly more often in agricultural, than managed or urban sites (effect size $\pm$ SE = 1.26 ± 0.33, p = 0.0002). Agricultural sites also had significantly higher visitation rates from shield-tipped small dark bees (effect size $\pm$ SE = 1.83 ± 0.78, p = 0.02) and medium striped hairy belly bees (effect size $\pm$ SE = 1.53 ± 0.62, p = 0.01). However, agricultural sites, compared to natural and urban sites, had significantly lower morphotype Shannon diversity (effect size $\pm$ SE = -0.488 ± 0.193, p = 0.009) and morphotype evenness (effect size $\pm$ SE = -0.264 ± 0.086, p = 0.002).

Visitation by native bees (here measured as visitation by non-honey bees, although there are a few other non-native species that may be included within the other morphotypes) was highest in urban sites (Table 3) compared to those in the other land use types (effect size $\pm$ SE = 1.389 ± 0.273, p<0.0001). Medium and small striped sweat bees were the most abundant groups after honey bees, which made up 12% and 7% of total bee observations

respectively. When analyzed by morphotype, urban sites had the highest visitation levels from medium striped sweat bees (effect size $\pm$ SE = 3.213 ± 0.268, p<0.0001), small striped sweat bees (effect size $\pm$ SE = 1.74 ± 0.53, p = 0.001), and small striped hairy belly bees (effect size $\pm$ SE = 1.055 ± 0.536, p = 0.04). Urban areas had higher morphotype richness (effect size $\pm$ SE = 0.369 ± 0.199, p = 0.06), but this effect was not significant.

None of the morphotypes were observed significantly most often in natural sites, although 2 of 3 sites where bumblebees were observed were natural sites.

To examine in more detail the effect of land use on bee visitation, we created a continuous variable for land use with an index ranging from agriculture to urban use based on proportional area of each type within a 500 m radius. We then used this measure of land use to assess the response of total bee visitation, native bee visitation, morphotype richness, evenness, and Shannon diversity using previously described mixed model techniques. We found that while there was no significant effect of land use on total bee visitation, native bee visitation observations increased with more surrounding urban area (effect size $\pm$ SE = 0.963 ± 0.22, p<0.001). We also saw the same effect of increasing surrounding urban area on morphotype richness (effect size $\pm$ SE = 0.27 ± 0.12, p = 0.02), Shannon diversity (effect size $\pm$ SE = 0.55 ± 0.144, p<0.01) and evenness (effect size $\pm$ SE = 0.237 ± 0.069, p<0.01) (Figure 3, Table 4).

Seed Set

Natural sites had the highest average rates of seed set, and urban areas had the lowest (effect size $\pm$ SE = -0.756 ± 0.371, p = 0.042, Figure 4), in direct contrast to the pattern found with floral visitation where urban sites had the highest rates of native bee visitation and natural sites had the lowest. In total 140 yellow starthistle seed heads were collected and dissected; 4 lost mesh bags in the field and were eliminated from the study. Of these, 43% of the collected seed heads experienced some type of seed damage, largely due to biological control efforts in the area involving tephritid flies and weevils. Seed predation decreased with amount of surrounding agricultural area (simple linear regression, p<0.01). Of the 79 seed heads that were intact, 73 had received

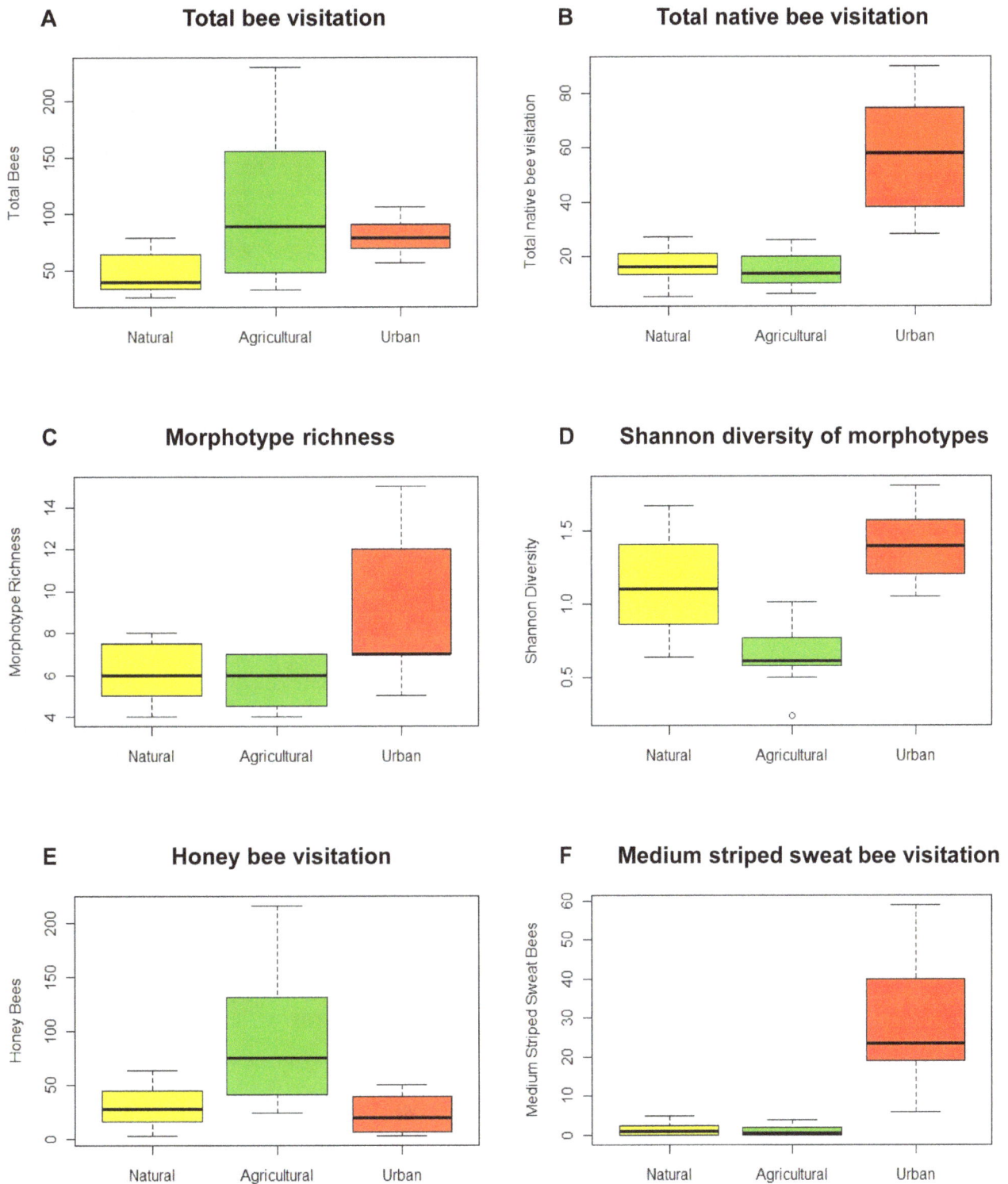

Figure 2. Box plots of bee visitation response variables in natural, agricultural, and urban sites. Bee morphotype visitation data and calculated community metrics were collected in East Contra Costa County, California.

the 4-hour treatment of being exposed to pollination. Only 6 flowers in the control group which were never exposed to pollinators experienced no predation. Of those, 3 had no viable seeds, and 3 had 4.7%, 8.3%, and 20% viable seeds respectively.

Bee Visitation and Seed Set

Of the 8 morphotypes that averaged at least one visit per 30 minute observation period, 3 exhibited significant relationships between visitation abundance at a site and seed set, although there was no significant relationship between site seed set and total bee

Table 2. Statistical output table for response variables having significant relationships with the agricultural land use type.

Response Variable	Effect size	Standard error	p- value
Total bee visitation	0.813	0.22	0.0002
Honey bees	1.26	0.33	0.0002
Shield-tipped small dark bees	1.83	0.78	0.02
Medium striped hairy belly bees	1.53	0.62	0.01
Morphotype Shannon diversity	−0.488	0.193	0.009
Morphotype evenness	−0.264	0.086	0.002

Bee morphotype visitation data and calculated community metrics were collected in East Contra Costa County, California. Significant relationships with the agricultural land use type were calculated based on generalized linear mixed models with land use type, bloom category of flowering patch, observation time period, wind, and temperature as fixed effects and site as a random effect. The natural land use type and morning (AM) observation time period were the model baselines for the categorical variables of land use type and observation time. Shannon diversity and evenness were fit with Gaussian distributions while all other variables were fit with Poisson distributions.

Table 3. Statistical output table for response variables having significant relationships with the urban land use type.

Response Variable	Effect size	Standard error	p- value
Total bee visitation	0.885	0.26	0.0007
Native bee visitation	1.389	0.273	<0.0001
Medium striped sweat bees	3.213	0.268	<0.0001
Small striped sweat bees	1.74	0.53	0.001
Small striped hairy belly bees	1.055	0.536	0.04
Morphotype richness	0.369	0.100	0.06

Bee morphotype visitation data and calculated community metrics were collected in East Contra Costa County, California. Significant relationships with the urban land use type were calculated based on generalized linear mixed models fit with Poisson distributions with land use type, bloom category of flowering patch, observation time period, wind, and temperature as fixed effects and site as a random effect. The natural land use type and morning (AM) observation time period were the model baselines for the categorical variables of land use type and observation time.

visitation, morphotype richness, or morphotype diversity. Increased seed set ratios correlated with sites that had more visitation from medium hairy leg bees (effect size±SE = 0.284±0.069, p<0.001) and to a smaller extent, round-tipped small dark bees (effect size±SE = 0.127±0.074, p = 0.04), despite there not being significant relationships between land use type and either of these bee groups. However, visitation by shield-tipped small dark bees (effect size±SE = −0.155±0.051, p = 0.002) had a significant negative effect on proportion of viable seeds (Figure 5).

Discussion

Our results show that rates of bee visitation and seed set vary among urban, agricultural, and natural landscapes, demonstrating the importance of land use in the dynamics of plant-pollinator interactions. We suggest that these effects are at least in part explained by floral availability, a vital bee resource, which can be highly variable among different land use types. For example, in August there are few plants in flower besides yellow starthistle in the natural areas of Contra Costa County, California, whereas in urban and agricultural areas there are many exotic plants and supplementary inputs available (personal observation). From pan-trapping of bee specimens in the region (Leong, unpublished data), we know that total bee abundance is highest in the spring in natural areas. However, towards the end of the summer when yellow starthistle is in flower, there is little difference in collected bee abundance between human-altered landscapes and natural areas, and human-altered areas may even exhibit overall higher bee abundance.

Our results of bee visitation to yellow starthistle support this pattern. Agricultural areas have large populations of managed honey bee colonies, so one would predict visitation to yellow starthistle by honey bees to be positively associated with surrounding agricultural land use. By contrast for native bees (total bee visitation excluding honey bees), the highest rates of visitation to yellow starthistle were in sites with more surrounding urban land use. Urban gardens have many exotic plants, often selected for aesthetic purposes, many of which are in flower later in the season than most California native plants. In addition, many of the plants in urban areas both directly and indirectly receive supplementary resources, particularly water, that further extend their flowering time. Even though agricultural areas also have

supplementary resources, the main crop in flower in East Contra Costa County later in the season is maize, which is wind-pollinated. There may be multiple impacts of exotic plants in urban areas. By filling the phenological flowering gap [48] noted above, they may help attract even larger populations of bees into the urban landscape. In addition, bees in urban sites may be behaviorally more likely to visit non-native plants due to the increased encounters they have with novel plants [49].

In agricultural and natural landscapes, a positive correlation between pollinator visitation and seed set is typical [50]. Surprisingly in our system, in human-altered landscapes, higher total observed bee visitation did not result in higher proportions of seed set, as would be expected. In fact, urban areas, despite receiving the highest rates of native bee visitation, exhibited the lowest rates of seed set. Conversely, natural areas, which received the lowest amount of total bee visitation, had the highest rates of seed set.

We suggest 2 possible explanations for this discrepancy between pollinator visitation and rates of seed set: 1) pollinator efficiency; and/or 2) the composition of the local flowering community. Depending on the plant, certain pollinator species are much more effective than others [51]. For example, *Osmia*, *Habropoda*, and *Apis*, have been found to produce varying amounts of seed set as a result of a single visit to blueberry, but these results vary slightly depending on the blueberry variety [52]. In the case of yellow starthistle, it is likely that the most frequent visitors are perhaps not the most efficient. When we directly compared average seed set at each site against visitation rates, we found a significant positive association with the medium hairy leg bees. The medium hairy leg bee morphotype includes those species which fall in both the Tribes Emphorini and Eucerini. Emphorini are known to largely be oligolectic (Michener 1999), meaning they specialize on certain plant groups, which theory suggests would make them more efficient pollinators than generalists [51].

The medium hairy leg bee morphotype was not significantly associated with any of the land use typesIt was also the only group that was observed most frequently during morning (AM) sampling, perhaps reflecting a difference in when yellow starthistle is most receptive to pollination. Despite the overwhelming abundance of honey bees in agriculture areas, we did not observe higher seed set in those regions, consistent with the observation that honey bees can be poorer pollinators than other species [53,54].

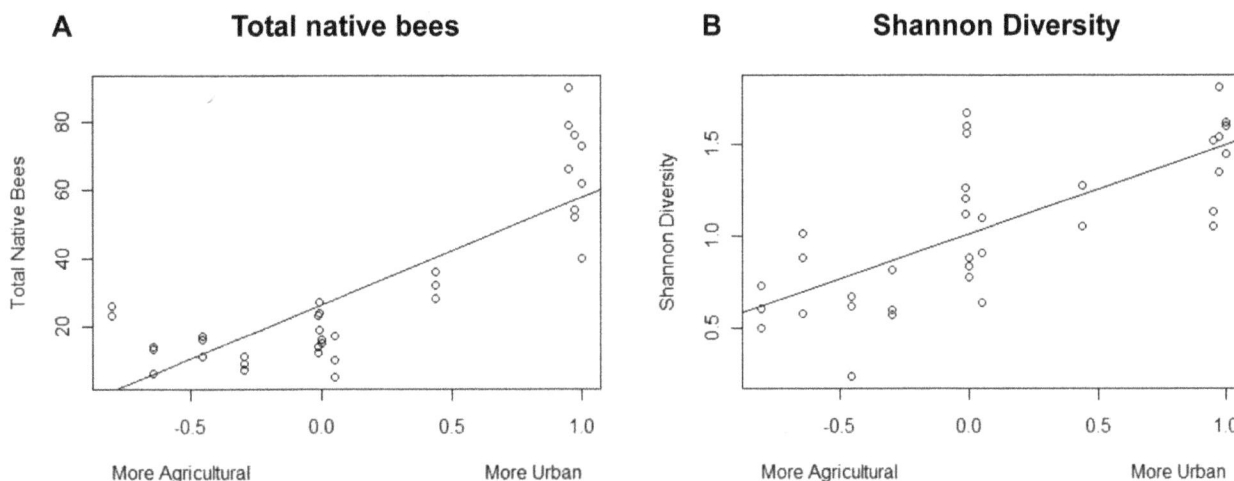

Figure 3. Bee visitation response variables as a function of surrounding anthropogenic land use. Bee morphotype visitation data and calculated community metrics were collected in East Contra Costa County, California. To examine in more detail the effect of anthropogenic land use on bee visitation, we created a continuous variable for land use with an index ranging from agricultural to urban land use based on proportional area of each type within a 500 m radius. As the x-axis moves from left to right, sites go from being more agricultural to more urban.

It is also important to note that this study used a morphotype classification, and there may be multiple species that fit within the same morphotype that provide varying degrees of pollination services [55]. It is possible there are rare, but highly efficient, pollinators that were rarely observed during the sampling period, or were lumped together with a more frequently observed morphotype.

An alternative explanation for the lack of an association between floral visitation and seed set is that higher plant diversity in urban and agricultural areas may decrease pollinator efficiency. Previous research has shown that invasive alien plants can have a negative effect on native plant communities by acting as attractors for pollinators, or decreasing pollinator efficiency by providing a wider range of resources for pollinators to visit, with the consequence that visitors transfer pollen from non con-specifics, potentially clogging stigmas and reducing pollination success [56,57,58]. In this case, our target plant, yellow starthistle is indeed considered an invasive alien plant, but the hypothesis of it being in a novel diverse community could lead to a similar effect on the frequency and quality of pollination services that it receives. In sites where there are many other potential plants to visit and accompanying decreased floral fidelity leading to diverse pollen loads, one predicts decreased pollinator efficiency. Abundant sources of exotic plant pollen could occur in areas where there is a greater diversity of nearby plants for pollinators to visit. This explanation might account for the observation that shield-tipped small dark bees were negatively correlated with seed set.

We selected yellow starthistle as the target plant for this study because of its ubiquitous distribution, reliance on pollination, and its attraction for a wide set of visitors; it is also a highly invasive and undesirable plant [59]. Previous research on yellow starthistle has found that its invasion can be facilitated other non-native pollinator species such as the honey bee, *Apis mellifera*, and the starthistle bee, *Megachile apicalis* [36,38], which is included in the medium striped hairy belly bee morphotype. However, the abundance of bees in both of these 2 morphotypes were most closely associated with agricultural areas, which did not have the highest rates of seed set as would be predicted by visitation alone.

Our results indicate clearly that bee visitation in human-altered landscapes can be higher than that in comparable natural areas, especially towards the end of the flowering season when there are few resources available in natural landscapes. Because the response of bee visitors to land use change depends on species-specific requirements and these pollinators also have variable effects on plants, understanding the effect of land use change on pollination services requires knowledge not only of which pollinator groups shift to the human-altered landscapes, but also the rate of pollination that those groups have on the plant species in those landscapes. Future research will benefit from looking at a wider range of plants with a different range of target pollinators and that flower earlier in the year to better tease out these hypotheses. If the patterns of bee visitation and seed set that we observed are indeed consistent across other plant species, the novel plant communities created in these human-altered landscapes and the generalist bee species that are favored in such landscapes will lead to a reduction in overall pollination services.

Table 4. Statistical output for response variables having significant relationships with the gradient of agricultural to urban land use.

Response Variable	Effect size	Standard error	p- value
Native bee visitation	0.963	0.22	<0.01
Morphotype richness	0.27	0.12	0.02
Morphotype Shannon diversity	0.55	0.144	<0.01
Morphotype evenness	0.237	0.069	<0.01

To examine in more detail the effect of anthropogenic land use on bee visitation, we created a continuous variable for land use with an index ranging from agriculture to urban land use based on proportional area of each type within a 500 m radius. Generalized linear mixed models were created with this calculated anthropogenic land use metric, bloom category of flowering patch, observation time period, wind, and temperature as fixed effects and site as a random effect. The morning (AM) observation time period was the model baseline for the categorical variable of observation time. Shannon diversity and evenness were fit with Gaussian distributions while all other variables were fit with Poisson distributions.

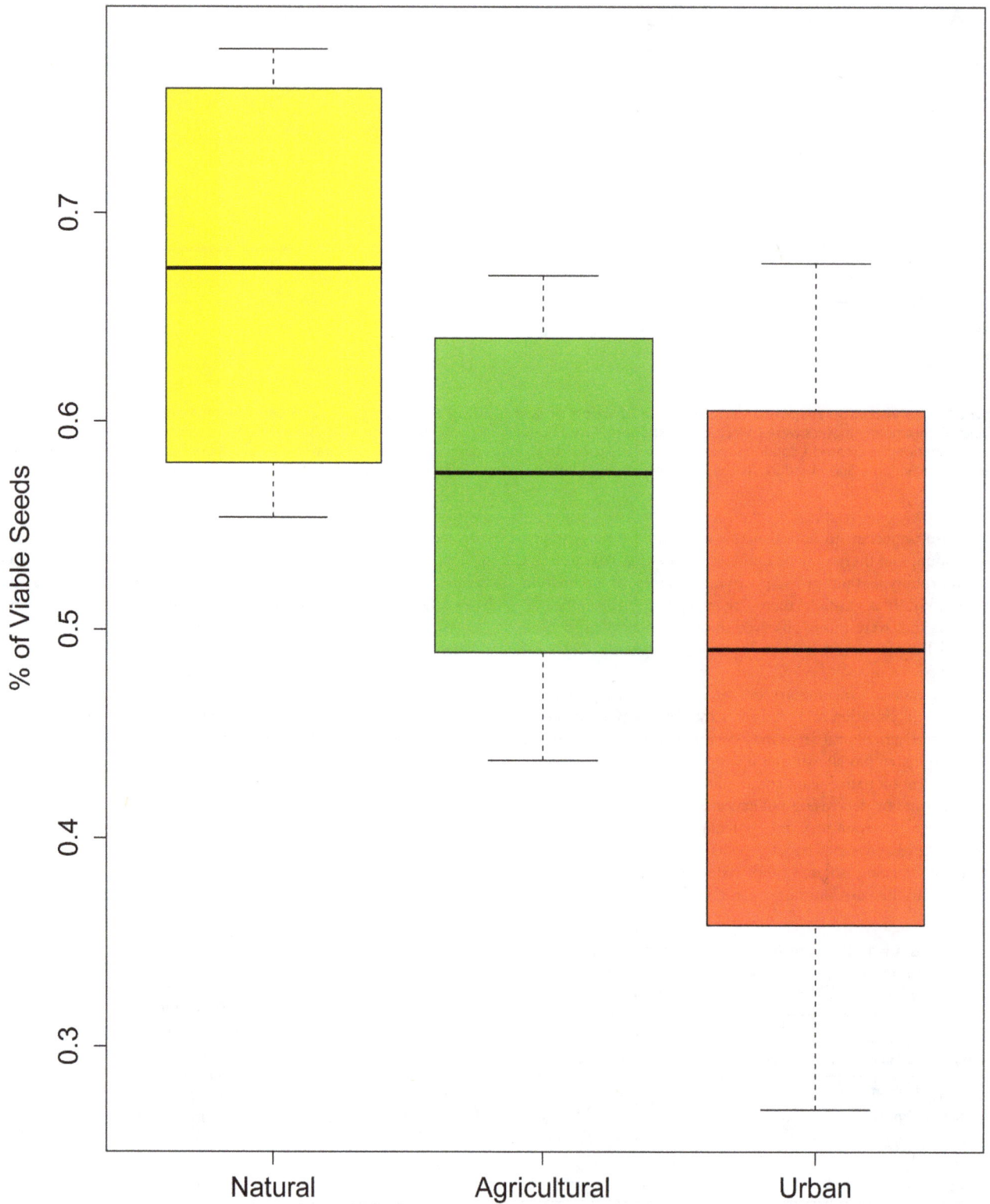

Figure 4. Box plot demonstrating the effect of land use type on percentage of viable seeds. Yellow starthistle seed heads were collected in East Contra Costa County, California and dissected in the lab after maturity. We calculated significance using a generalized linear mixed model fit with a binomial distribution, with land use type as a fixed effect and site as a random effect. With natural sites as the baseline, urban areas had significantly lower rates of seed set (effect size$\pm$SE = -0.756 ± 0.371, p = 0.042).

A

B

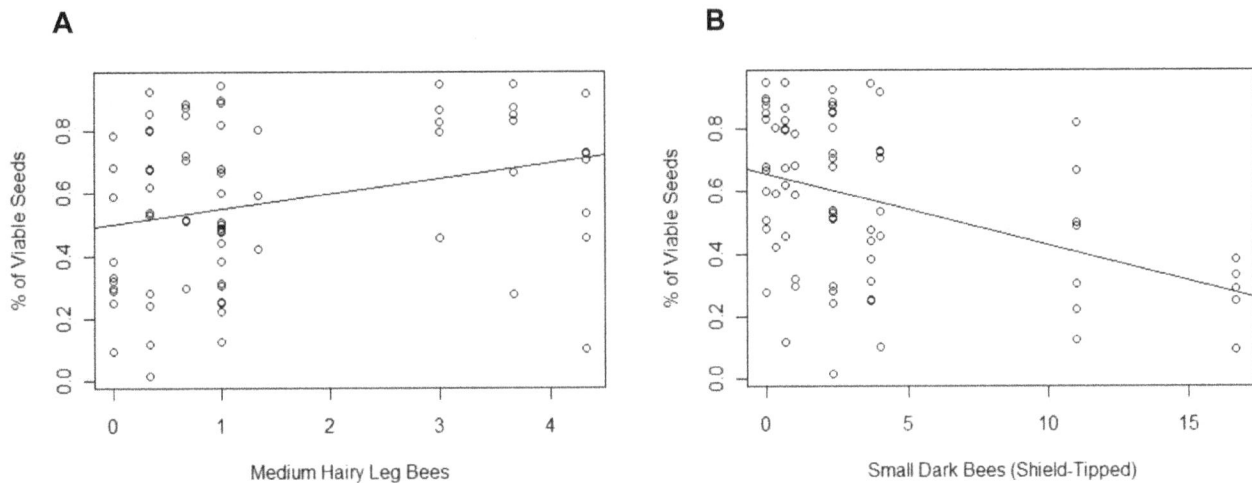

Figure 5. Correlation between the percentage of viable seeds in each yellow starthistle seed head and the average number of site visits by morphotype. Bee morphotype visitation data, calculated community metrics, and yellow starthistle seed heads were collected in East Contra Costa County, California. Bee morphotypes that averaged at least one visit per 30 minute observation window were included as fixed effects in a linear mixed model fit with a binomial distribution, with site as a random effect and the ratio of viable to total seeds as the response variable. Medium hairy leg bees (effect size±SE = 0.284±0.069, p<0.001) and shield-tipped small dark bees (effect size±SE = −0.155±0.051, p = 0.002) had significant effect sizes in the model. Regression lines were added to illustrate relationships.

Acknowledgments

We thank: Jennifer Imamura, Michael Leong, Kathy Leong, and Sebastien Renaudin for assistance in the field; Preston Chan and Felicia Han for lab help dissecting yellow starthistle seed heads; Lauren Ponisio, Steve Selvin, and Jack Kamm for statistical advice; The Los Vaqueros Watershed, Contra Costa Water District, and East Bay Regional Parks and Recreation for insights and access; and Linda Bürgi and Joanne Clavel for comments that greatly improved the manuscript.

Author Contributions

Conceived and designed the experiments: ML CK. Performed the experiments: ML. Analyzed the data: ML CK GR. Contributed reagents/materials/analysis tools: ML GR. Wrote the paper: ML CK GR.

References

1. Gill SE, Handley JF, Ennos AR, Pauleit S (2007) Adapting cities for climate change: the role of the green infrastructure. Built Environment 33: 115–133.
2. Ash C, Jasny BR, Roberts L, Stone R, Sugden AM (2008) Reimagining cities. Science 319: 739.
3. Grimm NB, Faeth SH, Golubiewski NE, Redman CL (2008) Global change and the ecology of cities. Science 319: 756–760.
4. Shochat E, Lerman S, Anderies J, Warren P, Faeth S, et al. (2010) Invasion, competition, and biodiversity loss in urban ecosystems. BioScience 60: 199–208.
5. Boone CG, Cook E, Hall SJ, Nation ML, Grimm NB, et al. (2012) A comparative gradient approach as a tool for understanding and managing urban ecosystems. Urban Ecosystems 15: 795–807.
6. McKinney ML (2008) Effects of urbanization on species richness: A review of plants and animals. Urban Ecosystems 11: 161–176.
7. Niemelä J, Kotze DJ, Venn S, Penev L, Stoyanov I (2002) Carabid beetle assemblages (Coleoptera, Carabidae) across urban-rural gradients: an international comparison. Landscape Ecology 17: 387–401.
8. Magura T, Tóthmérész B, Molnár T (2004) Changes in carabid beetle assemblages along an urbanisation gradient in the city of Debrecen, Hungary. Landscape Ecology 19: 747–757.
9. Winfree R, Griswold T, Kremen C (2007) Effect of Human Disturbance on Bee Communities in a Forested Ecosystem. Conservation Biology 21: 213–223.
10. Wojcik V, McBride J (2012) Common factors influence bee foraging in urban and wildland landscapes. Urban Ecosystems 15: 581–598.
11. Sax DF, Gaines SD, Brown JH (2002) Species invasions exceed extinctions on islands worldwide: A comparative study of plants and birds. American Naturalist 160: 766–783.
12. Hooper DU, Chapin FS, Ewel JJ, Hector A, Inchausti P, et al. (2005) Effects of biodiversity on ecosystem functioning: A consensus of current knowledge. Ecological Monographs 75: 3–35.
13. Barnosky AD, Hadly EA, Bascompte J, Berlow EL, Brown JH, et al. (2012) Approaching a state shift in Earth's biosphere. Nature 486: 52–58.
14. Tylianakis J, Didham R, Bascompte J, Wardle D (2008) Global change and species interactions in terrestrial ecosystems. Ecology Letters 11: 1351–1363.
15. Klein AM, Vaissiere BE, Cane JH, Steffan-Dewenter I, Cunningham SA, et al. (2007) Importance of pollinators in changing landscapes for world crops. Proceedings of the Royal Society B: Biological Sciences 274: 303–313.
16. Ollerton J, Winfree R, Tarrant S (2011) How many flowering plants are pollinated by animals? Oikos 120: 321–326.
17. Gallai N, Salles JM, Settele J, Vaissiere BE (2009) Economic valuation of the vulnerability of world agriculture confronted with pollinator decline. Ecological Economics 68: 810–821.
18. Kollin C (1991) On Balance: Weighing the benefits and costs of urban trees. In: Station USFSNFE, editor. Syracuse, NY.
19. Allen-Wardell G, Bernhardt P, Bitner R, Burquez A, Buchmann S, et al. (1998) The potential consequences of pollinator declines on the conservation of biodiversity and stability of food crop yields. Conservation Biology 12: 8–17.
20. Vanbergen AJ, Insect Pollinators Initiative (2013) Threats to an ecosystem service: pressures on pollinators. Frontiers in Ecology and the Environment 11: 251–259.
21. Kremen C, Williams NM, Aizen MA, Gemmill-Herren B, LeBuhn G, et al. (2007) Pollination and other ecosystem services produced by mobile organisms: a conceptual framework for the effects of land-use change. Ecology Letters 10: 299–314.
22. Errington A (1994) The Periurban Fringe - Europe Forgotten Rural-Areas. Journal of Rural Studies 10: 367–375.
23. Kremen C (2002) Crop pollination from native bees at risk from agricultural intensification. Proceedings of the National Academy of Sciences 99: 16812–16816.
24. Ricketts T, Regetz J, Steffan-Dewenter I, Cunningham S, Kremen C, et al. (2008) Landscape effects on crop pollination services: are there general patterns? Ecology Letters 11: 499–515.
25. Lonsdorf E, Kremen C, Ricketts T, Rachael W, Williams N, et al. (2009) Modelling pollination services across agricultural landscapes. Annals of Botany 103: 1589–1600.
26. Winfree R, Kremen C, Winfree R, Kremen C (2009) Are ecosystem services stabilized by differences among species? A test using crop pollination. Proceedings of the Royal Society B: Biological Sciences 276: 1–10.
27. Potts S, Biesmeijer J, Kremen C, Neumann P, Schweiger O, et al. (2010) Global pollinator declines: trends, impacts and drivers. Trends in Ecology & Evolution 25: 345–353.
28. Deguines N, Julliard R, Flores M, Fontaine C (2012) The Whereabouts of Flower Visitors: Contrasting Land-Use Preferences Revealed by a Country-Wide Survey Based on Citizen Science. PLoS One 7: e45822.
29. Hennig EI, Ghazoul J (2011) Plant–pollinator interactions within the urban environment. Perspectives in Plant Ecology, Evolution and Systematics 13: 137–150.

30. Pauw A (2007) Collapse of a pollination web in small conservation areas. Ecology 88: 1759–1769.

31. Hoehn P, Tscharntke T, Tylianakis J, Steffan-Dewenter I (2008) Functional group diversity of bee pollinators increases crop yield. Proceedings of the Royal Society B: Biological Sciences 275: 2283–2291.

32. Vazquez DP, Morris WF, Jordano P (2005) Interaction frequency as a surrogate for the total effect of animal mutualists on plants. Ecology Letters 8: 1088–1094.

33. Kearns C, Inouye D (1993) Techniques for Pollination Biology. BoulderCo: University Press of Colorado. 583 p.

34. Knight TM, Steets JA, Vamosi JC, Mazer SJ, Burd M, et al. (2005) Pollen limitation of plant reproduction: Pattern and process. Annual Review of Ecology, Evolution, and Systematics 36: 467–497.

35. Guise V (2011) 2011 Annual Crop and Livestock Report for Contra Costa County. ConcordCA: Director of Agriculture. 13 p.

36. McIver J, Thorp R, Erickson K (2009) Pollinators of the invasive plant, yellow starthistle (Centaurea solstitialis), in north-eastern Oregon, USA. Weed Biology and Management 9: 137–145.

37. Williams N, Cariveau D, Winfree R, Kremen C (2011) Bees in disturbed habitats use, but do not prefer, alien plants. Basic and Applied Ecology 12: 332–341.

38. Barthell JF, Randall JM, Thorp RW, Wenner AM (2001) Promotion of seed set in yellow star-thistle by honey bees: Evidence of an invasive mutualism. Ecological Applications 11: 1870–1883.

39. Gathmann A, Tscharntke T (2002) Foraging ranges of solitary bees. Journal of Animal Ecology 71: 757–757–764.

40. Zurbuchen A, Landert L, Klaiber J, Andreas M, Hein S, et al. (2010) Maximum foraging ranges in solitary bees: only few individuals have the capability to cover long foraging distances. Biological Conservation 143: 669–676.

41. Morandin L, Kremen C (2013) Hedgerow restoration promotes pollinator populations and exports native bees to adjacent fields. Ecological Applications 23: 829–839.

42. Kremen C, Ullmann K, Thorp R (2011) Evaluating the Quality of Citizen-Scientist Data on Pollinator Communities. Conservation Biology 25: 607–617.

43. Wilson LM, Jette C, Connett J, McAffrey J (2003) Biology and biological control of yellow starthistle.

44. Oksanen J, Blanchet F, Kindt R, Legendre P, Minchin P, et al. (2011) vegan: Community Ecology Package. R package version 20–2.

45. Dray S, Dufour AB (2007) The ade4 package: implementing the duality diagram for ecologists. Journal of Statistical Software 22: 1–20.

46. Bates D, Maechler M, Bolker B, Walker S (2011) lme4: Linear mixed-effects models using s4 classes. R package version 0999375–42.

47. Leong M, Kremen C, Roderick G (2013) Data from: Pollinator interactions with yellow starthistle (Centaurea solstitialis) across urban, agricultural, and natural landscapes.: Dryad Digital Repository.

48. Stout JC, Morales CL (2009) Ecological impacts of invasive alien species on bees. Apidologie.

49. Chittka L, Thomson JD, Waser NM (1999) Flower constancy, insect psychology, and plant evolution. Naturwissenschaften 86: 361–377.

50. Engel EC, Irwin RE (2003) Linking pollinator visitation rate and pollen receipt. American Journal of Botany 90: 1612–1618.

51. Schemske DW, Horvitz CC (1984) Variation among floral visitors in pollination ability - a precondition for mutualism specialization. Science 225: 519–521.

52. Sampson BJ, Cane JH (2000) Pollination efficiencies of three bee (Hymenoptera : Apoidea) species visiting rabbiteye blueberry. Journal of Economic Entomology 93: 1726–1731.

53. Garibaldi LA, Steffan-Dewenter I, Winfree R, Aizen MA, Bommarco R, et al. (2013) Wild Pollinators Enhance Fruit Set of Crops Regardless of Honey Bee Abundance. Science 339: 1608–1611.

54. Westerkamp C (1991) Honeybees are poor pollinators–why? Plant Systematics and Evolution.

55. Whittington R, Winston ML (2004) Comparison and examination of Bombus occidentalis and Bombus impatiens (Hymenoptera : Apidae) in tomato greenhouses. Journal of Economic Entomology 97: 1384–1389.

56. Grabas GP, Laverty TM (1999) The effect of purple loosestrife (Lythrum salicaria L,; Lythraceae) on the pollination and reproductive success of sympatric co-flowering wetland plants. Ecoscience 6: 230–242.

57. Bjerknes A-L, Totland O, Hegland S, Nielsen A (2007) Do alien plant invasions really affect pollination success in native plant species? Biological Conservation 138: 1–12.

58. Tscheulin T, Petanidou T (2013) The presence of the invasive plant Solanum elaeagnifolium deters honeybees and increases pollen limitation in the native co-flowering species Glaucium flavum. Biological Invasions 15: 385–393.

59. Swope SM, Parker IM (2012) Complex interactions among biocontrol agents, pollinators, and an invasive weed: a structural equation modeling approach. Ecological Applications 22: 2122–2134.

7

Simple Patchy-Based Simulators Used to Explore Pondscape Systematic Dynamics

Wei-Ta Fang[1]*, Jui-Yu Chou[2], Shiau-Yun Lu[3]

1 Graduate Institute of Environmental Education, National Taiwan Normal University, Taipei, Taiwan, ROC, **2** Department of Biology, National Changhua University of Education, Changhua, Taiwan, ROC, **3** Department of Marine Environment and Engineering, National Sun Yat-sen University, Kaohsiung, Taiwan, ROC

Abstract

Thousands of farm ponds disappeared on the tableland in Taoyuan County, Taiwan since 1920s. The number of farm ponds that have disappeared is 1,895 (37%), 2,667 ponds remain (52%), and only 537 (11%) new ponds were created within a 757 km² areain Taoyuan, Taiwan between 1926 and 1960. In this study, a geographic information system (GIS) and logistic stepwise regression model were used to detect pond-loss rates and to understand the driving forces behind pondscape changes. The logistic stepwise regression model was used to develop a series of relationships between pondscapes affected by intrinsic driving forces (patch size, perimeter, and patch shape) and external driving forces (distance from the edge of the ponds to the edges of roads, rivers, and canals). The authors concluded that the loss of ponds was caused by pond intrinsic factors, such as pond perimeter; a large perimeter increases the chances of pond loss, but also increases the possibility of creating new ponds. However, a large perimeter is closely associated with circular shapes (lower value of the mean pond-patch fractal dimension [MPFD]), which characterize the majority of newly created ponds. The method used in this study might be helpful to those seeking to protect this unique landscape by enabling the monitoring of patch-loss problems by using simple patchy-based simulators.

Editor: Bin Jiang, University of Gävle, Sweden

Funding: This work was supported by National Science Council grants (NSC99-2410-H-216-007 and NSC100-2628-H-003-161-MY2 to WF; NSC102-2311-B-018-001-MY2 to JC). The funders had no role in study design, data collection and analysis, decision to publish, or preparation of the manuscript.

Competing Interests: The authors have declared that no competing interests exist.

* E-mail: wtfang@ntnu.edu.tw

Introduction

Pond integrity is multifaceted with regard to the functions, structures, and variations of ponds based on a management perspective, particularly concerning local and regional biodiversity[1]. Regarding the functional paradigm of farm ponds, Smith et al.[2] examined millions of small water bodies throughout the USA. Compared with large water bodies (e.g., reservoirs and lakes), these small artificial ponds contain many more ecological organisms. In Norway, inland water bodies (e.g., ponds) are also regarded as one aspect of land cover and land use that can serve as a variable in land classification and delineation [3]. Numerous studies have explained the areal increase in species richness by citing an increase in microhabitat heterogeneity. Ponds are vulnerable to disturbances and vanish because of anthropogenic driving forces, which directly influence species distribution. The relative importance of pond configurations and site-specific structures to species is debatable. Several studies have reported that pond patterns (i.e., peripheral anthropogenic land uses) [4–6] and pond-size patterns [7,8] are related to species number and diversity. Pond size is the basic variable used to measure horizontal alterations. Historically, ecologists and geographers have attempted to determine the relationships between spatial size and species richness. As pond size increases, the likelihood of the animals that are new to this particular area will appear increases, thereby increasing species richness. Studies on a similar topic date back to 1921, when an increase in plant species number was observed as the same function of quadrat areas (i.e., the species-area

relationship was not just a phenomenon that occurred among plants) [9].

MacArthur and Wilson (1967) proposed an island biogeographic theory to explain spatial variances in avian richness values [10]. They developed the concept of island biogeography, which states that small islands have lower immigration rates and higher extinction rates, thereby resulting in the presence of fewer species compared with those of large islands. Alternatively, large islands, with a larger area and larger populations that are less prone to extinction, support more species. Based on this concept, MacArthur and Wilson suggested that more species inhabit large islands than small islands, and islands adjacent to a mainland contain more species than remote islands do because of higher immigration rates. The concept of island biogeography was also called the area-per-se hypothesis by Connor and McCoy [11]. The area-per-se hypothesis has since been applied to faunal communities, such as breeding birds on islands. This hypothesis was also the major concept used to explain species richness in several studies, which designed water regimes for vernal pools and refuges in earlier decades [12]. Ponds can differ in size for reasons such as drought, drainage, and drawdown of the water surface, which can reduce water regimes over time. Species richness is affected by both natural and anthropogenic influences on pond configuration [13]. The ecological significance of such disturbances might be crucial to aquatic communities. However, the edge effect between aquatic-terrestrial regimes has been completely ignored in species-area studies. An alternative hypothesis that was proposed to explain the area-per-se relationship stated that an increase in habitat heterogeneity accompanies an increase in area. Increased microhabitat diversity allows more species to find an increasing number of niches, resulting in greater species diversity.

Several studies have reported that the areal increase in species richness is caused by an increase in microhabitat heterogeneity. Recent studies have presented new theoretical and empirical methods aimed at creating theories based on transition processes at regional to global scales [14]. Several effects that could be beneficial to ecosystems were observed, such as the(a) area-per-se effect, (b) habitat effect, (c) distance effect, and (d) connectivity effect. The hypotheses require more detailed examinations of ecological theories to investigate anthropogenic influences. If sufficient mudflats and water regimes are maintained, ponds can support more species. Habitat effects can increase the diversity of ponds, and prevent negative anthropogenic influences at the edge of metropolitan areas. For example, if the curvilinear edge of a pond is limited to buildings (built-up areas) and/or highways, the pond is vulnerable to external pressures and damage. In addition, the smaller the distance is between ponds, the more beneficial it is for the dispersal of hydrophilic species.Cluster design should be considered to increase the connectivity of ponds and subsequently prevent the development of pondscapes that are located far

from each other, and/or prevent fragmentation that occurs when isolated patches develop in mosaics. Several of these aforementioned hypotheses must be carefully applied in the fields of landscape planning and design.

Forman [15] argued that ecological spatial form is not simply shaped by the effect of area-per-se and isolation patterns. This was only a theoretical hypothesis, which inferred that the pond size used in landscape design affects species richness within a heterogeneous terrain [16]. In addition to area, all variations in ponds can be condensed into four categories that address ecophysical identities: shape, edge, clustering, and connectivity [17,18]. Given the remarkable diversity and complex compositions of pond configurations, a pondscape is defined in this study as "a series of water surfaces of ponds associated with various surrounding landforms, including farms, creeks, canals, roads, houses, woodlands, and other open spaces," according to the reports of several researchers[19–21].

The field of landscape ecology was comprehensively developed based on the driving forces observed in pondscape studies, using the concepts of shape, edge, clustering, and connectivity. However, landscape ecology studies have seldom detected the anthropogenic influences of the aforementioned spatial driving forces (SDFs), compared with natural science studies [22–26].

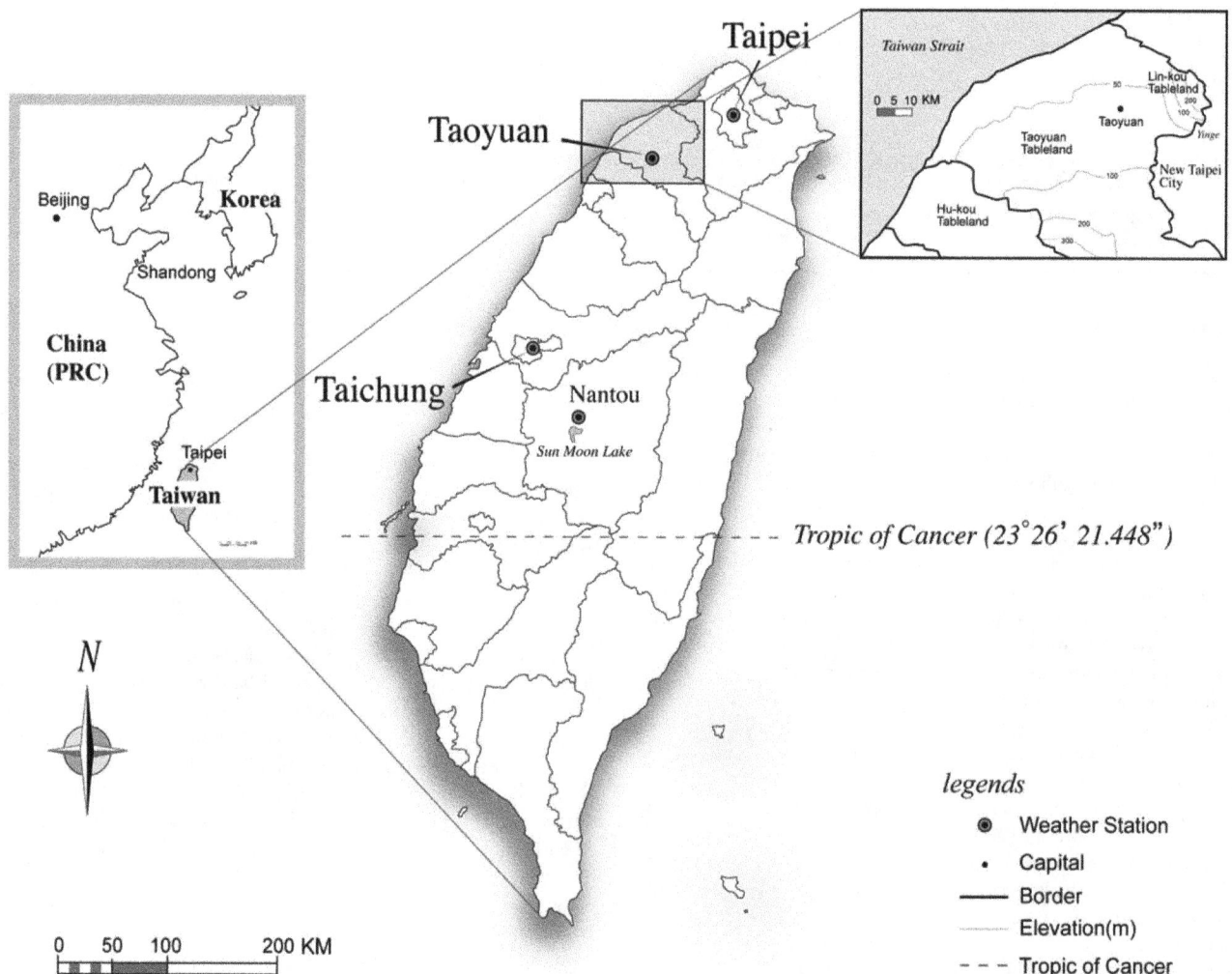

Figure 1. Study area.

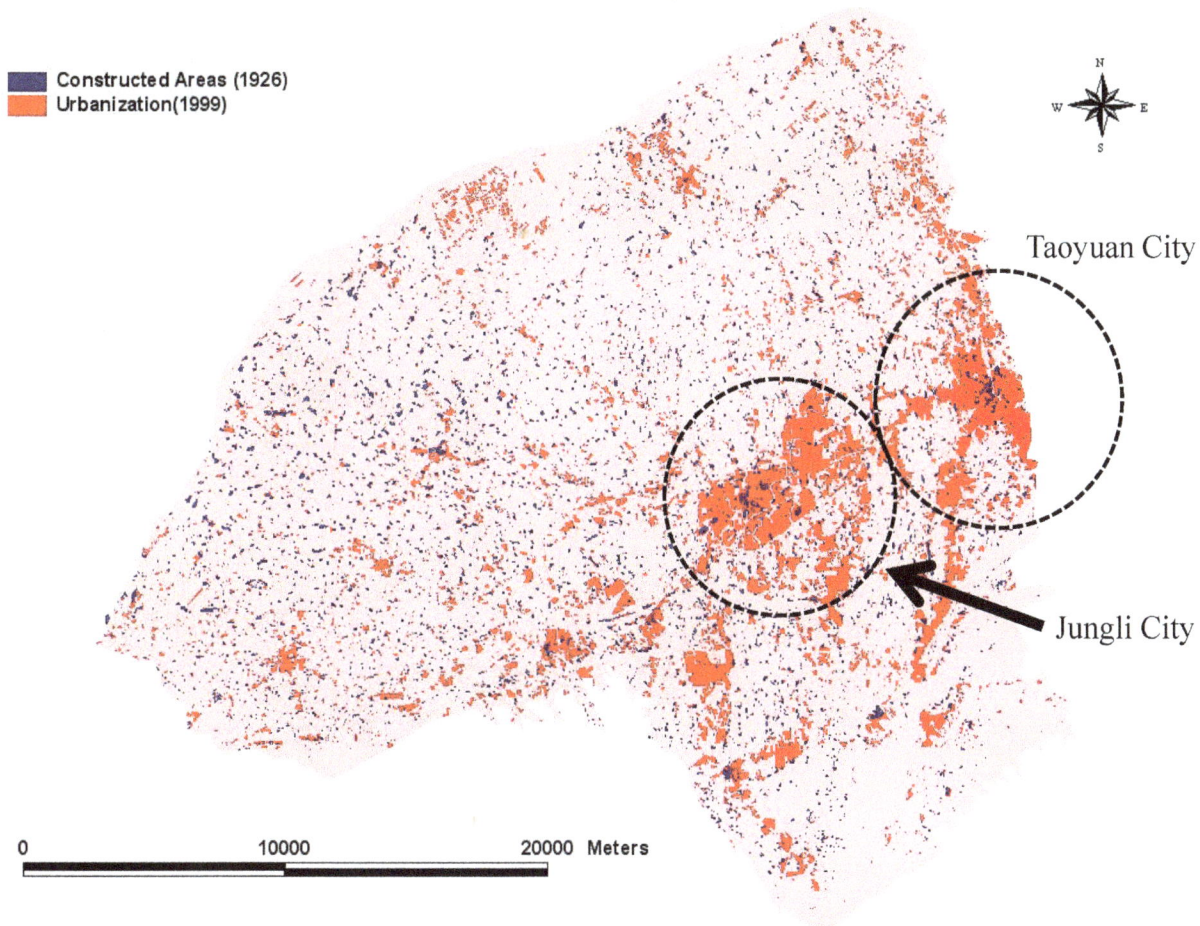

Figure 2. Trends of the progression of built-up areas in 1926.

Predicting the occurrence of anthropogenic factors was crucial in detecting human-induced land-use changes in closed depressions in Brussels [25]. In this study, we developed a series of relationships between pondscapes affected by intrinsic driving forces (patch size and patch shape) and external driving forces (distance from the edge of a pond to the edges of roads, rivers, and canals). This study focused on a unique area of pondscapes containing thousands of individual artificial ponds on the Taoyuan Tableland, which underwent dramatic pond losses during the past century [5,21,27]. Becausefew streams exist on the Taoyuan Tableland, the growth of water paddies could not be supported. However, the Taoyuan Tableland has an extremely flat slope (of 1/75–1/100) and the surface soil consists of impermeable red clay;consequently, residents dug ponds to hold water. Cultivation in Taoyuan began in 1680, and there are currently more than 8,000 ponds with a total area of 8,000 ha. At the beginning of the eighteenth century, the residents of Taoyuan constructed a network of ponds and canals for irrigation, which changed the pondscapes. Many ponds were merged into larger ponds and were connected by ditches. In 1964, the Shihmen main canal irrigation system was built to support the southern area of the Taoyuan farmlands through the Shihmen dam, which provided a more stable water supply for agricultural and industrial use. Thus, the surface area and number of ponds were again reduced in the southern region of the Taoyuan Tableland. In addition, because of the population growth and economic development that has

occurred in recent years, many of the ponds were filled up and became land used for human activities.

According to a review of studies on the thousands of farm ponds in Taoyuan County, Taiwan, the aforementioned information on pond-structure data has been seldom analyzed. Recently, spatially explicit land-coverdata derived from historical maps were directly linked to biological survey data exhibiting with spatial mapping by incorporating biophysical feedback in models of land-use changes[5,28]. In this study, the process of these changes and the leading causes of the disappearance of pondscapes were investigated.We first developed an exploratory analysis of pondscape changes during the twentieth century. Two geographic information systems (ESRI ArcGIS10 and ArcView 3.3) and a logistic regression model were then applied to analyze changes based on map-scanning data previously obtained during the period of Japanese colonial rule (1895–1945) and the modern period (the Republic of China's rule [1945 to the present]). Using these methods, we determined the causes of landscape changes that occurred after the Taoyuan main canal irrigation system, which supported the northern region of the Taoyuan farmlands, was constructed during the 1920s. The adoption of a Grand Canal irrigation system might result in an internal restructuring of agricultural land use from traditional farm-pond irrigation to more diversified water sources, such as ditches, canals, and reservoirs. The spatial dependency of pondscape changes associated with northern irrigation systems (Taoyuan main canal) constructed from 1926 to 1960 was

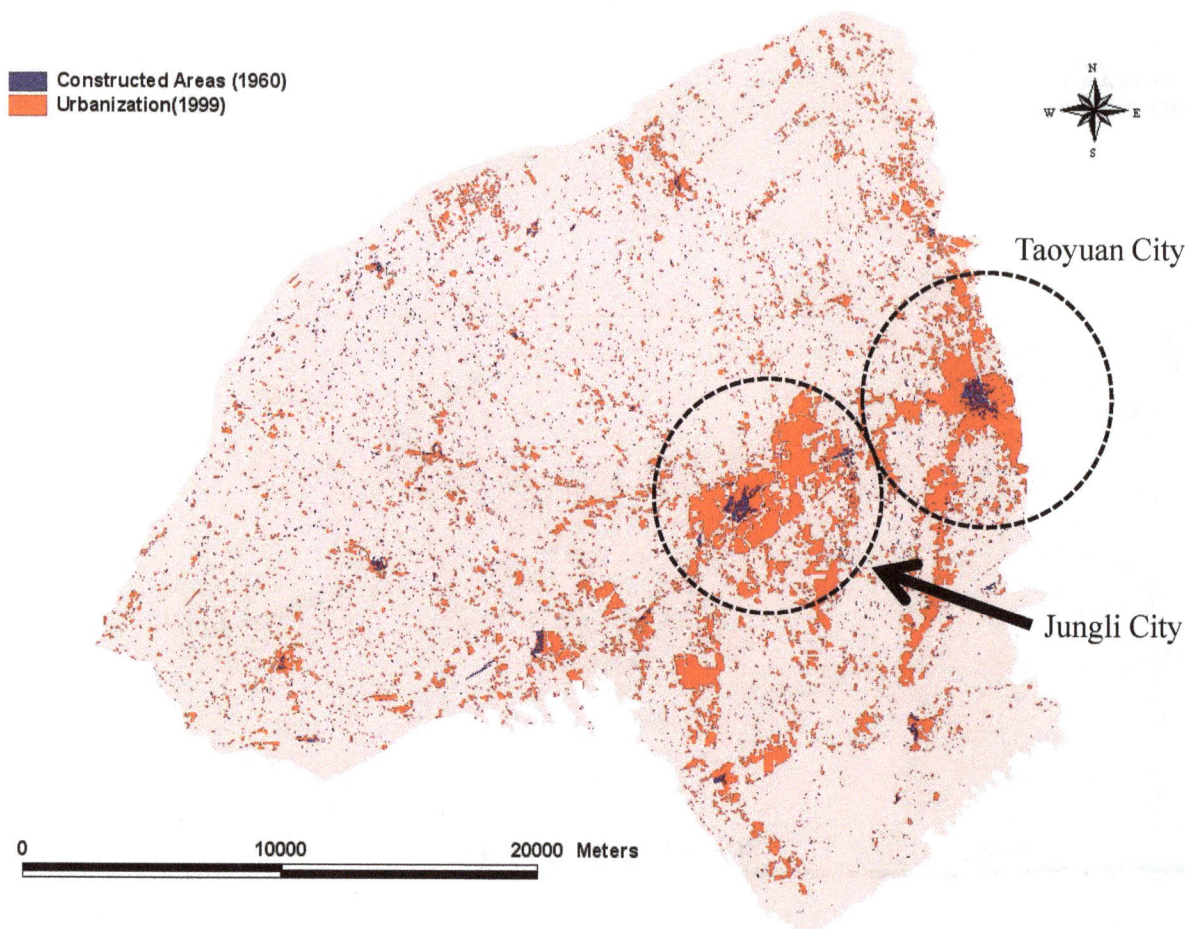

Figure 3. Trends of the progression of built-up areas in 1960.

identified. Furthermore, the process of urbanization that led to these changes was strategically analyzed using a computer-aided analytical model.

In this study, we (a) examined sensitive metrics, such as configuration metrics, which reduced the number of farm ponds; and (b) analyzed the quantitative limitations of these sensitive metrics. A statistical model was designed to examine how the perimeter effects of farm ponds caused losses on the Taoyuan Tableland. Land-use changes that occurred from 1926 to 1960 were analyzed to demonstrate how enforcing land-use policies might have influenced the direction and magnitude of pondscape changes.

Materials and Methods

Description of the Study Area

The Taoyuan Tableland, located in northwestern Taiwan at approximately 40 km southwest of Taipei, occupies an area of 757 km^2 based on the elevational contours (Figure 1). "Taoyuan," which means "peach garden," is situated in a rich agricultural area where numerous peach orchards existed during the nineteenth century. However, Taoyuan is now a regional market center for commercial and industrial activities and the transportation of goods,and is also considered a new potential aerotropolis. The extent of urban development has increased, and the Taoyuan

metropolitan area has become one of the fastest growing areas in Taiwan. Currently, the population of the Taoyuan Tableland is 2,038,000, according to the 2013 census data. This area has a population density of 2,692 persons/km^2, and the population is increasing at a rate of 2,000–3,000 persons/month. Urbanization influences the urban fabric and also causes changes in pondscape textures. Thus, the constant conversion of farm ponds into urban areas might partially be due to the proximity of Taoyuan County to the highly urbanized Taipei City. To understand the relationships between these phenomena, we set up areas of a matrix to analyze topographic maps by using ArcGIS 10. The total matrix area of the 757 km^2 area covers 11 administrative areas, comprising Shinwu Township, Guanyin Township, Dayuan Township, Jungli City, Yangmei Township, Pingjen City, Luchu Township, Taoyuan City, Bade City, Lungtan Township, and Dashi Township.

Description of Data Mining

We obtained authorization to scan 60 (4×15 pieces) original paper maps from four periods, using compact recordable disks with a total of 5,000 MB of memory. We digitized paper maps from 1904, 1926, and 1960, which were scanned at the Computer Graphic Center, Department of Geography, Chinese Cultural University (Taipei, Taiwan). The maps we digitized included the Taiwan fort maps from the Japanese colonial period at a scale of 1:

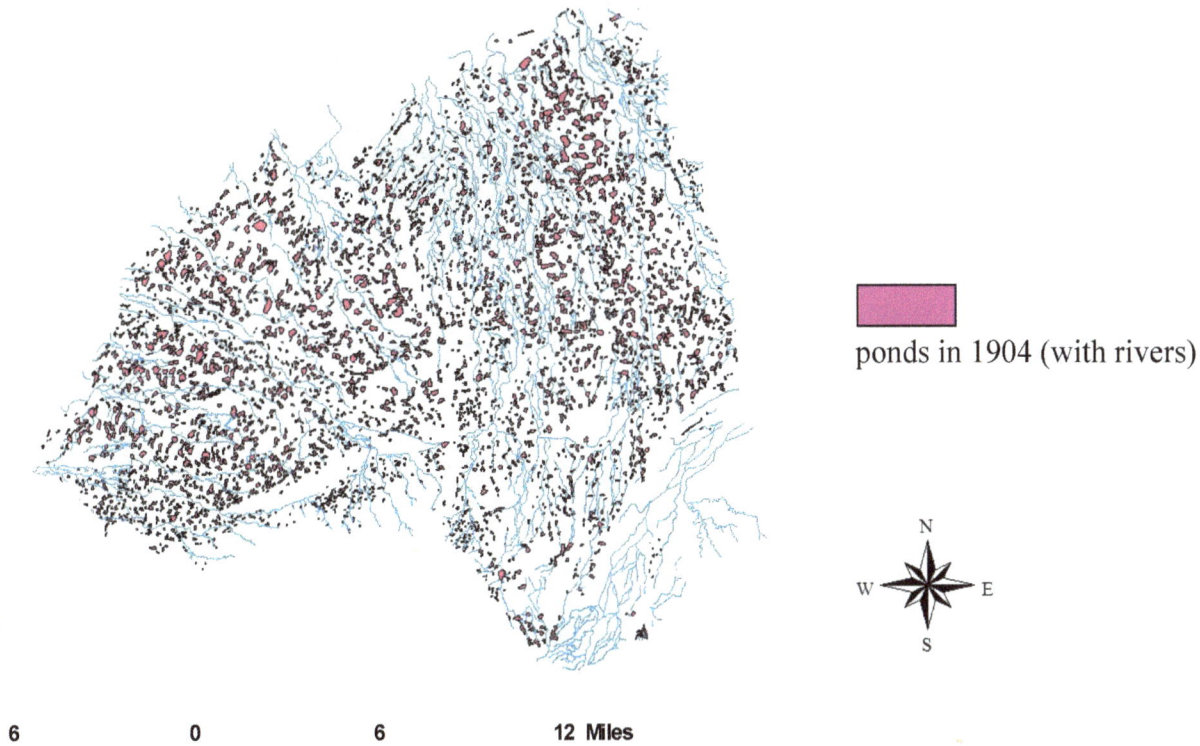

ponds in 1904 (with rivers)

6　　　　0　　　　6　　　　12 Miles

Figure 4. Farm ponds that existed on the Taoyuan Tableland in 1904 (overlapped with local rivers).

20,000, Taiwan topographic maps (sheet version of the 1926 maps, reprinted in 1998, at a scale of 1: 25,000), and Taiwan geographic maps (sheet version of the 1960 maps, at a scale of 1: 25,000) for three separate years during the twentieth century. Using the TM2 cartographic coordinate system, we established two or more control reference points to calibrate and convert the sheet data to true coordinates by using transformation tools. The digital maps comprised the following types of land use: farm ponds, canals, roads (national highways, provincial highways, and prefectural highways), railways, areas with buildings (built-up areas), and agricultural lands.

Logistic Regression Model

Landscape change processes were modeled using nonlinear techniques. A logistic regression is a binary classification algorithm determined using a discriminative parametric approach to predict the probability of the occurrence of an event [29,30]. Logistic regression also involves several predictor variables that are categorical and reflect binary data, such as "success" or "failure." Logistic regression models are widely used to analyze the driving forces underlying the proximate causes of various events, such as species recolonization [31,32] and landscape changes [33–44]. In this study, the binary value was the "existence" or "disappearance" of farm ponds. We then used the logarithm for calculating the ratio of the probability of "existence" to that of "disappearance" to form a linear regression equation. Because more than one independent variable can influence the probability of "existence" or "disappearance," the logistic regression model is

$$\pi_i = \frac{p_i}{1-p_i} = e^{\beta_0 + \sum_{i=1}^{m} B_j X_{ij}} \quad (1)$$

$$Z_i = \ln(\pi_i) = \ln\frac{p_i}{1-p_i} = \beta_0 + \beta_1 X_{i1} + \beta_2 X_{i2} + ... + \beta_m X_{im} \quad (2)$$

where ln is the natural logarithm, $\log_e$ (e = 2.71828...), P_i is the likelihood that pond loss occurs, $\frac{p_i}{1-p_i}$ is the "odds ratio," $\ln\frac{p_i}{1-p_i}$ is the log of the odds ratio, or "Logit," and all of the other components of the model are the same, such as

$$P_i = \frac{e^{zi}}{1+e^{zi}} = \frac{1}{1+e^{-zi}} \quad (3)$$

Whereas $y = \frac{\exp(\beta_0 + \beta_1 X_1 + \beta_2 X_2 + ... + \beta_m X_m)}{[1 + \exp(\beta_0 + \beta_1 X_1 + \beta_2 X_2 + ... + \beta_m X_m)]}$ or

$$y_i = \beta_0 + \beta_1 X_{i1} + \beta_2 X_{i2} + ... + \beta_m X_{im} \quad (4)$$

$X_{i1},...,X_{im}$ are m variables determined by the i th pond, and the likelihood is justified by internal driving forces ms (i.e., size, shape, etc.). The logistic regression coefficients were interpreted when the variables were correlated to explain the "vanished" conditions. We removed several variables from this analysis because they were only slightly correlated. Several sensitive variables (i.e., those highly correlated with ponds that had disappeared) were selected. Final variable coefficients for which $p \le .05$ were considered significant. In Equation (1), $p_i = P(y_i = 1 | x_{i1}, x_{i2},..., x_{im})$ is a series of

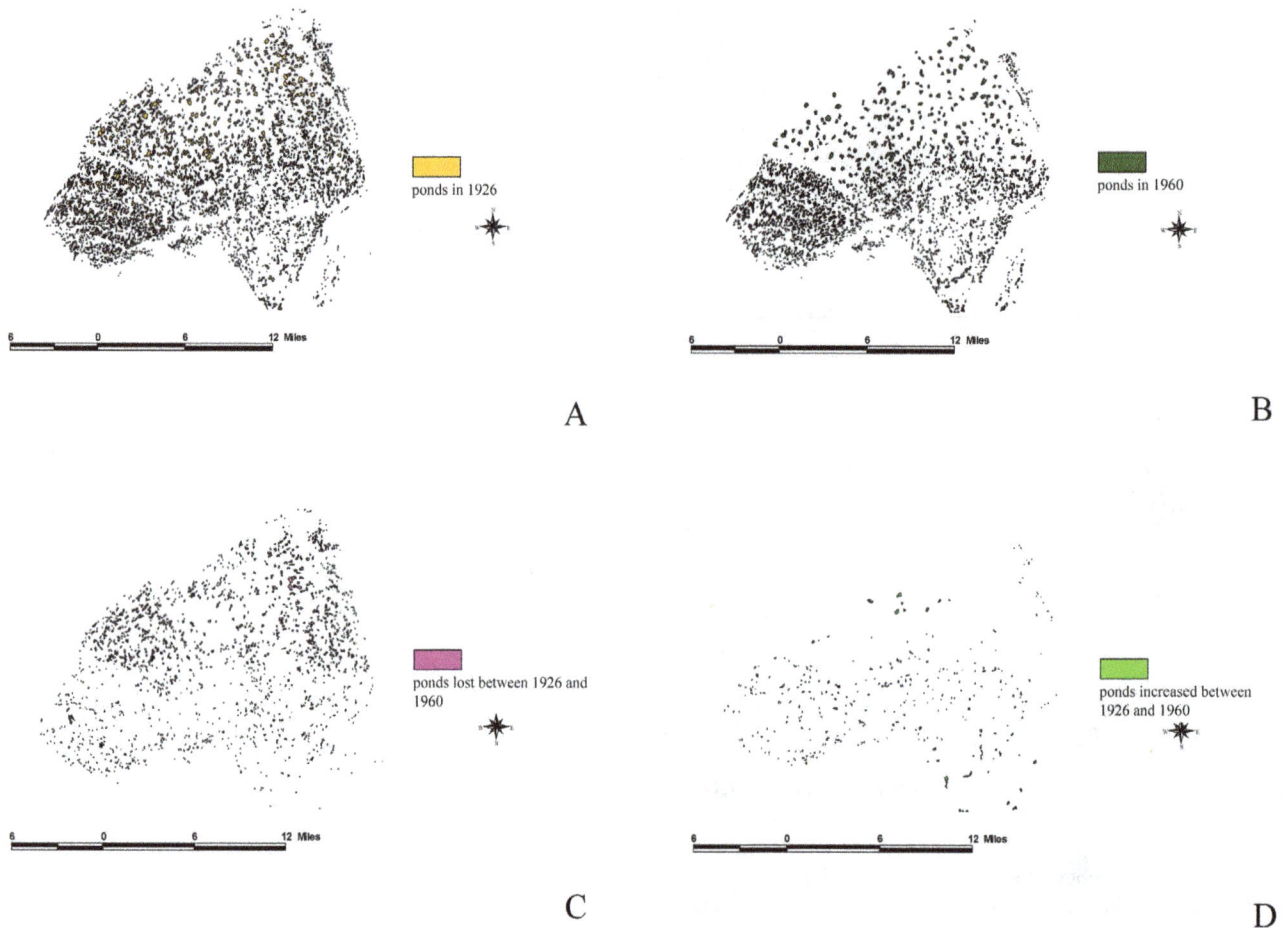

Figure 5. Pondscape changes from 1926 to 1960. (A) Farm ponds that existed on the Taoyuan Table land in 1926. (B) Farm ponds that remained on the Taoyuan Table land in 1960. (C) Trends of losses of farm ponds detected on the Taoyuan Table land between 1926 and 1960. It is obvious that more farm ponds in the northern region of theTaoyuan Tablel and disappeared than in the southern region. (D) Trends of new farm ponds detected on the Taoyuan Tableland between 1926 and 1960.

independent variables, x_{i1}, x_{i2},..., x_{im}, which represents the probability of occurrence. The following list gives the assumptions of the logistic regression model: (a) data should be derived from random samples; (b) the dependent variable, y_i, is assumed to be a function of m independent variables x_{im} (m = 1, 2, 3,...); (c) the logistic regression model is sensitive to multicollinearity, and the multicollinearity among independent variables influences the standard deviation; (d) the dependent variables used in the logistic regression model are binary variables, and these variables can be presented only as "0" or "1"; and (e) the relationships between independent variables and dependent variables are nonlinear.We randomly added variables to the model, based on the concepts developed by Turner [45], such as (a)nearest neighbor probabilities, (b) the amount of edge between land uses, and (c) the patches of ponds according to size class and land use, and then compared the results with a selected suitable model. IBM SPSS 21.0 software (IBM, Armonk, New York, USA) was used to produce the stepwise regression in this study.

The logistic models were selected using various methods to construct a variety of regression models from the same set of variables: (a) Enter: a process for selecting variables in which all of the variables in a block are entered in only a single step;(b) Forward Selection (Conditional): a stepwise selection method combined with entry testing, based on the significance of the score

statistic, and removal testing, based on the probability of a likelihood-ratio statistic and conditional parameter estimates; and (c) Forward Selection (Likelihood Ratio; LR): a stepwise selection method combined with entry testing, based on the significance of the score statistic, and removal testing, based on the probability of a likelihood-ratio statistic and the maximum partial likelihood estimates. Stepwise selection methods were applied in this study to identify covariables in the regression models. This model involves the use of likelihood ratio tests and is applied to prediction situations with a large number of variables to determine which variables are entered and in what order. The difference between Forward Selection (Conditional) and Forward Selection (Likelihood Ratio) is that Forward Selection (Conditional) can be used to perform fast but inaccurate calculations, and Forward Selection (Likelihood Ratio) can be used to perform time-consuming but accurate calculations. The conditional selection might also cause bias, which has been referred to as over fitting in combination with extreme values[46]. One of these methods can be adopted for performing forward stepwise selection to limit the number of covariables and subsequently select coefficients included in the model incrementally. We, therefore, added the final results by using the stepwise selection method based on the Likelihood Ratio approach.

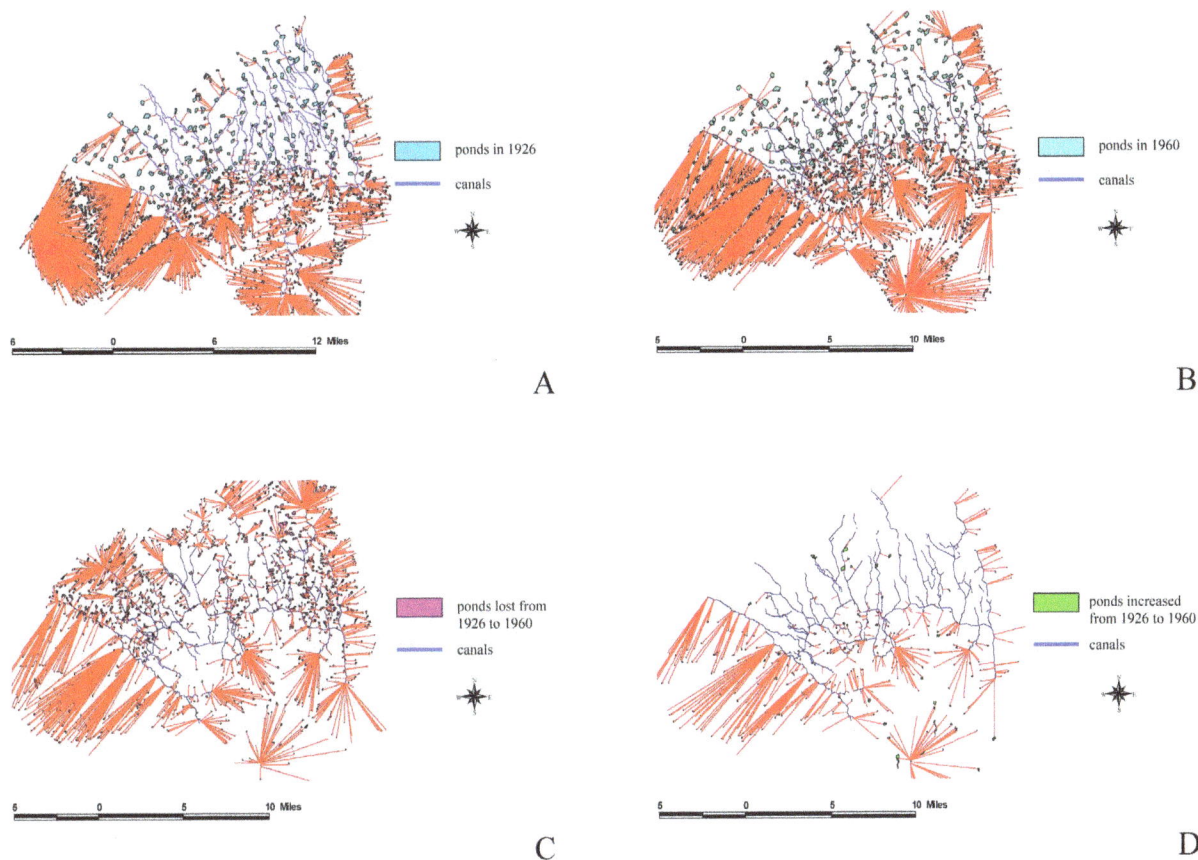

Figure 6. The influence of canals on pondscape changes from 1926 to 1960. (A) Distances between canals and existing ponds in 1926. (B) Distances between canals and remaining ponds in 1960. (C) Distances between canals and lost ponds from 1926 to 1960. (D) Distances between canals and new ponds from 1926 to 1960.

Results

"Orgware City" Trends

In this study, we constructed maps according to the topography of the Taoyuan Tableland from 1904 to 1926 and from 1926 to 1960. The location of cities on the Taoyuan Tableland was not obvious in 1904. By contrast, Taoyuan City and Jungli City had begun to form by 1926 (Figure 2), and two "real urban forms" had formed by 1960 because of population concentration (Figure 3). Similar to the scenario outlined in the urbanization model, the population moved from rural villages to cities, such as Taoyuan City and Jungli City, on the Taoyuan Tableland. These two cities are now densely populated areas where concentrated commercial activities occur. Thus, they have become centers of political, economic, and cultural activities. Various stages of concentration, dispersion, invasion, and separation of towns occurred in this area, and cities formed later. In particular, Taoyuan City and Jungli City competed and coexisted with each other, eventually forming an urban corridor along the main transportation lines. These changes followed the "Orgware City" concept and occurred through the interactions of human activities with nature, and nature with transportation (Figures 2 and 3).

Trends of Farm-Pond Losses

By comparing Figures 4 and 5A, we discovered that, despite the inconspicuous changes in farm ponds that occurred from 1904 to 1926, there was a decreasing trend in the number of farm ponds.

Because the scale of the 1904 maps (1: 20,000) differed from those of other periods (Figure 4), we analyzed only changes from 1926 (1: 25,000; Figure 5A) to 1960 (1: 25,000; Figure 5B). Because the various scales prevented the comparison between 1904 (Figure 4) and 1926 (Figure 5A), we used the map and determined that farm ponds have covered most of the study area since 1904, except for the southeastern corner. In addition, only sporadic villages and residences beyond the forms of cities developed in the early twentieth century.

Figure 5 shows the pattern of the disappearance of farm ponds in northern Taoyuan County, which differed from that of southern Taoyuan County. The number of farm ponds that disappeared was 1,895 (Figure 5C), 2,667 ponds remained, and only 537 new ponds were created (Figure 5D). We observed that the existence of farm ponds closely depended on the use of the Taoyuan main canal irrigation system on the Taoyuan Tableland. However, farm ponds with irregular shapes at the ends of the main canal in the northern region of the tableland disappeared first, whereas the smaller ponds disappeared later.

We used ArcGIS to analyze the overlap between the 1926 and 1960 maps to define dynamic patterns and subsequently understand the reasons for these changes. The results of this study indicated that the total area of farm ponds at the peak was 8,800 ha, and accounted for 11.8% of the tableland area. By contrast, the total area of the farm ponds has decreased to 2,898ha because of urbanization, accounting for only 3.8% of the tableland area. Using FRAGSTATS [47], we analyzed the driving forces behind

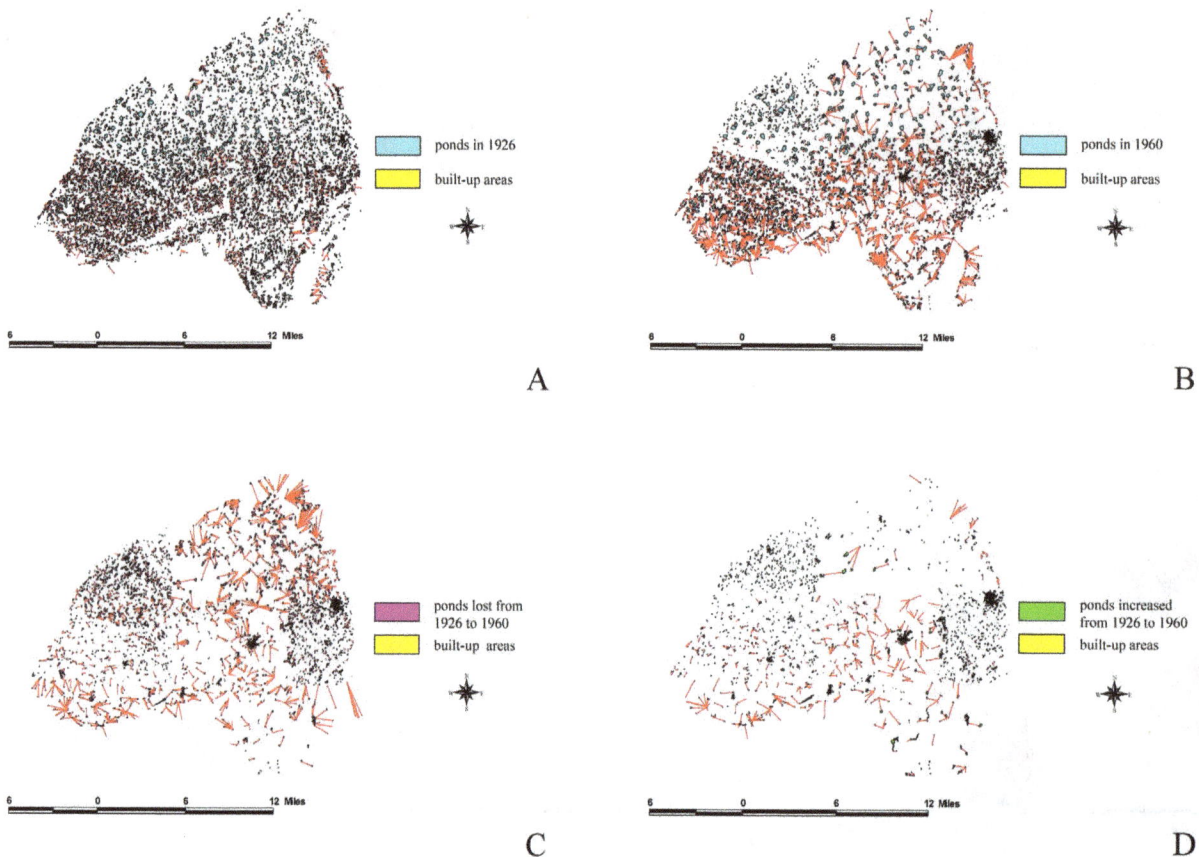

Figure 7. The influence of built-up areas on pondscape changes from 1926 to 1960. (A) Distances between built-up areas and existing ponds in 1926. (B) Distances between built-up areas and remaining ponds in 1960. (C) Distances between built-up areas and lost ponds from 1926 to 1960. (D) Distances between built-up areas and new ponds from 1926 to 1960.

the appearance and disappearance of farm ponds: (a) intrinsic driving forces, which include pond perimeter, pond size, and the mean pond-patch fractal dimension (MPFD); and (b) external driving forces, which included the minimal distance of a farm pond to a canal (Figures 6A, 6B, 6C, and 6D), building (Figures 7A, 7B, 7C, and 7D), river (Figures 8A, 8B, 8C, and 8D), road (Figures 9A, 9B, 9C, and 9D), or railway (Figures 10A, 10B, 10C, and 10D). We then used these relationships to further discuss the driving forces behind changes in farm ponds accompanied by the respective proximity effects.

The results based on the logistic regression model and using the independent variables indicated the possible reasons for pond changes, which included the area of farm ponds (AREA), the perimeter of farm ponds (PERI), the MPFD, and distances to the nearest artificial structure; specifically, canals (CANAL), buildings (built-up areas; BUILD), roads (ROAD), railways (RAIL), and rivers (RIVER)[45,48].

Ponds Lost from 1926 to 1960

The results were divided into 10 groups with approximately 456 cases in each group; see SAV S1 as original datasets. We divided all of the subjects into 10 groups based on their probabilities. Therefore, all of the subjects with a 0.1 probability or lower were placed in the lowest decile, and those with a 0.9 probability or higher were placed in the highest decile. The Hosmer and Lemeshow (HL) test values were determined at Step 2 (23.075, d.f. = 8, $p = .003 > .001$) of the Stepwise Forward Selection

(Likelihood Ratio) model. The chi-square value (23.075) with a higher P value (.003) was accepted for the goodness-of-fit test, compared with that of the other steps (Step1 [224.116, d.f. = 8, $p = .000$], Step 3 [74.601, d.f. = 8, $p = .000$], Step 4 [57.181, d.f. = 8, $p = .000$], and Step 5 [63.340, d.f. = 8, $p = .000$]). By using the HL statistic, we determined that high values indicated large differences between the actual and predicted values for a decile. Therefore, the lower the Hosmer-Lemeshow statistic is, or the higher the corresponding $p$value is, the lower the weakness of fit of the logistic regression model is.

At Step 2, we labeled the standardized regression coefficients as "Beta" (B). The B of PERI ($p = .000$) was equal to 0.002, and the constant was equal to 0.262 ($p = .000$). In this model, a unit of change in PERI led to a 0.002-fold change in the occurrence rate with significance. However, neither MPFD, AREA, CANAL, BUILD, RIVER, ROAD, nor RAILproduced a significant change in the respective occurrence rates. According to the data listed in Table 1, we present the logistic regression model as

$$\text{Logit } (\pi) = 0.262 + 0.002\text{PERI}$$

Based on this model, we concluded that the larger the perimeter of a pond was, the more easily it was lost, even if the cause was not obvious. By contrast, we determined that the shape of ponds significantly influenced the loss rate.

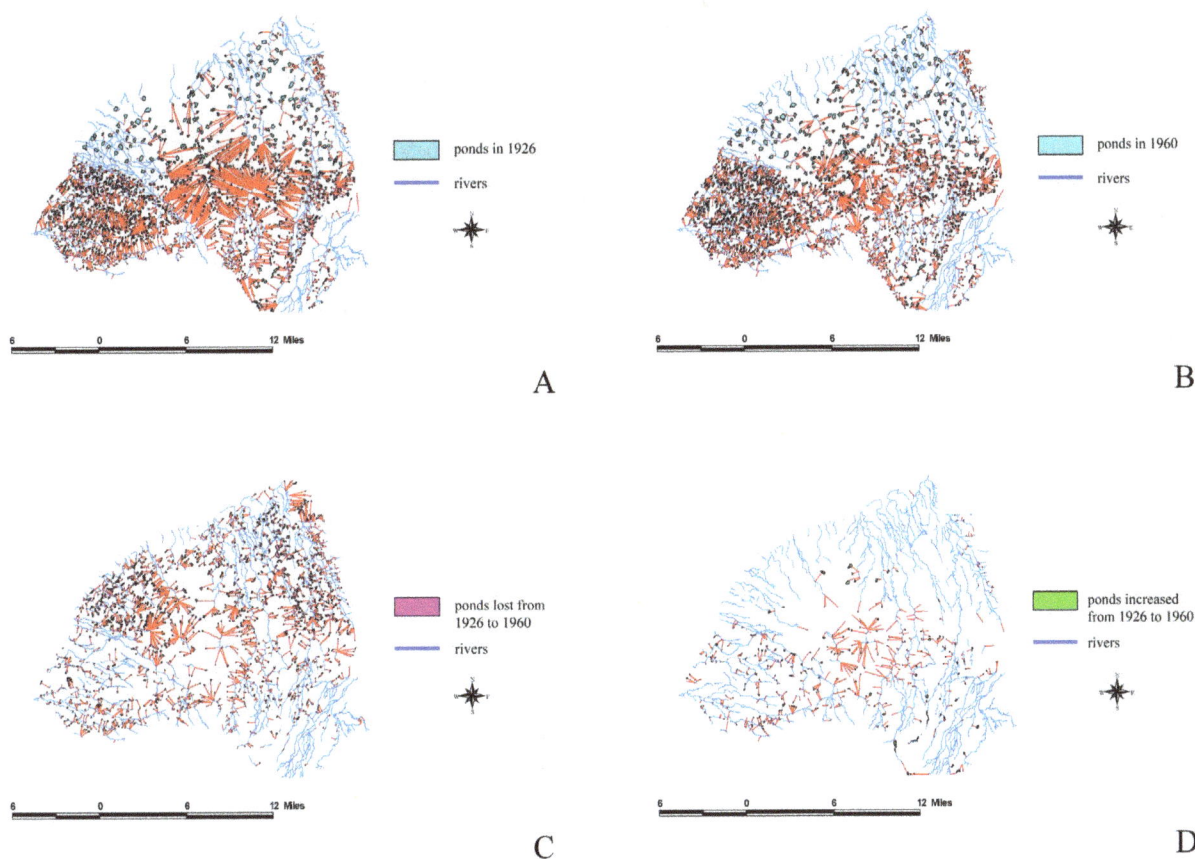

Figure 8. The influence of rivers on pondscape changes from 1926 to 1960. (A) Distances between rivers and existing ponds in 1926. (B) Distances between rivers and remaining ponds in 1960. (C) Distances between rivers and lost ponds from 1926 to 1960. (D) Distances between rivers and new ponds from 1926 to 1960.

Ponds Created from 1926 to 1960

The results were divided into 10 groups with approximately 310 cases in each group; see SAV S2 as original datasets. We divided all of the subjects into 10 groups based on their probabilities. The HL value was determined at Step 3 (12.420, d.f. = 8, $p = .133 > .001$) of the Stepwise Forward Selection (Likelihood Ratio) model. The chi-square value (12.420) with a higher $p$value was accepted for the goodness-of-fit test, compared with that of the other steps (Step 1 [56.017, d.f. = 8, $p = .000$] and Step 2 [16.291, d.f. = 8, $p = .000$]). As stated previously, by using the HL statistic, we determined that high values indicated large differences between the actual and predicted values for a decile. Therefore, the lower the HL statistic is, or the higher the corresponding P value is, the lower the weakness of fit of the logistic regression model is.

At Step 3, we labeled the standardized regression coefficients as B. The B of PERI ($p = .000$) was equal to 0.002, the B of MPFD was equal to -14.130 ($p = .000$), the B of RIVER was equal to 0.000 ($p = .006$), and the constant was equal to 20.261 ($p = .000$). Therefore, a unit change in PERI led to a 0.002-fold change in the occurrence rate, the MPFD led to a-14.130-fold change in the occurrence rate, And RIVER led to a 0-fold change in the occurrence rate. Neither AREA, CANAL, BUILD, ROAD, nor RAIL changed the occurrence rate. From Table 2, we present the logistic regression model as

Logit (π) = 20.261 + 0.002PERI − 14.130 MPFD

Based on this model, we concluded that the larger the perimeter of a pond was, the more likely the pond would be affected by human activities, even if the cause was not obvious. Consistent with the previous model, we determined that the shape of ponds significantly influenced the occurrence rate.

Discussion

Pond Loss as an Intrinsic Function

The number of farm ponds that disappeared was 1,895 (37%), 2,667 ponds remained (52%), and only 537 (11%) new ponds were created within a 757 km^2 area in Taoyuan, Taiwan between 1926 and 1960. According to the studies conducted by Heath and Whitehead[49], similar cases of pond loss were observed in the United Kingdom (U.K.). As an island nation, 55% of the ponds in the U.K.present in 1870 had disappeared by 1960, with the greatest loss occurring between 1920 and 1960. In addition, Western Europe also demonstrated a loss rate of 55% between 1900 and 1990 [50]. Systematic evidence has been presented for pond loss caused by infilling, which occurred mainly during land development periods in the twentieth century[51]. This type of land-form change, together with the extension of urban land uses and the extension of transportation land uses,negatively affected pond loss [52].

In this study, we examined large packages of mapped landscape metrics in pond loss by using multivariate statistical approaches to investigate phenomena that involved spatial interactions and neighborhood characteristics [52–54]. We determined proximate

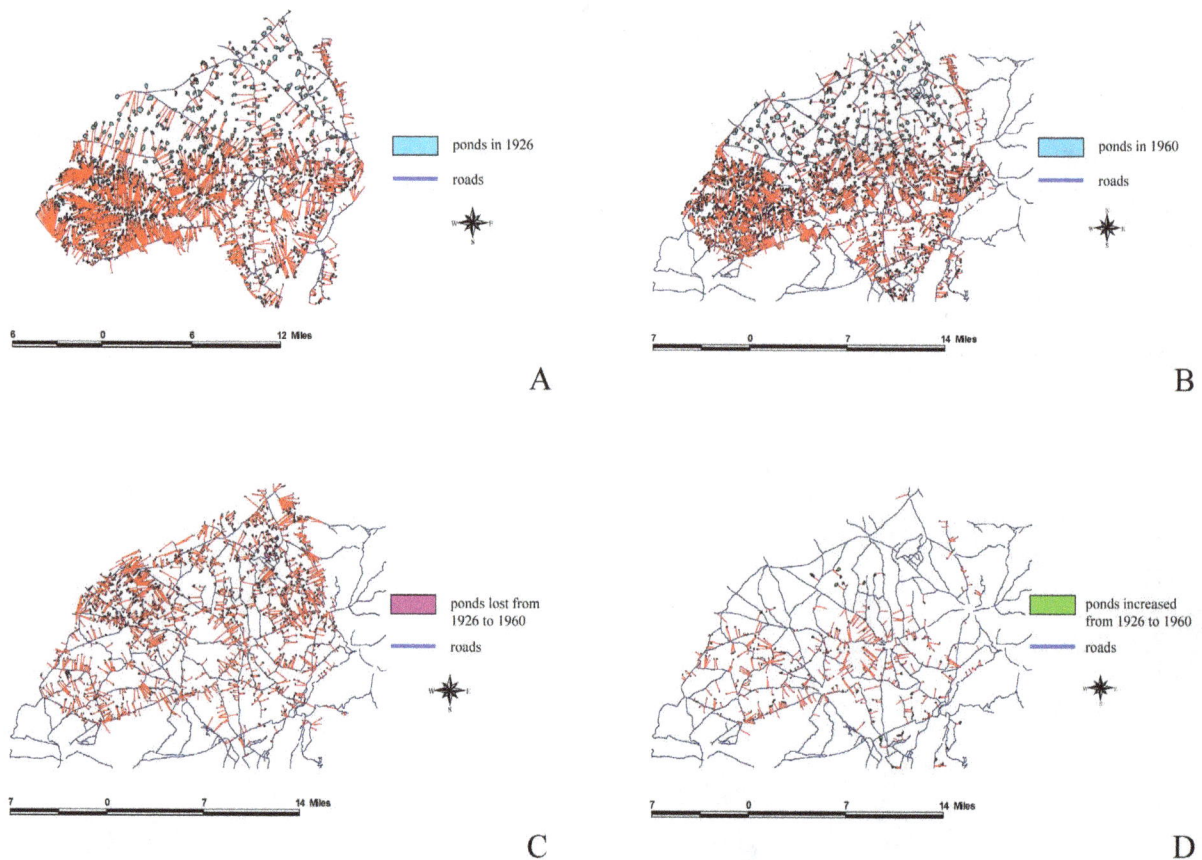

Figure 9. The influence of roads on pondscape changes from 1926 to 1960. (A) Distances between roads and existing ponds in 1926. (B) Distances between roads and remaining ponds in 1960. (C) Distances between roads and lost ponds from 1926 to 1960. (D) Distances between roads and new ponds from 1926 to 1960.

causes by using only intrinsic metrics, such as PERI and the MPFD, in addition to external driving forces (i.e., distance between the pond edge and road edge) based on historical images of pond losses. However, we determined that the main reasons for the loss of ponds or for the increase in the number of ponds were pond intrinsic factors, such as pond perimeter and pond shape, rather than external driving-force factors, such as the distance to roads, rivers, and canal edges. According to the first model, Logit $(\pi) = 0.262 + 0.002$PERI, an increase in pond perimeter led to the loss of ponds; however, according to the second model, Logit $(\pi) = 20.261 + 0.002$PERI $- 14.130$ MPFD, an increase in pond perimeter also led to an increase in the number of ponds.

In this study, pond losses and pond increases were considered to be independent phenomena that should be discussed separately. Generally, larger pond perimeters indicated increased likelihood of pond loss between 1926 and 1960; in addition, larger pond perimeters indicated increased likelihood that new ponds would be created between 1926 and 1960. However, a large perimeter was closely associated with circular shapes (low value of MPFD) beyond irregular and curvilinear shapes (high value of MPFD), which characterize the majority of ponds created during these periods.

Beyond External Driving Forces

This result of intrinsic functions beyond external driving forces was determined beyond our hypotheses, as claimed by several authors [34,35,39]. Studies conducted in an agricultural area of

northern France have reported that small, man-made ponds were more often affected and disappeared more rapidly compared with larger ponds[52]. Using the forward logistic regression approach, they also determined that pond persistence was consistently and positively affected by marsh, grasslands, and dune shrubs, and negatively affected by arable and urbanized lands. Braimoh and Onishi [39] claimed that rising demand for urban land and the construction of new roads changed the rural land located at the periphery of existing built-up areas[55]. This might have caused substantial losses of critical natural resources and increased landscape fragmentation. They discovered that one of the determinants (i.e., the proximity to roads) encouraged landform changes to original settlements and residential areas [34,35,55].

In comparing the results of this study to those of other related or similar studies, considering pond loss and the increase in the number of ponds in Taiwan as a land-use function driven by external driving-force factors, such as the distance to roads, rivers, canal edges, etc., is difficult to verify. Although the proposed model was developed using limited GIS maps from 1926 to 1960, it appeared to more suitably describe the function driven by intrinsic forces than that driven by landform changes based on the forward logistic regression. The results indicated that one of the major driving forces was pond shape. The final result also indicated that pond shape and pond perimeter were the major reasons that an increasing trend in the number of ponds from 1926 to 1960 was observed.

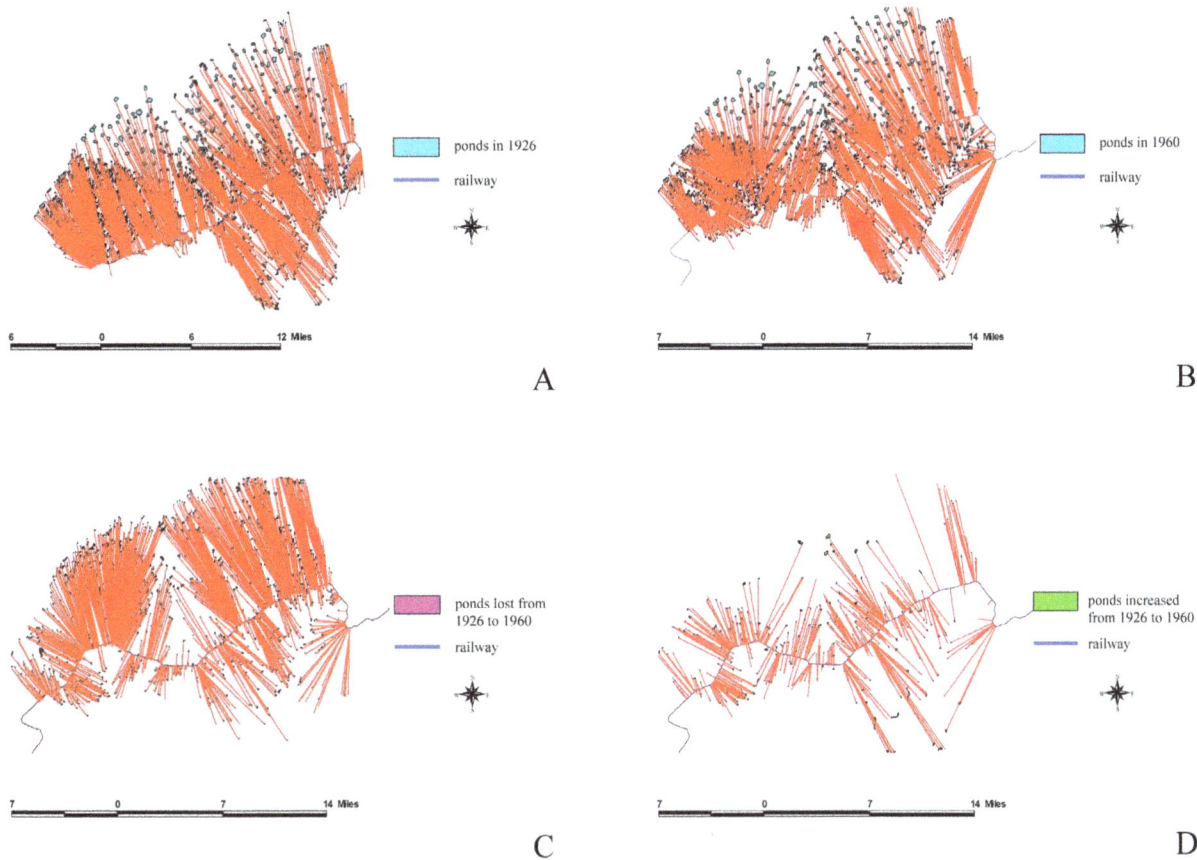

Figure 10. The influence of railways on pondscape changes from 1926 to 1960. (A) Distances between railways and existing ponds in 1926. (B) Distances between railways and remaining ponds in 1960. (C) Distances between railway and lost ponds from 1926 to 1960. (D) Distances between railway and new ponds from 1926 to 1960.

We also discovered that a curvilinear shape parameter was associated with several pond losses, in contrast with the increase in the number of ponds with a round shape. The more regular the surface of a pond patch was, the less likely the pond was lost. Therefore, curvilinear pond losses were more likely to occur during early agricultural periods because they were replaced by a canal irrigation system. This represents a crucial dynamic of irrigation types observed within the region.

By contrast, we discovered that SDFs, such as the distance of farm ponds from canals (CANAL), buildings (built-up areas; BUILD), rivers (RIVER), roads (ROAD), and railways (RAIL), were not major variables. The neighborhood characteristics of canals, buildings (built-up areas), rivers, roads, and railways presented in this paper were, therefore, not suitable for explaining pond-loss patterns on a regional scale. However, although land use from surrounding pondscapes became diverse and heterogeneous, the configuration of the lost ponds might not be statistically significant. Therefore, further ecological field research is necessary to interpret these pondscape dynamics more comprehensively. Finally, assumptions were made to validate the normality and linearity in this study for simulating land-use and land-cover changes [56,57].

Morphology Determined Landscape

Because of the characteristics of the mildly sloping gradients of the Taoyuan Tableland, morphology has long been at work in this region. In attempting to study the effects of retrospective influences on the increase in the number of ponds, we also discuss the morphologic influences determined by the landscape. For example, the results of this study suggested that the change in the condition of a pond, which can be either lost or increased, is related to the inherent morphology of the pond, such as size complexity and perimeter. These parameters are related; for example, if we consider two shapes with similar sizes, and one shape is characterized by a regular perimeter and the other is characterized by a fractal perimeter by a high MPFD value, the irregular perimeter is expected to be longer than that of the regular perimeter. Therefore, if the MPFD was greater than 1, the pond was highly formed as a curvilinear shape. Thus, we determined the conditions of the types of ponds that increased in number, which were characterized by a more circular and less curvilinear shape, and a lower value of MPFD. The reason that a farmer dug out a more circular pond than an irregularly shaped pond might have been that a circular pond was designed to avoid soil erosion from the pond fringe. In addition, the influences of an irregular pond fringe from potential nonpoint source pollution on nearby farm ponds should be considered. Because the original function of farm ponds was agricultural, farmers were more likely to secure water quality, quantity, and safety than focus on other functional uses between 1926 and 1960. Regarding these anthropogenic activities on the tableland, the results of this study indicated that the MPFD decreased as the number of ponds increased.

Table 1. Binomial logit model of pond losses for the period 1926~1960. Unit of observation: individual ponds in 1960.

		B	S.E.	Wals	df	Significance	Exp(B)	Exp(B) at a 95.0% confidence level	
								Lower level	Upper level
Step 1[a]	RAIL	.000	.000	284.656	1	.000	1.000	1.000	1.000
	Constant	1.102	.055	398.575	1	.000	3.011		
Step 2[b]	PERI	.002	.000	304.498	1	.000	1.002	1.002	1.003
	RAIL	.000	.000	374.463	1	.000	1.000	1.000	1.000
	Constant	.262	.072	13.316	1	.000	1.299		
Step 3[c]	PERI	.003	.000	373.709	1	.000	1.003	1.003	1.003
	CANAL	.000	.000	202.802	1	.000	1.000	1.000	1.000
	RAIL	.000	.000	325.626	1	.000	1.000	1.000	1.000
	Constant	−.519	.090	33.250	1	.000	.595		
Step 4[d]	PERI	.003	.000	369.346	1	.000	1.003	1.003	1.003
	CANAL	.000	.000	256.434	1	.000	1.000	1.000	1.000
	RIVER	.000	.000	69.760	1	.000	1.000	1.000	1.000
	RAIL	.000	.000	262.406	1	.000	1.000	1.000	1.000
	Constant	−.972	.105	85.276	1	.000	.378		
Step 5[e]	PERI	.003	.000	368.965	1	.000	1.003	1.003	1.003
	CANAL	.000	.000	247.843	1	.000	1.000	1.000	1.000
	RIVER	.000	.000	66.615	1	.000	1.000	1.000	1.000
	ROAD	.000	.000	15.229	1	.000	1.000	1.000	1.000
	RAIL	.000	.000	274.600	1	.000	1.000	1.000	1.000
	Constant	−1.059	.108	96.551	1	.000	.347		

Note 1:
a. Selected into a variable: railroad in Step 1.
b. Selected into a variable: PERI in Step 2.
c. Selected into a variable: CANALin Step 3.
d. Selected into a variable: RIVER in Step 4.
e. Selected into a variable: ROAD in Step 5.
Note 2:
1) AREA: the area of pond; PERI:the perimeter of pond; MPFD: Mean Pond Fractal Dimension;CANAL: distances of farm ponds from nearest canals;BUILD: distances of farm ponds from nearest buildings (built-up areas);RIVER: distances of farm ponds from nearest rivers;ROAD: distances of farm ponds from nearest roads;RAIL: distances of farm ponds from nearest railways.
2) The equation to Mean Pond Fractal Dimension (MPFD):

$$\text{MPFD} = \frac{\sum_{j=1}^{n}\left(\dfrac{2\ln p_{ij}}{\ln a_{ij}}\right)}{n_i} \tag{5}$$

a_{ij} = the area of pond ij (in m2).
n_i = the number of the pond ij.
p_{ij} = the perimeter of pond ij (in m).
Level: CLASS, LANDSCAPE
Units: None
Range:1<MPFD<2
Description: MPFD reflects shape complexity across a range of pond size. It equals 2 times the logarithm of pond perimeter (m) divided by the logarithm of pond area (m^2) (Li &Reynolds, 1994). MPFD approaches 1 for shapes with very simple perimeters such as circles or squares, and approaches 2 for shapes with highly convoluted and plane-filling perimeters.

Land Policies on Pond Conservation

Numerous factors affect farm-pond conversion. When the benefits off arm ponds do not exceed the benefits brought by changes in advanced land use, farm-pond conversion occurs. In addition to the aforementioned spatial parameters, we also considered the effect of land policy. Because the policies on farm land transformation were determined by private landowners, the farm-pond policies lifted the restrictions on farmland trades, thereby facilitating a more direct spatial transition of ponds. In this area, most farmland trades are associated with ponds located adjacent to urban-fringe lands. The reasons that the topic of farm-pond conversion should be emphasized are related to urbanization. Because the great demand for constructed land use in this situation decreases the number of ponds, the government should urge farmers to create more new ponds characterized by large, circular shapes because of the benefits associated with these types of ponds, according to the findings of this study. This could protect the integrity of the agricultural environment and subsequently secure the irrigation function, as well as provide efficiency and equity from a balanced perspective of natural resources.

Table 2. Binomial logit model of pond increases for the period 1926~1960. Unit of observation: individual ponds in 1960.

		B	S.E.	Wals	df	Significance	Exp(B)	Exp(B) at a 95.0% confidence level	
								Lower level	Upper level
Step 1[a]	MPFD	−17.735	1.206	216.370	1	.000	.000	.000	.000
	Constant	25.574	1.628	246.803	1	.000	127866882315.499		
Step 2[b]	PERI	.002	.000	24.436	1	.000	1.002	1.001	1.002
	MPFD	−13.807	1.424	94.077	1	.000	.000	.000	.000
	Constant	19.716	1.978	99.364	1	.000	365137138.357		
Step 3[c]	PERI	.002	.000	25.378	1	.000	1.002	1.001	1.002
	MPFD	−14.130	1.436	96.871	1	.000	.000	.000	.000
	RIVER	.000	.000	7.648	1	.006	1.000	1.000	1.000
	Constant	20.261	1.999	102.743	1	.000	629546300.185		

Note 1:
a. Selected into a variable: MPFD in Step 1.
b. Selected into a variable: PERI in Step 2.
c. Selected into a variable: RIVER in Step 3.
Note 2: Variables are defined in the footnotes to Table 1.

Conclusion

Pondsare defined as an artificial construction made to impound water through the use of a dam or an embankment, or by excavating a pit or a hole. This type of wetscape provides a unique scenic view of a landscape located between nature and humanity that has existed for over a hundred years in the Taoyuan Tableland of northwestern Taiwan. However, these wetscapes are being lost because of public construction, anthropogenic activities, and land use for economic development.

The relationship between pond loss and pond shape was a robust predictor for changes in landforms. We developed a series of models for pondscapes affected by intrinsic driving forces (patch size and patch shape) and external driving forces (distance from the edge of a pond to the edges of roads, rivers, and canals). The final results reflected the effect of various intensive driving forces on pond changes during specific periods in the history of the study area, and we concluded that these changes occurred because of pond intrinsic factors, such as pond size, and that pond-loss rates were affected by pond shape rather than by external driving forces. The models, based on a probabilistic method for calculations using binary logistic regression, were used to determine the effects of landform changes, which are represented by proximate causes underlying SDFs. In assessing the configurations of pondscape areas, however, assessing the loss rates and patterns of and identifying the effects of SDFs on land-use changes are crucial because they might indicate that the scientific rigor of generic theories of conceptual models must be reduced before these models can be used to explain land-use changes [14,58]. Several SDFs, however, are too simplified to explain their proximate

causes and consequences, and are difficult to use to support theories. In addition, a simple extrapolation of the trends possesses less predictive power because of the predictive uncertainties in space and/or time[59]. Therefore, the reasons for land-use changes should be discussed in both simple and general contexts[60]. This might be facilitated using logistic stepwise regression as a simple patchy-based approach to explore pondscape dynamics for monitoring pond-loss problems, once people begin to emphasize the importance of protecting this unique pondscape on the Taoyuan Tableland.

Supporting Information

Sav S1 The SPSS file indicated as pond disappeared (0) and pond remained (1) associated with parameters to support this study. (SPSS SAV, original datasets).

Sav S2 The SPSS file indicated as pond increased (0) and pond remained (1) associated with parameters to support this study. (SPSS SAV, original datasets).

Author Contributions

Conceived and designed the experiments: WTF JYC SYL. Performed the experiments: WTF JYC SYL. Analyzed the data: WTF JYC SYL. Contributed reagents/materials/analysis tools: WTF JYC SYL. Wrote the paper: WTF JYC SYL.

References

1. Lemmens P, Mergeay J, De Bie T, Van Wichelen J, De Meester L, et al. (2013) How to maximally support local and regional biodiversity in applied conservation? Insights from pond management. PLoS One 8: e72538.
2. Smith SV, Renwick WH, Bartley JD, Buddemeier RW (2002) Distribution and significance of small, artificial water bodies across the United States landscape. Sci Total Environ 299: 21–36.
3. Satrand GH (2011) Uncertainty in classification and delineation of landscapes: A probabilistic approach to landscape modeling. Environmental Modelling & Software 26: 1150–1157.
4. Houlahan JE, Findlay CS (2003) The effects of adjacent land use on wetland amphibian species richness and community composition. Canadian Journal of Fisheries and Aquatic Sciences 60: 1078–1094.
5. Fang W-T, Chu H-J, Cheng B-Y (2009) Modeling waterbird diversity in irrigation ponds of Taoyuan, Taiwan using an artificial neural network approach. Paddy and Water Environment 7: 209–216.
6. Hartel T, von Wehrden H (2013) Farmed areas predict the distribution of amphibian ponds in a traditional rural landscape. PLoS One 8: e63649.
7. Peaman PB (1995) Effects of pond size and consequent predator density on two species of tadpoles. Oecologia 102: 1–8.

8. Kadoya T, Suda S-I, Washitani I (2004) Dragonfly species richness on man-made ponds: effects of pond size and pond age on newly established assemblages. Ecological Research 19: 461–467.

9. Arrhenius O (1921) Species and area. Journal of Ecology 9: 95–99.

10. MacArthur RH, Wilson EO (1967) The Theory of Island Biogeography. Princeton, NJ: Princeton University Press: 203.

11. Conner EF, McCoy ED (1979) The statistics and biology of the species-area relationship. American Naturalist 113: 791–832.

12. Simberloff DS, Abele LG (1976) Island biogeographic theory and conservation practice. Science 191: 285–286.

13. Ward W, Blaustein L (1994) The overriding influence of flash floods on species-area curves in ephemeral Negev Desert pools: A consideration of the value of island biogeography theory. Journal of Biogeography 21: 595–603.

14. Lambin EF (1997) Modelling and monitoring land-cover change processes in tropical regions. Progress in Physical Geography 21: 375–393.

15. Forman RTT (1995) Land Mosaic: The Ecology of Landscape and Regions Cambridge, UK: Cambridge University: 632.

16. Oertli B, Joye DA, Castella E, Juge R, Cambin D, et al. (2002) Does size matter? The relationship between pond area and biodiversity. Biological Conservation 104: 59–70.

17. Linton S, Goulder R (2000) Botanical conservation value related to origin and management of ponds. Aquatic Conservation: Marine and Freshwater Ecosystems 10: 77–91.

18. Leitão AB, Ahern J (2002) Applying landscape ecological concepts and metrics in sustainable landscape planning. Landscape and Urban Planning 59: 65–93.

19. Froneman A, Mangnall MJ, Little RM, Crowe TM (2001) Waterbird assemblages and associated habitat characteristics of farm ponds in the Western Cape, South Africa. Biodiversity and Conservation 10: 251–270.

20. Francl KE, Schnell GD (2002) Relationships of human disturbance, bird communities, and plant communities along the land-water interface of a large reservoir. Environ Monit Assess 73: 67–93.

21. Fang W-T (2011) Creating pondscapes for avian communities: An artificial neural network experience beyond urban regions. In Hong, S.-K.; Wu, J.; Kim, J.-E.; Nakagoshi, N. (Eds.) Landscape Ecology in Asian Cultures. New York, NY: Springer: 187–200.

22. Geist HJ, Lambin EF (2002) Proximate causes and underlying driving forces of tropical deforestation. BioScience 52: 143–150.

23. Bürgi M, M HA, Schneeberger N (2005) Driving forces of landscape change - current and new directions. Landscape Ecology 19: 857–868.

24. Chowdhury RR (2006) Landscape change in the Calakmul Biosphere Reserve, Mexico: Modeling the driving forces of smallholder deforestation in land parcels. Applied Geography 26: 129–152.

25. Vanwalleghem T, Van Den Eeckhaut M, Poesen J, Govers G, Deckers J (2008) Spatial analysis of factors controlling the presence of closed depressions and gullies under forest: Application of rare event logistic regression. Geomorphology 95: 504–517.

26. Bhattarai K, Conway D, Yousef M (2009) Determinants of deforestation in Nepal's Central Development Region. J Environ Manage 91: 471–488.

27. Fang W-T, Huang Y-W (2012) Modelling Geographic Information System with Logistic Regression in Irrigation Ponds, Taoyuan Tableland. Procedia Environmental Sciences 12, part A: 505–513.

28. Veldkamp A, Lambin EF (2001) Predicting land-use change. Agriculture, Ecosystems & Environment 85: 1–6.

29. Berkson J (1944) Application of the Logistic Function to Bio-Assay. Journal of the American Statistical Association 39: 357–365.

30. Pampel FC (2000) Logistic Regression: A Primer (Quantitative Applications in the Social Sciences). Logistic Regression: Primer London, UK: Sage Publications.

31. Mladenoff DJ, Sickley TA, Wydeven AP (1999) Predicting gray wolf landscape recolonization: Logistic regression models vs. new field data. Ecological Applications 9: 37–44.

32. Stephenson CM, MacKenzie ML, Edwards C, Travis JMJ (2006) Modelling establishment probabilities of an exotic plant, Rhododendron ponticum, invading a heterogeneous, woodland landscape using logistic regression with spatial autocorrelation. Ecological Modelling 193: 747–758.

33. Ludeke AK (1990) An analysis of anthropogenic deforestation using logistic regression and GIS. Journal of Environmental Management 31: 247–259.

34. Schneider LC, Pontius Jr RG (2001) Modeling land-use change in the Ipswich watershed, Massachusetts, USA. Agriculture, Ecosystems & Environment 85: 83–94.

35. Gobin A, Campling P, Feyen J (2002) Logistic modelling to derive agricultural land use determinants: a case study from southeastern Nigeria. Agriculture, Ecosystems & Environment 89: 213–228.

36. Aspinall R (2004) Modelling land use change with generalized linear models–a multi-model analysis of change between 1860 and 2000 in Gallatin Valley, Montana. J Environ Manage 72: 91–103.

37. Verburg PH, de Nijs TCM, van Eck JR, Visser H, de Jong K (2004) A method to analyse neighbourhood characteristics of land use patterns. Computers, Environment and Urban Systems 28: 667–690.

38. Wimberly MC, Ohmann JL (2004) A multi-scale assessment of human and environmental constraints on forest land cover change on the Oregon (USA) coast range. Landscape Ecology 19: 631–646.

39. Braimoh AK, Onishi T (2007) Spatial determinants of urban land use change in Lagos, Nigeria. Land Use Policy 24: 502–515.

40. Machemer PL, Simmons C, Walker RT (2007) Refining landscape change models through outlier analysis in the Muskegon watershed of Michigan. Landscape Research 31: 227–294.

41. Serra P, Pons X, Saurí D (2008) Land-cover and land-use change in a Mediterranean landscape: A spatial analysis of driving forces integrating biophysical and human factors. Applied Geography 28: 189–209.

42. Vanwalleghem T, Van Den Eeckhaut M, Poesen J, Govers G, Deckers J (2008) Spatial analysis of factors controlling the presence of closed depressions and gullies under forest: Application of rare event logistic regression. Geomorphology 95.

43. Nong Y, Du Q (2011) Urban growth pattern modeling using logistic regression. Geo-spatial Information Science 14: 62–67.

44. Zheng X-Q, Zhao L, Xiang W-N, Li N, Lv L-N, et al. (2012) A coupled model for simulating spatio-temporal dynamics of land-use change: A case study in Changqing, Jinan, China. Landscape and Urban Planning 106: 51–61.

45. Turner MG (1988) A spatial simulation model of land use changes in a Piedmont county in Georgia. Applied Mathematics and Computation 27: 39–51.

46. Chatfield C (1995) Model Uncertainty, Data Mining and Statistical Inference. Journal of the Royal Statistical Society Series A (Statistics in Society) 158: 419–466.

47. McGarigal K, Cushman SA, Ene E (2012) FRAGSTATS v4: Spatial Pattern Analysis Program for Categorical and Continuous Maps. Computer software program produced by the authors at the University of Massachusetts, Amherst. Available at the following web site: http://www.umass.edu/landeco/research/fragstats/fragstats.html.

48. Helmer EH (2004) Forest conservation and land development in Puerto Rico. Landscape Ecology 19: 29–40.

49. Heath DJ, Whitehead A (1992) A survey of pond loss in Essex, South-east England. Aquatic Conservation: Marine and Freshwater Ecosystems 2: 267–273.

50. Hull AP (1997) The pond life project: a model for conservation and sustainability. In British Pond Landscape, Proceedings from the UK Conference of the Pond Life Projectt, Boothby J (ed) Pond Life Project: Liverpool: 101–109.

51. Boothby J (2003) Tackling degradation of a seminatural landscape: options and evaluations. Land Degradation & Development 14: 227–243.

52. Curado N, Hartel T, Arntzen JW (2011) Amphibian pond loss as a function of landscape change – A case study over three decades in an agricultural area of northern France. Biological Conservation 144: 1610–1618.

53. Verburg PH, van Eck JRR, de Nijs TCM, Dijst MJ, Schot P (2004) Determinants of land-use change patterns in the Netherlands. Environment and Planning B: Planning and Design 31: 125–150.

54. Verburg PH, Schot PP, Dijst MJ, Veldkamp A (2004) Land use change modelling: current practice and research priorities. GeoJournal 61: 309–324.

55. Serneels S, Lambin EF (2001) Proximate causes of land-use change in Narok District, Kenya: a spatial statistical model. Agriculture, Ecosystems and Environment 85: 65–81.

56. Parker DC, Manson SM, Janssen MA, Hoffmann MJ, Deadman P (2003) Multi-Agent Systems for the Simulation of Land-Use and Land-Cover Change: A Review. Annals of the Association of American Geographers 93: 314–337.

57. Pontius Jr RG, Boersma W, Castella JC, Clarke K, de Nijs T, et al. (2008) Comparing the input, output, and validation maps for several models of land change. The Annals of Regional Science 42: 11–37.

58. Hersperger AM, Gennaio MP, Verburg PH, Bürgi M (2010) Linking Land Change with Driving Forces and Actors: Four Conceptual Models. Ecology and Society 15: 1–17.

59. Pontius Jr RG, Spencer J (2005) Uncertainty in extrapolations of predictive land-change models. Environment and Planning B: Planning and Design 32: 211–230.

60. Lambin EF, Turner BL, Geist HJ, Agbola SB, Angelsen A, et al. (2001) The causes of land-use and land-cover change: moving beyond the myths. Global Environmental Change 11: 261–269.

8

Signals of Climate Change in Butterfly Communities in a Mediterranean Protected Area

Konstantina Zografou[1]*, Vassiliki Kati[2], Andrea Grill[3,4], Robert J. Wilson[5], Elli Tzirkalli[1], Lazaros N. Pamperis[6], John M. Halley[1]

1 Department of Biological Applications and Technologies, University of Ioannina, Ioannina, Greece, **2** Department of Environmental and Natural Resources Management, University of Patras, Seferi, Agrinio, Greece, **3** Department of Tropical Ecology and Animal Biodiversity, University of Vienna, Rennweg, Vienna, Austria, **4** Department of Organismic Biology, University of Salzburg, Hellbrunnerstraße, Salzburg, Austria, **5** Centre for Ecology and Conservation, University of Exeter Cornwall Campus, Penryn, United Kingdom, **6** P.O. Box 1220, Larissa, Greece

Abstract

The European protected-area network will cease to be efficient for biodiversity conservation, particularly in the Mediterranean region, if species are driven out of protected areas by climate warming. Yet, no empirical evidence of how climate change influences ecological communities in Mediterranean nature reserves really exists. Here, we examine long-term (1998–2011/2012) and short-term (2011–2012) changes in the butterfly fauna of Dadia National Park (Greece) by revisiting 21 and 18 transects in 2011 and 2012 respectively, that were initially surveyed in 1998. We evaluate the temperature trend for the study area for a 22-year-period (1990–2012) in which all three butterfly surveys are included. We also assess changes in community composition and species richness in butterfly communities using information on (a) species' elevational distributions in Greece and (b) Community Temperature Index (calculated from the average temperature of species' geographical ranges in Europe, weighted by species' abundance per transect and year). Despite the protected status of Dadia NP and the subsequent stability of land use regimes, we found a marked change in butterfly community composition over a 13 year period, concomitant with an increase of annual average temperature of 0.95°C. Our analysis gave no evidence of significant year-to-year (2011–2012) variability in butterfly community composition, suggesting that the community composition change we recorded is likely the consequence of long-term environmental change, such as climate warming. We observe an increased abundance of low-elevation species whereas species mainly occurring at higher elevations in the region declined. The Community Temperature Index was found to increase in all habitats except agricultural areas. If equivalent changes occur in other protected areas and taxonomic groups across Mediterranean Europe, new conservation options and approaches for increasing species' resilience may have to be devised.

Editor: Francesco de Bello, Institute of Botany, Czech Academy of Sciences, Czech Republic

Funding: This research has been co-financed by the European Union (European Social Fund - ESF) and Greek national funds through the Operational Program "Education and Lifelong Learning" of the National Strategic Reference Framework (NSRF) - Research Funding Program: Heracleitus II. Investing in knowledge society through the European Social Fund. The funders had no role in study design, data collection and analysis, decision to publish, or preparation of the manuscript.

Competing Interests: The authors have declared that no competing interests exist.

* E-mail: ntinazografou@yahoo.co.uk

Introduction

Major changes in climate worldwide have been identified as the cause of recent shifts observed in species' geographical distributions [1,2,3,4,5]. Many such shifts follow a poleward range expansion pattern [6,7,8]. Climate warming results in locations becoming generally more favourable for species near the "cool", high-latitude limits of their distributions, but it may be less favourable for species near their "warm", low-latitude limits [9], with consequent changes in relative species' abundance and community composition [10]. There is a documented pattern where widespread species (that are better able to expand their distributions through human-modified landscapes) or species associated with warm conditions are becoming more abundant due to warming, at the expense of habitat specialists or species restricted to higher latitudes or elevations [4,11,12]. Yet, different taxonomic groups and different regions have shown different levels of evidence of tracking changes to the climate [1,13].

Butterflies are known to be highly sensitive to climate change [6] and recent studies prove that they react faster than other groups such as birds [13]. A reason for this is because butterflies have relatively short generation times and are ectothermic organisms, meaning that their population dynamics may respond to temperature changes more directly and more rapidly [14]. Butterflies are among the most well-studied taxa in Europe, benefiting from a detailed dataset including relatively fine-resolution information on species' distributions and abundance [14], but they are still far less studied than vertebrates, although the latter comprise only a small fraction of global biodiversity. While further increases in the earth's temperature are anticipated [15] and are expected to lead to serious changes in diversity patterns worldwide, empirical evidence for such changes is still scarce for the Mediterranean biome [16] compared to temperate latitudes. Some evidence that the species composition of Mediterranean butterfly communities has not responded to climate warming as rapidly as expected based on the biogeographic associations of species [17] suggests that these communities may be comparatively resilient to climate change, but more research is needed to test this hypothesis. In addition, an urgent applied question related to climate-driven changes to ecological communities is whether European protected area networks may cease to be effective for conservation, if species are driven out of protected areas by climate warming [18]. So far, there is no empirical evidence on how climate change during the last decade has influenced species communities in Mediterranean nature reserves: precisely this kind of information is likely to be increasingly important for conservation planning in a global climate change scenario.

In this study, we assess if and how butterfly species richness and community composition have changed in response to climate change in the Greek nature reserve, Dadia-Leukimi-Soufli National Park. Greece is considered to be a biodiversity hotspot for butterflies, including more than 40% (234 species) of all European butterfly species (535) [19]. We selected Dadia-Leukimi-Soufli National Park (Dadia NP hereafter) as our study area, because its long conservation history has limited the scale of land use changes [20], and so differences in species composition can reasonably be attributed to factors other than land use change. In the case of Dadia NP, it has been acknowledged that in the absence of traditional activities (such as logging, livestock grazing), especially in the strictly protected core areas, forest encroachment at the expense of clearings and grasslands would have a negative impact on biodiversity, and particularly on species associated with open habitats [21,22]. Thus, the Specific Forest Management Plan of Dadia NP [23] considers the importance of landscape heterogeneity and open habitats, allowing controlled wood-cutting and grazing within the core areas. As a result, two of the most influential factors in the composition of butterfly communities, the intensity of livestock grazing and logging [24,25,26] have remained quite stable over the last decade (D. Vassilakis, Soufli Forest Department, *pers comm*). Moreover, preliminary data of an ongoing study on land cover changes in Dadia NP shows that forest cover remained quite consistent (72–74%) from 2001 to 2011 (K. Poirazidis, WWF Greece and P. Xofis, Inforest, unpublished data), implying that forest encroachment has been minimal during the period of study.

Sampling of butterfly communities was conducted in 2011 and 2012 and results were compared to an earlier study we carried out in 1998 [24]. The present paper is the first comparative study of community composition turnover in the light of climate change in Greece and the Balkan region. We investigate (a) if mean annual temperatures in the study area have increased since the 1990s and (b) if butterfly community composition and species richness have changed across a thirteen year period as a response to climate warming in a protected area, which is largely free of major changes to land use. Finally, we discuss how to implement our findings in a tangible conservation context for nature reserve management.

Materials and Methods

Ethics Statement

Specific permission for the field study described in Dadia-Leukimi-Soufli National Park was given by the Ministry of Environment Energy and Climate Change (Greece). Dadia forest has been owned and managed by the local government from 1980 when it was officially declared a Nature Reserve. The field observations included protected butterfly species but all individuals were released immediately after identification.

Study area

The study area of Dadia NP is situated in northeastern Greece (40°59'–41°15'N, 26°19'–26°36'E) (Fig. 1). It is a hilly area extending over 43000 ha with altitudes ranging from 20 to 650 m, including two strictly protected core areas (7290 ha), where only low-intensity activities such as periodic grazing and selective wood-cutting are allowed, under the control of the local Forest Service of Dadia NP. The core areas are surrounded by a buffer zone where certain human activities are also allowed such as domestic livestock grazing, small agriculture fields and controlled logging. The climate is sub-Mediterranean with an arid summer season (approximately July-September) and a mean annual rainfall ranging from 556 to 916 mm [27]. Mean annual temperature is 14.3°C with lowest values in January and the highest in July-August [27]. The forest is characterized by extensive pine and oak stands [28] and a heterogeneous landscape [29] supporting a high diversity of raptors [30], passerines [31], amphibians and reptiles [32], grasshoppers [33] orchids [34], vascular plants [28], beetles [35] and butterflies [24]. Dadia was established as a nature reserve in 1980 mainly due to its great variety of birds of prey and since then, it has become acknowledged as a region of interest for other groups of organisms as well.

Temperature data

Meteorological data (mean annual temperatures) were obtained from two stations, one located within the study area (Dadia NP station, functioning from 1994–2004) and a second one located 56 km away from the study area (the meteorological station in Alexandroupoli has been operating from 1964 until now [36]).

Butterfly sampling

To test for changes in community composition, the butterfly dataset recorded in 2011 followed exactly the same methodology as that used in 1998 [24], i.e. transects of 200m standard length at 3 locations per habitat type (7 habitats on the whole) were carried out, with transects in the same habitat type a minimum of 300 m distance and maximum of 1 km from one another (Fig. 1). Each transect was repeated 15 times, approximately every 10 days between May 14 and September 14. Habitat selection was representative of the predominant land use types in Dadia NP [37], containing 7 habitats which were: pine forest, oak forest, mixed forest (of mainly *Quercus spp.* and *Pinus brutia* stands), wet meadow, dry grassland, grazed pasture and agricultural fields. We conducted additional samplings in the broader Dadia NP area in 2011, to complete the NP species inventory, without considering them in the data analysis. Comparisons for the long-term period were conducted between the 21 transect sites for the years 1998–2011.

In addition, a third sampling was conducted in 2012, in order to clarify whether any long-term (1998–2011/12) community composition change can be attributed to long-term environmental changes such as climate change, or to short-term variation in community composition between successive years. To do so, a subset of six habitats out of seven (18 transects) was visited once (June 2012) at the same time and date as in 2011. Comparisons for the long and short-term period were conducted among these 18 transects for the years 1998–2011/2012 and 2011–2012 respectively.

Data analysis

Analysis of temperature. To estimate the temperature trend in Dadia NP during the last decades, a 22 year period (1990–2012) was considered. Because meteorological data for Dadia NP are

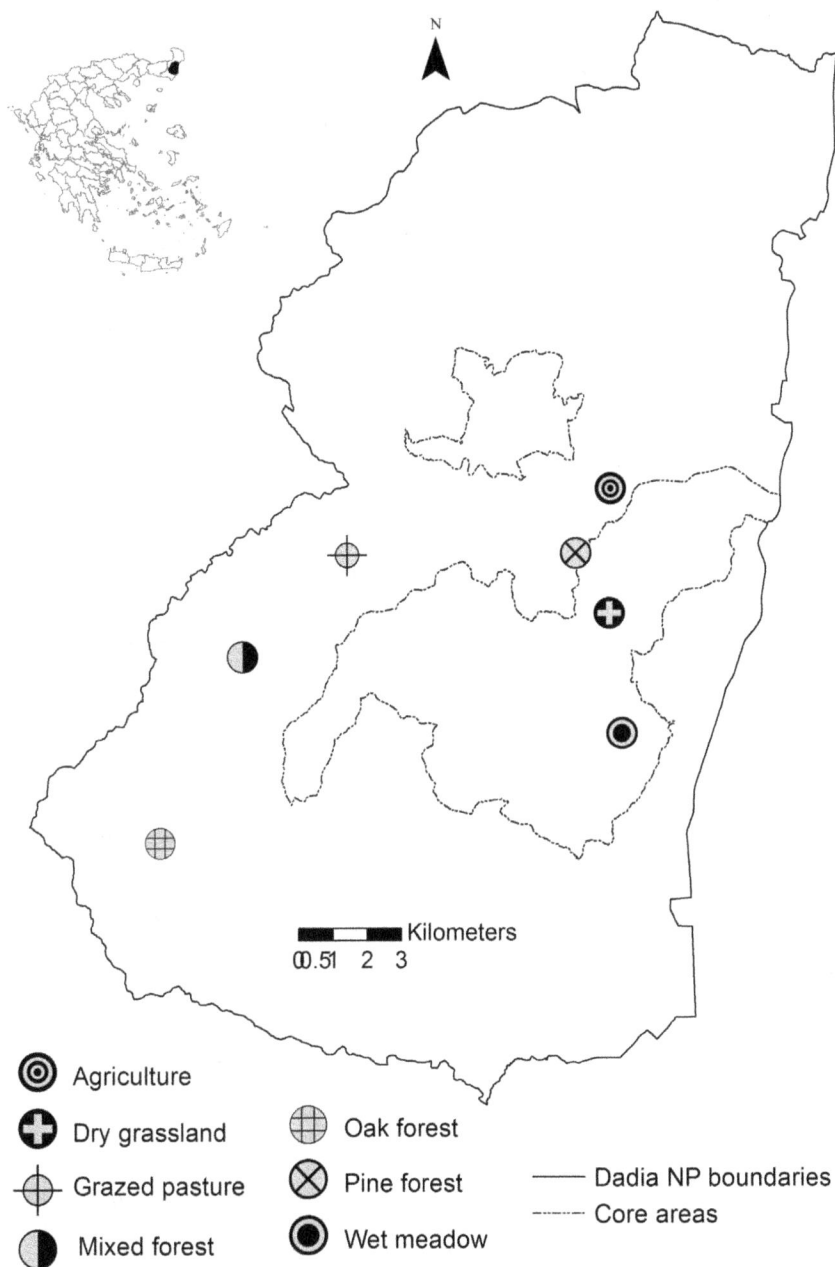

Figure 1. Map of the study area, Dadia National Park in NE-Greece. The map illustrates the geographic location of Dadia National Park where butterflies were sampled in seven habitat types (3 transects per habitat type) in 1998, 2011 and 2012.

only available for 1994–2004, a linear model (period 1994–2004) was run using Dadia NP station data as the response variable and Alexandroupoli station data as the independent variable. The obtained model was then used to estimate the temperature in the study area for all three butterfly surveys. Finally, a linear trend model with randomization (1000 times) was used to test for significant temperature change in the 22 year period. All these analyses were performed with Minitab® Statistical Software (ver.16.1.1).

Community composition change. To check the completeness of the sampling with respect to species detectable by each observer during 1998 and 2011, we assessed sampling efficiency in terms of proportion of species diversity sampled versus the species diversity estimated by non-parametric estimators (Chao 1) [38,39].

Based on this procedure, sampling efficiency was greater than 95% for both years (1998, 2011).

First, an Analysis of Similarities (ANOSIM test) was carried out to explore whether there was a significant change in community composition on a long-term (1998–2011) and short-term (2011–2012) period [40]. The ANOSIM test is based on the ranks of Bray-Curtis dissimilarity index and ranges from −1 to +1, where values greater than zero mean that community composition differs significantly between the years. We created two datasets, one for the long and one for the short-term periods, and we treated each one separately. We assessed the significance of the null hypothesis, namely equal similarity among replicates between groups (sampling periods) and within groups (21 transects) after conducting 999 permutations.

Secondly, the non-parametric method for multivariate analysis of variance based on permutation tests [41] was used, in order to determine the main influences on community composition changes. The permutation analysis of variance (PERMANOVA) for the 13 year period (1998–2011) was run for the 21 transects using species' abundances (counts during the 15 visits in each year) as the response variable, the year factor as a fixed effect and the repeated transects as the random effect in the model.

In order to create equivalent comparisons between the long and short-term periods, additional PERMANOVA were conducted for the 18 transects using the single June visit for (a) 1998 – 2011/ 2012 and (b) 2011 – 2012 respectively.

To pinpoint those species that contributed most to community composition changes, a separate univariate Poisson regression model was fitted for each species and the likelihood ratio statistic was used as a measure of change strength [42]. These analyses were carried out in R (R Development Core Team, 2009) using the *vegan* library [43] and *mvabund* package [44].

Measures of species' thermal associations. The first measure used for the regional thermal associations of butterflies was defined by three categories, in terms of their elevational distribution on Greek national territory, following the example of Wilson *et al.* [4] in Spain. We used the Greek Butterfly Atlas [45] and the 1260 actual localities (6'×6') recorded by the author or referred to in the bibliography, covering 61.19% of Greece. We classified species that occurred in more than 50% of these 1260 localities as "widespread". Species that occurred in fewer than 50% of the localities were classified according to their elevational associations. Those for which > 50% of the records came from localities with an elevation of more than 1000 m, were classified as "high-altitude". Those for which > 50% of the records came from localities with elevations below 1000 m were considered as "low-altitude" (Table S1). The elevation threshold of 1000 m was used for consistency with the four-grade scale provided in the Greek butterfly Atlas (0–500, 500–1000, 1000–1500 and >1500) [45]. Low and high-altitude species have been adequately sampled in the Greek butterfly Atlas in terms of sampling effort (number of localities) for the Greek territory below and above 1000 m. For each elevational zone, we took the ratio between the number of localities and the area covered by the Greek territory (km^2). The ratio ranged from 0.02 to 0.1, and a strong correlation emerged between the number of localities and the area at each elevational zone (Spearman rho = 1, n = 4, $P<0.001$) (Table S4).

The second measure for thermal associations of butterflies was the Species Temperature Index (STI), based on species' biogeographical associations in Europe. The STI is a species-specific value calculated as the average annual temperature across the 50×50 km grid squares where the species has been recorded in Europe [13,14,46,47,48]. At transect level, the average Species Temperature Index of all species was weighted by species' total abundance, in order to estimate a Community Temperature Index for each year. Then the respective transect community temperature indices for the years 1998 and 2011 were compared using a Wilcoxon rank sum test, to conclude whether there has been a significant change in butterfly community thermal structure.

European STI and our elevation-based measure of Greek butterfly thermal associations appeared to give a consistent measure of relative thermal associations of the species observed (Mann-Whitney U test for STI for high versus low-altitude species, n = 88 species, W = 1796, $P=0.02$).

Species diversity change. Considering the two butterfly surveys of 1998 and 2011 separately, alpha-diversity (Shannon–Wiener Index H′) was calculated for each of the 21 transects for all butterfly species and for high and low-altitude species separately.

Beta-diversity was also used to quantify species turnover within each habitat type (3 transects each), using Whittaker's formula, b = (S/ā)-1, where S is the total species number within each transect in each habitat type and ā is the average species number in that habitat type [39]. To test whether the values of alpha and beta-diversity differed between the sampling years we ran general linear models.

To pinpoint whether any significant differences between the two years for the high-altitude and low-altitude species were due to changes in species richness or abundance, Monte-Carlo permutation tests were used. Assuming for the null hypothesis that both years were equivalent and that high-altitude and low-altitude species had the same probability of occurrence in a given sample, the following test statistics for species richness (T_{sp}) and abundance (T_{ab}) were used:

$$T_{sp} = \frac{L_2}{L_2 + H_2} - \frac{L_1}{L_1 + H_1},$$

$$T_{ab} = \frac{l_2}{l_2 + h_2} - \frac{l_1}{l_1 + h_1}$$

where L is the number of low-altitude species and H the number of high-altitude species for the years (1) 1998 and (2) 2011, and l is the abundance of the low-altitude species and h the abundance of the high-altitude species for the years (1) 1998 and (2) 2011. Thus, if the relative proportion of low-altitude species increases, we expect T_{sp} or T_{ab} to be positive. These steps were repeated 1000 times with no replacement. If the observed value (T_{sp} or T_{ab}) falls within the range of the randomly generated values (two-tailed test for $P< 0.025$) we cannot reject the null hypothesis, namely that both high and low-altitude species have the same probability to occur in the sampling years (in terms of species richness or abundance). We carried out these analyses in Minitab and R using libraries *vegan* and *nlme* [49].

Results

Butterfly diversity of Dadia NP

A total of 78 species (3248 individuals) were recorded in 2011, 35 species (427 individuals) in 2012 and 75 (2855) in 1998. The number of species and the number of species of European conservation concern (SPEC) [50] per habitat type for each sampling period (1998–2011–2012) are given in the supporting information (Fig. S1).

Community composition change

A significant difference in community composition over the long-term period (1998–2011) and a non-significant difference over the short-term period (2011–2012) was found, according to ANOSIM results (R = 0.32, n = 42, $P=0.006$ and R = 0.02, n = 42, $P=0.4$ respectively). The PERMANOVA analysis for the 13 year period indicated a significant effect of the year x transect interaction on community composition ($F_{1,168}=1.2$, $P=0.01$, Table S2). A *posteriori* test among levels of the factor 'year', within levels of the factor 'transect', showed significant differences in time only for five transect sites (Table S3). Contrasting results of the single repetition in June between the long and short-term period were found with an additional PERMANOVA. A significant year x transect interaction emerged for the long-term period (1998–2011: $F_{1,24}=4.63$, $P=0.001$; 1998–2012: $F_{1,24}=3.42$, $P=0.001$), indicating that differences among transects affected the response of community composition to different years over the longer period, while a non-significant year x transect interaction emerged for the short-term period (2011–2012: $F_{1,24} = 0.56$, $P = 0.9$). This result suggests that the lack of difference between 2011–12, in contrast to the difference between 1998 versus both 2011 and 2012, is not simply due to a lack of power in using the single June transect counts for comparisons involving 2012. A *posteriori* test among levels of the factor year, within levels of the factor transect, showed no significant differences.

Nineteen species which contributed most importantly to the difference between the years 1998 and 2011 (Table 1) were pinpointed, out of which 10 species had decreased in abundance. The species with the strongest changes in abundance were the widespread species *Aporia crataegi* (decrease), and *Argynnis paphia* (decrease). *Arethusana arethusa* has become totally extinct in all study sites since 1998, *Melitaea trivia* considered to be a low-altitude species showed a strong decline (over 90% of its abundance compared to 1998), while species like *Hipparchia fagi*, *Kirinia roxelana* and *Aricia agestis* almost doubled their abundance.

Temperature trend

A significant increase of mean annual temperature in Dadia NP was found between 1990 and 2012, of 0.95°C (Fig. 2). The null hypothesis (no significant change in temperature) was rejected after conducting 1000 randomizations ($P = 0.003$).

Table 1. Results from univariate Poisson regression models fitted to each taxon.

Species names	LR	SC	PC
Arethusana arethusa	215.01	HA	−100
Melitaea trivia	405.52	LA	−95
Argynnis paphia	1125.59	HA	−85
Aporia crataegi	2662.47	HA	−85
Pieris mannii	293.70	HA	−84
Vanessa cardui	258.51	W	−83
Brenthis daphne	461.67	HA	−74
Brintesia circe	91.47	HA	−56
Issoria lathonia	243.78	HA	−29
Coenonympha pamphilus	126.65	HA	−28
Maniola jurtina	1395.40	LA	+5
Colias crocea	361.90	W	+8
Melitaea didyma	238.92	HA	+11
Polyommatus icarus	615.09	W	+25
Satyrium ilicis	455.98	LA	+34
Thymelicus sylvestris	200.98	HA	+79
Hipparchia fagi	303.96	HA	+109
Kirinia roxelana	151.81	LA	+187
Aricia agestis	126.65	LA	+511

LR: Likelihood ratio test statistic used as a measure of species strength of between-years effect, SC: species categories (HA: high-altitude, LA: low-altitude, W: widespread) created using species elevational distributions in Greece, PC: proportional change (%) of species abundance among 1998 and 2011 (formula used N_{2011} / N_{1998}).
Only statistically significant species ($P < 0.05$) are shown, while species are ranked from those with the greatest declines to those with the greatest increases in abundance between 1998 and 2011 (%).

Changes in species diversity and thermal associations

Using the first measure of species' regional thermal associations, 40 high-altitude species were observed in both 1998 and 2011 (1557 individuals in 1998, versus 1161 in 2011), whereas 25 low-altitude species (913 ind.) were observed in 1998, versus 31 (1657 ind.) observed in 2011. Only 7 (1998) and 5 (2011) species were classified as "widespread" (Table S1). A significant increase in alpha-diversity for the low-altitude species and respectively a significant decrease for high-altitude species was found. The alpha-diversity increase was not significant, when considering all species regardless of whether they were high or low-altitude (Table 2). None of the beta-diversity changes between 1998 and 2011 were significant (Table 2), with slight increases for the overall butterfly community and the low-altitude species, versus a slight decrease for the high-altitude species. According to the Monte-Carlo permutations, the changes in species diversity were due to species abundance differences ($T_{ab} = 0.2$, $P < 0.025$) and not to species richness ($T_{sp} = 0.05$, $P = 0.086$).

Using the second measure of the species' European thermal associations, the community temperature index was found to change significantly between the years 1998 and 2011 (Wilcoxon rank sum test W = 344, n = 42, $P = 0.0036$). In fact, a significant increase of community temperature indices was found in all habitats except for the agricultural areas where the community temperature index had decreased (Fig. 3). To ensure that the CTI change did not result from phenological change, we repeated the process of index calculation for all visits during the summer except for the first in 1998 and the last in 2011. CTI again showed a significant increase between time periods, implying that changes in butterfly community composition were independent of any advancement in mean flight dates by the constituent species.

Discussion

Signals of climate change

Butterfly community composition changed significantly over the 13-year period in conjunction with a recent temperature increase. We found significant changes in the abundance of regionally high versus low-altitude species, as well as a significant increase of the Community Temperature Index based on the thermal associations of species' distributions in Europe. In the later recording period, species associated with warm conditions (i.e. low-altitude species) came to dominate over species associated with cool conditions (i.e. high-altitude species). This suggests that butterfly communities in the study area may have responded to climate warming, even in as short a period as 13 years. Of course, it is well known that there are changes over all timescales in temperature time-series due to local or regional changes that need not be attributed to a prevailing global-warming trend [51]. It is also well established that the expansion of forest owing to land abandonment in the Mediterranean region during the last century may threaten open habitat species [21,22,52]. However, the protected status of Dadia NP and the subsequent stability of land use regimes over the last decade (see Introduction) suggest that our results are nonetheless consistent with the global warming interpretation.

We found marked changes in butterfly community composition over a 13 year time period, but on the other hand our analysis gave no evidence of significant short-term year-to-year variability in butterfly community composition. Butterfly community composition was most influenced by the factors year and transect, when comparing datasets over the long-term period (1998–2011 and 1998–2012). Different habitat types naturally host different butterfly communities [53,54], explaining the transect factor effect. On the other hand, the long conservation history of Dadia NP,

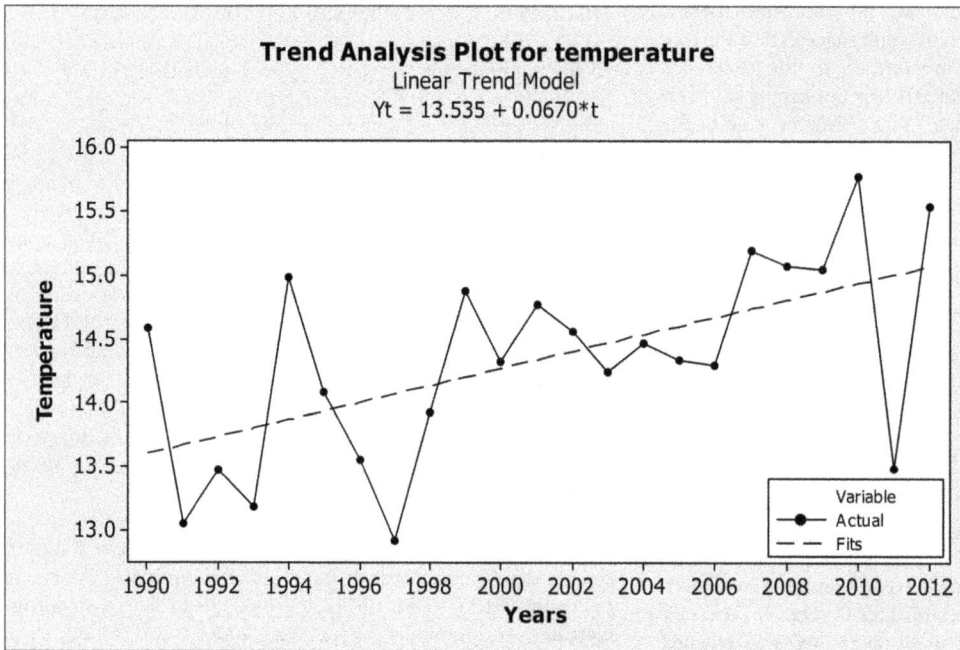

Figure 2. Temperature trend analysis plot for temperature in Dadia National ark. The solid line illustrates the mean annual temperatures from 1990 to 2012 in Dadia National Park, and the dotted line the fitted trend line after 1000 repetitions (randomization). The mean annual temperatures show a general upward trend.

where habitat quality and land use have been kept quite consistent, support our hypothesis that changes in community composition between the sampling periods might be attributable to climate change rather than land use change and therefore explaining the factor of year. A *posteriori* test showed that when a specific habitat type is considered, butterfly communities seem to remain the same between years, suggesting minor changes within the same habitat type (Table S3). Small changes within the same habitat type could be due to more than just a direct impact of climate on the butterflies. Climate can influence the relative abundance of species through direct effects on physiology, growth or survival (e.g. [50,55]), or through indirect effects on the insects by influencing the availability of larval foodplants (e.g. [56,57]). Further investigation into how climate may influence butterfly population dynamics and community structure in Mediterranean terrestrial habitats is needed.

Low-altitude species showed a significant increasing trend in terms of alpha-diversity (see Table 2). This suggests a community response to climate warming, where a shift towards a dominance of lower-elevation species is expected [4,8,10]. Only one species, *M. trivia*, a Near Threatened species at the European Union (EU27) level [58], was a distinct exception. It is a low-altitude species but suffered a dramatic population decline of over 95% (based on its abundance in 1998). Similarly, other *Melitaea* species such as *M.cinxia* have experienced a significant population decline in the Mediterranean (NE Spain) from 1994 to 2008 [17].

High-altitude species showed a significant decreasing trend in terms of alpha-diversity. This represents further evidence of a change in the distribution and abundance of such species towards cooler locations at higher latitudes or elevations [1,6,7]. Two high-altitude species that contributed much to the between-year-difference declined over 80% over the 13-year period, *A. crataegi*

Table 2. Alpha-diversity (mean Shannon index at transect level) and beta-diversity (Whittaker index at habitat level) for (a) all butterfly species, (b) high-altitude species and (c) low-altitude species and respective general linear models testing their significant change between the years 1998 and 2011.

		Year	(a) All species	(b) HA species	(c) LA species
Transects	α-diversity	1998	2.5	1.94	1.73
		2011	2.7	1.68	1.95
	GLM	F	1.26	5.61	4.67
		p-value	0.26	**0.02**	**0.03**
Habitats	β-diversity	1998	0.45	0.62	0.45
		2011	0.51	0.57	0.48
	GLM	F	0.37	0.14	0.07
		p-value	0.55	0.71	0.78

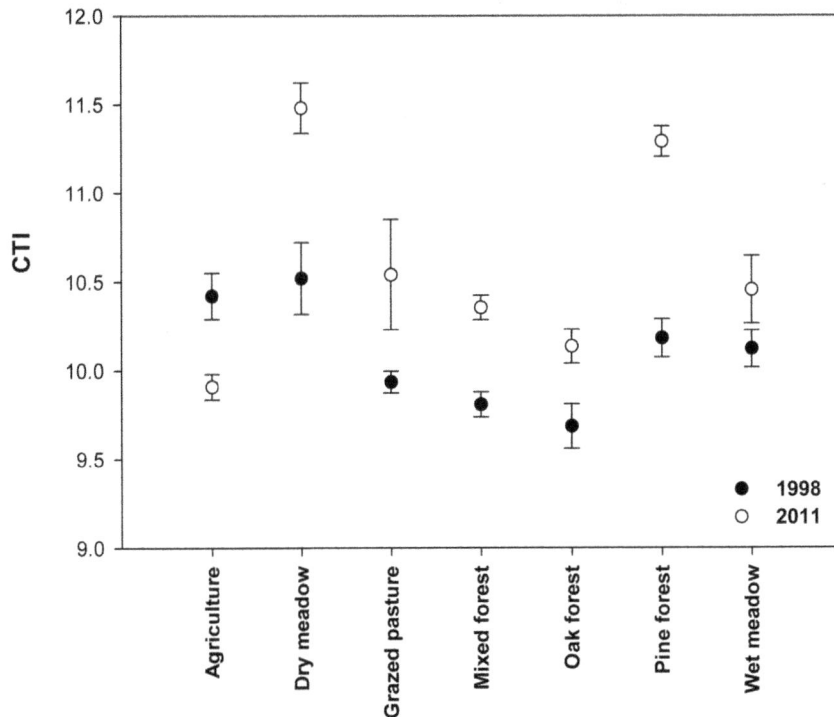

Figure 3. Community Temperature Index (CTI) among the sampled habitats in 1998 and 2011. A Community Temperature Index (CTI, y-axis) was calculated for each one of the seven habitats (x-axis) as the average Species Temperature Index (calculated after the average temperature of each species' geographical range in Europe, see [13,14]) weighted by species' total abundance, sampled in 1998 (filled circle) and 2011 (empty circle) in each of the habitats. Figure shows significant increase of CTI in all habitats except for the agricultural areas.

and *Pieris mannii*. Recent changes in the distribution of *A. crataegi* in Europe appear to reflect effects of both climate and land use change [59]: in central Spain the species has declined at low elevations, leading to an upward altitudinal shift [55]; in Scandinavia it has expanded its range, whereas in central Europe it has suffered serious declines [59]. *P. mannii* is known to have expanded its northern range limit in Switzerland and Germany in association with climate warming [60]. Regional warming cannot, however, explain the significant decline of *Vanessa cardui*, a migrant and 'widespread' species, whose population size is largely regulated by climatic conditions in its overwintering habitat in Africa [61]. Finally, two more high-altitude species in our study area, *Melanargia galathea* and *Coenonympha leander*, were only recorded 6.7 km away to the north-west (800m altitude) from their site of observation in 1998 (mixed forest, 350m altitude), suggesting maybe the first signals of some species' movement to higher altitudes.

Our results showed a significant increase in the butterfly Community Temperature Index of sample sites (see Fig. 3). In contrast with the non significant trends observed in NE Spain [17], our findings suggest that butterfly communities may indeed have responded to regional warming in the Eastern Mediterranean basin, even during a relatively short period of 13 years. Our findings are consistent with similar patterns of increasing Community Temperature Index observed in northern Europe [13,14,62]. Agricultural habitats were the exception to the above general pattern. Here, the butterfly community changed from hotter (1998) to cooler (2011) thermal associations. We attribute this pattern to both the presence of natural hedges and tree lines providing shade at field edges, as well as to irrigation systems, which have recently been found to buffer butterfly communities

against the effects of drought in the Mediterranean [63]. Water availability is a key factor determining the distribution of butterflies and many other taxa in dry, low latitude, ecosystems [64,65] prolonging the "green season" of the field margins and therefore the food resources until late summer. Despite their anthropogenic origin, our evidence suggests that cool or moist microhabitats provided by mosaic agricultural landscapes may play a role in supporting butterfly populations under the increased thermal stress of the summer over a period of climate warming. These anthropogenic features may have enabled populations of butterfly species associated with relatively cool or moist conditions to "bounce back" from the effects of preceding hot years during the relatively cool conditions of the field survey in summer 2011 (see Fig. 2).

Conservation implications

New approaches for species conservation in existing protected areas may be needed as the climate warms [18]. Our study showed that artificially cooled or moist habitats such as in traditional agriculture can support species associated with cooler conditions (low temperature index), through possible effects of irrigation during the dry and hot summers of the South-east Mediterranean (see Fig. 3). Perhaps, preserving traditional small agricultural plots with hedges and tree lines and maintaining the current irrigation system could be a useful approach for increasing resilience to climate change [66]. In addition, in order to accommodate the possible distributional movement of species towards higher altitudes (we observed this for two species, *M. galathea*, *C. leander*, that formerly occurred in the study transects), we propose the future expansion of the existing reserve's borders to the west, towards the South-Eastern hills of the Rhodopi mountains.

Our results demonstrate that a 13 year period of assessment may be adequate to detect responses of butterfly communities in terms of species abundance and thermal structure. Although it is possible that a longer time period may be needed to detect changes in species richness or communities in cold ecosystems of higher latitudes [67], the documented signals even in this relatively short period underline the necessity for systematic research into hotter, low latitude, Mediterranean ecosystems.

The buffer zone of Dadia NP is of greater conservation importance for butterflies than the core areas constituted mainly by pinewoods and designed for the needs of raptors and the black vulture in particular. More than 55% of the regional butterfly species of European conservation concern were recorded in the park's buffer zone. Likewise, the most species-rich sites with the highest conservation importance for Orthoptera [33], orchids [34], passerines, amphibians and reptiles [32] as well as butterflies in 1998 [24] are situated in the buffer zone. Importantly, this research provides further evidence that 'buffer zones' are not only transition zones to unprotected areas, but essential parts of a reserve, contributing to its value for nature conservation. Considering that only a small proportion of total land area can ever be realistically protected in the form of nature reserves, conservation efforts must also comprise the surrounding area of nature reserves considering all components of biodiversity [29]. This becomes particularly important in a changing climate scenario, when species – as we have shown here for butterflies – may leave existing nature reserves or alter their habitat associations in search of more climatically-suitable habitats [18,68].

Supporting Information

Figure S1 Number of species and number of SPEC (Species of European conservation concern) per habitat type (7), per sampling year (1998-2011-2012).

Table S1 Presence absence data of all butterfly species for the 7 habitat types (21 transects) per sampling year (1998-2011-2012).

Table S2 Results of permutational multivariate analysis of variance (PERMANOVA).

Table S3 Results of pair-wise a posteriori test of permutational multivariate analysis of variance (PERMANOVA).

Table S4 Distribution of the 1260 actual localities (corresponding to 5193 points observed by the author or referred to the bibliography on Greek butterfly Atlas) among 4 elevation zones of Greek territory.

Acknowledgments

We are grateful to Dadia's forestry and WWF Greece for supporting the field research. Also, G.C. Adamidis is thanked for valuable advice and comments on the manuscript as well as Pistolas, Christina and Johanne for their support in field. We thank John Haslett (University of Salzburg) for initiating the survey in 1998. Our special thanks to Chris Van Swaay and Josef Settele for allowing us to use the Species Temperature Data (STI). Finally we thank an anonymous reviewer, Constantí Stefanescu and Martin Wiemers for their valuable comments on the manuscript.

Author Contributions

Conceived and designed the experiments: JMH VK KZ. Performed the experiments: KZ. Analyzed the data: KZ JMH RJW. Contributed reagents/materials/analysis tools: AG ET RJW LNP JMH. Wrote the paper: KZ VK AG RJW JMH.

References

1. Hickling R, Roy DB, Hill JK, Fox R, Thomas CD (2006) The distributions of a wide range of taxonomic groups are expanding polewards. Global Change Biology 12: 450–455.
2. Lenoir J, Gégout JC, Marquet PA, De Ruffray P, Brisse H (2008) A significant upward shift in plant species optimum elevation during the 20th century. Science 320: 1768–1771.
3. Root TL, Price JT, Hall KR, Schneider SH, Rosenzweig C, et al. (2003) Fingerprints of global warming on wild animals and plants. Nature 421: 57–60.
4. Wilson RJ, Gutiérrez D, Gutiérrez J, Monserrat VJ (2007) An elevational shift in butterfly species richness and composition accompanying recent climate change. Global Change Biology 13: 1873–1887.
5. Huntley B, Collingham YC, Willis SG, Green RE (2008) Potential impacts of climatic change on European breeding birds. PLoS One 3.
6. Parmesan C, Ryrholm N, Stefanescu C, Hill JK, Thomas CD, et al. (1999) Poleward shifts in geographical ranges of butterfly species associated with regional warming. Nature 399: 579–583.
7. Parmesan C, Yohe G (2003) A globally coherent fingerprint of climate change impacts across natural systems. Nature 421: 37–42.
8. Warren MS, Hill JK, Thomas JA, Asher J, Fox R, et al. (2001) Rapid responses of British butterflies to opposing forces of climate and habitat change. Nature 414: 65–69.
9. Opdam P, Wascher D (2004) Climate change meets habitat fragmentation: Linking landscape and biogeographical scale levels in research and conservation. Biol Conserv 117: 285–297.
10. Beaugrand G, Reid PC, Ibañez F, Lindley JA, Edwards M (2002) Reorganization of North Atlantic marine copepod biodiversity and climate. Science 296: 1692–1694.
11. Dapporto L, Dennis RLH (2013) The generalist-specialist continuum: Testing predictions for distribution and trends in British butterflies. Biol Conserv 157: 229–236.
12. Barry JP, Baxter CH, Sagarin RD, Gilman SE (1995) Climate-related, long-term faunal changes in a California rocky intertidal community. Science 267: 672–675.

13. Devictor V, Van Swaay C, Brereton T, Brotons L, Chamberlain D, et al. (2012) Differences in the climatic debts of birds and butterflies at a continental scale. Nat Clim Chang 2: 121–124.
14. Van Swaay CAM, van Strien A, Harpke A, Fontaïne B, Stefanescu C, et al. (2012) The European Butterfly Indicator for Grassland species 1990–2011. Report VS2012.019. De Vlinderstichting, Wageningen.
15. IPCC (2007) The Third Assessment Report of the Intergovernmental Panel on Climate Change. http://www.ipcc.ch/publications_and_data/ar4/wg1/en/ spmsspm-projections-of.html.
16. Peñuelas J, Filella I, Comas P (2002) Changed plant and animal life cycles from 1952 to 2000 in the Mediterranean region. Glob Change Biol 8: 531–544.
17. Stefanescu C, Torre I, Jubany J, Páramo F (2011) Recent trends in butterfly populations from north-east Spain and Andorra in the light of habitat and climate change. J Insect Conserv 15: 83–93.
18. Araújo MB, Alagador D, Cabeza M, Nogués-Bravo D, Thuiller W (2011) Climate change threatens European conservation areas. Ecol Lett 14: 484–492.
19. Werner U, Buszko J (2005) Detecting biodiversity hotspots using species-area and endemics-area relationships: The case of butterflies. Biodivers Conserv 14: 1977–1988.
20. Catsadorakis G, Källander H (2010) The Dadia-Lefkimi-Soufli Forest National Park, Greece: Biodiversity, Management and Conservation. Athens: WWF Press. 316 p.
21. Debussche M, Lepart J, Dervieux A (1999) Mediterranean landscape changes: Evidence from old postcards. Glob Ecol Biogeogr 8: 3–15.
22. Gerard F, Petit S, Smith G, Thomson A, Brown N, et al. (2010) Land cover change in Europe between 1950 and 2000 determined employing aerial photography. Prog Phys Geogr 34: 183–205.
23. Gatzogiannis S (1999) Guidlines for the preparation of forest management plans. -Forest Research Institute, N.AG.RE.F., Thessaloniki. (In Greek).
24. Grill A, Cleary DFR (2003) Diversity patterns in butterfly communities of the Greek nature reserve Dadia. Biol Conserv 114: 427–436.
25. Stefanescu C, Peñuelas J, Filella I (2009) Rapid changes in butterfly communities following the abandonment of grasslands: A case study. Insect Conserv Diver 2: 261–269.

26. Verdasca MJ, Leitão AS, Santana J, Porto M, Dias S, et al. (2012) Forest fuel management as a conservation tool for early successional species under agricultural abandonment: The case of Mediterranean butterflies. Biol Conserv 146: 14–23.
27. Maris F, Vasileiou A (2010) Hydrology and torrential environment. In: Catsadorakis G, Kälander H, editors. The Dadia-Lefkimi-Soufli Forest National Park, Greece: Biodiversity, Management and Conservation. Athens: WWF Greece. pp. 41–45.
28. Korakis G, Gerasimidis A, Poirazidis K, Kati V (2006) Floristic records from Dadia-Lefkimi-Soufli National Park, NE Greece. Flora Mediterranea 16: 11–32.
29. Kati V, Poirazidis K, Dufrêne M, Halley JM, Korakis G, et al. (2010) Towards the use of ecological heterogeneity to design reserve networks: A case study from Dadia National Park, Greece. Biodivers Conserv 19: 1585–1597.
30. Schindler S (2010) Dadia National Park, Greece - An integrated study on landscape biodiversity, raptor populations and conservation management. Vienna: University of Vienna. 318 p.
31. Kati V, Sekercioglu CH (2006) Diversity, ecological structure, and conservation of the landbird community of Dadia reserve, Greece. Divers Distrib 12: 620–629.
32. Kati V, Foufopoulos J, Ioannidis Y, Papaioannou H, Poirazidis K, et al. (2007) Diversity, ecological structure and conservation of herpetofauna in a Mediterranean area (Dadia National Park, Greece). Amphibia Reptilia 28: 517–529.
33. Kati V, Dufrêne M, Legakis A, Grill A, Lebrun P (2004) Conservation management for Orthoptera in the Dadia reserve, Greece. Biol Conserv 115: 33–44.
34. Kati V, Dufrêne M, Legakis A, Grill A, Lebrun P (2000) Les Orchidées de la reserve de Dadia (Grèce), leurs habitats et leur conservation. Les Naturalistes Belges 81: 269–282.
35. Argyropoulou MD, Karris G, Papatheodorou EM, Stamou GP (2005) Epiedaphic Coleoptera in the Dadia forest reserve (Thrace, Greece): The effect of human activities on community organization patterns. Belg J Zool 135: 127–133.
36. tutiempo website. Available: http://www.tutiempo.net/en/Climate/Alexandroupoli_Airport/166270.htm. Accessed 2011 Dec 15.
37. Adamakopoulos T, Gatzogiannis S, Poirazidis K (1995) Specific Environmental Study of the Dadia Forest Special Protection Area. Parts A+B, C. Ministry of Environment, Ministry of Agriculture, ACNAT. WWF Greece, Athens (In Greek).
38. Colwell RK, Chang XM, Chang J (2004) Interpolating, extrapolating, and comparing incidence-based species accumulation curves. Ecology 85: 2717–2727.
39. Magurran AE (2004) Measuring biological diversity: MA: Blackwell, Malden.
40. Clarke KR (1993) Non-parametric multivariate analyses of changes in community structure. Aust J Ecol 18: 117–143.
41. Anderson MJ (2005) PERMANOVA: A FORTRAN Computer Program for Permutational Multivariate Analysis of Variance. Department of Statistics, University of Auckland, New Zealand.
42. Warton DI, Wright ST, Wang Y (2012) Distance-based multivariate analyses confound location and dispersion effects. Methods in Ecology and Evolution 3: 89–101.
43. Oksanen J, Blanchet G, Kindt R, Legendre P, O'Hara RB, et al. (2010) vegan: Community Ecology Package. R package version 1. 17–2 ed.
44. Wang Y, Naumann U, Wright ST, Warton DI (2012) Mvabund- an R package for model-based analysis of multivariate abundance data. Methods in Ecology and Evolution 3: 471–474.
45. Pamperis LN, Stavridis SK (2009) The Butterflies of Greece. Athens: Pamperis. 766 p.
46. Kampichler C, van Turnhout CAM, Devictor V, van der Jeugd HP (2012) Large-Scale Changes in Community Composition: Determining Land Use and Climate Change Signals. PLoS One 7 (4), art. no. e35272.

47. Filz KJ, Wiemers M, Herrig A, Weitzel M, Schmitt T (2013) A question of adaptability: Climate and habitat change lower trait diversity in butterfly communities in south-western Germany. Eur J Entomol: 633–642.
48. Schweiger O, Harpke A, Wiemers M, Settele J (2014) CLIMBER: Climatic niche characteristics of the butterflies in Europe. ZooKeys, 367: 65–84.
49. Pinheiro J, Bates D, DebRoy S, Sarkar D, Team tRDC (2003) nlme: Linear and Nonlinear Mixed Effects Models. R package version 3. 1–108.
50. Van Swaay C, Cuttelod A, Collins S, Maes D, López Munguira M, et al. (2010) European Red List of Butterflies. Luxembourg: Publications Office of the European Union.
51. Halley JM, Kugiumtzis D (2011) Nonparametric testing of variability and trend in some climatic records. Clim Change 109: 549–568.
52. Matter SF, Doyle A, Illerbrun K, Wheeler J, Roland J (2011) An assessment of direct and indirect effects of climate change for populations of the Rocky Mountain Apollo butterfly (Parnassius smintheus Doubleday). Insect Sci 18: 385–392.
53. New TR (1997) Are Lepidoptera an effective 'umbrella group' for biodiversity conservation? J Insect Conserv 1: 5–12.
54. Robinson N, Armstead S, Deane Bowers M (2012) Butterfly community ecology: The influences of habitat type, weather patterns, and dominant species in a temperate ecosystem. Entomol Exp Appl 145: 50–61.
55. Merrill RM, Gutiérrez D, Lewis OT, Gutiérrez J, Díez SB, et al. (2008) Combined effects of climate and biotic interactions on the elevational range of a phytophagous insect. J Anim Ecol 77: 145–155.
56. Schweiger O, Settele J, Kudrna O, Klotz S, Kühn I (2008) Climate change can cause spatial mismatch of trophically interacting species. Ecology 89: 3472–3479.
57. Pateman RM, Hill JK, Roy DB, Fox R, Thomas CD (2012) Temperature-dependent alterations in host use drive rapid range expansion in a butterfly. Science 336: 1028–1030.
58. Van Swaay C, Maes D, Collins S, Munguira ML, Šašić M, et al. (2011) Applying IUCN criteria to invertebrates: How red is the Red List of European butterflies? Biol Conserv 144: 470–478.
59. Asher J, Warren M, Fox R, Harding P, Jeffcoate G, et al. (2001) The Millennium Atlas of Butterflies in Britain and Ireland. OxfordUK: Oxford University Press. 433 p.
60. Settele J, Kudrna O, Harpke A, Kühn I, van Swaay C, et al. (2008) Climatic Risk Atlas of European Butterflies: BioRisk. 712 p.
61. Stefanescu C, Páramo F, Åkesson S, Alarcón M, Ávila A, et al. (2012) Multi-generational long-distance migration of insects: Studying the painted lady butterfly in the Western Palaearctic. Ecography 36: 474–486.
62. Devictor V, Julliard R, Couvet D, Jiguet F (2008) Birds are tracking climate warming, but not fast enough. P Roy Soc Lond B Bio 275: 2743–2748.
63. González-Estébanez FJ, García-Tejero S, Mateo-Tomás P, Olea PP (2011) Effects of irrigation and landscape heterogeneity on butterfly diversity in Mediterranean farmlands. Agric Ecosyst Environ 144: 262–270.
64. Stefanescu C, Herrando S, Páramo F (2004) Butterfly species richness in the north-west Mediterranean Basin: The role of natural and human-induced factors. J Biogeogr 31: 905–915.
65. Stefanescu C, Carnicer J, Peñuelas J (2011) Determinants of species richness in generalist and specialist Mediterranean butterflies: The negative synergistic forces of climate and habitat change. Ecography 34: 353–363.
66. Morecroft MD (2012) Adapting conservation to a changing climate. J Appl Ecol 49: 546.
67. Menéndez R, Megías AG, Hill JK, Braschler B, Willis SG, et al. (2006) Species richness changes lag behind climate change. P Roy Soc Lond B Bio 273: 1465–1470.
68. Suggitt AJ, Stefanescu C, Páramo F, Oliver T, Anderson BJ, et al. (2012) Habitat associations of species show consistent but weak responses to climate. Biol Lett 8: 590–593.

9

Long Term Effect of Land Reclamation from Lake on Chemical Composition of Soil Organic Matter and Its Mineralization

Dongmei He, Honghua Ruan*

Faculty of Forest Resources and Environmental Science, and Key Laboratory of Forestry and Ecological Engineering of Jiangsu Province, Nanjing Forestry University, Nanjing, Jiangsu, China

Abstract

Since the late 1950s, land reclamation from lakes has been a common human disturbance to ecosystems in China. It has greatly diminished the lake area, and altered natural ecological succession. However, little is known about its impact on the carbon (C) cycle. We conducted an experiment to examine the variations of chemical properties of dissolved organic matter (DOM) and C mineralization under four land uses, i.e. coniferous forest (CF), evergreen broadleaf forest (EBF), bamboo forest (BF) and cropland (CL) in a reclaimed land area from Taihu Lake. Soils and lake sediments (LS) were incubated for 360 days in the laboratory and the CO_2 evolution from each soil during the incubation was fit to a double exponential model. The DOM was analyzed at the beginning and end of the incubation using UV and fluorescence spectroscopy to understand the relationships between DOM chemistry and C mineralization. The C mineralization in our study was influenced by the land use with different vegetation and management. The greatest cumulative CO_2-C emission was observed in BF soil at 0–10 cm depth. The active C pool in EBF at 10–25 cm had longer (62 days) mean residence time (MRT). LS showed the highest cumulative CO_2-C and shortest MRT comparing with the terrestrial soils. The carbohydrates in DOM were positively correlated with CO_2-C evolution and negatively correlated to phenols in the forest soils. Cropland was consistently an outlier in relationships between DOM chemistry and CO_2-evolution, highlighting the unique effects that this land use on soil C cycling, which may be attributed the tillage practices. Our results suggest that C mineralization is closely related to the chemical composition of DOM and sensitive to its variation. Conversion of an aquatic ecosystem into a terrestrial ecosystem may alter the chemical structure of DOM, and then influences soil C mineralization.

Editor: Han Y.H. Chen, Lakehead University, Canada

Funding: This study was supported by the MOST of China (973 Program No. 2012CB416904), the National Science Foundation of China (No. 31170417) and partially supported by PAPD and Collaborative Innovation plan of Jiangsu higher education, Doctorate Fellowship Foundation of Nanjing Forestry University. The funders had no role in study design, data collection and analysis, decision to publish, or preparation of the manuscript.

Competing Interests: The authors have declared that no competing interests exist.

* E-mail: hhruan@njfu.edu.cn

Introduction

Microbial respiration of soil organic matter (SOM) causes a large flux of CO_2 to the atmosphere, and understanding the controls on this flux is a critical component of society's effort to cope with increasing atmospheric CO_2 and climate change [1–2]. Many abiotic and biotic factors affect the global carbon (C) cycle, especially, human activities, such as the fossil fuel combustion and land use change [3]. Of these factors, land use changes, such as deforestation and afforestation, draining of wetlands, converting grassland to arable cropping and reclaiming lakes, directly affect microbial respiration [4–5]. Many studies on the influence of land use change have shown it results in drastic changes to soil C cycling [6–8]. Land use determines the vegetation type grown on the soil, and therefore the type of organic matter input to the ecosystem [5,9]. Land use change can also increase microbial C mineralization, which may exacerbate the trend of global warming and other effects related to climate change [7].

In the late 1950s, reclaiming land from lakes became a common type of land use in China. China is a traditional agricultural country with a growing population, and it was deemed necessary to expand cultivated land to develop agriculture as well as feed an increasing population. Thus, after several decades more than 1.4×10^4 km^2 of land was reclaimed from lakes [3,10]. Taihu Lake as one of the five largest freshwater lakes in China also experienced land reclamation. Taihu Lake is situated in the south of the Yangtze Delta among 30°55′–31°33′N and 119°55′–120°36′E and with a land area of 2.3×10^3 km^2. From the 1950s to 1980s, more than 160 lakes around the Taihu Lake basin were disappeared and the area of water surface had reduced by 13.6% [11]. In recent years, many studies have focused on the effects of reclamation on ecosystem. They observed that many negative consequences caused by reclaiming land from lakes, such as species decline for habitat loss, eliminating the natural buffers and flood control capacity, water pollution and eutrophication caused by agriculture measures and so on [12]. However, there are few studies about the effect of lake reclamation on the properties of soil organic matter in the terrestrial ecosystem and little is known about its impact on soil C cycling.

Solid soil organic matter must pass through the dissolved phase to be decomposed by microbes [13–14]. Although dissolved organic C (DOC) makes up only a small portion of total soil organic carbon (mostly less than 1%), it represents the active or labile C pool [15–17] and turn over more than 4000 times annually [18]. Thus, the DOM pools is closely correlated to the labile fraction of soil organic matter and is a sensitive indicator of its overall dynamics [16,19]. Also, many studies observed the close relationships between biodegradability of DOC and C mineralization [14,16,20].

Land use is considered as the factor with the greatest influence on soil DOM because it determines the quantity and quality of organic matter input into the soil [5]. However, the mechanism of how land use influences dynamics of DOM is still not clear. Several authors have found that biodegradability of DOM largely depends on its chemical composition [21–23]. It is generally assumed that the easily degradable DOM consists mainly of simple carbohydrate monomers (i.e., glucose, fructose), low molecular organic acids (i.e., citric, oxalic, succinic acid), amino acids, amino sugars and low molecular weight proteins [13,18]. The stable DOM fraction generally contains polyphenolic, aromatic structures and other complex macromolecules [24]. So, the chemical structure of DOM and the complexity of its molecules were considered to correlate with C mineralization. The chemical structure of DOM often described with UV and fluorescence spectroscopy [25–26]. Coupling and understanding of the chemical composition of DOM and C mineralization could be a useful tool for evaluating the influence of land use change on soil C dynamics.

In a previous study, Wang et al. [3] reported the labile SOC concentrations were different from each other under four types of land use from land reclaimed from Taihu Lake, China. Here we endeavored to link the C mineralization potential to the chemical makeup of DOM in soils of land that had been converted from lake sediments to forests and cropland. We hypothesized that this conversion of aquatic ecosystems to terrestrial ecosystems resulted in increased recalcitrance in chemical structure of DOM and ultimately affected potential soil C mineralization, because terrestrial organic matter is usually rich in aromatic and phenolic structures while DOM from aquatic environments has a lower content of aromatic C and more aliphatic C [27–28]. Our objectives were to (1) test for differences in C mineralization in lake sediments and terrestrial soils under different land use on reclaimed land; (2) compare the chemical composition of DOM in lake sediments and soils under different land use; (3) relate C mineralization and chemical properties of DOM in order to explore the role of DOM in C cycling in reclaimed soils.

Materials and Methods

Site description

The study was conducted at the Xiaodian Lake Forest (31°10′N, 120°48′E), located in the northeast of Wujiang City, Jiangsu province, southeast of China (fig. 1) (The Department of Agriculture & Forestry of Jiangsu Province is the authority responsible for it. There were no specific permissions required for the study area. Our field studies did not involve endangered or protected species). The climate in this region is humid north subtropical monsoon with a mean annual temperature of 16°C, annual rainfall of approximately 1100 mm (mainly during the summer months), annual average relative humidity of 78%, and an annual non-frost period of up to 240 days [3]. This area used to be a part of the Taihu Lake, and the lake area was converted to farmland in the early 1960s. The soil properties of the reclaimed land were unsuitable for food crops and so, afforestation projects were carried out in 1969 in much of the region. After more than 40 years of forest management, it has developed into a forest park

with 1.33 km² forest-covered area. The dominant forest types in the park are dominated by coniferous forest (*Metasequoia glyptostroboides*), evergreen broadleaf forest (*Cinnamomum camphora*), bamboo forest (*Phyllostachys heterocycla*) and all of them are the planted tree species monoculture. Additionally, there is approximately 7.0 ha of cropland with rotations of rice and canola in parcels around the park and with the common tillage practices of ploughing, irrigation, mineral and organic fertilization. Soils in this region are all derived from lake sediment and were similar following reclamation. The general characteristics of each site were described by Wang et al. [3].

Soil sampling

We chose four sites within coniferous forest (CF), evergreen broadleaf forest (EBF), bamboo forest (BF) and cropland (CL), respectively. Four 10 m×10 m plots were randomly established within each site. Soil samples were collected randomly using a 3 cm diameter soil sampler at 0–10 cm and 10–25 cm depths from each plot in October 2011. After removal of the litter layer, ten soil cores were taken inside each plot and pooled to form a composite soil sample. Visible roots, residues and stones were immediately removed after sampling and then field-moist soil samples were sieved through a 2 mm mesh. Soil samples were divided into two parts: one stored at 4°C for analysis of soil microbial biomass, DOM and C mineralization incubation experiments. The other part was air dried and analyzed for pH, soil total carbon (TC) and total nitrogen (TN). Lake sediments (LS) were sampled with plexiglass tube (11 cm I.D., 50 cm in length) by cylindrical sampler (Rigo Co. Φ110 mm×500 mm) at four points in the water area of the lake. Sediments in the tube were push-back and sliced at 10 cm for 0–10 cm depth and then 15 cm for 10–25 cm depth using a stainless steel spatula [29–30]. A total of 40 sediment samples were collected and kept in room temperature of 4°C.

Soil physical and chemical characteristics

Soil bulk density was measured with the core method in each layer by using a 5 cm×5 cm metal cylinder [31]. Soil moisture was determined gravimetrically by oven drying at 105°C for 24 h. Soil pH was measured on air-dried soil in a 1:2.5 (w/w) soil-water suspension using a glass electrode. Soil TC and TN were determined via dry combustion and thermal conductivity detection using an elemental analyzer (Vario EL III, Elementar, Germany). Microbial biomass carbon (MBC) was determined by chloroform fumigation as described by Vance et al. [32].

Incubation experiment and analytical methods

Carbon mineralization. C mineralization was measured in each soil sample during incubation in the dark at 25°C for 360 days [33]. From each incubation sample, 100 g fresh moist soil was incubated in a 1000 ml sealed Mason jar. A 50 ml beaker containing 20 ml 1 M NaOH solution was placed in each jar to trap the evolved CO_2. Three additional jars containing 20 ml 1 M NaOH solution but no soil were used as a control. During incubation, the evolved CO_2 trapped in NaOH was determined by titration with 1 M HCl after precipitating the carbonate with 1 M $BaCl_2$ solution. After the CO_2 traps were taken out, the jars were left open for 2 h to maintain aerobic conditions in each jar, and resealed for further incubation. Water loss in the jars was monitored by weighing the jars and replenished by adding ultrapure water after opening [34]. The cumulative C mineralization was expressed as g CO_2-C kg^{-1} soil.

DOM extraction and characterization. DOM was extracted from each soil sample before and after soil incubation. 10 mM

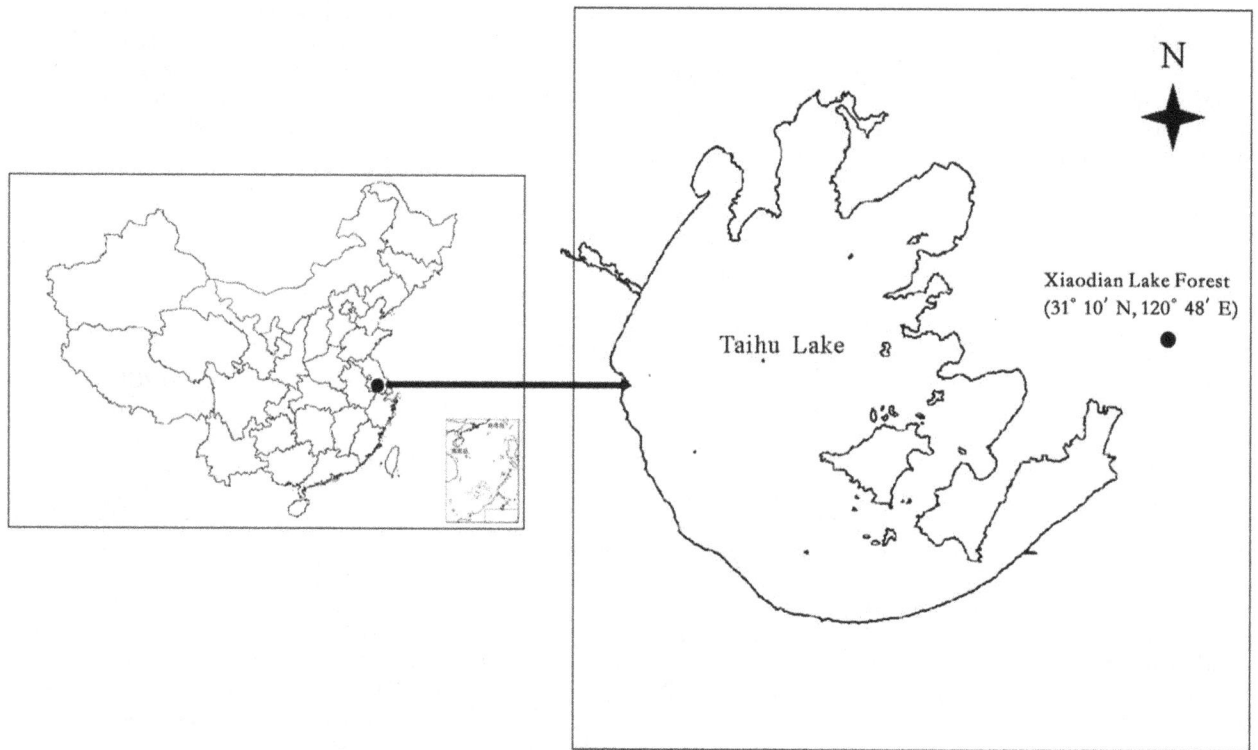

Figure 1. The map of study site in Xiaodian Lake area of the Taihu Lake basin China.

CaCl$_2$ solution was added to an aliquot of soil samples at a soil-solution ratio of 1:2 (w/v) and shaken for 30 min in a horizontal shaker at room temperature. The suspension was spun in a centrifuge for 20 min at 4000 rpm and filtered through a 0.45 μm pore-size cellulose acetate membranes (Schleicher & Schuell, OE 67). The filtered solutions were stored frozen (−20°C) for additional analyses.

DOC concentration was determined by Shimadzu TOC-5050 total organic carbon analyzer. Hydrophilic and hydrophobic fractions in DOM were separated by Amberlite XAD-8 resin (Rohm and Haas, Philadelphia, PA) [35]. Briefly, DOM solutions were acidified to pH 2 with HCl, and then eluted through glass columns filled with Amberlite XAD-8 resin. The column effluent representing the hydrophilic fraction of DOM (Hi) was analyzed for organic C and the amount of organic C in hydrophobic fraction was calculated by the difference between organic C in DOM solution and in hydrophilic fraction. Total carbohydrates in DOM (CH) were measured by phenol-sulfuric acid method [36], using glucose as a standard. Total soluble phenolic compounds (Phe) were analyzed by Folin-Ciocalteau method [37], using tannic acid as a standard.

For spectroscopic measurements, all samples were diluted to 10 mg C L^{-1} to avoid concentration effects and were brought to a constant pH of 7.7 [21] by adding NaOH. UV absorption at 254 nm of DOM was measured using a UV/Vis spectrophotometer (Shimadzu UV-2550, Japan). The specific UV absorbance values (SUVA$_{254}$) were determined as the ratio of the UV absorbance at 254 nm to the DOM concentration and multiplying the value of 100, i.e. (UV$_{254}$/DOC)×100. SUVA$_{254}$ can serve as an indicator of the aromaticity of DOM.

Fluorescence emission spectra were obtained with a Varian Cary Eclipse Fluorescence Spectrophotometer (λ_{ex} 254 nm, slit 10 nm, λ_{em} 300–480 nm, slit 10 nm, and scan speed 1200 nm

min^{-1}) using 1 cm cuvettes. Fluorescence efficiency index (FE) proposed by Ewald et al. [38] was considered to be proportional to the quantum efficiency and was defined as the ratio of maximum fluorescence intensity (F$_{max}$) divided by UV absorption at the excitation wavelength of 254 nm (F$_{max}$/Abs). The emission humification index (HIX$_{em}$, dimensionless) was determined as the ratio between the area in the upper quarter (Σ435–480 nm) of the usable fluorescence emission spectrum and the area in the lower usable quarter (Σ300–345 nm) [39–41]. The HIX$_{em}$ can show the degree of complexity and condensation of the DOM.

Statistical analysis

The double exponential model [20,21,42], separating the mineralized organic C into an active C pool and slow C pool can be presented as:

$$C_m(\%) = a \times [1 - \exp(k_1 t)] + (100 - a) \times [1 - \exp(-k_2 t)]$$

where, t is incubation days; C_m is cumulative value of mineralized C at t time presented as percent of initial C in the soil; a is the portion of organic C that is readily decomposed (in % of initial C in the soil = labile C); $(100 - a)$ is the part of organic C that is slowly decomposed (stable C); k_1 and k_2 are mineralization rate constants for active and stable C pools (day^{-1}). Mean residence time (MRT) for each pool is calculated as the reciprocal of the decomposition rate constant in the double exponential model.

This double decomposition equation for two distinct C pools with different mineralization rate constants was fitted with the nonlinear regression that was used in the Marquardt algorithm and an iterative process to find the parameter values that could minimize the residual sum of squares. The model that gave the least squared error was chosen to be the best. For a model to be chosen, all the parameters had to generate plausible results in realistic ranges. For example, the rate constants could not be negative, and the sum of active and stable C pools should not exceed initial C in the soil.

All results were reported as the mean ± SE of the four field replicates. Differences in soil properties, mineralization parameters and DOM chemical properties among land use types for each site were tested using ANOVA followed by least significant difference (LSD). T-test was used to assess the differences between soil depths. Statistical significance was determined at $p < 0.05$ level. Linear regression analysis was used to determine relationships between C mineralization and chemical properties of DOM. We calculated Pearson correlation coefficients among properties of soil, C mineralization parameters and the chemical characteristics of DOM to discuss the influence of selected parameters on C mineralization. All statistical analyses were performed with SPSS 13.0.

Results

Soil properties

We observed variation in several soil characteristics under different land uses and soil depth (Table 1). Soil bulk density was lower at 0–10 cm soil depth than 10–25 cm depth and BF was significantly lower than all other land uses ($p < 0.05$). Soil pH in all soils and sediments were ranging from 4.13 to 6.30 with the highest value in LS. TC and TN concentrations from the terrestrial soils were significantly higher at 0–10 cm soil depth (TC 14.95–35.35 g C/kg soil, TN 1.77–3.41 g N/kg soil), but both of these two factors for LS were lower at upper layer. BF soils were observed remarkably higher TC and TN than other land uses. There was significant difference in MBC and DOM between the land use types and soil depth (Table 1). MBC and DOM were significantly higher at 0–10 cm soil depth in terrestrial land uses. Under the four land uses, MBC showed highest value in CL and lowest value in CF at both soil depths, while DOM was significantly greater in CF at 0–10 cm depth and significantly lower in EBF at 10–25 cm depth. MBC and DOM in LS showed an inverse trend compared with the other land uses, which was greater at 10–25 cm depth.

Chemical characteristics of DOM

The chemical composition of DOM was compared for each sample before and after incubation (Table 2). Before incubation, the fraction of CH at 0–10 cm soil depth ranged from 11.25 to 28.39% of total DOM, with the highest value in BF. The Phe at 0–10 cm soil depth ranging from 1.06 to 3.74% of total DOM was significantly larger than at 10–25 cm depth, with the highest value in CL. The Hi fraction at 0–10 cm depth was varied in the land uses with the highest in CL (61.75%) and lowest in BF (21.22%). At 10–25 cm soil depth, the fraction of CH in BF and LS was significantly larger than the other soils. The largest Phe concentration was observed in LS (3.49%). The proportion of hydrophilic C at 10–25 cm soil depth was no significant difference between each site, but it is significantly higher than at 0–10 cm depth except for CL soil.

From the initial samples before incubation, the spectral indicators were found to be significantly different among the land uses and between soil depths (Table 2). At 0–10 cm depth, $SUVA_{254}$ was significantly greater in CF while FE was significantly greater in CL. At 10–25 cm soil depth, $SUVA_{254}$ and HIX_{em} were both lower in CL. Difference in FE at 10–25 cm was statistically significant between each site. There were no significant differences between soil depths in $SUVA_{254}$ for EBF and LS, but FE for all the

Table 1. Soil total C concentration (TC), total N concentration (TN), dissolved organic M (DOM), microbial biomass C (MBC), pH and the bulk density of different land uses soils collected on reclaimed land from Taihu Lake, China.

Site	Land use	TC (g C/kg Soil)	TN (g N/kg Soil)	DOM (mg C/kg Soil)	MBC (mg C/kg Soil)	pH	Bulk density (g cm^{-3})
Depth							
0–10 cm	CF	19.12±1.64b*	2.24±0.15b*	126.63±2.75a*	214.50±5.77d*	4.45±0.07d*	1.22±0.03a*
	EBF	16.05±0.69bc*	1.77±0.04b*	93.50±2.56d*	370.40±9.64c*	4.78±0.08c	1.18±0.03a
	BF	35.35±1.79a*	3.41±0.22a*	110.50±2.40b*	423.01±5.13b*	4.13±0.06e*	0.89±0.06b
	CL	14.95±0.96c*	1.84±0.10b*	105.80±2.81bc*	521.98±7.72a*	5.67±0.08b*	1.16±0.03a*
	LS	16.14±1.38bc	2.26±0.23b	99.95±2.11cd*	350.03±7.64c*	6.19±0.02a	Nd
10–25 cm	CF	8.34±0.53c	1.18±0.07b	97.72±4.50b	112.77±4.43a	4.89±0.15bc	1.31±0.01a
	EBF	8.81±0.77c	1.16±0.08b	78.39±3.06c	309.94±3.90c	4.98±0.07b	1.25±0.02a
	BF	23.76±0.94a	2.59±0.25a	95.88±2.46b	232.28±4.65d	4.30±0.02c	0.99±0.06b
	CL	9.30±1.16c	1.23±0.13b	91.40±3.13b	455.99±5.20b	6.18±0.16a	1.31±0.03a
	LS	17.51±1.03c	2.34±0.13c	115.48±4.21a	558.82±10.28e	6.30±0.07a	Nd

Values are mean ± SE (n = 4). Means within a column of the corresponding depth followed by different letters are significantly different and * indicates the significant difference between the soil depth. (Significance at $p < 0.05$). CF: coniferous forest; EBF: evergreen broadleaf forest; BF: bamboo forest; CL: cropland; LS: lake sediment; Nd: Not determined.

Table 2. Chemical properties of soil DOM at two soil depths under different land uses reclaimed from Taihu Lake, China.

Site Depth	Land use	DOM	SUVA$_{254}$ (l mg C^{-1} m^{-1})	FE	HIX$_{em}$	CH (% of total DOM)	Phe	Hi
0-10 cm	CF	Initial	2.30±0.06a*	321.28±2.28b*	1.76±0.11ab	23.86±1.21b	2.50±0.04b*	32.71±1.38c*
		Δ	2.65±0.05a*	29.62±1.95c	4.17±0.11a*	32.87±2.02a*	4.68±0.22a*	12.15±2.20b
	EBF	Initial	1.00±0.09d	301.18±3.40b*	1.64±0.03b	21.83±0.49b*	2.08±0.04c	37.47±1.35b*
		Δ	2.38±0.12b*	95.03±5.61b*	4.50±0.69a*	25.40±0.51b*	2.81±0.17b*	8.72±1.91b
	BF	Initial	2.07±0.07b*	252.31±3.26d*	1.94±0.07a	28.39±1.13a	2.14±0.06c*	29.22±1.14c*
		Δ	1.05±0.06c	216.67±6.46a*	5.59±0.70a*	4.70±0.37c*	1.37±0.10c*	19.90±2.17a
	CL	Initial	1.20±0.04c*	430.40±2.29a*	1.11±0.08c*	18.98±1.02c*	3.74±0.12a*	61.75±0.70a
		Δ	0.55±0.06d*	-59.55±6.02e*	2.78±0.43b	6.09±0.37c*	-2.61±0.11e*	-3.61±1.35c*
	LS	Initial	1.15±0.06c	289.30±17.80c*	1.68±0.11b	11.25±0.72d*	1.06±0.07d*	59.58±1.19a*
		Δ	0.47±0.07d*	-4.65±1.88d	0.79±0.01c	-4.00±0.31d*	0.25±0.07d*	10.87±3.92b*
10-25 cm	CF	Initial	0.98±0.05b	253.32±1.30d	1.69±0.02a	21.37±0.46b	1.33±0.13c	47.77±1.12a
		Δ	0.60±0.07b	24.40±2.49bc	1.84±0.23c	6.74±0.17a	1.72±0.06a	7.31±1.48c
	EBF	Initial	1.20±0.06a	408.93±4.16a	1.07±0.02b	15.35±0.79c	2.18±0.09b	52.48±0.70b
		Δ	0.18±0.06c	11.25±2.46c	2.24±0.15b	2.78±0.38b	0.60±0.03a	6.98±2.83c
	BF	Initial	1.30±0.06a	170.44±2.17e	1.88±0.02a	26.14±0.81a	1.51±0.02c	46.18±0.49a
		Δ	1.10±0.04a	87.96±2.75a	2.85±0.45a	-2.14±0.85c	0.42±0.05a	17.15±0.69b
	CL	Initial	0.78±0.05c	316.91±2.46c	0.48±0.02c	11.28±0.55d	0.92±0.04d	56.06±1.65a
		Δ	0.15±0.05c	25.32±6.10b	3.52±0.16a	3.61±0.50b	-0.08±0.02c	18.33±3.58b
	LS	Initial	1.23±0.05a	350.29±11.95b	1.60±0.27a	24.81±0.37a	3.49±0.12a	40.76±1.08a
		Δ	0.20±0.04c	-48.88±11.18d	0.40±0.23d	-13.39±0.80d	0.38±0.09b	43.26±3.23a

Results shown are characteristics of initial DOM and the variation of each characteristic at the end of 360 days incubation. Values are mean ± SE (n = 4). Means within a column of the corresponding depth followed by different letters are significantly different and * indicates the significant difference between the soil depth. (Significance at $p < 0.05$).

CF: coniferous forest; EBF: evergreen broadleaf forest; BF: bamboo forest; CL: cropland; LS: lake sediment; SUVA$_{254}$: specific UV absorbance at 254 nm; FE: fluorescence efficiency ($F_{max}/A254$); HIX$_{em}$: humification index using emission fluorescence spectra (ratio of areas: 435–480 nm/300–345 nm); CH: carbohydrate C; Phe: phenol C; Hi: hydrophilic C; initial: properties of initial samples (soils before incubation); Δ: variation of DOM chemical properties between the initial and final value during incubation.

sites were observed significantly different between soil depths. Differences in HIX_{em} between soil depths were only found in CL.

After 360 days incubation, there were significant amounts of variation in DOM chemical properties among the land uses and lake sediments and most of the variation between soil depths was also significant (Table 2). The proportion of CH at 0–10 cm soil depth increased 17%–138% in the land uses with significantly greater value in CF. The proportion of Phe and Hi respectively increased 23%–188% and 18%–68% for CF, EBF, BF, and LS at 0–10 cm depth, but the values of these two factors for CL showed decreased trend after incubation. The increased values of Phe were significantly different from each site with the greatest variation in CF, while increased amount of Hi was lager in BF, at 0–10 cm depth. The proportion of CH fraction had the largest increase in CF, but it was observed reducing in BF and LS, at 10–25 cm soil depth. At this layer, both the proportion of Phe and Hi increased after incubation, and the largest increase was found in LS. There was no significant difference in the variation of Phe among the three forest sites. $SUVA_{254}$ and HIX_{em} from the samples at the end of incubation were increased at both 0–10 and 10–25 cm soil depth (Table 2). CF had the greatest variation in $SUVA_{254}$ and HIX_{em} at 0–10 cm depth and the variation of FE in BF was larger the other sites. There was also no significant difference in the variation of HIX_{em} among the three forest sites. The variation of FE values was significantly different between each other site and the FE values in CL and LS decreased after incubation. At 10–25 cm depth, the variation of $SUVA_{254}$ and FE in BF was significantly larger than other sites.

Carbon mineralization

The cumulative CO_2 production under different land uses and sediments ranged from 0.88 to 7.72 g CO_2-C kg^{-1} soil and there were significant difference between soil depths (Fig. 2a). At 0–10 cm soil depth, cumulative C mineralization in BF soils (2.87 g CO_2-C kg^{-1} soil) was significantly greater than other land uses and CL had the lowest amounts of cumulative CO_2-C (1.62 g CO_2-C kg^{-1} soil). The cumulative CO_2-C for the three forest sites decreased with the increasing soil depth. At 10–25 cm soil depth, C mineralization was found to be significantly different in the following order of CL>BF>CF>EBF (Table 3). Representing the precursor of the four terrestrial soils, LS had different C mineralization patterns than the four terrestrial soils, and its evolution of cumulative CO_2-C (4.59 and 7.72 g CO_2-C kg^{-1} soil, respectively) was significantly larger than other soils, at 0–10 and 10–25 cm soil depth. After 360 days of incubation, the percentage of cumulative mineralized C (as % of TC) ranged from 5.11% to 47.19% with the highest C mineralization in LS at 10–25 cm depth (Fig. 2b). At 0–10 cm soil depth, the percentage of mineralized C was significantly different between each other site with largest value in EBF (15.23%) and lowest value in BF (8.11%). At 10–25 cm soil depth, CL had the highest percentage of cumulative mineralized CO_2-C (21.07%) among the four land uses, approximately four times larger than BF, which had the lowest percentage of mineralized C (5.11%).

The rate of CO_2-C evolution from all land use types were highest at the beginning and then decreased progressively with the advancement of time (Fig. 3). The CO_2-C evolution patterns in all land uses and lake sediments were best described by the double exponential model, with the model fits resulting in r^2 values between 0.96 and 0.99 (Table 3). The size of the labile C pool in different soils was comprised of 0.2%–7.09% with the highest value in LS at 10–25 cm depth. The k_1 values (mineralization rate constant) of labile C at 0–10 cm depth ranged from 0.02 to 0.19 with the mean residence time (MRT_1) of 21 days (average of all

soils), varying from 0.02 to 0.12 at 10–25 cm depth with an average MRT_1 of 33 days. Determination of the size and turnover of the stable C pool showed values at 0–10 cm soil depth were larger than those at 10–25 cm soil depth (Table 3). The mineralization rate constants of the stable C pool (k_2) were 2–3 orders of magnitude lower than k_1, ranging from 0.00022 to 0.00098 in upper soil layer with an average of MRT_2 of 9 years and ranging from 0.00011 to 0.00129 in lower soil layer with an average of MRT_2 of 12 years. At 0–10 cm soil depth, the mean residence time of active and stable C pools in CL sites was larger than the others, whereas at 10–25 cm soil depth, the largest mean residence time of stable C pools was found in BF. The mean residence time of active and stable C pools in LS was lower than all other land use types.

C mineralization in relation to DOM characteristics

C mineralization was significantly correlated with DOM characteristics and its change (the variation of DOM properties between beginning and ending of incubation) under the four land uses (Table 4, Fig. 4). The CO_2-C evolution at the 0–10 cm soil depth had a positive correlation with the proportion of CH, variation of the Hi fraction (ΔHi), fluorescence efficiency variation (ΔFE) and HIX_{em} variation (ΔHIX_{em}), while had a negative correlation with initial Phe, Hi, FE and variation of CH (ΔCH), Phe (ΔPhe) and $SUVA_{254}$ ($\Delta SUVA_{254}$) (Fig. 4). The high proportion of labile C pool (a value of the double exponential model) from upper soil was related to the high proportion of Phe, Hi fraction and high values of FE. In contrast, it was negatively related to most of the parameters including CH fraction, ΔPhe, ΔHi, HIX_{em}, $SUVA_{254}$, ΔFE, ΔHIX_{em} and $\Delta SUVA_{254}$ (Table 4). The k_1 values of upper soils were correlated very well with proportion of Phe and Hi, the values of FE, HIX_{em}, ΔFE and ΔHIX_{em}, whereas the k_2 values were only correlated with Phe, ΔCH, ΔPhe and $\Delta SUVA_{254}$ (Table 4).

In 10–25 cm soil layer, cumulative CO_2-C positively correlated with proportion of CH and the values of HIX_{em}, ΔHi, ΔFE, ΔHIX_{em}, $\Delta SUVA_{254}$, while negatively correlated with proportion of Phe and Hi, as well as the values of FE and $SUVA_{254}$ (Table 4). The proportion of labile C pool showed positive correlations with Phe and FE, and showed the negative relationships with ΔHi, HIX_{em} and ΔFE. The k_1 values highly related to proportion of CH, Phe and Hi, as well as values of $SUVA_{254}$ and ΔHIX_{em}. The relationships between k_2 values and the properties of DOM were similar with the k_1 values (Table 4).

Discussion

Chemical characteristics of DOM

The decomposition rate of organic matter is in part controlled by its chemical fractions and structures [13,21]. The highest portion of CH fraction in BF was due to a lot of litter and root exudate input. For the four land uses, the percentage of CH was observed lower in CL, which may be explained by two reasons: (i) soil microbial community may deplete CH fraction to a greater degree because fertilization can increase the microorganism activity [9] and (ii) the type of crop residue input may influence the chemical composition of DOM [9]. Different from the land uses, the CH for LS was greater at 10–25 cm, probably due to the deeper sediments in anoxic condition would limit the microbial activity [43]. The lowest proportion of phenol C of the DOM from the un-incubated soils was found in LS at 0–10 cm soil depth, possibly because most of the DOM is derived from algae, bacteria and macrophytes, which contain lower phenolic groups [44]. The highest percentage of Hi fraction in CL agreed with Chantigny [5]

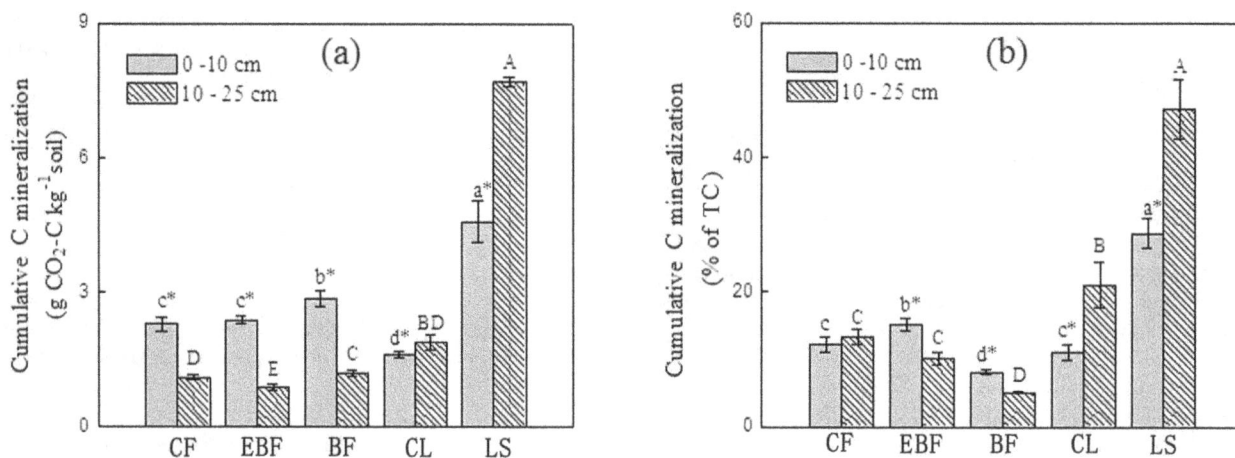

Figure 2. Carbon mineralized at the end of the 360 days incubation period for different land uses and lake sediments. Bars represent standard errors (n = 4). Different letters above bars indicate significant differences ($p<0.05$) of mean values of different sites. * indicates the significant difference ($p<0.05$) between the soil depth.

who reported that DOM from crop residue contained more hydrophilic fractions. During the incubation period, the quality and quantity of organic matter in soils changed, which should lead to variation in the chemical structure of DOM. The proportion of Hi fraction, CH and Phe in DOM under most of the sites were observed to increase at the end of incubation, indicating that some insoluble components were converted to soluble fractions, possibly due to desorption of adsorbed organic matter or other biotic regulatory mechanisms [13,45]. Kalbitz et al. [24] reported that relative increase in polysaccharides after incubation is likely caused by microbial formation and that many bacteria and fungi can release diverse polysaccharides. The relative enrichment in phenols was possibly due to oxidation and fragmentation of lignin in soils [46]. Although both of the percentage of labile and stable fractions in DOM increased at the end of the incubation due to complement of soil organic matter [20], there was a greater increase in the proportion of Phe than CH and Hi fractions. The larger increase in the proportion of Phe at the end of incubation suggested that phenols were the most stable fractions, even though these compounds can experience partial degradation [47]. Our results suggested that CH and Hi fractions were preferentially utilized by microorganisms while phenols were resistant to degradation during incubation. Unlike the other sites, both the proportion of Phe and Hi fractions in CL DOM decreased slightly, probably due to the influence of management practices such as fertilization which changed the soil environment and accelerated both of them degradation [9,48].

The specific UV absorbance at 254 nm ($SUVA_{254}$), humification indices (HIX_{em}) and fluorescence efficiency indices (FE) have been used to characterize the content of aromatic structures and complexity degree in DOM [49–51]. In our study, $SUVA_{254}$ and HIX_{em} decreased with soil depth, consistent with Corvasce et al. [52] and Bu et al. [53], suggesting that the aromatic fraction of partially degraded lignin-derived compounds might be gradually adsorbed by upper mineral soil and protected from microbial degradation [54]. CF and BF soils had the highest $SUVA_{254}$ and HIX_{em} values, indicating that there were more complex and condensed polyaromatic structures in the DOM due to the presence of ligninolitic compounds [55]. Our results agree with Khomutova et al. [56] who also found values of specific UV absorbance at 260 nm were larger in coniferous forest soils compared to deciduous forest and pasture soils, suggesting that

DOM under conifer soils is enriched in hydrophobic aromatic compounds [57]. Just like $SUVA_{254}$ and HIX_{em}, fluorescence efficiency (FE) was also used to express the degree of condensation of DOM. The low FE in BF soils may be partially attributed to its highly substituted aromatic structural features and its inter- or intra-molecular bonding which could result in self-quenching within macromolecules [26]. Ewald et al. [38] drew a similar conclusion that fluorescence efficiency had an inverse linear correlation with molecular weight of the fulvic acid fractions, which was due to internal quenching in macromolecules caused by energy transfers. The smaller values of $SUVA_{254}$ and HIX_{em} as well as larger value of FE were in CL consistent with Chantigny [5] who reported there were more lignin and other recalcitrant compounds in forest litter compared to crop residues. The increase of $SUVA_{254}$ and HIX_{em} under different sites suggested aromatic structures were accumulating with the decomposition of the labile components of DOM during incubation [21,50]. Hur et al. [27,44] also found the increase of SUVA values for all the DOM samples after microbial incubation and attributed this result to preferential microbial utilization of non-aromatic fractions and/or to microbial transformation of labile compounds into aromatic C structures. At 0–10 cm soil depth, the increase of $SUVA_{254}$ and HIX_{em} in CL were significantly smaller than the forest soils, indicating the same conclusion that DOM in CL contains smaller complex compounds. Although the changes in spectra parameters during incubation exhibited increasing trends under most land use types, the increased DOM for LS was smaller than other land uses. This result suggests that the accumulation of condensed aromatic structures was relatively small during incubation, supporting the previous statement that accumulation of phenols was small compared to hydrophilic fractions for LS DOM. For the soil DOM samples, a pronounced increase in FE value was inconsistent with the changes in HIX_{em} or $SUVA_{254}$, which couldn't be explained by fluorescent quenching. This inconsistency may be attributed to the characterization method of FE influenced by the changes in DOM properties.

Soil carbon mineralization in relation to DOM characteristics

The C mineralization was significantly higher at 0–10 cm soil depth regardless of vegetation types for all land uses except CL.

Table 3. C mineralization kinetics of soil after 360 days incubation at 25°C: cumulative CO_2-C, sizes of the labile and stable C pools, mineralization rate constants and mean residence times of the labile and the stable C pools.

Site		C_m	Labile C	Stable C	k_1	k_2	MRT_1	MRT_2	r^2
Depth	Land use	(%)	(%)	(%)	(day^{-1})	(day^{-1})	(days)	(years)	
0–10 cm	CF	12.24	0.75	99.25	0.03	0.00036	30.48	7.68	0.993
	EBF	15.23	0.72	99.28	0.13	0.00048	7.82	5.80	0.979
	BF	8.11	0.20	99.80	0.10	0.00023	10.17	12.13	0.998
	CL	10.95	2.62	97.38	0.02	0.00022	53.13	12.60	0.999
	LS	28.71	1.13	98.87	0.19	0.00098	5.15	2.85	0.983
10–25 cm	CF	13.45	3.07	96.93	0.03	0.00032	33.27	8.67	0.995
	EBF	10.11	4.81	95.19	0.02	0.00016	61.71	17.80	0.999
	BF	5.11	1.42	98.58	0.02	0.00011	48.62	26.11	0.999
	CL	21.07	1.70	98.30	0.07	0.00065	13.53	4.26	0.993
	LS	47.19	7.09	92.91	0.12	0.00128	8.11	2.16	0.957

CF: coniferous forest; EBF: evergreen broadleaf forest; BF: bamboo forest; CL: cropland; LS: lake sediment; C_m: cumulative mineralized C as percentage of initial C in the soil. Labile C: rapidly mineralizable C (calculated using a double exponential model); Stable C: slowly mineralizable C (calculated using a double exponential model); k_1: mineralization rate constant of the labile C pool (double exponential model); k_2: Mineralization rate constant of the stable C pool (double exponential model); MRT_1: Mean residence times of the labile C pool ($MRT_1 = 1/k_1$); MRT_2: Mean residence times of the stable C pool ($MRT_1 = 1/k_2$); r^2: coefficient of determination of the double exponential model.

Values are mean (n = 4).

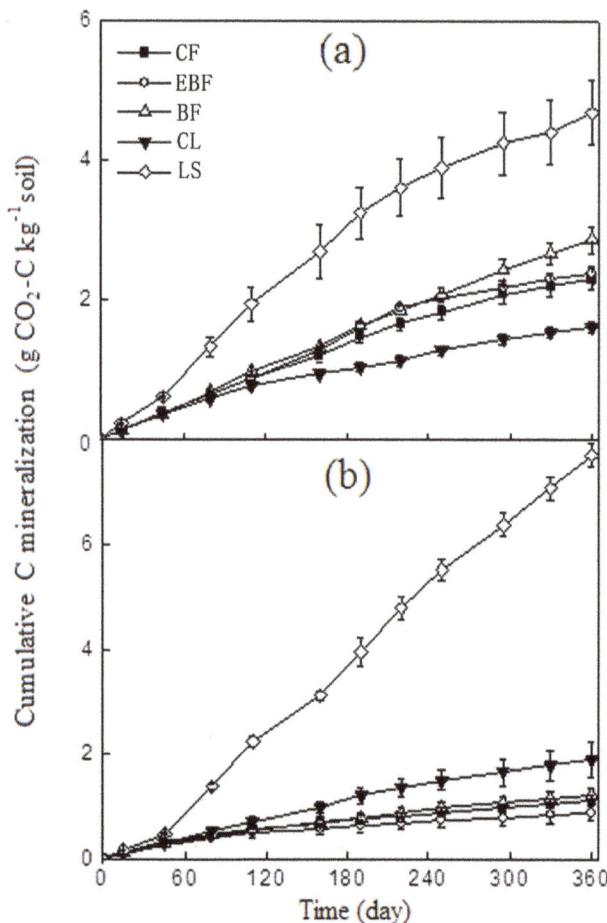

Figure 3. Dynamics of C mineralization of soils under different land use types reclaimed from Taihu Lake, China. (a): C mineralization at 0–10 cm depth; (b): C mineralization at 10–25 cm depth. Bars indicate the standard errors (n = 4).

The higher C mineralization observed at 10–25 cm soil depth in CL was likely due to the tillage practice of ploughing which stimulated the C mineralization in deeper layers [58]. At 0–10 cm soil depth, cumulative C mineralization in BF soils was greater than other land uses, also possibly due to plenty of SOC deriving from degradation of bamboo litter and rhizomatous. While cumulative C emission for CL was significantly lower than the forest soils at 0–10 cm and Zhao et al. [20] observed the similar result that the average cumulative mineralized CO_2-C in forest sites and arable sites was 382 and 279 mg C/kg soil, respectively. The amount of cumulative mineralized CO_2-C during 360 days incubation was significantly higher in lake sediments compared to the terrestrial soils. This result was consistent with our primary hypothesis. Mora et al. [33] observed that during 10 days incubation 0.5% of the sediment C, as well as 0.25% of soil C was mineralized.

The analysis of double exponential model showed that emission of CO_2-C increased as the labile C pool (a) decreased, which was inconsistent with the results of Kalbitz et al. [21] and Bu et al. [59]. During the incubation period, lake sediments had a higher percentage of C mineralized (as % of TC), a larger pool of labile C, and shorter residence times for labile and stable C. This suggested that lake sediments were enriched in less complex, more easily mineralized forms of C [60–62]. The smaller fraction of

mineralized C and lower mineralization rate constants for the stable C pool found in BF soils suggested that organic C in BF soils was composed mainly of polyphenols and aromatic structures deriving from the decomposition of bamboo rhizomes and recalcitrant to biodegradation [63–65]. Compared to forest soils, both labile and stable C pools in CL soils had lower mineralization rate constants, which was different from the result of Kalbitz et al. [21]. This discrepancy was likely explained by DOM representing a small fraction of labile C in CL soils, and mineralization of soil C can't be explained only by consumption of DOM. Zhao et al. [20] also found no correlations between the decrease in quantity of DOM and soil C mineralization. At 0–10 cm depth, mineralization rate constants for both the labile and stable C pool of CF soils were significantly smaller than EBF soils, consistent with previous studies which found C mineralization rate in a conifer forest to be lower than in an evergreen broadleaf forest [66].

In our study, we found the chemical compositions and spectra characteristics of DOM were correlated with C mineralization, but the relationship varied across land uses. The cumulative CO_2-C emission showed a significant positive correlation with proportion of CH at the beginning of the incubation and a significant negative relationship with the proportion of initial Phe. The results suggest that CH and Phe respectively represent the labile and refractory fractions in DOM, and they may regulate the mineralization potential of SOM. Soil carbohydrates are readily available for biodegradation and often have short half-lives due to rapid uptake and assimilation by soil microbial communities [22]. At the end of incubation, the increase in proportion of CH at 0–10 cm soil depth had a negative relationship with cumulative CO_2-C emission, similar to Tian et al. [34], and emphasizing that carbohydrates are utilized preferentially by microorganisms [13,21]. With the consumption of labile fractions, the increasing proportion of Phe was closely related to the less CO_2-C emission, explained by inhibiting the activity of various enzymes [34,46]. The correlation between ΔCH (0–10 cm), ΔPhe (0–10 cm), CH (10–25 cm) and cumulative CO_2-C emission in CL statistically deviated from the regression line (fig. 4a, 4b and 4g), suggesting that the contribution rate to C mineralization made by Phe or CH in CL was different from other land uses. Hi fractions of DOM separated by XAD-8 resin can be characterized as labile soluble organic moieties, particularly carbohydrates, amino sugars and low-molecular-weight organic acids and show a higher degree of biodegradation [13,23,35,57]. In our study, the increase of Hi fraction was strongly tied to increasing C mineralization, which was inconsistent with the general results of previous studies [21,44,67]. This result attributed to the fact that part of the hydrophilic fractions such as hydrophilic neutral (HiN) is composed mainly of carbohydrates that are bound to aromatic compounds and are converted into typical hydrophobic fractions during long term incubation [21]. However, the correlation of Hi and C mineralization at 10–25 cm depth in CL also deviated from the regression line.

Recently, fluorescence and UV spectroscopy has been used successfully to describe the chemical properties of DOM [21,53,68–69]. In many studies, the extent of C mineralization was inversely related to the specific UV absorbance of DOM [13,53,70–71]. This was confirmed by our results observed in 10–25 cm soil depth, indicating that UV-inactive substances were degraded preferentially. The relative increase in $SUVA_{254}$ at the end of incubation was positively related to cumulative CO_2-C emission, which suggests that more aromatic structures were cumulated during the C mineralization. Fluorescence spectroscopy can provide additional information relating to structure, functional groups, conformation, and heterogeneity, as well as dynamic

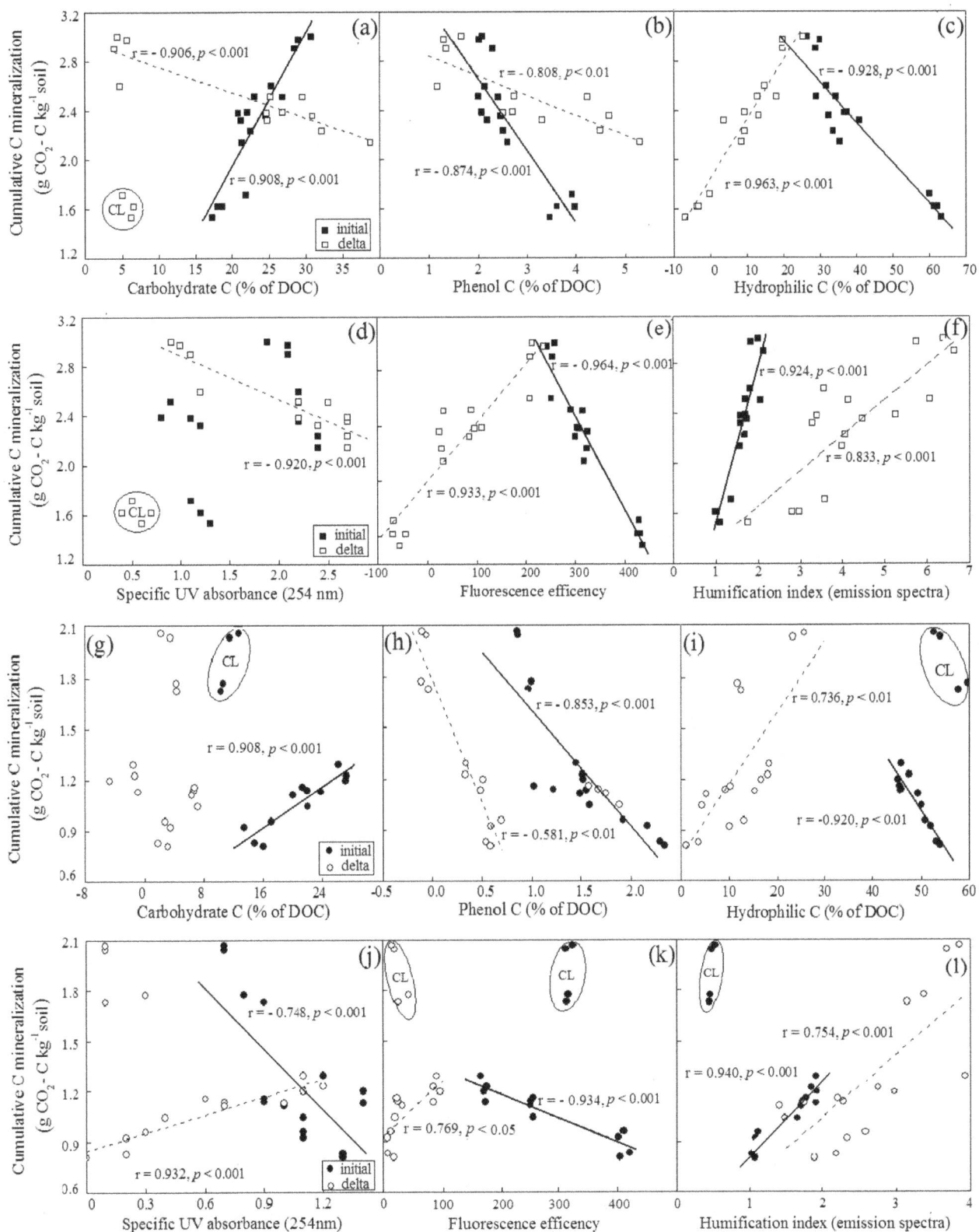

Figure 4. Relationships between cumulative C mineralization and DOM chemical properties of the four land use types. 0–10 cm soil depth: a–f; 10–25 cm soil depth: g–l. (a and g) between carbohydrate carbon percentages and cumulative CO_2-C, (b and h) between phenol carbon percentages and cumulative CO_2-C, (c and i) between hydrophilic carbon and cumulative CO_2-C, (d and j) between SUVA$_{254}$ values and cumulative CO_2-C, (e and k) between the fluorescence efficiency and cumulative CO_2-C, (f and l) between the humification index (emission fluorescence spectra) and cumulative CO_2-C. The curves for the figures are generated with the data of the CL samples excluded.

Table 4. Pearson correlation coefficients (r-value) between C mineralization and soil properties and chemical characteristics of DOM under four terrestrial land use types reclaimed from Taihu Lake, China.

		DOC	CH	ΔCH	Hi	ΔHi	Phe	ΔPhe	SUVA$_{254}$	ΔSUVA$_{254}$	FE	ΔFE	HIX$_{em}$	ΔHIX$_{em}$
0-10 cm	a	-0.140	-0.762**	-0.315	0.976**	-0.893**	0.951**	-0.782**	-0.504*	-0.538*	0.982**	-0.871**	-0.902**	-0.702**
	k_1	-0.586*	0.373	0.048	-0.555*	0.499*	-0.792**	0.339	-0.263	0.275	-0.737**	0.752**	0.510*	0.532*
	k_2	-0.285	-0.148	0.803**	-0.346	0.090	-0.548*	0.666**	-0.289	0.849**	-0.283	0.043	0.210	0.106
	MRT$_1$	0.270	-0.600*	-0.203	0.822**	-0.742**	0.948**	-0.604*	-0.101	-0.447	0.928**	-0.876**	-0.755**	-0.665**
	MRT$_2$	0.130	0.099	-0.886**	0.414	-0.145	0.562*	-0.762**	0.142	-0.921**	0.305	-0.031	-0.268	-0.117
	CO$_2$-C	0.085	0.908**	0.044	-0.928**	0.963**	-0.873**	0.573*	0.432	0.279	-0.964**	0.934**	0.925**	0.833**
10-25 cm	a	-0.600*	-0.263	0.390	0.135	-0.696**	0.755**	0.368	0.200	-0.499	0.776	-0.688**	-0.954**	-0.534*
	k_1	0.170	-0.653**	0.276	0.642**	0.466	-0.796**	-0.483	-0.819**	-0.456	0.090	-0.230	0.575	0.586*
	k_2	-0.420	0.408	-0.383	-0.394	-0.410	0.916**	0.189	0.825**	0.235	0.172	0.125	0.079	0.459
	MRT$_1$	0.142	-0.710**	0.469	0.662**	0.318	-0.758**	-0.322	-0.886**	-0.560*	0.199	-0.404	-0.700*	-0.436
	MRT$_2$	-0.088	0.694**	-0.794**	-0.578*	0.028	0.581*	-0.074	0.885**	0.661**	-0.360	0.680**	0.191	-0.122
	CO$_2$-C	0.310	-0.446	0.014	0.423	0.736**	-0.853**	-0.581*	-0.748**	-0.223	-0.122	0.012	0.940**	0.754**

n = 16; abbreviations see Table 2 and Table 3;
*Significance at $p < 0.05$;
**Significance at $p < 0.01$.

properties of DOM [25–26,49]. In our study, the FE and HIX$_{em}$ deduced from fluorescence emission spectra exhibited strong correlation with CO_2 emission. The FE and HIX$_{em}$ for initial DOM respectively showed negative and positive correlation with C mineralization, suggesting that in the long-term incubation not only labile C but also stable C contributed to the C mineralization. The ΔFE and ΔHIX$_{em}$ during incubation were positively correlated with C mineralization, supporting expectation that with the oxidation or degradation of labile fractions, more aromatic compounds and other complex stable molecules are accumulated [68]. The relationship between part of spectroscopy parameters and cumulative CO_2 emission in CL also deviated from the regression line (fig. 4d, 4k and 4l) possibly due to the effluence of land use and management.

Conclusions

Soils from varied land uses under reclaimed area from Taihu Lake had significant differences in soil DOM chemical properties and C dynamics. The C mineralization of lake sediments was much larger than that in terrestrial ecosystems, confirming our hypothesis. For the terrestrial soils, C mineralization in the upper soil layer was higher in BF and lower in CL. We have found that C mineralization in our study was fitting the double exponential model and the kinetics parameters were closely related to the chemical properties of DOM and sensitive to its variation.

Cumulative CO_2-C was positively correlated with the carbohydrates while negatively correlated with phenols and the aromaticity of DOM, indicating labile compounds are preferentially utilized by microbes. Moreover, the variation of Phe (ΔPhe) and humificaton (ΔHIX$_{em}$) at the end of incubation showed reverse correlation with cumulative CO_2, suggesting in the long-term incubation stable compounds also make contribution to C mineralization.

Overall, our study suggests that in conversion from aquatic ecosystem to terrestrial ecosystem, different land use types with different vegetation cover and varied management practices lead to significant changes in the chemical structure of DOM, which can influence the C dynamics. Inconsistent trend observed in CL soils may be attributed to the tillage practices (e.g. fertilization, crop rotation).

Acknowledgments

We gratefully acknowledge Yiling Luan, Juhua Yu, Jiaojiao Guo, Zilong Ma, Yupeng Zhao and zhiqin Hua for help with field and lab work. Special thanks go to Dr. Jason Vogel for his helpful comments.

Author Contributions

Conceived and designed the experiments: HR DH. Performed the experiments: DH. Analyzed the data: DH. Contributed reagents/materials/analysis tools: DH. Wrote the paper: DH.

References

1. Leinweber P, Jandl G, Baum C, Eckhardt K, Kandeler E (2008) Stability and composition of soil organic matter control respiration and soil enzyme activities. Soil Biology & Biochemistry 40: 1496–1505.
2. Hansson K, Kleja DB, Kalbitz K, Larsson H (2010) Amounts of carbon mineralized and leached as DOC during decomposition of Norway spruce needles and fine roots. Soil Biology & Biochemistry 42: 178–185.
3. Wang Y, Ruan HH, Huang LL, Feng YQ, Zhou JZ, et al. (2010) Soil labile organic carbon with different land uses in reclaimed land area from Taihu lake. Soil Science 175: 624–630.
4. Lal R (2002) Soil carbon dynamics in cropland and rangeland. Environmental Pollution 116: 353–362.
5. Chantigny MH (2003) Dissolved and water-extractable organic matter in soils: a review on the influence of land use and management practices. Geoderma 113: 357–380.
6. Caravaca F, Masciandaro G, Ceccanti B (2002) Land use in relation to soil chemical and biochemical properties in a semiarid Mediterranean environment. Soil & Tillage Research 68: 23–30.
7. Martin D, Lal T, Sachdev CB, Sharma JP (2010) Soil organic carbon storage changes with climate change, landform and land use conditions in Garhwal hills of the Indian Himalayan mountains. Agriculture, Ecosystems and Environment 138: 64–73.
8. Tobiašová E (2011) The effect of organic matter on the structure of soils of different land uses. Soil & Tillage Research 114: 183–192.
9. Kalbitz K, Solinger S, Park JH, Michalzik B, Matzner E (2000) Controls on the dynamics of dissolved organic matter in soils: a review. Soil Science 165: 277–304.
10. Ge QS, Zhao MC, Zheng JY (2000) Land use change of China during the 20th century. Acta Geographica Sinica 55: 698–706.
11. Qin BQ, Xu PZ, Wu QL, Luo LC, Zhang YL (2005) Environmental issues of Lake Taihu, China. Hydrobiologia 581: 3–13.
12. Søndergaard M, Jeppesen E (2007) Anthropogenic impacts on lake and stream ecosystems, and approaches to restoration. Journal of Applied Ecology 44: 1089–1094.
13. Marschner B, Kalbitz K (2003) Controls of bioavailability and biodegradability of dissolved organic matter in soils. Geoderma 113: 211–235.
14. Li ZP, Han CW, Han FX (2010) Organic C and N mineralization as affected by dissolved organic matter in paddy soils of subtropical China. Geoderma 157: 206–213.
15. Cookson WR, Abaye DA, Marschner P, Murphy DV, Stockdale EA, et al. (2005) The contribution of soil organic matter fractions to carbon and nitrogen mineralization and microbial community size and structure. Soil Biology & Biochemistry 37: 1726–1737.
16. Marinari S, Liburdi K, Fliessbach A, Kalbitz K (2010) Effects of organic management on water-extractable organic matter and C mineralization in European arable soils. Soil & Tillage Research 106: 211–217.
17. Kowalczuk P, Durako MJ, Young H, Kahn AE, Cooper WJ, et al. (2009) Characterization of dissolved organic matter fluorescence in the South Atlantic

18. Bight with use of PARAFAC model: Interannual variability. Marine Chemistry 113: 182–196.
18. Boddy E, Hill PW, Farrar J, Jones DL (2007) Fast turnover of low molecular weight components of the dissolved organic carbon pool of temperate grassland field soils. Soil Biology & Biochemistry 39: 827–835.
19. Marschner B, Bredow A (2002) Temperature effects on release and ecologically relevant properties of dissolved organic matter in sterilised and biologically active soils. Soil Biology & Biochemistry 34: 459–466.
20. Zhao MX, Zhou JB, Kalbitz K (2008) Carbon mineralization and properties of water-extractable organic carbon in soils of the south Loess Plateau in China. European Journal of Soil Biology 44: 158–165.
21. Kalbitz K, Schmerwitz J, Schwesig D, Matzner E (2003) Biodegradation of soil-derived dissolved organic matter as related to its properties. Geoderma 113: 273–291.
22. Don A, Kalbitz K (2005) Amounts and degradability of dissolved organic carbon from foliar litter at different decomposition stages. Soil Biology & Biochemistry 37: 2171–2179.
23. Said-Pullicino D, Kaiser K, Guggenberger G, Gigliotti G (2007) Changes in the chemical composition of water-extractable organic matter during composting: Distribution between stable and labile organic matter pools. Chemosphere 66: 2166–2176.
24. Kalbitz K, Schwesig D, Schmerwitz J, Kaiser K, Haumaier L, et al. (2003) Changes in properties of soil-derived dissolved organic matter induced by biodegradation. Soil Biology & Biochemistry 35: 1129–1142.
25. Chen J, Gu BH, LeBoeuf EJ, Pan HJ, Dai S (2002) Spectroscopic characterization of the structural and functional properties of natural organic matter fractions. Chemsophere 48: 59–68.
26. Chen J, LeBoeuf EJ, Dai S, Gu BH (2003) Fluorescence spectroscopic studies of natural organic matter fractions. Chemosphere 50: 639–647.
27. Hur J, Kim G (2009) Comparison of the heterogeneity within bulk sediment humic substances from a stream and reservoir via selected operational descriptors. Chemosphere 75: 483–490.
28. Teixeira MC, Azevedo JCR, Pagioro TA (2011) Spatial and seasonal distribution of chromophoric dissolved organic matter in the Upper Paraná River floodplain environments (Brazil). Acta Limnologica Brasiliensia 23: 333–343.
29. You BS, Zhong JC, Fan CX, Wang TC, Zhang L, et al. (2007) Effects of hydrodynamics processes on phosphorus fluxes from sediment in large, shallow Taihu Lake. Journal of Environmental Sciences 19: 1055–1060.
30. Yin HB, Fan CX, Ding SM, Zhang L, Zhong JC (2008) Geochemistry of iron, sulfur and related heavy metals in metal-Polluted Taihu Lake sediments. Pedosphere 18: 564–573.
31. Blake GR, Hartge KH (1986) Bulk density. In: Klute A ed. Methods of soil analysis. Part I. Physical and mineralogical methods. Madison, WI: American Society of Agronomy and Soil Science Society of America 363–375.

32. Vance ED, Brookes PC, Jenkinson DS (1987) An extraction method for measuring soil microbial biomass carbon. Soil Biology & Biochemistry 19: 703–707.

33. Mora JL, Guerra JA, Armas CM, Rodríguez-Rodríguez A, Arbelo CD, et al. (2007) Mineralization rate of eroded organic C in Andosols of the Canary Islands. Science of the Total Enviroment 378: 143–146.

34. Tian L, Dell E, Shi W (2010) Chemical composition of dissolved organic matter in agroecosystems: Correlations with soil enzyme activity and carbon and nitrogen mineralization. Applied Soil Ecology 46: 426–435.

35. Simonsson M, Kaiser K, Danielsson R, Andreux F, Ranger J (2005) Estimating nitrate, dissolved organic carbon and DOC fractions in forest floor leachates using ultraviolet absorbance spectra and multivariate analysis. Geoderma 124: 157–168.

36. Chantigny MH, Angers DA, Kaiser K, Kalbitz K (2007) Extraction and characterization of dissolved organic matter. In: Carter MR, Gregorich EG eds. Soil sampling and methods of analysis, chap 48. CRC Press. 617–635.

37. Kalbitz K, Meyer A, Yang R, Gerstberger P (2007) Response of dissolved organic matter in the forest floor to long-term manipulation of litter and throughfall inputs. Biogeochemistry 86: 301–318.

38. Ewald M, Berger P, Visser SA (1988) UV-visible absorption and fluorescence properties of fulvic acids of microbial origin as function of their molecular weights. Geoderma 43: 11–20.

39. Zsolnay Á, Baigar E, Jimenez M, Steinweg B, Saccomandi F (1999) Differentiating with fluorescence spectroscopy the sources of dissolved organic matter in soil subjected to drying. Chemosphere 38: 45–50.

40. Kalbitz K, Geyer W (2001) Humification indices of water-soluble fulvic acids derived from synchronous fluorescence spectra-effects of spectrometer type and concentration. Journal Plant Nutrition and Soil Science 164: 259–265.

41. Kalbitz K, Geyer W, Geyer S (1999) Spectroscopic properties of dissolved humic substances-areflection of land use history in a fen area. Biogeochemistry 47: 219–238.

42. Yang LX, Pan JJ, Yuan SF (2006) Predicting dynamics of soil organic carbon mineralization with a double exponential model in different forest belts of China. Journal of Forestry Research 17: 39–43.

43. Wang XW, Li XZ, Hu YM, Lv JJ, Sun J, et al. (2010) Effect of temperature and moisture on soil organic carbon mineralization of predominantly permafrost peatland in the Great Hing'an Mountains, Northeastern China. Journal of Environmental Sciences 22: 1057–1066.

44. Hur J, Lee B, Shin H (2011) Microbial degradation of dissolved organic matter (DOM) and its influence on phenanthrene–DOM interactions. Chemosphere 85: 1360–1367.

45. Kemmitt SJ, Lanyon CV, Waite IS, Wen Q, Addiscott TM, et al. (2008) Mineralization of native soil organic matter is not regulated by the size, activity or composition of the soil microbial biomass-a new perspective. Soil Biology & Biochemistry 40: 61–73.

46. Rovira P, Vallejo VR (2007) Labile, recalcitrant, and inert organic matter in Mediterranean forest soils. Soil Biology & Biochemistry 39: 202–215.

47. Wieder WR, Cleveland CC, Townsend AR (2008) Tropical tree species composition affects the oxidation of dissolved organic matter from litter. Biogeochemistry 88: 127–38.

48. Majcher EH, Chorover J, Bollag JM, Huang PM (2000) Evolution of CO_2 during birnessite-induced oxidation of ^{14}C-labeled catechol. Soil Science Society of America Journal 64: 157–163.

49. Zsolnay Á (2003) Dissolved organic matter: artefacts, definitions, and functions. Geoderma 113:187–209.

50. Akagi J, Zsolnay Á, Bastida F (2007) Quantity and spectroscopic properties of soil dissolved organic matter (DOM) as a function of soil sample treatments: Air-drying and pre-incubation. Chemosphere 69: 1040–1046.

51. Wang LY, Wu FC, Zhang RY, Li W, Liao HQ (2009) Characterization of dissolved organic matter fractions from Lake Hongfeng, Southwestern China Plateau. Journal of Environmental Sciences 21: 581–588.

52. Corvasce M, Zsolnay A, D'Orazio V, Lopez R, Miano TM (2006) Characterization of water extractable organic matter in a deep soil profile. Chemosphere 62: 1583–1590.

53. Bu XL, Wang LM, Ma WB, Yu XN, McDowell WH, et al. (2010) Spectroscopic characterization of hot-water extractable organic matter from soils under four different vegetation types along an elevation gradient in the Wuyi Mountains. Geoderma 159: 139–146.

54. Traversa D, D'Orazio V, Senesi N (2008) Properties of dissolved organic matter in forest soils: influence of different plant covering. Forest Ecology and Management 256: 2018–2028.

55. Fuentes M, González-Gaitano G., García-Mina JM (2006) The usefulness of uv-visible and fluorescence sepectroscopies to study the chemical nature of humic substances from soils and composts. Organic Geochemistry 37: 1949–1959.

56. Khomutova TE, Shirshova LT, Tinz S, Rolland W, Richter J (2000) Mobilization of DOC from sandy loamy soils under different land use (Lower Saxony, Germany). Plant and Soil 219: 13–19.

57. Kiikkilä O, Kitunen V, Smolander A (2006) Dissolved soil organic matter form surface organic horizons under birch and conifers: Degradation in relation to chemical characteristics. Soil Biology & Biochemistry 38: 737–746.

58. Cookson WR, Murphy DV, Roper MM (2008) Characterizing the relationships between soil organic matter components and microbial function and composition. Soil Biology & Biochemistry 40: 763–777.

59. Bu XL, Ding JM, Wang LM, Yu XN, Huang W, et al. (2011) Biodegradation and chemical characteristics of hot-water extractable organic matter from soils under four different vegetation types in the Wuyi Mountains, southeastern China. European Journal of Soil Biology 47: 102–107.

60. Jacinthe PA, Lal R, Owens LB, Hothem DL (2004) Transport of labile carbon in runoff as affected by land use and rainfall characteristics. Soil and Tillage Research 77: 111–123.

61. Rodríguez-Rodríguez A, Guerra A, Arbelo C, Mora JL, Gorrín SP, et al. (2004) Forms of eroded soil organic carbon in andosols of the Canary Islands. Geoderma 121: 205–219.

62. Juarez S, Rumpel C, Mchnu C, Vincent C (2011) Carbon mineralization and lignin content of eroded sediments from a grazed watershed of South-Africa. Geoderma 167–168: 247–253.

63. Faikd HA (2000) Primary studies on lactone in bamboo leaf. MS degree Thesis. Zhejiang University, China.

64. Zhang LY (2005) The application and Study of Flavone in Bamboo Leaf. Modern Food Science and Technology 22: 247–249 (in Chinese).

65. Zhou Y (2009) Soil organic carbon pools and the characteristics of mineralization along an elevation gradient in Wuyi Mountain, China. Ph.D. Dissertation. Nanjing Forestry University, China.

66. Rey A, Jarvis P (2006) Modelling the effect of temperature on carbon mineralization rates across a network of European forest sites (FORCAST). Global Change Biology 12: 1894–1908.

67. Nguyen HV, Hur J (2011) Tracing the sources of refractory dissolved organic matter in a large artificial lake using multiple analytical tools. Chemosphere 85: 782–789.

68. Glatzel S, Kalbitz K, Dalva M, Moore T (2003) Dissolved organic matter properties and their relationship to carbon dioxide efflux from restored peat bogs. Geoderma 113: 397–411.

69. Matilainen A, Gjessing ET, Lahtinen T, Hed L, Bhatnagar A, et al. (2011) An overview of the methods used in the characterisation of natural organic matter (NOM) in relation to drinking water treatment. Chemosphere 83: 1431–1442.

70. Embacher A, Zsolnay A, Gattinger A, Munch JC (2007) The dynamics of water extractable organic matter (WEOM) in common arable topsoils: I. Quantity, quality and function over a three year period. Geoderma 139: 11–22.

71. Embacher A, Zsolnay A, Gattinger A, Munch JC (2008) The dynamics of water extractable organic matter (WEOM) in common arable topsoils: II. Influence of mineral and combined mineral and manure fertilization in a Haplic Chernozem. Geoderma 148: 63–69.

10

Does Habitat Heterogeneity in a Multi-Use Landscape Influence Survival Rates and Density of a Native Mesocarnivore?

Eric M. Gese[1]*, Craig M. Thompson[2]¤

1 United States Department of Agriculture, Wildlife Services, National Wildlife Research Center, Department of Wildland Resources, Utah State University, Logan, Utah, United States of America, **2** Department of Wildland Resources, Utah State University, Logan, Utah, United States of America

Abstract

The relationships between predators, prey, and habitat have long been of interest to applied and basic ecologists. As a native Great Plains mesocarnivore of North America, swift foxes (*Vulpes velox*) depended on the historic disturbance regime to maintain open grassland habitat. With a decline in native grasslands and subsequent impacts to prairie specialists, notably the swift fox, understanding the influence of habitat on native predators is paramount to future management efforts. From 2001 to 2004, we investigated the influence of vegetation structure on swift fox population ecology (survival and density) on and around the Piñon Canyon Maneuver Site, southeastern Colorado, USA. We monitored 109 foxes on 6 study sites exposed to 3 different disturbance regimes (military training, grazing, unused). On each site we evaluated vegetation structure based on shrub density, basal coverage, vegetation height, and litter. Across all sites, annual fox survival rates ranged from 0.50 to 0.92 for adults and 0.27 to 0.78 for juveniles. Among sites, population estimates ranged from 1 to 7 foxes per 10 km transect. Fox density or survival was not related to the relative abundance of prey. A robust model estimating fox population size and incorporating both shrub density and percent basal cover as explanatory variables far outperformed all other models. Our results supported the idea that, in our region, swift foxes were shortgrass prairie specialists and also indicated a relationship between habitat quality and landscape heterogeneity. We suggest the regulation of swift fox populations may be based on habitat quality through landscape-mediated survival, and managers may effectively use disturbance regimes to create or maintain habitat for this native mesocarnivore.

Editor: Jesus E. Maldonado, Smithsonian Conservation Biology Institute, United States of America

Funding: Funding and logistical assistance was provided by the United States Army, Directorate of Environmental Compliance and Management, Fort Carson, Colorado, through the United States Fish and Wildlife Service, Colorado Assistance Office, Golden, Colorado, and the Utah Cooperative Fish and Wildlife Research Unit at Utah State University. Additional support was provided by the United States Department of Agriculture, Wildlife Services, National Wildlife Research Center at Utah State University, Logan, Utah. The funders had no role in study design, data collection and analysis, decision to publish, or preparation of the manuscript.

Competing Interests: The authors have declared that no competing interests exist.

* Email: eric.gese@usu.edu

¤ Current address: United States Department of Agriculture, Forest Service, Pacific Southwest Research Station, Fresno, California, United States of America

Introduction

Historically, North American grasslands and shrub-steppe systems were maintained through the interactions of frequent, low intensity disturbances such as fire, native herbivore grazing, drought, and soil disturbances [1,2]. These interactions resulted in a mosaic of different-aged grasslands across the landscape [3], which benefited native wildlife [4], and enhanced community richness and diversity [5]. However, during the 1900s natural grassland systems in the Great Plains of North America were altered through processes such as the conversion of prairie into ranchland and cropland, fire suppression, and predator control programs [6]. The alterations interacted to create a variety of landscape changes including the conversion of native grassland to shrubland [7] and the homogenization of the landscape [5]. Concurrently, swift fox (*Vulpes velox*) populations declined and by 1950 they were believed to be absent from much of their historic range [8]. In 1978 the swift fox was declared extirpated in the Canadian prairies [9].

While the direct effects of disturbances on native species are often limited, indirect effects mediated through changes in vegetation structure are thought to have a much greater effect [2]. Since the mid-1970s, extensive research has focused on swift fox distribution and demographics [8,10–14]. However, much of this has focused on the characteristics of individual populations, leaving a large gap in the understanding of landscape-level influences [14,15]. Lately, researchers have investigated the influence of landscape variation on swift fox ecology, or compared spatial ecology and demographics across habitat types [11,13,14,16–18]. Viewed as shortgrass specialists, swift foxes have been shown to be capable of exploiting a variety of habitats and prey [8,19,20].

In 1982 the United States Army purchased 1,040 km² of southeastern Colorado grassland for the purpose of mechanized infantry training. Since then, livestock have been excluded from the area, and fire suppression increased. Military training activity commenced in 1985 on the site, primarily in the form of mechanized infantry [21]. Due to the scale of training maneuvers, some areas of the base were underutilized resulting in some areas

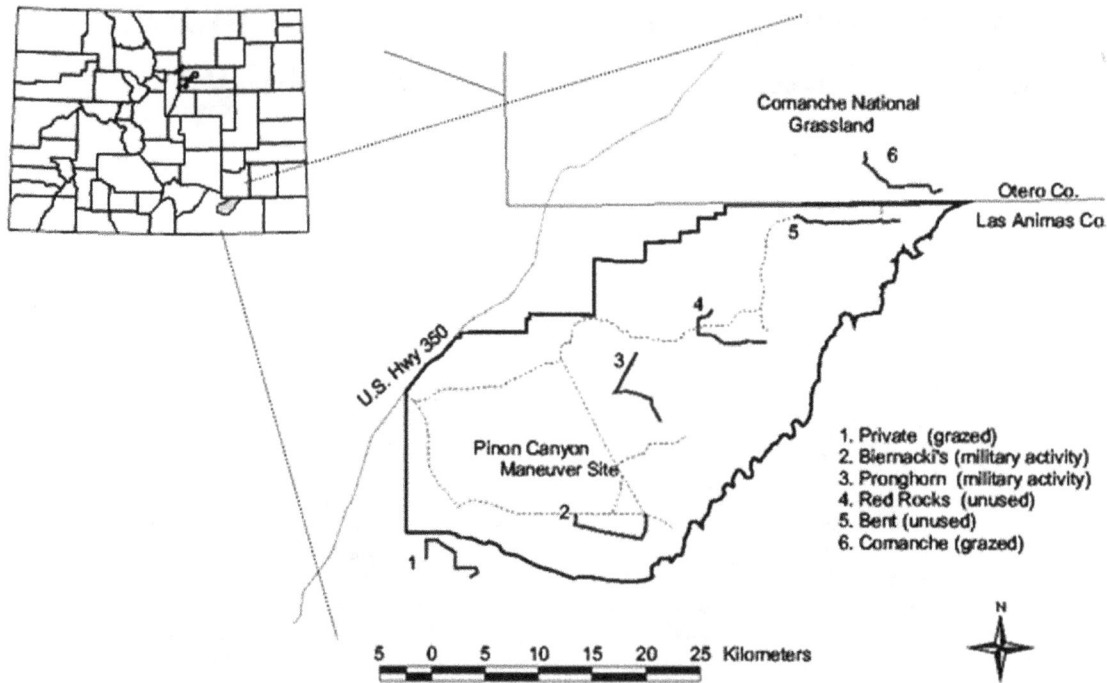

Figure 1. Six study sites on and around the Piñon Canyon Maneuver Site, southeastern Colorado, USA. Locations of the 6 transects are indicated, as well as the associated dominant land use.

being disturbance-free. Research on the response of the vegetative community to this change in ownership and land use has identified two interacting landscape trajectories: an increase in basal cover and grass height following the release from grazing and a reduction in basal cover, shrub height, and shrub density associated with military training [21,22].

The objective of our research was to investigate the influence of vegetation structure on swift fox population ecology, principally survival rates and density, on and around the U.S. Army Piñon Canyon Maneuver Site, southeastern Colorado, USA. The abrupt shift in land ownership, the discrete boundaries of the training area, and the patterns of land use within the military parcel coalesced into a natural experiment on the influence of landscape heterogeneity and vegetation structure on swift fox ecology. While there was no true experimental control of treatments in our study, due to the temporal and spatial scale of terrestrial vertebrate research, observational studies following landscape-level changes are often the only available option. We therefore use the term 'natural experiment' cautiously; our research was observational yet capitalized on a well-defined change in land use practices and the resulting changes in landscape structure and swift fox demographics.

Methods

Ethics Statement

Fieldwork was approved and sanctioned by the United States Department of Agriculture's National Wildlife Research Center, the United States Army – Directorate of Environmental Compliance and Management, and the United States Forest Service. Permission to access land on the Piñon Canyon Maneuver Site was obtained from the United States Army, permission to access land of the Comanche National Grassland was obtained from the United States Forest Service, and permission to access private land was obtained from the landowner.

Capture and handling protocols were reviewed and approved by the Institutional Animal Care and Use Committees (IACUC) at the United States Department of Agriculture's National Wildlife Research Center (QA-930) and Utah State University (#1060). Permits to capture and handle swift foxes and small mammals were obtained from the Colorado Division of Wildlife (state license numbers 01-TR001, 02-TR001, 03-TR001, 04-TR001). Data were archived with the United States Department of Agriculture's National Wildlife Research Center (QA-930) and is available with permission from the authors.

Study Area

The study area was on and around the 1,040-km^2 Piñon Canyon Maneuver Site (PCMS) located in Las Animas County, Colorado, USA, plus areas on the United States Forest Service Comanche National Grassland, and private ranchland (Fig. 1). The region was classified as semi-arid grassland steppe, with approximately 60% categorized as shortgrass prairie dominated by blue grama (*Bouteloua gracilis*), western wheatgrass (*Pascopyrum smithii*), and galleta (*Hilaria jamesii*) [23]. Shrublands interspersed throughout the area included four-winged saltbrush (*Atriplex canescens*) and greasewood (*Sacrobatus vermiculatus*), plus prickly pear cactus (*Opuntia phaeacantha*), cholla (*Opuntia imbricata*), and yucca (*Yucca glauca*). The remaining landscape was dominated by pinyon-juniper woodlands (*Pinus edulis*, *Juniperus monosperma*). Elevation varied between 1,310 to 1,740 m, average temperatures ranged from 1°C in January to 23°C in July, and precipitation averaged 30 cm [21]. Monthly precipitation was highest in July with an average of 4.3 cm of rain, though the 35% of the annual precipitation that fell during the cool-season (March-May) had a proportionally greater impact on productivity [22].

Study Design

In order to deal with the range of spatial scales used by predators and prey, we developed a hierarchical study design. We identified 6 study sites in areas subjected to 3 land use regimes: livestock grazing, mechanized military training, and unused. Unused areas were considered controls despite the fact that 'no disturbance' is an unnatural state for grassland ecosystems; these were sites on military property and protected from grazing, yet were not used for training purposes. Two study sites were located in each land use regime and sites were named according to landmarks or historical owners: Private (PRV), Biernacki's (BTS), Pronghorn (PRN), Red Rocks (RRK), Bent (BNT), and Comanche (COM). Each study site was centered around a 10-km trapping transect [24,25], and the outer boundaries were defined by the home ranges of resident swift foxes [24]. Within each site, we randomly placed 50×70 m sampling grids within 1 km of the trapping transect. We used a random number generator to create a distance along the trapping transect, a direction (right or left), and a distance from the transect. This point became the northwest corner of the grid. These grids served as sampling units for both small mammal trapping and vegetation structure surveys. Each study site was considered to be spatially independent (i.e., home ranges of foxes did not overlap adjacent transects, nor did foxes travel beyond one transect during a season).

Swift fox populations on each site were evaluated based on density and survival rates [24,25]. Each year was divided into 3 seasons based on fox behavior: breeding/gestation: 15 December – 14 April; pup-rearing: 15 April – 14 August; dispersal: 15 August – 14 December [25,26]. We calculated both overall and seasonal estimates of population density and survival rates [24,25]. Small mammal and vegetation surveys were also conducted seasonally at a rate of 4 grids per site per season, resulting in 12 grids sampled/site/year [27]. While we assumed differences in vegetation structure resulted primarily from differences in land use, each study site was considered an experimental unit due to the intrinsic small scale variation between them. We attempted to minimize the effect of within-site heterogeneity through replication and the distribution of sites; however additional uncontrollable and confounding factors such as disturbance intensity, cattle stocking rates, and the degree of fire suppression precluded the use of a treatment – control design.

Swift Fox Capture and Radio-Telemetry

We captured foxes using box traps baited with chicken [28]. Traps were placed 500 m apart along each 10 km trapping transect resulting in 21 trap locations per study site. Each trap was oriented and covered with brush to provide protection from exposure. Traps were set in the late afternoon, checked early the following morning, and left closed throughout the day. Each site was trapped for 4 consecutive nights 3 times per year. For recollaring or targeting animals, a trap-enclosure system was used at den sites [29]. We used subsequent home range analyses to identify gaps between resident swift fox territories, and we trapped these gaps to assure full population monitoring. Captured foxes were handled without anesthesia, weighed, sexed, and aged through tooth wear (adult, juvenile). Foxes were considered juvenile until the pup-rearing season following their birth (15 April). Foxes were ear-tagged and radio-collared with 30–50 g radio transmitters (Advanced Telemetry Systems, Isanti, MN, USA). Attempts were made to remove transmitting radio-collars at the end of the study.

We located foxes a minimum of 3 times per week, twice during nighttime hours when animals were actively hunting and once during daylight hours to locate den sites. Locations were considered independent when separated by >4 hours [30]; more than sufficient time for a fox to cross its home range. Nocturnal locations were estimated using triangulation of 2–3 bearings within 5 minutes and separated by at least 40°. Triangulation was done using Program Locate II (Pacer, Truro, Nova Scotia); telemetry error on the study area was ±8° as determined from reference transmitters [24,25]. Diurnal locations were collected visually by approaching the animal until either a den could be identified, or the animal was seen. Mortality sensors within transmitters indicated when a collar had been stationary for 4–6 hours. When a mortality signal was detected, the transmitter was recovered immediately and the location was recorded. Efforts to determine the cause of death included searching the area for tracks and other sign, as well as necropsy of any remains [31].

Vegetation Structure

Vegetation structure has been defined as the "height, density, biomass, and dispersion of herbaceous and woody vegetation" [32]. For each of the 6 study sites, we evaluated vegetation structure based on the 50×70 m sampling grids randomly located within 1 km of the trapping transect [27]. Four grids were sampled each season, and new grids were selected each subsequent season. Each grid consisted of seven 50-m line-transects oriented north-south and spaced 10 m apart. On each line transect, vegetation type and height was evaluated by dropping a measuring pin every 1 m and recording the type and height of the tallest vegetation encountered [33]. For each grid, point measurements were combined into estimates of percent basal cover, percent bare ground, percent litter (dead material), and mean shrub and grass height. Shrub density was calculated by counting all woody plants >20 cm high within the grid. Grid estimates were combined into seasonal and annual averages for each study site. Standard deviations of grid estimates for each study site were used to represent the homogeneity of vegetation characteristics across each study site.

Prey Base

Following vegetation sampling, we placed 35 Sherman live traps with 10 m spacing throughout the 50×70 m sampling grid. Traps were baited with equine sweet feed (corn, oats, molasses). Trapping grids were run for 4 consecutive nights; checked and closed each morning and reset each afternoon. Captured rodents were marked with Sharpie pens on the tail and abdomen allowing for identification of recaptures over the 4-day trapping period [2,34,35]. Relative abundance for each species was estimated based on the number of individuals captured. We calculated community richness as the number of species captures and we estimated community diversity using the Shannon-Weaver index [36].

Data Analysis

We estimated average seasonal survival rates for juvenile and adult swift foxes, as well as an overall survival rate for each of the 6 sites using the known fate model in Program MARK [37]. The model was age-structured, allowing juveniles to graduate into the adult cohort after surviving through April of the year following their birth. Individuals not located during a season were censored for that season.

We estimated the number of foxes in each site using the robust model in Program MARK [37] and Huggin's estimator. Seasonal survival estimates for each site were taken from the known fate model. Estimates of the number of foxes were converted into density estimates by calculating the 'effective trapping area' associated with each transect [24,25]. The radius of the average seasonal 95% kernel home range for foxes associated with each transect was used to buffer the transect in ArcView GIS. The resulting polygon was considered the 'effective trapping area' for that transect. The estimated number of foxes per transect could then be converted into a density estimate for each site. Density estimates were consolidated into seasonal averages as well as an overall estimate for each site. Chi-square analysis was used to test for differences in capture rates among sites.

We used Pearson correlation coefficients to identify vegetation variables for final analysis. We selected variables based on their independence and ability to discriminate among study sites. We evaluated seasonal differences in vegetation structure among sites using the GLM procedure carried out in SAS v9.2 and separated sites into statistically significant groupings. Due to the large number of models generated, Tukey's studentized range was used to control for the experiment-wise error rate.

We compared seasonal swift fox population parameters to seasonal vegetation variables using both univariate and multivariate techniques. We used linear regression to compare seasonal fox survival and population density with seasonal vegetation variables. We also constrained the above mentioned MARK models using combinations of grass height, shrub density, percent basal cover, and percent litter in order to further evaluate the effect of vegetation structure on fox demographics. The logit link function was used to run constrained models. We used likelihood ratio tests and AIC statistics [38] to evaluate whether the inclusion of vegetation data improved the explanatory power of the original, unconstrained known fate and robust models.

Results

Between 20 November 2001 and 27 November 2004, 116 swift foxes were captured 238 times; 109 foxes were fitted with radio-collars. Captures were not distributed equally among sites ($\chi^2 = 26.6$, df = 5, $P < 0.001$), with 86% of all captures occurring on the grazed or military sites (Table 1); trapping effort was equal across all study sites. Fewer foxes and a greater proportion of juvenile foxes were captured in unused sites compared to military or grazed sites (Table 1). Throughout the study, 7595 locations were recorded on the 109 collared foxes. The mean number of days a fox was monitored, from radio-collaring to either death, loss of signal, or radio-collar removal, was 299 days (SD = 284.5). A total of 55 swift foxes died during the study (38 adult, 17 juvenile). Of these deaths, 24 (44%) were suspected coyote predation, 22 (40%) were confirmed coyote predation, 3 (5%) were badger predation, 3 (5%) were vehicle collision, 2 (4%) were golden eagle predation, and 1 (2%) was bobcat predation. Many of the suspected coyote predation events were when we recovered a torn, bloody, or buried radio-collar and were unable to conduct a necropsy. Thus, suspected and confirmed predation by coyotes accounted for 84% of the swift fox deaths with predation being the main cause of death across all study sites.

Prey Base and Swift Fox Survival and Density

Small mammal communities were sampled on 185 trapping grids. Northern grasshopper mice (*Onychomys leucogaster*), Ord's kangaroo rat (*Dipodomys ordii*), silky pocket mice (*Perognathus flavus*), western harvest mice (*Reithrodontomys megalotis*), white-footed mice (*Peromyscus leucopus*), southern plains woodrat (*Neotoma micropus*), thirteen-lined ground squirrels (*Spermophilus tridecemlineatus*), deer mice (*Peromyscus maniculatus*), and spotted ground squirrels (*Spermophilus spilosoma*) accounted for >99% of all captures. Three species, Northern grasshopper mice, deer mice, and Ord's kangaroo rat, accounted for 76% of all captures. Only one small mammal

parameter, relative abundance of Northern grasshopper mice, differed significantly between sites ($F = 2.62$, df = 5, 179, $P = 0.03$) (Table 2) and it was unrelated to either fox density ($F = 0.002$, df = 5, 179, $P = 0.96$), or fox survival ($F = 1.60$, df = 4, $P = 0.29$).

Vegetation Structure and Swift Fox Survival and Density

Between 2001 and 2004, 185 vegetation grids were sampled across the 6 study sites. Mean vegetation height ranged from 7.7 cm in the PRV site to 21.5 cm in the COM site and shrub density ranged from 0.03 (PRN) to 2.7 shrubs/100 m^2 (COM) (Table 2). Only one vegetation parameter, percent basal cover, was not significantly different among sites ($F = 1.38$, df = 5, 179, $P = 0.23$). The remaining 6 vegetation parameters evaluated differed significantly among sites ($P < 0.01$ in all cases). Groupings varied and did not correspond to the dominant land use (Table 2).

Swift fox survival estimates did not differ significantly between seasons ($F = 0.01$, $P = 0.99$), by year ($F = 0.98$, $P = 0.386$), by age ($F = 0.02$, $P = 0.891$), or by site ($F = 0.57$, $P = 0.721$) (Table 3). Only one vegetation variable, shrub density, was significantly related to swift fox survival (Fig. 2). However, this relationship depended on a single outlying point. When this point was removed from the analysis, the R^2 value dropped to 0.004 and the associated P value rose to 0.796. Fox population density was negatively related to all 4 vegetation variables, but only the fox density - mean grass height relationship was both statistically significant and significantly different from zero (Fig. 3).

Constraining the known fate survival model, using data on shrub density, consistently improved model performance (Table 4). However, only two single-variable models, the interaction of shrub density and grass height ($\chi^2 = 4.38$, $P = 0.036$) and shrub density alone ($\chi^2 = 4.19$, $P = 0.041$), showed statistically significant improvement over the null, age-structured model based on likelihood ratio tests. These two models were roughly equivalent with ΔAICc values differing by only 0.19, and their combined AICc weight equaled 0.328. Only one additional known fate model, constrained by the interaction of shrub density and percent basal cover, had a ΔAICc value <2. No models incorporating the standard deviation of vegetation variables outperformed the null model.

Population density estimates differed by season and site (Table 5). Site most strongly influenced estimates ($F = 5.78$, df = 5, $P = 0.004$, R$^2 = 0.385$). Season was marginally significant ($F = 3.07$, df = 5, $P = 0.057$); however its inclusion in the model raised the R^2 to 0.467. One robust population density model, constrained by shrub density and percent basal cover, significantly outperformed all other models as well as the null model ($\chi^2 = 39.32$, $P < 0.001$; Table 6). This model had an AICc weight of 0.98 and the next best performing model, constrained by shrub density alone, had a ΔAICc value of 9.68.

Discussion

The swift fox survival rates we recorded were similar to those previously reported on the PCMS and elsewhere. On Piñon Canyon, estimated annual adult survival rates have ranged from 0.52 [39] to 0.88 [25]. In Wyoming, swift fox survival estimates ranged from 0.40 to 0.69 [40]. In general, our results were similar with one exception. On the Comanche site we recorded an adult survival rate of 0.92. This is one of the highest survival rates reported for swift foxes, and was based on a sample of 17 animals monitored for >3 years. On this site, population density was low, survival was high, and resident animals were larger and heavier than on other sites. While we do not have sufficient data to explain this, we speculated that effective management of the Comanche National Grassland during drought conditions resulted in the best

Table 1. Number, age, and sex ratios of swift foxes captured in 6 sites in southeastern Colorado, USA, 2001–2004.

Dominant land use	Site	# animals captured	Males: Females	Proportion adults
Grazed	PRV	32	18/14	0.44
	COMM	17	7/10	0.41
Military	BTS	23	9/14	0.57
	PRN	28	14/14	0.36
Unused	RRK	9	3/6	0.22
	BNT	7	5/2	0.29

of both worlds for swift foxes: an average grass height to allow predator detection and high shrub density to maintain prey density. While population density on the Comanche National Grassland was lower than on other sites, we believe this reflected a stable population with long-term residents and low turnover.

In contrast, estimates of juvenile survival have ranged widely. Rongstad et al. [39] estimated annual juvenile survival on PCMS at only 0.05. On the same landscape, Karki et al. [28] reported a range of survival estimates from 0.41 to 0.60. Reports have varied on whether juvenile swift foxes experience higher or lower survival than adults. Kamler et al. [41] found juvenile swift foxes had higher survival rates than adults, while Sovada et al. [16] and Schauster et al. [25] reported the opposite. Our estimates ranged from 0.27 in an 'unused' site to 0.78 in a site exposed to military training. The high variation in juvenile survival rates indicated fluctuating environmental conditions may play a role. For example, annual precipitation and the resulting growing season may influence juvenile survival during dispersal due to vegetation height but we did not consider a 3-year study sufficient to evaluate climate-related effects.

Density estimates on PCMS have averaged 0.22 [25] and 0.26 foxes/km^2 [28]. These estimates were based on telemetry studies of known populations during a time when swift fox populations were believed to be at their peak. Our estimates in the same area, averaged 0.10 foxes/km^2, were based on mark-recapture data during drought conditions. In northern Colorado, swift fox densities ranged from 0.2/km^2 in poor habitat to 1.1/km^2 in good habitat [42]. Our estimates ranged from 0.03 foxes/km^2 on an 'unused' site to 0.18 foxes/km^2 on a grazed site.

It is important to note that our results varied considerably among sites. While results on military-used lands were fairly consistent, swift fox population parameters on grazed lands varied widely and may have been influenced by finer scale heterogeneity. In one grazed site (COM), we recorded above average survival and below average population estimates. In the other grazed site (PRV), we recorded the highest population estimate and average survival rates. This may be related to the variation in vegetation structure among sites. In general, military sites were more homogeneous while grazed sites showed greater variation in structural measurements between vegetation grids (Table 4). 'Grazed' appears to be a far too simplistic category for a wildlife/landscape interaction study such as ours: individual management practices resulted in different landscape conditions and shifts between grassland and shrubland often depended on seasonal effects. During our study, drought conditions prompted the U.S. Forest Service to reduce stocking rates on the Comanche

Table 2. Mean (± SD) vegetation structure and small mammal population parameters for 6 study sites in southeastern Colorado, USA, 2001–2004.

	Grazed		Military		Unused	
	PRV	COM	BTS	PRN	RRK	BNT
% basal cover	38.0±19.9	45.1±18.9	40.3±17.2	35.3±12.8	43.8±18.1	43.7±21.7
% bare ground	42.6±19.5[a]	26.1±13.9[b]	40.0±12.9[a]	37.0±11.7[a]	29.5±11.4[b]	30.6±13.8[b]
% litter	18.0±8.9[a]	24.9±14.1[b]	18.2±13.5[a]	26.4±13.8[b]	23.4±13.2[b]	16.9±9.3[a]
Mean veg. ht	7.7±5.0[a]	21.5±43.4[b]	10.2±5.1[a]	9.6±4.4[a]	16.9±24.3[b]	16.3±8.8[b]
Mean grass ht	6.7±3.9[a]	10.6±3.1[b]	9.5±5.1[b]	8.9±4.3[b]	9.4±3.0[b]	12.8±4.9[c]
Mean shrub ht	12.9±20.9[a]	59.1±62.8[b]	18.8±17.2[a]	17.1±15.4[a]	53.4±79.1[b]	42.4±65.1[b]
Shrubs/100 m^2	0.9±2.2[a]	2.7±2.0[b]	0.7±1.0[a]	0.3±0.3[c]	0.7±0.8[a]	1.2±0.9[a]
Total Captures	1.0±1.3	2.6±3.5	2.0±3.0	2.2±2.8	3.4±8.2	4.2±6.0
NGM	0.2±0.4[a]	0.1±0.4[a]	1.0±1.7[b]	0.7±1.1[b]	0.5±1.1[a,b]	0.6±1.3[b]
DM	0.1±0.4	1.0±1.9	0.3±0.6	0.1±0.3	1.9±6.0	1.3±2.5
OKR	0.1±0.4	0.9±1.4	0.5±1.2	1.0±2.4	0.3±0.7	1.1±2.5
Richness	0.8±1.1	1.2±1.6	0.9±1.2	1.1±1.0	1.0±1.3	1.6±1.9
Diversity	0.2±0.4	0.3±0.5	0.2±0.4	0.2±0.4	0.2±0.3	0.4±0.6

Values are averages of 36 sampling grids/site. Heights are given in centimeters. Letters refer to statistically significant ($P \leq 0.05$) groupings for each parameter. NGM: Northern grasshopper mouse; DM: deer mouse; OKR: Ord's kangaroo rat.

Table 3. Estimates of survival rates (± SE) for adult and juvenile swift foxes on 6 study sites located in southeastern Colorado, USA, 2001–2004.

Dominant land use	Site	Age class (n)	Seasonal survival rate			Annual survival rate
			Dispersal	*Breeding*	*Pup rearing*	
Grazed	PRV	Adult (15)	0.81 (0.09)	0.81 (0.09)	0.83 (0.08)	0.54
		Juvenile (12)	0.89 (0.10)	0.88 (0.12)		0.65
	COM	Adult (6)	1.0 (0.0)	0.92 (0.08)	1.0 (0.0)	0.92
		Juvenile (5)	1.0 (0.0)	0.50 (0.25)		0.50
Military	BTS	Adult (12)	0.82 (0.08)	0.84 (0.08)	0.78 (0.10)	0.54
		Juvenile (3)	1.0 (0.0)	1.0 (0.0)		0.78
	PRN	Adult (11)	0.79 (0.09)	0.73 (0.11)	0.94 (0.06)	0.54
		Juvenile (11)	0.91 (0.10)	0.60 (0.15)		0.51
Unused	RRK	Adult (2)	0.75 (0.22)	0.80 (0.18)	0.83 (0.15)	0.50
		Juvenile (7)	0.33 (0.27)	1.0 (0.0)		0.27
	BNT	Adult (1)	–	–	–	–
		Juvenile (5)	–	0.75 (0.22)		–

Sample size indicates the age at capture. Juveniles surviving into April of their second year graduated into the adult cohort for analysis.

National Grassland (COM), most likely resulting in greater annual plant production compared to the PRV site where stocking rates remained constant.

At this point, extensive information exists on individual swift fox populations scattered throughout their historic range. However, there is a scarcity of information regarding the variation between these populations and what habitat factors contribute to differences in densities or demographic rates. Our results indicated a strong link between vegetation structure and swift fox population ecology, yet this link was not related to prey abundance in our

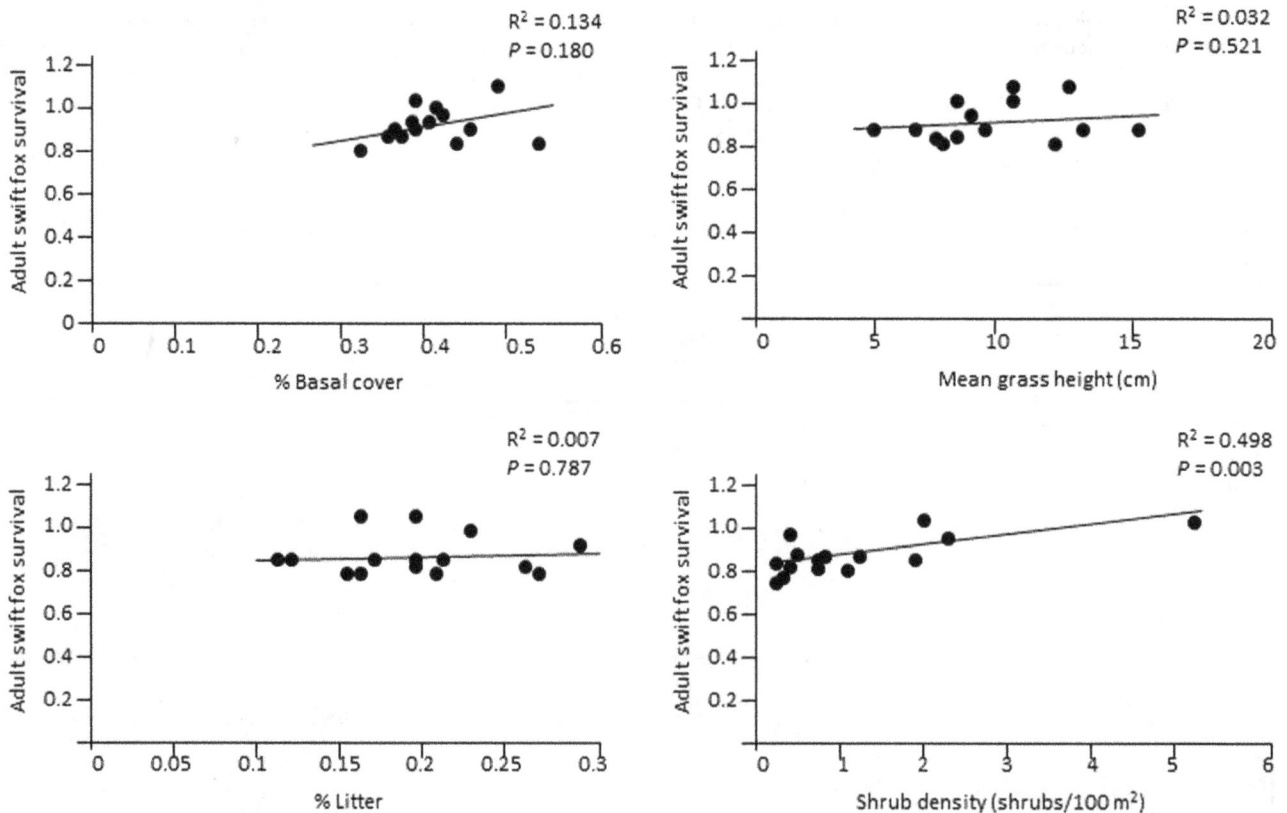

Figure 2. Linear regressions showing the relationships between swift fox survival rates and vegetation structure, southeastern Colorado, USA, 2001–2004.

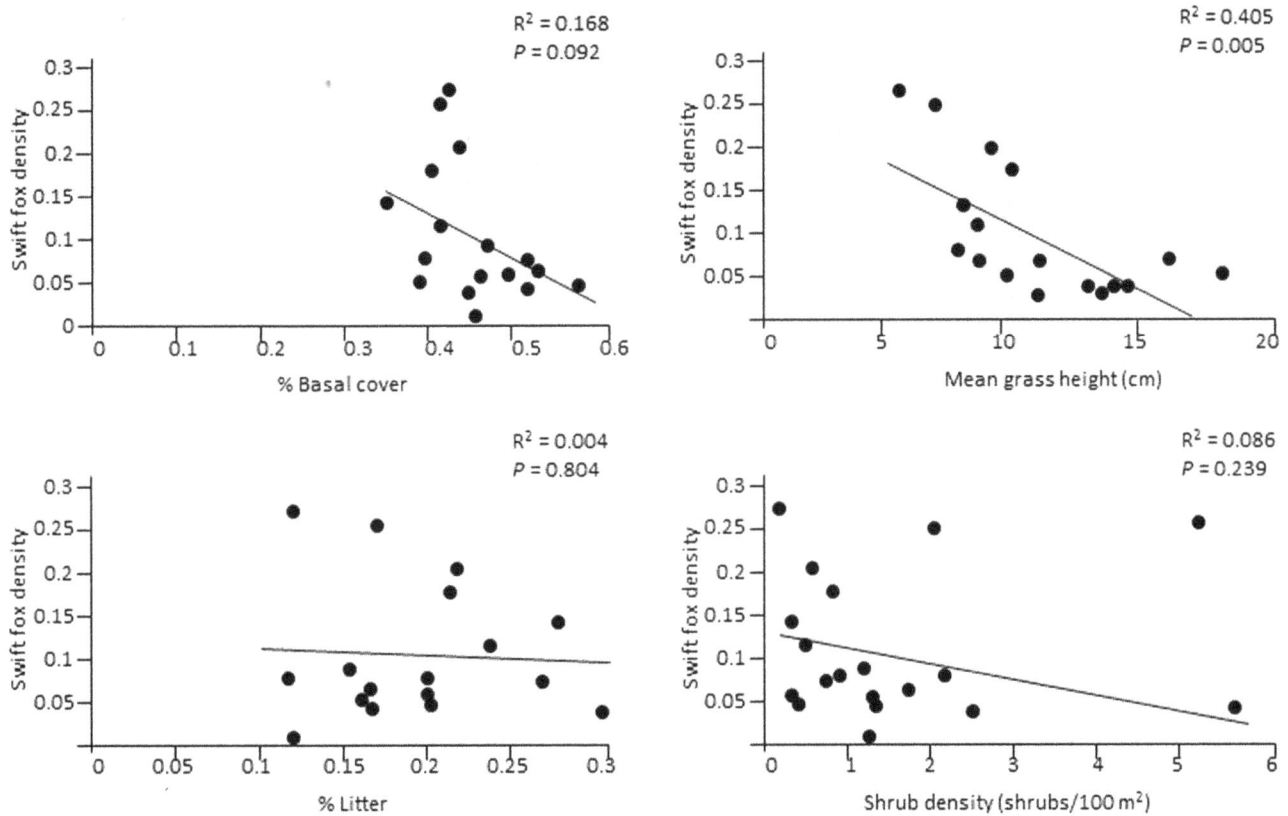

Figure 3. Linear regressions showing the relationships between swift fox population density and vegetation structure, southeastern Colorado, USA, 2001–2004.

study area. Population estimates were negatively related to mean grass height and adult survival was slightly positively related to shrub density. The relationship with grass height has been hinted at but not documented in previous work. For example, Kamler et al. [11] suggested the lack of swift fox activity on ungrazed Conservation Reserve Program grasslands was related to the presence of taller vegetation. They noted that even inexperienced juveniles showed an almost complete avoidance of these areas. Similarly, in our unused Bent site where mean grass height was the highest, only transient foxes were captured which were predominantly young foxes attempting to establish a home range in less habitable areas. We found that no radio-collared foxes established

home ranges despite the lack of competition; all radio-collared foxes either died or left the site. The lack of resident animals in this site hindered our ability to accurately estimate survival. Our results indicated that while swift foxes were capable of exploiting a range of habitats, they showed a higher probability of population persistence in areas where disturbances kept vegetation short.

White et al. [43] documented the transition from grassland to shrubland can be accompanied by a shift from relatively rare, large bodied rodents to more abundant, small-bodied species that have fewer anti-predatory defenses. Mesocarnivores such as swift foxes may benefit from this shift due to the more abundant, vulnerable prey base, and this may explain the slight positive

Table 4. Results from age-structured known fate survival models, constrained by vegetation characteristics.

Model	AICc	ΔAICc	AICc weight	Model likelihood
shrub*grass	257.753	0.00	0.17208	01.0000
shrub	257.944	0.19	0.15643	0.9091
shrub*basal	258.620	0.87	0.11156	0.6483
shrub + shrub*grass	259.835	2.08	0.06076	0.3531
shrub + basal	259.924	2.17	0.05811	0.3377
shrub + litter	259.960	2.21	0.05708	0.3317
shrub + grass	259.968	2.22	0.05684	0.3303
null	260.047	2.29	0.05466	0.3176

Models shown are those that outperformed the null (age-structured) model.

Table 5. Swift fox population density estimates (foxes/km^2) on 6 study sites exposed to 3 land use practices in southeastern Colorado, USA, 2001–2004.

Dominant land use	Site	Seasonal density estimates			Total density estimate (SD)
		Dispersal	Breeding	Pup-rearing	
Grazed	PRV	0.15	0.22	0.21	0.18 (0.10)
	COM	0.03	0.03	0.06	0.04 (0.05)
Military	BTS	0.06	0.17	0.07	0.11 (0.08)
	PRN	0.04	0.11	0.09	0.09 (0.06)
Unused	RRK	0.04	0.06	0.04	0.05 (0.03)
	BNT	0.05	0.04	0.0	0.03 (0.05)

relationship between fox survival and shrub density. However, this benefit comes with increased risk; more shrubs generally mean reduced visibility and more jackrabbits (*Lepus californicus*), leading to increased risk of coyote predation [44]. Throughout the study, we documented dispersing swift foxes avoiding these areas of dense vegetation and assumed this indicated either an innate or learned avoidance of intraguild predation risk. At the same time, foxes living in heavily grazed areas with high shrub density and low mean grass height (COM site) were larger, heavier, and survived longer [14]. Alone, an increase in shrub density appears to carry both costs and benefits for swift foxes; increased predation risk as well as increased prey availability. The addition of increased basal cover and/or grass height, such as found in undisturbed grassland systems, appears to tip the balance and make the landscape unsuitable, presumably by increasing the risk of intraguild predation.

Recent experiments have tested the hypothesis that coyote control will result in increased swift fox survival and density with mixed results. Kamler et al. [45] reported coyote control resulted in increased swift fox survival, density, and recruitment. On the PCMS, Karki et al. [28] found coyote control resulted in increased juvenile fox survival but did not increase fox density due to compensatory dispersal, and suggested coyote control was not an effective means of increasing swift fox densities. Our results suggested an alternative, non-lethal, means of increasing swift fox population viability. Management practices oriented toward reintroducing more complex disturbance regimes such as the combination of an infrequent, intense physical disturbance and

periodic prescribed burning would reduce vegetation density and grass height, and could increase the quality of habitat for swift foxes. Besides the use of prescribed burning, other disturbance regimes that might also reduce vegetation height include controlled grazing during appropriate times of the year, mechanical reduction of woody vegetation (i.e., brush management), or using crop management to reduce crop stubble. No research examining these approaches has been conducted, but warrant future consideration for managing or enhancing swift fox populations.

The relationship between grassland vegetation structure and disturbance regimes has been well established. Swift foxes evolved in grassland systems and as a result depend on grassland disturbance dynamics to maintain habitat quality in our region. Disruptions in grassland disturbance regimes have the potential to degrade swift fox habitat through long-term changes in vegetation structure [8]. A similar scenario was presented by List and Macdonald [46] for kit foxes (*Vulpes macrotis*) on Mexican grasslands, where prairie dog (*Cynomys ludovicianus*) eradication programs risk long term, indirect harm due to shrubland expansion. Our results support the evidence that swift foxes in our region are a shortgrass prairie specialist despite being capable of exploiting sub-optimal habitats [8]. We also found strong evidence of a relationship between habitat quality and landscape heterogeneity, though additional information is needed on exactly how vegetation structure influences swift fox ecology through shifts in prey base or predation pressure. We suggest the regulation of swift fox populations may be based on habitat quality through a

Table 6. Results from age-structured robust design models using Huggin's estimator to derive population size and constrained by vegetation characteristics.

Model	AICc	ΔAICc	AICc weight	Model likelihood
shrub + basal	1722.043	0.00	0.98171	1.0000
shrub	1731.718	9.68	0.00778	0.0079
grass + shrub	1733.196	11.15	0.00372	0.0038
shrub + litter	1733.855	11.81	0.00267	0.0027
shrub + sdshrub	1733.935	11.89	0.00257	0.0026
shrub + sdshrub + shrub*sdshrub	1736.055	14.01	0.00089	0.0009
grass	1736.635	14.59	0.00067	0.0007
null	1775.135	53.09	0.00000	0.0000

Models shown are those that outperformed the null (age-structured) model and resulted in an AICc weight greater than zero.

type of landscape-mediated survival (i.e., mostly predation), and therefore managers may effectively use disturbance regimes to create or maintain habitat.

Acknowledgments

We thank T. Warren, B. Rosenlund, G. Belew, R. Bunn, and M. Klavetter for logistical assistance; E. Joyce, J. White, E. Cleere, M. Watkins, D. Degeranno, A. Larkins, C. Roemer, D. Fletcher, W. Ulrey, S. Schopman, C. Gazal, A. Knipps, J. Garner, and C. Briggs for field assistance; and M. Conner, S. K. M. Ernest, J. A. MacMahon, P. S. Gipson, and F. D. Provenza for reviews of the manuscript.

Author Contributions

Conceived and designed the experiments: EMG CMT. Performed the experiments: EMG CMT. Analyzed the data: EMG CMT. Contributed reagents/materials/analysis tools: EMG CMT. Wrote the paper: EMG CMT.

References

1. Hartnett DC, Hickman KR, Fischer Walter LE (1996) Effects of bison grazing, fire, and topography on floristic diversity in tallgrass prairie. J Range Manag 49: 413–420.
2. Valone TJ, Nordell SE, Ernest SKM (2002) Effects of fire and grazing on an arid grassland ecosystem. Southwest Nat 47: 557–565.
3. Wright HA (1980) The role and use of fire in the semidesert grass-shrub type. US Forest Service, General Technical Report INT-85, Ogden, Utah.
4. Lyon LJ, Huff MH, Smith JK (2000) Fire effects on fauna at landscape scales. In: Smith JK, editor. Wildland fire in ecosystems: effects of fire on fauna. US Forest Service, Rocky Mountain Research Station, General Technical Report RMRS-GTR-42-vol.1, Fort Collins, Colorado. pp. 43–49.
5. Valone TJ, Kelt DA. (1999) Fire and grazing in a shrub-invaded grassland community; independent or interactive ecological effects? J Arid Environments 42: 15–28.
6. Barbour MG, Burk JH, Pitts WD, Gillman FS, Schwartz MW. (1999) Terrestrial plant ecology. Menlo Park, California: Benjamin/Cummings.
7. Brown JR, Archer S. (1999) Shrub invasion of grassland: recruitment is continuous and not regulated by herbaceous biomass or density. Ecology 80: 2385–2396.
8. Sovada MA, Woodward RO, Igl LD (2009) Historical range, current distribution, and conservation status of the swift fox, Vulpes velox, in North America. Can Field Nat 123: 346–367.
9. Scott-Brown JM, Herrero S, Reynolds J (1987) Swift fox. In: Novak M, Baker JA, Obbard ME, Malloch B, editors. Wild furbearer management and conservation in North America. Ontario Trappers Association, North Bay, Ontario. pp. 433–441
10. Sovada MA, Carbyn L, editors (2003) Ecology and conservation of swift foxes in a changing world. Canadian Plains Research Center, University of Regina, Saskatchewan.
11. Kamler JF, Ballard WB, Fish EB, Lemons PR, Mote K, et al (2003) Habitat use, home ranges, and survival of swift foxes in a fragmented landscape: conservation implications. J Mammal 84: 989–995.
12. Finley DJ, White GC, Fitzgerald JP (2005) Estimation of swift fox population size and occupancy rates in eastern Colorado. J Wildl Manag 69: 861–873.
13. Nicholson KL, Ballard WB, McGee BK, Surles J, Kamler JF, et al (2006) Swift fox use of black-tailed prairie dog towns in northwest Texas. J Wildl Manag 70: 1659–1666.
14. Thompson CM, Gese EM (2007) Food webs and intraguild predation: community interactions of a native mesocarnivore. Ecology 88: 334–346.
15. Moehrenschlager A, Cypher BL, Ralls K, List R, Sovada MA (2004) Swift and kit foxes. In: Macdonald DW, Sillero-Zubiri C, editors. Biology and conservation of wild canids. Oxford: Oxford University Press. pp. 185–198.
16. Sovada MA, Roy CC, Bright JB, Gillis JR (1998) Causes and rates of mortality in swift foxes in western Kansas. J Wildl Manag 62: 1300–1306.
17. Nicholson KL, Ballard WB, McGee BK, Whitlaw HA (2007) Dispersal and extraterritorial movements of swift foxes (Vulpes velox) in northwestern Texas. Western North Amer Nat 67: 102–108.
18. Martin DJ, White GC, Pusateri FM (2007) Occupancy rates by swift foxes (Vulpes velox) in eastern Colorado. Southwest Nat 52: 541–551.
19. Jackson VL, Choate JR (2000) Dens and den sites of the swift fox, Vulpes velox," Southwest Nat 45: 212–220.
20. Matlack RS, Gipson PS, Kaufman DW (2000) The swift fox in rangeland and cropland in western Kansas: relative abundance, mortality, and body size. Southwest Nat 45: 221–225.
21. Shaw RB, Diersing VE (1990) Tracked vehicle impacts on vegetation at the Piñon Canyon Maneuver Site, Colorado. J Environmental Quality 19: 234–243.
22. Milchunas DG, Schulz KA, Shaw RB (1999) Plant community responses to disturbance by mechanized military maneuvers. J Environmental Quality 28: 1533–1547.
23. Shaw RB, Anderson SL, Schulz KA, Diersing VE (1989) Plant communities, ecological checklist, and species list for the U.S. Army Pinon Canyon Maneuver Site, Colorado. Colorado State University Science Series 37, Fort Collins.
24. Schauster ER, Gese EM, Kitchen AM (2002) An evaluation of survey methods for monitoring swift fox abundance. Wildl Soc Bull 30: 464–477.
25. Schauster ER, Gese EM, Kitchen AM (2002) Population ecology of swift foxes (Vulpes velox) in southeastern Colorado. Can J Zool 80: 307–319.
26. Kitchen AM, Gese EM, Schauster ER (1999) Resource partitioning between coyotes and swift foxes: space, time, and diet. Can J Zool 77: 1645–1656.
27. Thompson CM, Gese EM (2013) Influence of vegetation structure on the small mammal community in a shortgrass prairie ecosystem. Acta Theriologica 58: 55–61.
28. Karki SM, Gese EM, Klavetter ML (2007) Effects of coyote population reduction on swift fox demographics in southeastern Colorado. J Wildl Manag 71: 2707–2718.
29. Kozlowski AJ, Bennett TJ, Gese EM, Arjo WM (2003) Live capture of denning mammals using an improved box-trap enclosure: kit foxes as a test case. Wildl Soc Bull 31: 630–633.
30. Swihart RK, Slade NA (1985) Influence of sampling interval on estimates of home range size. J Wildl Manag 49: 1019–1025.
31. Disney M, Speigel LK (1992) Sources and rates of San Joaquin kit fox mortality in western Kern County, California. J Wildl Manag 64: 388–400.
32. Guthery FS (1996) Upland gamebirds. In: Krausman PR, editor. Rangeland wildlife. Society for Range Management, Denver. pp. 59–69.
33. Dale MR (1999) Spatial pattern analysis in plant ecology. Cambridge: Cambridge University Press.
34. Thibault KM, Ernest SKM, Brown JH (2010) Redundant or complementary? Impact of a colonizing species on community structure and function. Oikos 119: 1719–1726.
35. Allington GRH, Koons DN, Ernest SKM, Schutzenhofer MR, Valone TJ (2013) Niche opportunities and invasion dynamics in a desert annual community. Ecology Letters 16: 158–166.
36. Morin PJ (1999) Community ecology, Malden, Massachusetts: Blackwell Science.
37. White GC, Burnham KP (1999) Program MARK: survival estimation from populations of marked animals. Bird Study Supplement 46: 120–138.
38. Akaike H (1973) Information theory and an extension of the maximum likelihood principle. In: Petran BN, Csaki F, editors. International symposium on information theory, Akadéniai Kiadi, Budapest, Hungary. pp. 267–281.
39. Rongstad OJ, Laurion TR, Andersen DE (1989) Ecology of the swift fox on the Piñon Canyon Maneuver Site, Colorado. Final report, Wisconsin Cooperative Wildlife Research Unit, University of Wisconsin, Madison, Wisconsin.
40. Olson TL, Lindzey FG (2002) Swift fox survival and production in southeastern Wyoming. J Mammal 83: 199–206.
41. Kamler JF, Ballard WB, Gilliland RL, Mote K (2003) Spatial relationships between swift foxes and coyotes in northwestern Texas. Can J Zool 81: 168–172.
42. Fitzgerald J, Roell B, Dent L, Schafer M, Irby L, et al (1996) Population dynamics of the swift fox (Vulpes velox) in northern Colorado. Final report, University of Northern Colorado, Greeley, Colorado.
43. White EP, Ernest SKM, Thibault KM (2004) Trade-offs in community properties through time in a desert rodent community. Amer Nat 164: 670–676.
44. Germano DJ, Hungerford R, Martin SC (1983) Responses of selected wildlife species to the removal of mesquite from desert grasslands. J Range Manag 36: 309–311.
45. Kamler JF, Ballard WB, Gilliland RL, Lemons PR, Mote K (2003) Impacts of coyotes on swift foxes in northwestern Texas. J Wildl Manag 67: 317–323.
46. List R, Macdonald DW (2003) Home range and habitat use of the kit fox (Vulpes macrotis) in a prairie dog (Cynomys ludovicianus) complex. J Zool 259: 1–5.

11

Estimating Landholders' Probability of Participating in a Stewardship Program, and the Implications for Spatial Conservation Priorities

Vanessa M. Adams[1,2]*, **Robert L. Pressey**[1], **Natalie Stoeckl**[3]

[1] Australian Research Council Centre of Excellence for Coral Reef Studies, James Cook University, Townsville, Queensland, Australia, [2] Research Institute for the Environment and Livelihoods and Northern Australia National Environmental Research Program Hub, Charles Darwin University, Darwin, Northern Territory, Australia, [3] School of Business and Cairns Institute, James Cook University, Townsville, Queensland, Australia

Abstract

The need to integrate social and economic factors into conservation planning has become a focus of academic discussions and has important practical implications for the implementation of conservation areas, both private and public. We conducted a survey in the Daly Catchment, Northern Territory, to inform the design and implementation of a stewardship payment program. We used a choice model to estimate the likely level of participation in two legal arrangements - conservation covenants and management agreements - based on payment level and proportion of properties required to be managed. We then spatially predicted landholders' probability of participating at the resolution of individual properties and incorporated these predictions into conservation planning software to examine the potential for the stewardship program to meet conservation objectives. We found that the properties that were least costly, per unit area, to manage were also the least likely to participate. This highlights a tension between planning for a cost-effective program and planning for a program that targets properties with the highest probability of participation.

Editor: Zoe G. Davies, University of Kent, United Kingdom

Funding: VMA acknowledges funding for field work through The Nature Conservancy and The Thomas Foundation. RLP acknowledges support from the Australian Research Council. The funders had no role in study design, data collection and analysis, decision to publish, or preparation of the manuscript.

Competing Interests: The authors have declared that no competing interests exist.

* E-mail: vanessa.adams@cdu.edu.au

Introduction

Private land conservation is becoming more prominent and important as expansion of strict protected areas is increasingly constrained by reduced availability of land, insufficient budgets for acquisition, and escalating management costs of small, isolated reserves [1–3]. Longstanding conservation programs on private land include the US Conservation Reserve Program [4] and, in Australia, the Victorian Bush Tender Program [5].

Farmers, Indigenous owners and other private landholders manage approximately 77% of Australia's land area. This statistic alone indicates that conservation on private land is integral to Australia's biodiversity conservation strategy [6]. All Australian states and territories have legislation for conservation covenanting on private properties, although some state programs are longer established and cover larger areas than others [7]. Several states have competitive tendering for conservation contracts including the Victorian Bush Tender Program [5], the New South Wales Environmental Services Scheme, and the Queensland Nature Assist program.

Understanding landholders' willingness to participate has two important implications for private land conservation. First, this understanding will shape policy for the design of incentives. For example, factors specific to program design, such as proposed land management, constraints on land title, and delivery of incentives, will influence willingness to participate [8]. Typical approaches to assess the design and viability of stewardship programs include methods such as choice modelling and auctions [9,10]. While these approaches will reveal expected participation levels and provide insights into the effective design of programs, they are not typically structured to assess whether a program is likely to achieve spatial conservation objectives.

The second implication of information on willingness to participate is that identifying willing landholders is vital to identifying areas that are both valuable for achieving objectives and feasible for conservation action. For example, a map of landholders' willingness can be used to design a configuration of protected areas that will, at least in theory, be more easily implemented because it selects those properties owned or managed by people more likely to engage in formal protection [11]. Alternatively, a map of landholders' willingness can be used to assess the likely spatial configuration of voluntary, private protected-area management resulting from a conservation auction, demonstrating the scope for an auction program to achieve conservation outcomes.

Combining willingness and spatial conservation priorities will allow for conservation programs to enhance the likelihood that spatial conservation objectives are met. One example of the potential to incorporate spatial conservation priorities into the auction process is the Western Australian Conservation Auction, in which assessment of the benefits offered by properties accounted for complementarity of conservation values between bids [12]. This process demonstrated the potential to integrate well-developed auction processes with spatial planning to establish sets of private conservation areas that maximized the achievement of conservation objectives within budgets. The conservation outcomes of a program might also depend on aspects of spatial configuration of the properties selected [13–15], for example to achieve objectives related to connectivity and buffering from surrounding land uses. The potential to consider configuration as well as representation of ecosystems and species has been demonstrated in applications to protect and restore private lands [16,17]. Furthermore, configuration of properties might have important social implications in addition to ecological benefits. For example, landholders might be more inclined to participate in a program if their neighbours are participating, one reason being the added certainty that benefits associated with improved management would not be at risk from unmanaged threats nearby (e.g. spread of unmanaged weeds or fire) [18,19].

Of the Australian states and territories, the Northern Territory's policies and funding for conservation on private lands are the least developed, with financial support for conservation covenants and management agreements under consideration. Therefore, we undertook a pilot study in the Daly Catchment to assess the potential for such programs to meet conservation objectives. The program under consideration is for stewardship payments to leverage already extensive routine land management by altering or extending land management practices to meet conservation objectives on private lands. The program would include covenants, which are perpetual titles on private land, as well as management agreements, which are legal agreements between the Government and landholders. We have examined aspects of designing the program such as costs and payment structures [18]. Here, we report on landholders' willingness to participate in such a program.

Our study had three aims. The first was to assess landholders' willingness to participate and inform the design of a stewardship-payment program in the Daly Catchment, Northern Territory. We used a choice experiment to estimate the probability of participation in the program relative to: 1. contract type (covenant versus management agreement); 2. payment amount; and 3. required change in proportion of property managed for conservation. These factors have been identified as important in influencing participation in programs in other regions (e.g., participants relying on production for income may require higher levels of compensation) [20,21]. Choice modelling can estimate the effects of combinations of factors on participants' choices and is therefore useful for designing policies [22,23] and has been used in other regions to explore the influence of attributes such as compensation and duration of contract on willingness to participate [21,24]. The choice model allowed us to estimate the expected level of participation in a program in the Daly, which can indicate the viability of the program more broadly and provide guidelines to the Government about adequate budgets to meet desired participation levels. The choice model also allowed us to examine landholders' preferences for the two mechanisms presented (covenants and management agreements) and how these preferences varied with respect to payment amount and required change in proportion of property to be managed for conservation.

Our second aim was to assess the potential of the stewardship program to meet spatial conservation objectives. We therefore predicted spatially landholders' willingness to participate at the resolution of individual properties and incorporated these predictions into conservation planning software. Predicting willingness to participate for individual properties allowed us to consider the potential spatial distribution of participating properties and

therefore the likely conservation outcomes relative to vegetation types mapped across the catchment. Understanding whether a stewardship program would have the desired impacts of achieving adequate protection for spatially variable conservation features is an important step in scoping a program that has been underutilized.

The third aim of our study was to analyze how the interactions between willingness to participate and conservation costs can influence solutions identified in spatial conservation planning. The potential correlations between conservation costs and willingness to participate have not yet been examined, although they could determine the success of a conservation program. For example, if costs and willingness are negatively correlated then an incentive program would probably be feasible: the properties most likely to be included would also be the most cost-efficient to engage. However, a positive correlation would mean that the most willing landholders also have the least cost-efficient properties, posing difficulties for the design of an incentive scheme. We examined the implications of interactions between costs and willingness to participate in our study region. Our study is the first to incorporate both spatially variable willingness to participate and spatially variable costs. Therefore, this is the first study to elucidate how these two components of the planning problem interact and potentially enhance or constrain capacity to meet conservation objectives.

Materials and Methods

Ethics Statement

This study was approved by James Cook University's Human Ethics Committee (H3283).

Study Area

The study area was the whole of the Daly River catchment in the Northern Territory, covering approximately 5.2 million ha and extending from the coastline south-west of Darwin to 250 km inland (Figure 1A). The Daly River is one of the major river systems in the Top End. Riparian strips in the Daly catchment contain some of the most extensive gallery (rainforest) vegetation in the Northern Territory. Five of the sixty-seven sites of conservation significance identified in the Northern Territory occur in the Daly Catchment [25]. Approximately 10% of the catchment is protected by national parks, such as Nitmiluk Gorge, and Indigenous Protected Areas, such as Fish River. However, protection is not representative across the 105 mapped vegetation types, with 48 having at least 10% area protected and the remaining 57 having less than 10% area protected. Therefore, considerable effort is still needed to ensure adequate and representative protection of the vegetation types in the Daly catchment. In addition, the Daly catchment area is regarded as a highly prospective region for further development. The potential for future pressure to clear native vegetation makes the area a high priority for conservation to ensure valued areas are adequately protected.

The many conservation priorities in the catchment are unlikely to be addressed with further acquisition for national parks because of the large property sizes and correspondingly large acquisition and management costs. Instead, the region is suitable for off-reserve programs involving stewardship payments in conjunction with conservation agreements between the Government and landholders. The mean size of private properties in the Daly is ~10,500 ha (median size 90 ha). Properties larger than 5,000 ha represent approximately 13% of landholders but about 90% of the catchment's private land (Figure 1B). Therefore, engaging with

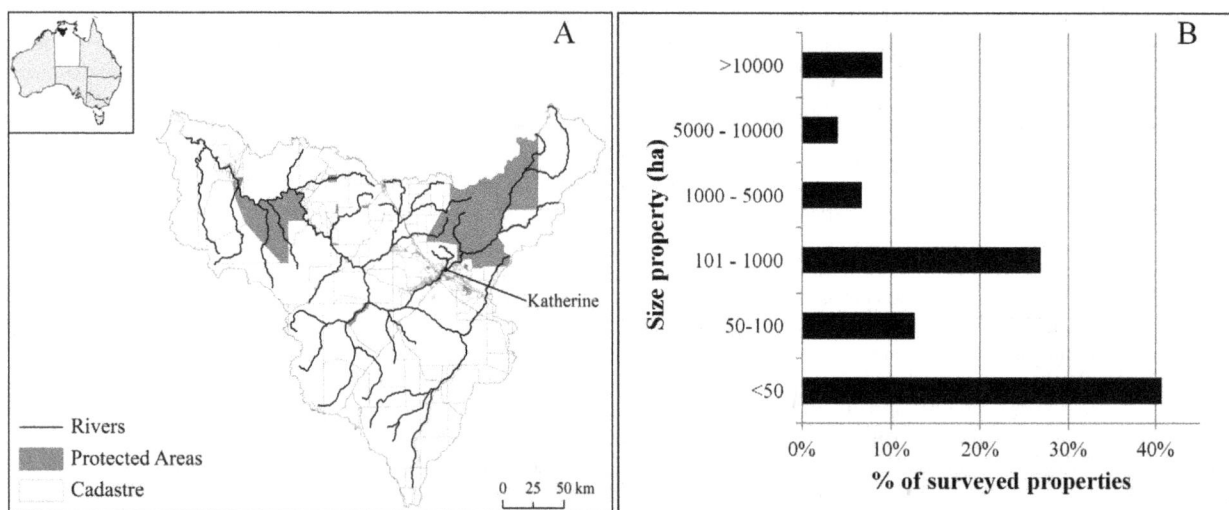

Figure 1. The Daly catchment and pastoral and horticultural properties. The map inset shows the Northern Territory with pale shading and the Daly catchment in black. A) Rivers, protected areas and boundaries of properties (cadastre). Two large national parks extend into the north-east corner of the catchment: Nitmiluk National Park and the southern portion of Kakadu National Park. Fish River Indigenous Protected Area is in the north-west portion of the catchment. B) Size distribution of the 440 properties included in our survey.

relatively few landholders has the potential for extensive conservation benefits. In addition to engaging with private landholders, the Government is interested in funding new Indigenous Protected Areas. These are agreements between traditional owners and the Australian Government, considered to be similar to national parks. Funding both Indigenous Protected Areas and a stewardship program on non-Indigenous properties would equitably provide opportunities for all Daly residents to access financial support for conservation management.

Discussions with the Northern Territory Government

The Northern Territory is the only Australian jurisdiction without well-established arrangements for covenants and conservation management agreements. Therefore, we used the structure of the Queensland Nature Refuge program, which supports establishment of covenants on freehold and leasehold land, as the basis for designing our survey questions. The state of Queensland has more private land under covenant (referred to as Nature Refuges) than any other Australian jurisdiction [26] and has recently implemented legislation, called the Delbessie Agreement, to encourage participation in the program by landholders on extensive leasehold properties [27]. Under the Delbessie Agreement, lessees with properties identified as having conservation value must either enter into a Nature Refuge agreement and be rewarded with a 10-year lease extension or elect to have their properties acquired if they do not wish to participate.

We undertook a series of conversations with the Northern Territory Government that indicated that the relevant agency would consider a scheme similar to Queensland's Nature Refuge program and that the Daly catchment was a high priority area for trialling such a program. We therefore designed our survey with the assumptions that the Northern Territory would model its covenant program for private land on Queensland's and that legislation similar to Queensland's Delbessie Agreement would be considered to support the environmentally sustainable, productive use of rural leasehold land.

Based on a pilot survey, below, and discussions with the Northern Territory Government, including staff working on private protected-area initiatives and spatial conservation plan-

ning, we identified three realistic parameters of a stewardship program. First, the Government would pay a premium to engage landholders in conservation covenants in preference to management agreements because of the perceived benefits of permanent title for conservation (payments of 150% of actual stewardship costs for covenants as opposed to 100% of actual stewardship costs for management agreements). Second, most landholders are currently not managing any areas for conservation, over and above routine property management. Third, landholders participating in the stewardship program would be required to manage several small patches on their property for conservation.

Officers of the Northern Territory Government also indicated that they would consider equal funding for Indigenous Protected Areas alongside funding for a stewardship program on private land to ensure that funds were available for conservation across tenures.

Choice Modelling Experiment and Survey Methods

The survey included questions about the characteristics of landholders and properties, current expenditures on land management and conservation management, and other information specific to the choice experiments. For the choice experiment, respondents were asked to consider the hypothetical scenario of a stewardship program with three alternatives for landholders: conservation covenant, conservation management agreement, or sell property. Choice experiments typically include a status-quo or default option. In our design, we did not include an 'opt-out' option because we wanted to mirror legislation similar to the Delbessie Agreement. Under that arrangement, 'sell property' could be considered the opt-out or status quo because it is the only option for landholders unwilling to place portions of their properties under covenant. Not all on-farm conservation programs have similar 'conserve or sell' clauses, so the results of this experiment are not transferrable to those situations. Indeed, the probabilities of participation estimated here will likely exceed those obtained in situations where neither sale nor participation is necessary. Our results are therefore optimistic estimates of environmental outcomes from a stewardship program and we would expect larger shortfalls in meeting conservation objectives in situations with the default option of not participating.

Based on landholders' attitudes and responses to the Nature Refuge program, we hypothesized that willingness to participate in a program would depend on the type of agreement (covenant or management agreement), the proportion of property already set aside for conservation, the additional proportion of property to be set aside for the program, and the financial payment relative to costs of conservation management above day-to-day land management costs [8,20]. Adams et al. [18] estimated the additional costs of conservation management above day-to-day land management costs for landholders in the Daly and we term these costs 'stewardship costs'. We assumed for our study that landholders would receive stewardship payments as a function of their additional costs to achieve conservation objectives.

In a pilot study, we tested different attributes of a stewardship program to ensure they were cognitively accessible to respondents. Based on the pilot study, we represented financial payment as a percentage of stewardship costs because these costs will vary with current management activities and characteristics of properties, including size. We assumed that financial payments would range from 0% to 150% of stewardship costs (presented to survey respondents as a payment relative to actual costs incurred, Figure 2), and used incremental amounts across that range to allow interpolation between points in our model (Figure 3). We represented the required change in proportion of property set aside for conservation with five representative combinations identified from the pilot study (Figure 3). We constructed the choice sets using a full factorial design, resulting in 80 different combinations (4 covenant payments×4 management agreement payments×5 changes in proportion of property set aside). Because 80 choice sets would be too demanding for a respondent, we chose a blocked full factorial design; blocking is a common way to handle the trade-off between maximising the data collected from each respondent and fatigue of the respondent [28]. The choice sets were blocked into 8 versions of the choice experiment. Each participant was randomly assigned a block of ten choice sets and we ensured that the received responses were evenly distributed across the 80 choice sets. Respondents were given a set of definitions for alternative stewardship arrangements or sale of property using an information box and then asked to choose the preferred option in each choice set (example in Figure 2).

For our survey, we considered only land parcels in the catchment of 10 ha or larger and excluded properties within the town of Katherine (Figure 1). Properties in the town or smaller than 10 ha are probably not good candidates for conservation agreements because they are predominantly residential and unmanaged. We sent surveys to all landholders eligible for private land stewardship agreements, defined here as all 440 pastoralist landholders [see 18 for more details]. We used the Dillman tailored design method [29]. Of the 440 landholders contacted, 25 requested to be removed from the survey and 50 addresses were no longer active, leaving a total of 365 possible respondents. The response rate to the survey (about 25%, or 92 of 365 landholders, with 710 choice sets completed) was in line with similar surveys in the region [30,31]. Responses were also representative of property size and types across the catchment [18] (Survey S1). Based on Orme's rule of thumb [32,33] the target number of respondents was 125. While our achieved response (92) was approximately 25% less than the target and we acknowledge that a small sample is likely to be high risk, our estimated coefficients were all statistically significant suggesting that our sample size was adequate.

Choice Experiment Analysis

We analyzed the choice sets using a conditional mixed-effects logit model in STATA version 9. Based on the choice experiment, the probability of an individual i choosing an alternative m is given by

$$P(y_i = m | x_i, z_i) = \frac{\exp(z_{im}\gamma + x_i\beta_m)}{\sum_j \exp(z_{ij}\gamma + x_i\beta_j)}$$

where alternative specific variables for individual i for alternative m are given by z_{im} and coefficients are denoted by γ, case-specific variables for individual i are given by x_i, and coefficients are denoted by β. In our choice experiment, conservation payments were alternative-specific while conservation configuration was case-specific and landholder-specific variables were included as case-specific variables. We explored a range of landholder-specific variables including size of property, engagement in conservation efforts, land use, number of years on property, and natural characteristics of properties [18]. In our final model we included the only two statistically significant landholder-specific variables: size of property (ln(property size, ha)); and a binary flag indicating whether the landholder was currently engaged in conservation management (conservation flag) (Table 1). Ideally, we would have also tested whether sale values of properties influenced landholders' choices to sell, but reliable sales data were not available for the region.

Application of Choice Model

We used our final choice model (Table 1) for two purposes. First, we explored how the probability of participation was affected by different payment levels, to understand how to maximize participation. Using the survey sample averages, we estimated the catchment-wide average probability of participation in covenants and management agreements based on three payment scenarios: 50% of stewardship costs for both conservation covenants and conservation management agreements; 100% of stewardship costs for both conservation covenants and conservation management agreements; and 150% of stewardship costs for conservation covenants and 100% of stewardship costs for conservation management agreements. For these scenarios, we assumed configuration 2 (no patches currently set aside for conservation and landholders would be required to set aside several small patches for conservation in the future). This was the most likely configuration across the properties in the catchment. These scenarios reflect discussions held with the NT Government about the likely design of the payment program (see section on discussions, above, for further detail).

The second use of the choice model was to create a map of expected probability of participation of individual properties so that our planning scenarios could preferentially select properties with higher probabilities of participation. For each property, we therefore estimated probability of participation assuming payments of 150% for covenants and 100% for management agreements and configuration 2 (no patches currently set aside for conservation and several patches to be set aside for conservation in the future).

Spatial Planning Using Marxan with Zones

We conducted a spatial planning exercise to demonstrate how a map of estimated probability of participation can be used to design configurations of properties that contribute to conservation objectives. In many settings, any one spatial configuration identified during a planning process will change dynamically as

Imagine that the government would like you to set aside two more 'patches' of land on your property for conservation purposes. This would change the configuration of your land	
From: ● **where there is one small patch set aside for conservation**	***To:*** ● ● ■ **where there are several patches set aside for conservation**

Would you choose to	**Payment (as a % of Total Costs)**	**Choice** ✓
Accept a **Conservation covenant**, that would require you to pay for the survey costs, and then spend 1-2 days per month 'managing' the extra conservation areas	*and* receive compensation for **50%** of all costs	☐
Or		
Accept a **Conservation management agreement**, that would require you to purchase some extra supplies (e.g. fencing) and labour (to put the fences in) and that would require you to spend an extra 1-2 days per month 'managing' the areas	*and* receive compensation for **50%** of all costs	☐
Or		
Sell your entire property at market value	Market value	☐

Figure 2. Example choice set presented to respondents in survey.

planners engage with stakeholders and the actual, as opposed to estimated, willingness of landholders to participate is revealed [34]. If landholders identified in the initial configuration refuse to participate, the configuration would be iteratively updated until the full budget was exhausted. For our study, we assumed that the identified configuration of properties would be used to direct first engagement with landholders, so we compared initial configurations between several scenarios (Table 2 and below).

We chose to explore how conservation objectives would be met across the catchment, subject to constraints on funds and area dedicated to conservation, by both Indigenous Protected Areas and stewardship agreements collectively. Indigenous Protected Areas, although very different mechanisms from covenants and management agreements, were important to consider because of the Government's preparedness to consider them as part of an overall approach to nature conservation. Using Marxan with Zones [35] we planned for five zones: 1. national parks; 2. Indigenous Protected Areas (IPAs); 3. stewardship agreements; 4. riparian buffer areas and other sites that are protected under clearing guidelines for the Daly catchment, termed here the 'never clear' zone; and 5. un-engaged areas used for production but not conservation management, termed here the 'available' zone. We chose these zones to account for existing formal conservation areas (zone 1) or parts of the catchment protected from clearing through other measures (zone 4) and to explicitly plan for new conservation areas through IPAs (zone 2) and stewardship agreements (zone 3). Because of the size of the optimization problem (large number of planning units, features, and zones), we chose to select properties for a generalized 'stewardship agreement' zone without differentiating between covenants and management agreements. For the planning process, we assumed a single time step in which areas were identified for engagement for stewardship or Indigenous Protected Areas and that engagement and conservation management would continue. It was beyond the scope of our study to predict the vagaries of iterative adjustments to configurations [34]

as individual landholders are engaged and some decline participation.

Marxan, a widely used conservation planning tool, uses the simulated annealing algorithm to minimize the objective function score:

$$\sum_{i=1}^{m} c_i x_i + \sum_{i=1}^{m} \sum_{h=1}^{m} x_i * (1 - x_h) cv_{ih}$$

subject to the constraint that objectives are met:

$$\sum_{i=1}^{m} x_i r_{ij} \geq T_j, \forall j$$

For m planning units, n features, r_{ij} is the occurrence level of feature j in site i and x_i is the control variable that indicates which planning unit is in, or out of, the reserve system. Marxan with Zones generalizes this approach by increasing the number of states or zones to which a planning unit can be assigned.

We used Marxan with Zones to examine, for four scenarios, possible spatial configurations of Indigenous Protected Areas and stewardship agreements. The scenarios (Table 2) were designed to examine the influence of variable costs and variable probability of participation on: 1. the spatial configuration of properties selected for a stewardship program; and 2. the capacity to meet conservation objectives within budget constraints.

Scenario 1 (uniform costs). Cost of each planning unit is equal to its area; not considering probability of participation.

Scenario 2 (variable costs). Cost of each planning unit is equal to the expected cost of participation in a stewardship program; not considering probability of participation.

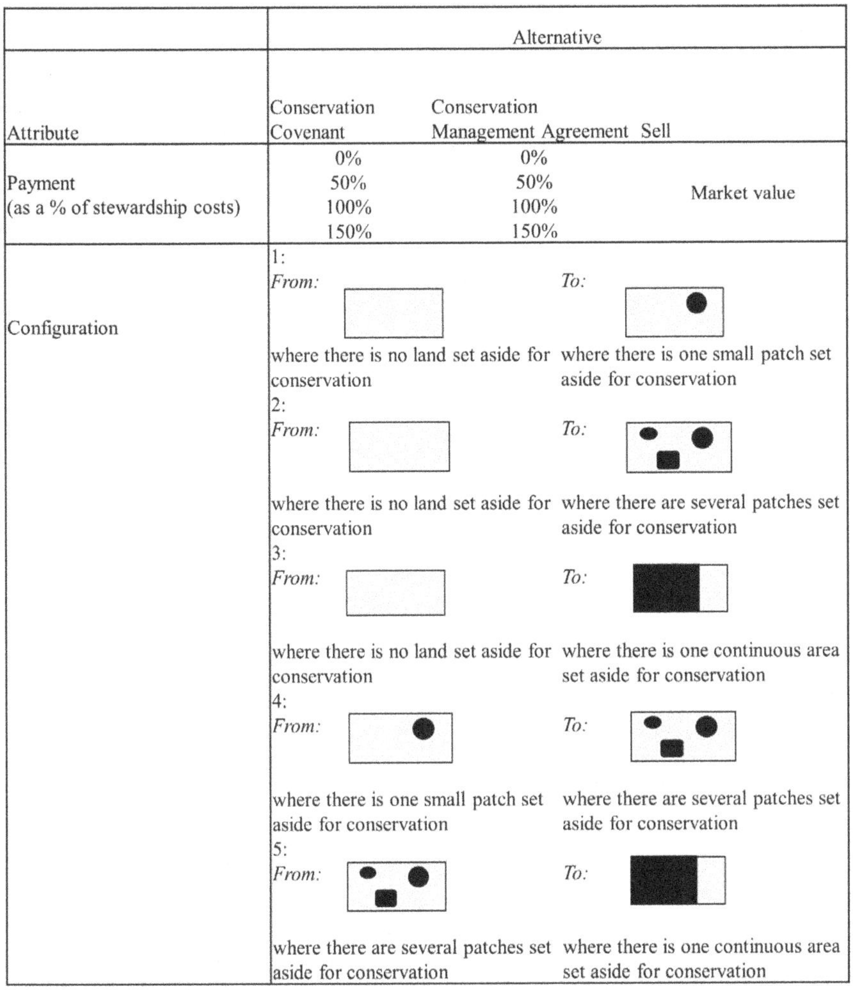

	Alternative		
Attribute	Conservation Covenant	Conservation Management Agreement	Sell
Payment (as a % of stewardship costs)	0% 50% 100% 150%	0% 50% 100% 150%	Market value
Configuration	1: *From:* [ ] where there is no land set aside for conservation 2: *From:* [ ] where there is no land set aside for conservation 3: *From:* [ ] where there is no land set aside for conservation 4: *From:* [●] where there is one small patch set aside for conservation 5: *From:* [● ● ■] where there are several patches set aside for conservation	*To:* [●] where there is one small patch set aside for conservation *To:* [● ● ■] where there are several patches set aside for conservation *To:* [█] where there is one continuous area set aside for conservation *To:* [● ● ■] where there are several patches set aside for conservation *To:* [█] where there is one continuous area set aside for conservation	

Figure 3. Attribute levels and changes for the choice experiment. The survey provided respondents with three alternatives: conservation covenant, conservation management agreement, or sell property. The choice experiment explored two attributes that might influence respondents' choices: payment level as a percentage of stewardship costs (defined here as the additional costs of managing land for conservation, over and above routine property management) and change in extent and configuration of conservation management, defined relative to current configuration (*From*) and future configuration (*To*). We considered four payment levels and five changes in configuration.

Scenario 3 (uniform costs + probability of participation). Cost of each planning unit is equal to its area; considering probability of participation.

Scenario 4 (variable costs + probability of participation). Cost of each planning unit is equal to the expected cost of participation in a stewardship program; considering probability of participation.

We divided all properties in the catchment into planning units of square 25 ha grids (n = 212,173). Relatively small planning units allowed us to capture already protected areas in 'never clear' zones within properties and to identify spatial heterogeneity of conservation priority within properties (as opposed to identifying only whole properties as priorities). To control the aggregation of selected areas [35], we identified the zone boundary cost for each scenario with the method of Stewart & Possingham [36]. Because properties were divided into multiple planning units we checked that the percentage of each property selected for stewardship agreements was in line with the configuration assumptions made for calculating probability of participation. We included quantitative objectives for 105 vegetation types and the 5 sites of conservation significance in the Daly catchment, to give a total of

110 conservation features. The sites of conservation significance within the Daly River catchment have been assessed as either nationally or internationally significant and include features such as the Daly River, Anson Bay and Floodplains and Western Arnhem Plateau [25]. Based on discussions with the Northern Territory Government Department for Natural Resources, Environment, The Arts and Sport (NRETAS), our objectives were 30% of the current extent of each vegetation type (because pre-clearing data were not available) and 100% of each site of conservation significance. The Northern Territory has clearing guidelines for the Daly that allocate buffers around sensitive vegetation or other features that cannot be cleared (e.g. a required 250 meter buffer around all streams) [37]. Therefore, we locked all required buffers (the 'never clear' zone in Marxan with Zones) into the selected conservation configuration for all scenarios. In addition, we locked in all existing national parks and Indigenous Protected Areas. We assumed that the different zones contributed differentially to conservation objectives (Table 2), reflecting different levels of commitment of management to conservation.

Marxan with Zones minimizes the total cost of the zoning plan C:

Table 1. Conditional mixed-effects logit model.

Variable	Coefficient		SE
CC intercept	0.3704		0.3861
CMA intercept	0.3788		0.3492
Payment	0.0133	***	0.0012
Configuration 2, CC	−1.1400	***	0.3420
Configuration 3, CC	−1.3841	***	0.3562
Configuration 4, CC	−1.0396	**	0.3442
Configuration 5, CC	−1.1116	***	0.3336
Configuration 2, CMA	−0.6737	**	0.3123
Configuration 3, CMA	−1.0710	***	0.3216
Configuration 4, CMA	−0.4958	*	0.3091
Configuration 5, CMA	−0.8896	**	0.3162
Conservation flag, CC	2.2508	***	0.2935
Conservation flag, CMA	1.3770	***	0.2625
ln(property size), CC	−0.6335	***	0.1230
ln(property size), CMA	−0.4577	***	0.1004
N (Choice sets)	710		
Log L	−654.32		
rho^2	0.16		

CC indicates conservation covenant; CMA indicates conservation management agreement. Configuration was coded as a set of dummy variables (corresponding to alternative changes in configuration in Figure 2) with configuration 1 chosen as the status quo.
*$p<0.05$,
**$p<0.005$,
***$p<0.001$.

$$C = \sum_{i=1}^{M} \sum_{j=1}^{N} c_{ij} x_{ij}$$

where $x_{ij} = 1$ if the ith planning unit is included in the jth zone, subject to the constraint that a planning unit can only be placed in one zone. For scenarios 1 and 3, a uniform cost was used for planning units (i.e. cost was equal to the area of the planning unit). For scenarios 2 and 4 a spatially variable cost was used for

Table 2. Scenarios compared using Marxan with Zones.

Scenario	Zones included (proportional contribution of zones to objectives in parentheses)	Cost IPA	Cost stewardship
Scenario 1 - Uniform costs	1 – National Park (1)	Area	Area
	2 – IPA (1)		
	3 – Stewardship (1)		
	4 – Never Clear (0.7)[a]		
	5 – Available (0)[b]		
Scenario 2 - Variable costs	As above	$2.25 per ha	Estimated expected stewardship costs per ha
Scenario 3 - Uniform costs + probability of participation	As above	Area	Area
Scenario 4 - Variable costs + probability of participation	As above	$2.25 per ha	Estimated expected stewardship costs per ha

We defined scenarios in terms of zones considered, proportional contribution of zones to conservation objectives, costs of management in Indigenous Protected Areas (IPAs), and costs of stewardship (private pastoral zone).
[a]Areas covered by legislation that prevents clearing, assuming that this legislation is fully effective for ensuring that the area will not be cleared but contributes less than a protected area managed for conservation. For example these areas may be grazed or have invasive weeds or feral animals present, which may result in lower biodiversity compared to conserved land [49].
[b]Currently not managed for conservation but available for management either with IPA or stewardship.

planning units. For properties under consideration for Indigenous Protected Areas, we used an expected conservation management cost per ha of $2.25 [25]. Because we did not distinguish between covenants and management agreements in our optimization problem, we calculated an expected stewardship cost per property based on: 1. covenants receiving a premium over management agreements (150% of total costs as compared to 100% for management agreements); 2. the estimated probability of participation in each mechanism; and 3. the explanatory model of stewardship costs from Adams et al. [18] to estimate the per-ha costs of stewardship payments. Therefore, we calculated the total expected cost per ha of stewardship payments per property as:

$$E(C) = prob_{cma|part}c_{cma} + prob_{cc|part}c_{cc}$$

where $prob_{cma}$ is the probability of landholder i selecting a conservation management agreement calculated with the choice model, $prob_{cc}$ is the probability of landholder i selecting a conservation covenant calculated with the choice model, $prob_{cma|part}$ is the probability of landholder i selecting a conservation management agreement given the landholder has agreed to participate in the program (equal to $prob_{cma}/(1-prob_{sell})$), $prob_{cc|part}$ is the probability of landholder i selecting a conservation covenant given the landholder has agreed to participate in the program (equal to $prob_{cc}/(1-prob_{sell})$), c_{cma} is the cost of stewardship payment to landholder i based on Adams et al [18], and c_{cc} is 150% of c_{cma}. We then calculated the management cost of each planning unit from the calculated per-ha expected cost of stewardship.

For scenarios 2 and 4 we ran Marxan with Zones to achieve objectives within a constrained budget of $1.5 million to fund Indigenous Protected Areas and stewardship agreements in the catchment. This figure was based on the non-spatial financial estimate of $1 million required for stewardship agreements across pastoral properties in the catchment [18] and a pro-rated estimate of $0.5 million for Indigenous Protected Areas over about 1.5 million ha of Indigenous land. To ensure that scenarios 1 and 2 were directly comparable, we selected a budget for scenario 1 equal to the average area selected in scenario 2 under the constrained budget of $1.5 million (740,000 ha). Similarly, for comparability of scenarios 3 and 4, we selected a budget for scenario 3 equal to the average area selected in scenario 4 under the constrained budget of $1.5 million (620,000 ha).

In scenarios 3 and 4 we include the estimated probability of participation in the stewardship program in the optimization problem to demonstrate how these data might be used in spatial planning. We wanted to select properties with the highest probability of participation while still meeting our objectives within a constrained budget. To do this we included the estimated probability of participation as a conservation feature for each pastoral property and set a catchment-wide objective of 15% of the total probability (which is computationally similar to the approach used by other studies) [11]. The 15% objective was selected to reflect the non-spatial findings of Adams et al. [18] that a $1 million budget would be sufficient to support participation of the most cost-efficient properties (i.e. the largest 15% of properties). Importantly, this approach also allowed probability of participation to be separated from stewardship costs in the software analyses.

We ran Marxan with Zones with 100 runs for each scenario and recorded best solutions and selection frequency for each scenario. For the best solution in each scenario we summarized the total cost and area as well as the cost and area for the stewardship zone (a subset of the areas in the solution for each scenario, Table 2). For

properties selected for stewardship we also calculated the average and median property size, percentage of properties selected for stewardship (out of 535 properties), average probability of participation of properties selected, and percentage of total probability. To isolate the effects of including variable costs, we compared scenarios 1 and 2 and scenarios 3 and 4, respectively. To isolate the effects of including probability of participation, we compared scenarios 2 and 4. Scenarios 1 and 3 were not directly comparable because their area budgets were different, having been calibrated, respectively, from scenarios 2 and 4. Lastly, we compared scenarios 3 and 4 to examine the combined effects of including variable costs and probability of participation.

Results

The final conditional mixed-effects logit model for the choice experiment (Table 1) included the two significant landholder-specific variables - ln(property size, ha) and conservation flag (p<0.001) - in addition to the two design variables being investigated (configuration and payment). The coefficient for property size was negative, indicating that owners of larger properties were less likely to participate. The coefficient for conservation flag was positive, indicating that owners already engaged in conservation management were more likely to participate. The coefficients for configuration levels were negative, and increasingly so with the extent of change in required proportion of property to be managed for conservation. Accordingly, configuration 3, requiring landholders to change from no patches to one large continuous patch set aside for conservation, had the largest negative coefficient. This trend was similar for both covenants and management agreements. However, the coefficients for covenant configurations were more strongly negative, indicating that landholders were less likely to select a covenant than a management agreement. The coefficient for payment level was positive, indicating that probability of participation increased with payment amount.

For our three payment scenarios, in which payment levels were varied but configuration was held constant, the predicted probabilities of participation in stewardship arrangements increased from 42% to 64% as payment levels increased (Table 3). Respondents always preferred conservation management agreements to covenants. However, the payment premium for covenants substantially increased the probability of participating through a covenant (29% for 150% payment, 18% for 100% payment, Table 3). The design of our choice experiment, lacking an alternative for 'opting-out' of negotiations without selling, probably produced absolute probabilities of participating that were higher than an alternative survey design with an 'opt-out' choice. However, we expect that the relative probabilities between payment levels and stewardship arrangements reliably indicate the preferences of landholders in the Daly. In fact, the preference of management agreements over covenants was supported by qualitative results from in-person interviews and unsolicited comments provided in survey responses. In addition, if the design of the stewardship program in the Northern Territory comes to reflect the constraints of the Queensland program, coupled with the Delbessie Agreement, then our probabilities will be directly applicable.

In all scenarios, there was approximately 0.5 million ha in existing national parks and Indigenous Protected Areas and an additional 1.5 million ha in buffer areas (the 'never clear' zone). Eighty-nine of the 110 objectives were fully achieved in these buffers and protected areas.

For the annual budget of $1.5 million to support management of Indigenous Protected Areas and stewardship agreements, not all

Table 3. Estimated probabilities of participation for three payment scenarios.

	Payment scenarios		
	50% CC, 50% CMA	100% CC, 100% CMA	150% CC, 100% CMA
Conservation Covenant (CC)	0.13	0.18	0.29
Conservation Management Agreement (CMA)	0.29	0.40	0.35
Stewardship arrangement (CC + CMA)	0.42	0.58	0.64
Sell Property	0.58	0.42	0.36

CC indicates conservation covenant; CMA indicates conservation management agreement.

conservation objectives could be met (number of missed objectives ranged from 8 to 11 across scenarios, Table 4). In all cases, the 100% targets for the five sites of conservation significance could not be met. Other shortfalls were for rarer vegetation types.

Including variable costs reduced the number of properties engaged in stewardship agreements by selecting larger properties, a consequence of strong economies of scale for stewardship costs (compare scenarios 1 and 2 and scenarios 3 and 4, respectively, Table 4). Including variable costs also lowered the overall probability of participation across selected properties because landholders on larger properties were less likely to participate (Table 4). The effects of considering variable costs, in terms of number and size of properties, were more dramatic when probability of participation was not considered (compare scenarios 1 and 2, Table 4). With probability of participation also included, these effects of variable costs were tempered (compare scenarios 3 and 4, Table 4) by the inverse relationship between cost and probability of participation, below.

Including variable costs and probability of participation (scenario 4) shifted spatial selections to smaller properties with higher probability compared to using variable costs only (scenario 2), and this increased the number of missed objectives marginally to 11, compared to 9 in scenario 2. In relation to scenario 2, the average probability of participation of selected properties in scenario 4 almost doubled, the percentage of total probability in selected properties increased five-fold, and the median property size dropped from 3,717 ha to 579 ha (Table 4).

Compared to probability of participation with uniform costs (scenario 3), including both probability of participation and variable costs (scenario 4) decreased overall probability of participation. This was because including variable costs slightly increased the average size of properties, due to economies of scale, but also reduced the average probability of participation, due to the negative relationship between property size and probability of participation (Table 1). This reflects a tension between two key considerations in the Daly: cost-effectiveness requires that larger properties should be targeted for stewardship, but overall probability of participation is thereby lowered.

Discussion

Choice modelling has been applied to management of protected areas or design of conservation incentives [22,23] but, to our knowledge, it has not previously been combined with conservation planning for optimal spatial design of a stewardship program. Our choice analysis provides several insights for designing and implementing a stewardship program in the Northern Territory. We estimated that a large percentage of landholders – between 42% to 64%, depending on payment levels - would be willing to participate in stewardship agreements (i.e. a covenant or a management agreement). We found that landholders were financially motivated in their preferences between conservation management agreements and conservation covenants. All else being equal, landholders preferred management agreements, reflecting their reported concerns over the title implications of covenants and potential negative effects on sale values. This is consistent with previous reports of respondents' concerns over agreements impinging upon their rights to use and manage land [20] and previous findings that shorter or less restrictive management agreements are preferred [24]. However, this preference for management agreements in the Daly can apparently be weakened with a payment premium for covenants. Covenants have benefits for the Government. The first is the security of permanent titling [7]. Second, titling allows covenants to be classified as IUCN-recognized protected areas (Class VI in the case of Nature Refuges, however private protected areas may qualify for all classes) [38] so that covenants then contribute to national conservation goals such as the 2020 17% target under the Convention on Biological Diversity [39].

The stewardship payment model developed by Adams et al. [18] found strong economies of scale with the largest properties being the most cost-efficient. However, in our choice model, the negative coefficient associated with *ln(property size, ha)* indicated that the most cost-efficient properties were also the least likely to participate. This finding could reflect the tendency for larger properties to be more likely associated with production land uses, with production landholders more concerned about lost income from stewardship agreements than non-production landholders on smaller properties [20]. Our spatial zonings supported the findings of Adams et al. [18] that including variable stewardship costs to select the most cost-efficient implementation of the stewardship program resulted in engaging with larger properties. However, our analyses here also demonstrated that the budget level of $1.5 million per annum was insufficient for all conservation objectives to be met. Furthermore, if the stewardship program were implemented as a closed-bid auction, probably even fewer conservation objectives would be met because landholders on the most cost-efficient properties would be less likely to submit bids. Rather, the more willing participants would be more likely to have smaller properties that are more costly to manage per ha, and a larger budget would therefore be needed to meet conservation objectives while engaging these landholders. Therefore, our analysis indicates that, to meet the kinds of conservation objectives used here, the design of the stewardship program would need to encourage participation of large properties by addressing their managers' specific concerns about enrolling [8,20] or involve a

Table 4. Summary results from Marxan with Zones for the four scenarios, including number of objectives met, total cost, total cost of stewardship agreements, total area, total area selected for stewardship, average and median property size, percentage of properties selected for stewardship, average probability of participation of selected properties (average probability across all properties is 32.7% with a range from 0.9% to 78.7%), and percentage of total probability of participation in selected properties (objective was 15% for scenarios 3 and 4).

	Objectives met (of 110)	Total cost ($)	Total Stewardship cost ($)	Total area (ha)	Total area – Stewardship (ha)	Average property size (ha)	Median property size (ha)	Percentage of properties selected for stewardship	Average Probability	Percentage of total probability
Scenario 1 - Uniform cost	103	1,962,437	1,125,664	740,000	371,899	11,274	392	45.7%	22.2%	23.4%
Scenario 2 - Variable cost	101	1,500,000	623,172	740,000	389,701	18,610	3,717	27.5%	12.6%	2.8%
Scenario 3 - Uniform cost + PoP*	103	1,650,119	1,019,919	620,000	280,089	11,794	481	43.6%	21.6%	20.0%
Scenario 4 - Variable cost + PoP*	99	1,500,000	822,245	620,000	301,225	12,778	579	40.3%	20.2%	15.0%

*Probability of participation.

larger budget than $1.5 million to engage small properties. Alternatively the Government might fund an outreach campaign prior to starting the stewardship program to increase the probability of larger properties participating.

Our choice experiment sought to mimic a key characteristic of the Delbessie Agreement by explicitly not offering landholders a choice to 'opt-out'. We believe, however, that our experimental design would not have exaggerated one of our key conclusions: that property size was inversely related to probability of participation, creating a tension between selecting properties that are cost-efficient and selecting properties with landholders who are willing to participate. Our design would have exaggerated the negative association between property size and probability of participating only if probability of selling and property size were positively related, that is, if owners of larger properties were more likely to sell than those of smaller properties. In that case, having the option to sell rather than to engage in stewardship would be more appealing to owners of larger properties. However, we found that property size and number of years of ownership, admittedly an imprecise proxy for propensity to sell, were uncorrelated. We conclude that it is unlikely that an alternative experimental design would have changed our observed negative association between property size and probability of participation.

The number of studies considering variable conservation costs has increased recently, demonstrating the benefits associated with incorporating costs into priority setting [40]. Recent advances have included more sophisticated dynamics such as land-market feedbacks [41,42]. However, studies of variable costs typically assume uniform availability of land. Specifically, they fail to consider that some landholders will be more or less willing to engage in conservation management, whether by selling their land or participating in stewardship programs. Progress on incorporating costs parallels advances in integrating other social considerations in systematic planning, such as measures of willingness or social indicators of feasibility. These studies have demonstrated ways of making plans more readily implemented [43,44]. Variation in landholders' willingness to participate, similar to our approach here, has been considered in two other spatial prioritizations [11,45], but those studies included costs (unrealis-

tically) as uniform, average sales prices across properties. To our knowledge, no previous study has included both spatially variable costs and spatial variation in willingness to participate in spatial optimization.

By selecting areas with spatially variable data on both costs and willingness to participate, our study demonstrated that, with a constrained budget, spatially variable costs can be more important than willingness in determining conservation priorities. This is likely to be the case more generally, where economies of scale apply to costs such as those of acquisition and management [46–48]. This result provides an important insight into the potential interactions between the spatial distribution of conservation features, costs of conservation, and willingness of landholders to engage in conservation. These interactions will be important to consider for future studies concerned with opportunities for and constraints on implementation. Our analyses highlight important design and policy issues associated with implementing a stewardship program in the Northern Territory and other parts of the world. If planners understand the spatial drivers of both costs and probability of participation, then trade-offs can be addressed proactively with engagement strategies or arguments for adequate budgets.

Acknowledgments

VMA thanks D. Witzke for field assistance and those landholders who responded to the survey.

Author Contributions

Conceived and designed the experiments: VMA RLP NS. Performed the experiments: VMA. Analyzed the data: VMA. Wrote the paper: VMA RLP NS.

References

1. Bruner AG, Gullison RE, Balmford A (2004) Financial costs and shortfalls of managing and expanding protected-area systems in developing countries. Bioscience 54: 1119–1126.
2. James AN, Gaston KJ, Balmford A (1999) Balancing the Earth's accounts. Nature 401: 323–324.
3. Joppa LN, Loarie SR, Pimm SL (2008) On the protection of "protected areas". Proc Natl Acad Sci USA 105: 6673–6678.
4. Claassen R, Cattaneo A, Johansson R (2008) Cost-effective design of agri-environmental payment programs: US experience in theory and practice. Ecol Econ 65: 737–752.
5. Stoneham G, Chaudhri V, Ha A, Strappazzon L (2003) Auctions for conservation contracts: an empirical examination of Victoria's BushTender trial. The Australian Journal of Agricultural and Resource Economics 47: 477–500.
6. Commonwealth of Australia (2008) Caring for our Country Outcomes 2008–2013. Canberra.
7. Fitzsimons JA, Wescott G (2004) The classification of lands managed for conservation: existing and proposed frameworks, with particular reference to Australia. Environmental Science & Policy 7: 477–486.
8. Moon K, Cocklin C (2011) A landholder-based approach to the design of private-land conservation programs. Conserv Biol 25: 493–503.
9. Ferraro PJ (2008) Asymmetric information and contract design for payments for environmental services. Ecol Econ 65: 810–821.
10. Jack BK, Leimona B, Ferraro PJ (2009) A Revealed Preference Approach to Estimating Supply Curves for Ecosystem Services: Use of Auctions to Set Payments for Soil Erosion Control in Indonesia. Conserv Biol 23: 359–367.
11. Guerrero AM, Knight AT, Grantham HS, Cowling RM, Wilson KA (2010) Predicting willingness-to-sell and its utility for assessing conservation opportunity for expanding protected area networks. Conserv Lett 3: 332–339.
12. Hajkowicz S, Higgins A, Williams K, Faith DP, Burton M (2007) Optimisation and the selection of conservation contracts. Australian Journal of Agricultural and Resource Economics 51: 39–56.
13. Lombard AT, Cowling RM, Vlok JHJ, Fabricius C (2010) Designing conservation corridors in production landscapes: Assessment methods, implementation issues, and lessons learned. Ecol Soc 115: [online] Available: http://www.ecologyandsociety.org/vol15/iss13/art17/.
14. Nicholson E, Westphal MI, Frank K, Rochester WA, Pressey RL, et al. (2006) A new method for conservation planning for the persistence of multiple species. Ecol Lett 9: 1049–1060.
15. Rouget M, Cowling RM, Lombard AT, Knight AT, Graham IHK (2006) Designing large-scale conservation corridors for pattern and process. Conserv Biol 20: 549–561.
16. Bryan BA, Crossman ND (2008) Systematic regional planning for multiple objective natural resource management. J Environ Manage 88: 1175–1198.
17. Seddon JA, Barratt TW, Love J, Drielsma M, Briggs SV, et al. (2010) Linking site and regional scales of biodiversity assessment for delivery of conservation incentive payments. Conserv Lett 3: 229–235.
18. Adams VM, Pressey RL, Stoeckl N (2012) Estimating land and conservation management costs: the first step in designing a stewardship program for the Northern Territory. Biol Conserv 148: 44–53.
19. Adams VM, Setterfield SA (2013) Estimating the financial risks of Andropogon gayanus to greenhouse gas abatement projects in northern Australia. Environ Res Lett 8: doi:10.1088/1748-9326/1088/1082/025018.
20. Moon K, Cocklin C (2011) Participation in biodiversity conservation: Motivations and barriers of Australian landholders. Journal of Rural Studies 27: 331–342.

21. Broch S, Vedel S (2012) Using Choice Experiments to Investigate the Policy Relevance of Heterogeneity in Farmer Agri-Environmental Contract Preferences. Environmental and Resource Economics 51: 561–581.

22. Jacobsen JB, Thorsen BJ (2010) Preferences for site and environmental functions when selecting forthcoming national parks. Ecol Econ 69: 1532–1544.

23. Horne P, Boxall PC, Adamowicz WL (2005) Multiple-use management of forest recreation sites: a spatially explicit choice experiment. For Ecol Manage 207: 189–199.

24. Horne P (2006) Forest owners' acceptance of incentive based policy instruments in forest biodiversity conservation-A choice experiment based approach. Silva Fennica 40: 169–178.

25. NRETAS (2009) Recognising sites of conservation significance for biodiversity values in the Northern Territory. Palmerston: Biodiversity Conservation Unit, Department of Natural Resources, Environment, The Arts and Sport.

26. Adams VM, Moon K (2013) Security and equity of conservation covenants: contradictions of private protected area policies in Australia. Land Use Policy 30: 114–119.

27. DERM (2007) Delbessie Agreement (State Rural Leasehold Land Strategy). Available: http://www.nrw.qld.gov.au/land/state/rural_leasehold/strategy.html. Accessed 2010 Feb.

28. Hensher DA, Rose JM, Greene WH (2006) Applied Choice Analysis A Primer. Cambridge: Cambridge University Press.

29. Dillman DA (2007) Mail and internet surveys: the tailored design method - 2007 update. Hoboken, NJ: John Wiley. 565 p.

30. Zander KK, Garnett ST, Straton A (2010) Trade-offs between development, culture and conservation - Willingness to pay for tropical river management among urban Australians. J Environ Manage 91: 2519–2528.

31. Zander KK, Straton A (2010) An economic assessment of the value of tropical river ecosystem services: Heterogeneous preferences among Aboriginal and non-Aboriginal Australians. Ecol Econ 69: 2417–2426.

32. Orme B (1998) Sample size issues for conjoint analysis studies. Sawthooth Software Research paper Series Squim, WA, USA: Sawthooth Software Inc.

33. Rose J, Bliemer MJ (2013) Sample size requirements for stated choice experiments. Transportation 40: 1021–1041.

34. Pressey RL, Mills M, Weeks R, Day JC (2013) The plan of the day: managing the dynamic transition from regional conservation designs to local conservation actions. Biol Conserv 166: 155–169.

35. Watts ME, Ball I, Stewart RS, Klein CJ, Wilson K, et al. (2009) Marxan with Zones - software for optimal conservation-based land- and sea-use zoning. Environ Model Software: 1–9.

36. Stewart RR, Possingham HP (2005) Efficiency, costs and trade-offs in marine reserve system design. Environmental Modeling and Assessment 10: 203–213.

37. NRETAS (2010) Department of Natural Resources, Environment, The Arts and Sport land clearing guidelines. Palmerston NT: Northern Territory Department of Natural Resources, Environment, The Arts and Sport.

38. Fitzsimons JA (2006) Private Protected Areas? Assessing the suitability for incorporating conservation agreements over private land into the National Reserve System: A case study of Victoria. Environmental and Planning Law Journal 23: 365–385.

39. UNEP (2010) Report of the Tenth Meeting of the Conference of the Parties to the Convention on Biological Diversity (UNEP/CBD/COP/10/27).

40. Polasky S (2008) Why conservation planning needs socioeconomic data. Proc Natl Acad Sci USA 105: 6505–6506.

41. Armsworth PR, Daily GC, Kaireva P, Sanchirico JN (2006) Land market feedbacks can undermine biodiversity conservation. Proc Natl Acad Sci USA 103: 5403–5408.

42. Armsworth PR, Sanchirico JN (2008) The effectiveness of buying easements as a conservation strategy. Conserv Lett 1: 182–189.

43. Ban NC, Mills M, Tam J, Hicks CC, Klain S, et al. (2013) A social–ecological approach to conservation planning: embedding social considerations. Front Ecol Environ 11: 194–202.

44. Game ET, Lipsett-Moore G, Hamilton R, Peterson N, Kereseka J, et al. (2011) Informed opportunism for conservation planning in the Solomon Islands. Conserv Lett 4: 38–46.

45. Knight AT, Grantham HS, Smith RJ, McGregor GK, Possingham HP, et al. (2011) Land managers' willingness-to-sell defines conservation opportunity for protected area expansion. Biol Conserv 144: 2623–2630.

46. Armsworth PR, Cantú-Salazar L, Parnell M, Davies ZG, Stoneman R (2011) Management costs for small protected areas and economies of scale in habitat conservation. Biol Conserv 144: 423–429.

47. Ban NC, Adams VM, Pressey RL, Hicks J (2011) Promise and problems for estimating management costs of marine protected areas. Conserv Lett 4: 241–252.

48. McCrea-Strub A, Zeller D, Rashid Sumaila U, Nelson J, Balmford A, et al. (2010) Understanding the cost of establishing marine protected areas. Marine Policy 35: 1–9.

49. Woinarski J, Green J, Fisher A, Ensbey M, Mackey B (2013) The effectiveness of conservation reserves: Land tenure impacts upon biodiversity across extensive natural landscapes in the tropical savannahs of the Northern Territory, Australia. Land 2: 20–36.

12

Bird Community Conservation and Carbon Offsets in Western North America

Richard Schuster[1]*, Tara G. Martin[1,2], Peter Arcese[1]

1 Department of Forest and Conservation Sciences, University of British Columbia, Vancouver, British Columbia, Canada, **2** Ecosciences Precinct, CSIRO Ecosystem Sciences, Brisbane, Queensland, Australia

Abstract

Conservation initiatives to protect and restore valued species and communities in human-dominated landscapes face huge challenges linked to the cost of acquiring habitat. We ask how the sale of forest carbon offsets could reduce land acquisition costs, and how the alternate goals of maximizing α or β-diversity in focal communities could affect the prioritization land parcels over a range of conservation targets. Maximizing total carbon storage and carbon sequestration potential reduced land acquisition costs by up to 48%. Maximizing β rather than α-diversity within forest and savannah bird communities reduced acquisition costs by up to 15%, and when these solutions included potential carbon credit revenues, acquisition cost reductions up to 32% were achieved. However, the total cost of conservation networks increased exponentially as area targets increased in all scenarios. Our results indicate that carbon credit sales have the potential to enhance conservation outcomes in human-dominated landscapes by reducing the net acquisition costs of land conservation in old and maturing forests essential for the persistence of old forest plant and animal communities. Maximizing β versus α-diversity may further reduce costs by reducing the total area required to meet conservation targets and enhancing landscape heterogeneity. Although the potential value of carbon credit sales declined as a fraction of total acquisition costs, even conservative scenarios using a carbon credit value of \$12.5/T suggest reductions in acquisition cost of up to \$235 M, indicating that carbon credit sales could substantially reduce the costs of conservation.

Editor: Mark S. Boyce, University of Alberta, Canada

Funding: The Natural Sciences and Engineering Research Council Canada Discovery Grant (PA, http://www.nserc-crsng.gc.ca/), W. and H. Hesse (RS, http://hesse-award.sites.olt.ubc.ca/) and the University of British Columbia (RS, PA, http://ubc.ca/) kindly funded our work. The funders had no role in study design, data collection and analysis, decision to publish, or preparation of the manuscript.

Competing Interests: The authors have declared that no competing interests exist.

* E-mail: mail@richard-schuster.com

Introduction

There is a pressing need to develop mechanisms to promote biodiversity conservation in the face of climate and land use change and the competing needs of humans [1–4]. This challenge is particularly severe in human-dominated landscapes, where private ownership prevails and the cost of purchasing properties or compensating land holders for lost opportunity incurred as a result of conservation can be substantial [5,6]. There are multiple routes available to land conservation, such as land purchase or private land conservation initiatives [7–10]. Here we focus on land purchase rather than conservation agreements on private lands, to avoid having to estimate the long-term costs of monitoring and enforcement or the probability that private conservation agreements are challenged in future [7,11,12]. One way of making conservation via land purchase more affordable is by offsetting those costs via payments for ecosystem services. The use of carbon markets to pay for carbon sequestration is an ecosystem service gaining global attention [13–15], in part because public concerns about the consequences of climate change have motivated 35 nations and 13 sub-national jurisdictions to put a price on carbon [16]. To the degree that carbon and biodiversity values overlap, carbon offsets could therefore be used to protect forests that would otherwise be logged [17,18] or to restore those still supporting valued old forest communities [19].

Several outstanding issues arise when considering the role of carbon markets in forest restoration. One issue is that biodiversity values may be lower in stands with the highest returns from carbon sequestration sales because sequestration rates typically peak in stands of intermediate age [20]. In contrast, older forests act as carbon sinks and continue to accumulate carbon over time, but at lower rates on average [21], resulting in high initial returns on the sale of carbon storage credits, where one carbon credit represents the offset of greenhouse gas emissions by one tonne of carbon dioxide equivalent (CO_2-e). A second issue is whether to develop conservation plans that maximize species richness (α-diversity) within habitats or maximize dissimilarities in community composition (β-diversity) to accommodate landscape complexity and species that utilize multiple habitats [22,23]. Under climate change, it has been suggested that an emphasis on community dissimilarity (β-diversity) may deliver more robust conservation plans than those based on species richness [23,24]. Here we examine the potential value of carbon credit sales to offset land acquisition costs by developing conservation area designs that maximize β or α-diversity in native old forest and savannah bird communities in relation to forest structure and human land use. Specifically, we ask how protecting forests with high carbon storage versus high carbon uptake is likely to affect conservation outcomes.

Carbon and Biodiversity in the Georgia Basin

The Coastal Douglas Fir (CDF) ecozone of the Georgia Basin (British Columbia, Canada [25]) is a classic example of an endangered but extraordinarily diverse region that has been rapidly converted to exclusive human use ($\geq$60%) [26] and thus retains $\leq$0.3% of historic old forests (>250 years) [27] and $\leq$10% of oak woodland and savannah [28], both of which provide habitat for 117 species at risk of extirpation, which represents the highest density of species of global and provincial concern to conservation of any ecozone in BC [26]. Because regional, provincial and federal authorities own <20% of the region and only ~9% is already conserved, cost-efficient routes to conservation are urgently needed to help reduce the risk of extirpation for those species and related ecosystems.

Prior to European colonization the CDF occurred as uneven-aged forest (often >300 years) dissected by shallow and deep-soil meadow and woodland communities [25,29] maintained in part by aboriginal land management practices to enhance hunting opportunities and root and fruit harvests [30–33]. In addition to recent human-caused disturbances, oak woodland and savannah community distributions are predicted to shift under future climate conditions, and only a small fraction of the current protected areas have the potential to accommodate this shift [34]. The resulting land use heterogeneity within the region and potential for humans to directly or indirectly affect native species richness [19,35–37] make this system ideal for studying trade-offs involved when attempting to maximize α- versus β-diversity in conservation plans, while simultaneously maximizing ecosystem service values represented as total carbon stored or sequestration potential. To do so, we compared systematic conservation scenarios that maximized old forest and savannah bird biodiversity (α-diversity) or their dissimilarity (β-diversity), and then quantified their relative costs given alternate carbon markets, and in relation to increasing targets for the total area conserved [38] (Table 1).

Materials and Methods

Ethics Statement

Permits or permission for the use of bird point count locations were obtained from Parks Canada (locations in National Park Reserves), private land owners (locations on private land), or did not require specific permission as they occurred on public right of ways (e.g., roadsides, regional parks). As private land owners did not want their information posted publically please contact the authors for contact details. The field studies did not involve endangered or protected species. This study did not require approval from an Animal Care and Use Committee because it was a non-invasive observational field study, and did not involve the capture and handling of wild animals.

Biodiversity data

We used trained observers to conduct 1,770 point counts on mainland BC and 53 islands from 30 Apr–11 Jul, 2005–2011 (Fig.1, 48.7° N, 123.5° W) to record all birds detected in 10 min, 50 m radius counts between 5 AM–12 PM at 713 sample locations (>100 m apart). Locations were re-visited 1–12 times and geo-referenced via a GPS (GPS60, Garmin Ltd, Kansas, USA). We extended the approach of Schuster & Arcese [19] geographically (from 1560 km^2 to 2520 km^2) by adding 601 counts to create predictive distribution models for 47 bird species and 25 covariates based on remote-sensed data and models incorporating imperfect detectability [39]. To estimate detectability we used one site specific (crown closure) and three observation specific (time of date, Julian date and observer identity) covariates.

We associated bird species indicators with the habitats they were expected to occupy by using 11 experts to rank the likelihood of observing 47 species in 10 focal habitat types using photographic and text descriptions of herbaceous, shrub, woodland, wetland, four forest types (pole, young, mature and old), and 2 human-dominated habitats (rural, urban), to create two community metrics indicating Old Forest (OF, [19]) and Savannah (SAV) habitats standardized between 0 and 1 by dividing through the maximum value possible (details in Appendix S1), where:

$$OF = \frac{-2 * Herb - 1 * Shrub - 0.5 * Pole + 0.5 * YFor + 1 * MFor + 2 * OFor}{7}$$

$$SAV = \frac{2 * Wood + 2 * Herb + 1 * Shrub}{5}$$

These metrics match our goals given the region's history and focus on Old Forest and Savannah community conservation (see Introduction). Specifically, each species contributed to the cumulative Old Forest or Savannah community score, weighted by its expert opinion score for the given sub-type, summed across species to create community specific association scores from 0 to 1, and corresponding to none versus all members of the community expected to be present. The metrics were then projected spatially as predictive maps of community occurrence over the entire study area (2520 km^2, Fig. 1) as 1 ha hexagonal polygons (Fig. S1–S2 in Appendix S1,see also [19]).

Table 1. Summary of diversity features, land cost metrics, conservation targets and carbon prices used in 144 Marxan scenarios.

Diversity features (n = 2)	Property cost metrics (n = 4)	Conservation Targets [%] (n = 9)	Carbon credit value (CC) (n = 3)
α-diversity (maximize Old Forest + Savannah individually)	Total Land value (TLV)	10 to 50 (in 5% steps)	9 $/T (lowest price PCT has paid for credits so far)
β-diversity (maximize β-score)	TLV – StC	-	12.5 $/T (half the cost PCT charges, as well as roughly the average price PCT is paying for credits)
	TLV – SeqC		25 $/T (the price that PCT is charging for credits)
	TLV – TotC		

PCT = Pacific Carbon Trust; StC = Carbon Storage * CC; SeqC = Carbon.
Sequestration potential * CC; TotC = StC+SeqC.

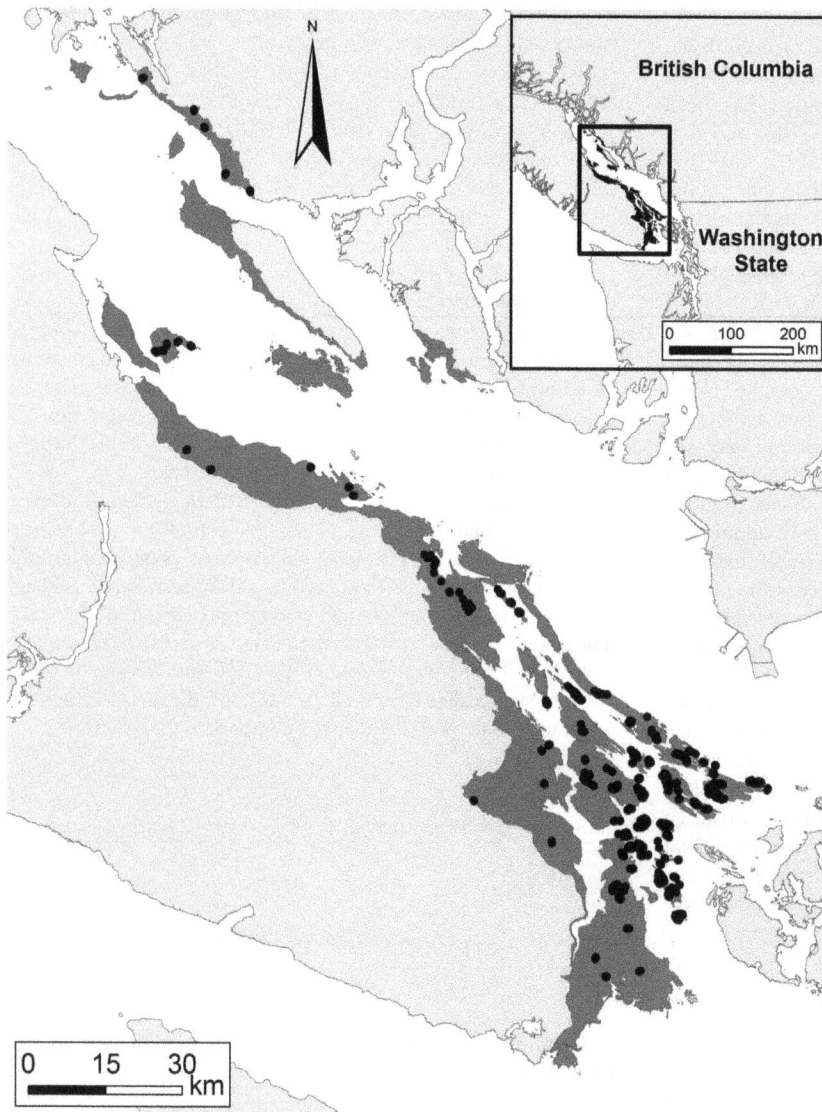

Figure 1. Georgia Basin study area including bird point count locations. Dark grey area indicates the extent of the study region and black dots represent bird point count locations.

Carbon estimates

Forest carbon storage and sequestration rates were estimated for all forested land in the study area using terrestrial ecosystem mapping (TEM; [27]) and FORECAST [40]. FORECAST is a stand-level forest ecosystem simulator that is one of two models approved by the BC Ministry of Forests for carbon budget assessments [41], and the only model calibrated for use in the CDF (Blanco et al. 2007) and linked to TEM [42]. To facilitate carbon analysis TEM polygons were stratified into homogenous analysis units based on site series. Net ecosystem carbon storage was limited to: above and below-ground tree biomass, deadwood biomass, and dead below-ground biomass. Each analysis unit was simulated for a period of 300 years with results reported for annual time steps to create carbon storage curves. FORECAST results were subsequently assigned to individual TEM polygons by estimating the age of each polygon subsection based upon the current assigned structural stage and estimated productivity class [42]. These age estimates were derived from ranges provided by Meidinger et al. [43] for regional forest ecosystems. Ages of old

stands (structural stage 7) were set at 200 to be conservative. Age estimates were verified against a subset of TEM polygons (Southern Gulf Islands of Southwestern BC) for which direct age estimates were available (n = 254). For conservation prioritization analysis we used predicted net ecosystem carbon storage and net ecosystem carbon sequestration estimates for 20 years from now due to uncertainty about fire frequency in the future. Further details on this analysis are provided in Appendix S2.

Cadastral layer and property costs

We incorporated spatial heterogeneity in land values [5,44–46] in our plans by using cadastral data and 2012 land value assessments (Integrated Cadastral Information Society of BC, ICIS). However, because there is no centralized entity curating cadastral data for British Columbia, we combined data from ICIS, the BC Assessment agency and the Integrated Cadastral Fabric. Doing so required processing to remove stacked and overlapping polygons and slivers. The combined cadastral layer included 193,623 polygons. Current assessments were available for 187,139 polygons, but missing for 3,281 polygons or reduced relative to market value due to taxation or administrative reasons unrelated to our work (e.g farm or managed forest land, 3,203 ploygons). For these 6,484 polygons we applied an inverse distance weighted interpolation to estimate land values by splitting cadastral polygons into 10 groups based on polygon size to accommodate high size related heterogeneity in assessed cost using R v.2.15.2 [47] and packages gstat v.1.0-14 [48] and sp v.1.0-1 [49].

We used tax assessment land values to estimate acquisition costs because they are revised annually in the region, and because more 'realistic' strategies would require speculation on how purchase cost may be affected by location of existing reserves, evolving zoning plans, the willingness of owners to sell, or other effects. In particular, there is as yet no consensus on the effect of conservation agreements on land values [50–52].

Marxan inputs

We used Marxan [53] to prioritize cadastral polygons for inclusion in conservation area designs by using them as planning units (n = 193,623). We calculated biodiversity and carbon estimates for each planning unit using ArcGIS v.10.1 [54] and area weighted sums in Geospatial Modelling Environment v.0.7.2.1 [55].

To determine whether maximizing β- versus α- diversity affected conservation outcomes we created two sets of diversity features as inputs to Marxan. First we included the diversity features individually in the analysis and set conservation targets for Old Forest and Savannah scores as the percentage of total old forest or savannah habitat existing within the study region. The second approach we used was to pre-specify a β-diversity metric to combine biodiversity features with the goal to specifically maximize highly diverse habitat patches. For this purpose we created the following metric:

$$\beta - score = \frac{2 * OF * SAV}{OF + SAV}$$

This represents the Old Forest and Savannah community dissimilarity, using a scaling factor of 2 to create β-scores between 0 and 1 (Fig. S3 in Appendix S1). In Marxan analyses we set targets for the β-score, while still including Old Forest and Savannah metrics (without setting a target) to keep track of individual community representation. We used a total of four property cost metrics per diversity scenario: i) Total land value (TLV) for each property, which is the sum of the assessed property value and any improvement on that parcel; ii) TLV minus the current carbon storage (T) times the carbon credit value ($/T). Here we used $12.5 Canadian per credit, which is half the amount that Pacific Carbon Trust, a crown corporation established in 2008 to deliver greenhouse gas offsets in the province of British Columbia (http://pacificcarbontrust.com, date accessed: 2013-12-10), sells credits for and about the average amount they pay for credits. iii) TLV minus the amount of potential carbon sequestration over 20 years times the carbon credit value; iv) TLV minus ii and iii combined (Table 1).

Marxan scenarios

We used the two diversity scenarios α (Old Forest + Savannah) and β (β-score) in combination with the four cost scenarios (Table 1). An important consideration for this study was what level to set the required conservation target to, in order to ensure the study system will maintain viable populations of native species and be resilient to predicted environmental change in the future. As there is debate about what constitutes appropriate conservation goals [38] we used a range of conservation targets (10–50%) to investigate the potential trade-offs of different targets. We calibrated each diversity scenario to ensure robust analysis by initially setting the diversity target to 50% (the most costly to reach) and the number of restarts to 100, as we were not so much interested in the spatial representation of the reserve design but rather its cost effectiveness [56]. For the same reason we also refrained from setting boundary length modifiers. For each diversity scenario we created Marxan solutions for combinations of the following species penalty factors (SPF's): 1–10,15,20 and number of iterations: 10 k, 50 k, 100 k, 500 k, 1 M, 5 M, 10 M, 25 M, 50 M, 100 M, for a total of 65 calibration analyses per diversity scenario. We created cumulative distribution functions using number of solutions on the y-axis, solution cost on the x-axis for SPF and Marxan score for number of iterations [56]. Based on the results we used the following values for SPF and number of iterations respectively: Old Forest + Svannah (3/10 M); β-score (3/10 M). We also investigated summed solutions to make sure every restart met its targets, excluding ones that missed the target by >5%. For ease of computation we created an R function to batch run Marxan (Appendix S3).

We held the calibrated values constant in subsequent analyses and ran Marxan scenarios for the two diversity metrics in combination with the four cost metrics, using the baseline carbon credit value of $12.5. For each combination we further varied the conservation target from 10–50%. From each run we recorded the cost of the total reserve system averaged over the number or restart (100), while ensuring conservation targets were met. To examine the amount of remaining Old Forest and Savannah communities protected by maximizing β-diversity we compiled community scores as Marxan features in these scenarios without setting targets, allowing us to keep track of Old Forest and Savannah representation without affecting the analysis. We used the results from these analyses to compare the reserve prices within each diversity scenario as well as across scenarios. In addition we calculated the potential cost savings between fee simple acquisition scenarios (TLV only) and ones that utilize the sale of carbon credits. As market prices of carbon credits are highly variable we extended our approach to include variation in carbon credit value, by repeating the entire analysis for two additional carbon credit values: i) $9 per credit (the lowest rate PCT has ever paid for credits), and ii) $25 per credit (the price PCT sells credits for). In total 144 Marxan scenarios were investigated (Table 1). All results presented here relate to the baseline carbon credit value of $12.5 unless otherwise stated.

Results

Land acquisition cost, diversity and planning goals

Acquisition costs of conservation networks increased from $180 M to $2.45 B as targets increased from 10 to 50% of remaining Old Forest and Savannah bird communities when maximising α-diversity (Fig. 2), but reduced slightly when maximizing β-diversity ($172 M to $2.16 B, 10–50% target; Fig. 2), representing savings of 4–15% as compared to equivalent α-diversity scenarios depending on conservation target (Fig. 3a). Savings were due in part to a reduction in total area needed to reach a given target when maximising β versus α-diversity (mean = 7%, range = 5–11%; Fig. 3b). The amount of standing and sequestration potential carbon in conserved landscapes also declined slightly when maximizing β-diversity (mean = 2%, range = 0.7–5%; Fig. 3c). In contrast, representation of Old Forest communities slightly increased (1–2.5%) and representation of

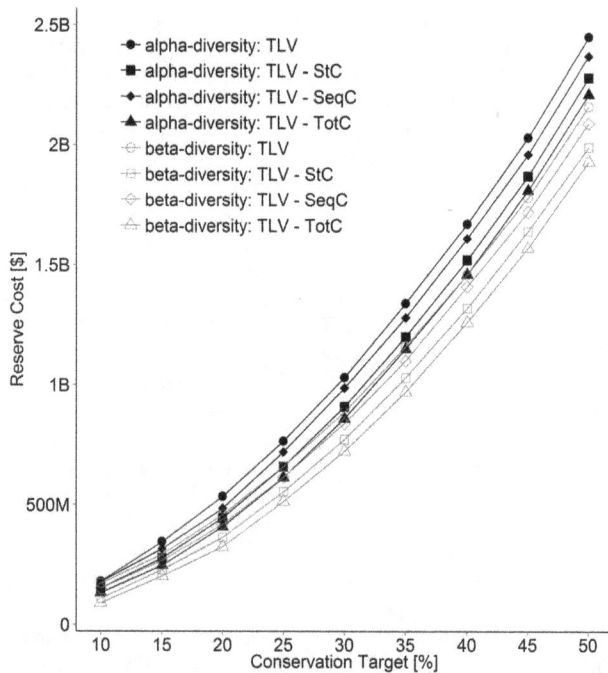

Figure 2. Reserve costs using alpha and beta diversity and a carbon credit value of $12.5 across a range of conservation targets (term definitions in Table 1).

Savannah declined (−2.0–−5.7%) when maximizing β versus α-diversity (Table 2).

Cost savings given carbon credits

Maximizing total (standing + sequestered) carbon resulted in the largest cost savings in both α and β-diversity scenarios aimed at protecting Old Forest and Savannah habitats. Acquisition costs increased from $133 M to $2.21 B as target increased from 10 to 50% when maximizing α-diversity, which represent potentials offset of $47 M–235 M, equivalent to a 10–28% cost reduction via carbon credit sales (Fig.2). In comparison, acquisition costs were lower for scenarios that maximised β-diversity ($90 M to $1.93 B), in part because implied carbon credit sales ($82–227 M) contributed slightly more to cost reduction (e.g., 11–48%; Fig. 2). Maximising carbon storage and carbon sequestration potential individually reduced acquisition costs to a smaller extent, but carbon storage offered superior savings (Fig. 2). Overall, maximizing total carbon returned networks that were 17.5% cheaper on average when maximizing β versus α-diversity compared to 12.3% without using carbon storage and sequestration values (Fig. 3a).

Conservation targets and carbon price

The cost of conservation networks increased exponentially with increasing targets for all scenarios (Fig. 2). In β-diversity scenarios the total area that needed to be acquired to reach a conservation target was 11–5% lower and acquisition costs 32–13% less than scenarios that maximized α -diversity (Fig. 3). The percent reduction in total acquisition costs due to carbon value also declined as conservation targets increased in α and β-diversity scenarios (Fig. 4a,b). The magnitude by which acquisition costs were reduced by carbon value was similar across prices considered but maximized at $25/T in most scenarios (Fig. 4a,b). Relative

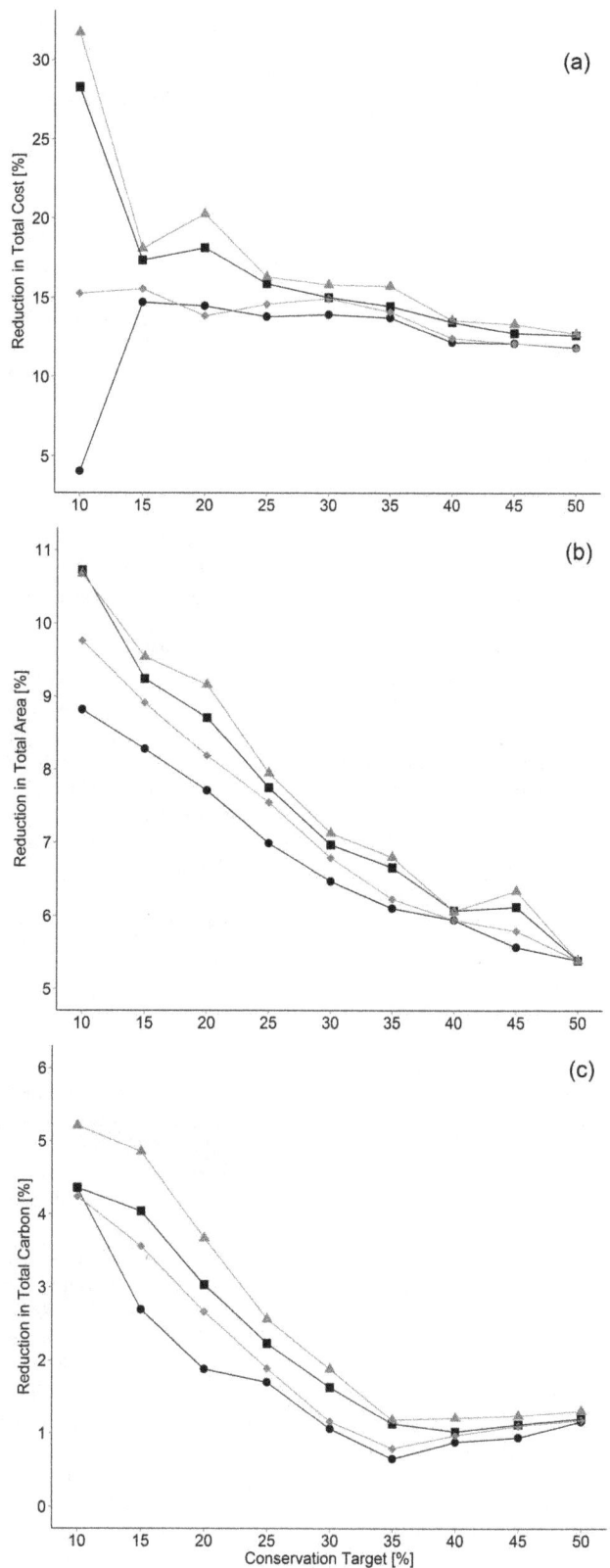

Figure 3. Comparison of α and β-diversity scenario results. Presented are the % reductions when using a β-diversity approach for: a) reserve network cost, b) reserve network area, c) total carbon included in the reserve networks. Circles represent TLV, squares StC, diamonds SeqC and triangles TotC (term definitions in Table 1) (c).

Table 2. Conservation targets (bold) and % actually included in β-diversity scenarios (BETA) using carbon offset value of $12.5/T.

Scenario		Percent of Target Bird Community Protected [%]								
		10	15	20	25	30	35	40	45	50
Total Land Value (TLV)	BETA	10.00	15.00	20.00	25.00	30.00	35.00	40.00	45.00	50.00
	OF	11.12	16.31	21.60	26.81	32.07	37.01	42.18	47.29	52.41
	SAV	8.80	13.57	18.09	22.87	27.59	32.49	37.22	42.07	46.85
TLV - StC	BETA	10.00	15.00	20.00	25.00	30.00	35.00	40.00	45.00	50.00
	OF	11.18	16.40	21.73	26.89	32.07	37.13	42.23	47.51	52.47
	SAV	8.72	13.38	17.93	22.68	27.45	32.34	37.14	41.78	46.78
TLV - SeqC	BETA	10.00	15.00	20.00	25.00	30.00	35.00	40.00	45.00	50.00
	OF	10.98	16.31	21.67	26.83	32.04	37.10	42.14	47.35	52.43
	SAV	8.96	13.56	18.07	22.85	27.57	32.42	37.30	42.02	46.84
TLV - TotC	BETA	10.00	15.00	20.00	25.00	30.00	35.00	40.00	45.00	50.00
	OF	11.17	16.43	21.72	26.95	31.95	37.18	42.27	47.50	52.48
	SAV	8.72	13.36	17.94	22.70	27.63	32.26	37.10	41.79	46.73

β-targets were met in each case and Old Forest (OF) was generally over and Savannah (SAV) underrepresented. StC = carbon storage credits; SeqC = sequestration potential credits; TotC = StC+SeqC.

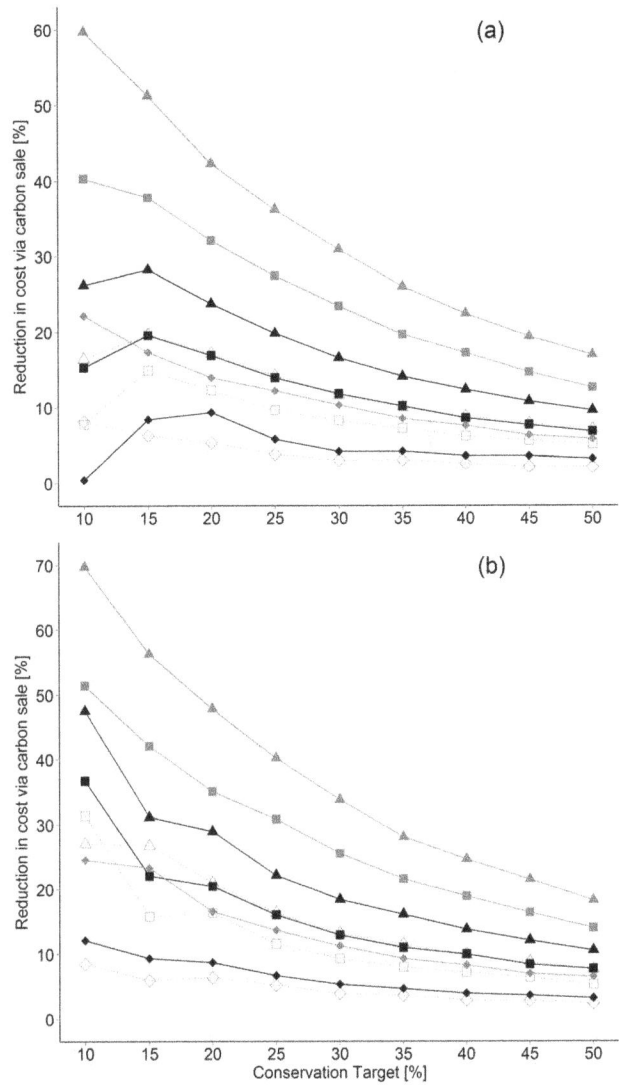

Figure 4. Cost savings when using carbon credit sales in relation to outright acquisition (TLV scenarios). Carbon credit values used are shown in parenthesis for a) α-diversity and b) β-diversity. Rectangles represent StC, diamonds SeqC and triangles TotC. Carbon credit values are as follows: dotted ($9), black ($12) and grey ($25) (term definitions in Table 1).

reduction in cost due to carbon was maximized at the 10% target in all β-diversity scenarios (Fig. 4b).

Discussion

Carbon credit sales have the potential to reduce land acquisition costs by up to 48% in Coastal Douglas fir forest and woodland communities of western North America given values already paid in the region ($12.5/T: [57]; Fig. 4b). The largest benefits were realized in scenarios that maximized total carbon and β-diversity in native woodland and old forest bird communities of the region, because those scenarios achieved their targets by selecting cheaper and slightly smaller networks than scenarios maximizing α-diversity in these communities independently. We now develop these points in light of literature on ecosystem services, land acquisition and conservation applied to threatened plant and animal communities in human-dominated landscapes of the Georgia Basin of western North America.

Maximizing β versus α diversity

Prioritizing β over α-diversity in Old Forest and Savannah bird communities reduced acquisition costs by up to 15%, or up to 32% including carbon values (Fig.3a). One reason for these savings is that the fraction of Old Forest bird habitat included in conservation networks was larger in β as compared to α-diversity scenarios (Table 2), resulting in more carbon stored per unit area conserved (Fig. 3b,c). However, over-representation of Old Forest relative to Savannah communities also reduced acquisition costs in scenarios not including carbon value, indicating that Old Forest habitat was on average less valuable than Savannah habitat in the region, perhaps due to high human amenity values (Table 2; e.g., [58]). Scenarios maximizing β-diversity also met conservation targets by protecting less total area and up to an 11% cost savings compared to scenarios maximizing α-diversity (Fig.3b), and this finding was largely independent of carbon value or conservation target (Fig.3b). Prior studies of the consequence of emphasizing α- versus β-diversity in conservation planning have concluded that a focus on β-diversity is likely to enhance long term persistence in diverse species assemblages and reserve networks [22,59,60]. Our results broaden these conclusions by showing that scenarios that maximize β-diversity may also reduce the cost of conservation by reducing the area required to meet realistic targets for land acquisition.

Conservation cost and carbon

Our results indicate that carbon credit sales should be considered as an option to maximize return on conservation investments in regions where land cost is high and old or growing forests offer habitat for valued focal communities. Our results therefore compliment suggestions that carbon credit sales have the potential to advance conservation and mitigate the impacts of climate change [13,61,62] but extend those suggestions by providing a spatially explicit, empirical example applied to a landscape with high conservation and cultural values [26,58,63]. The largest reductions in cost due to carbon credit sales were obtained by including carbon storage and carbon sequestration potential (Fig. 4), indicating that flexibility in carbon credit sales with respect to forest age can also increase economic efficiency. Although our results are based on a 20 year time-frame due to uncertainty about fire frequency, versus more typical 100 year time-frame for such projects [41,61], they could easily be revised with new data.

Our finding that carbon storage reduced costs more than sequestered carbon (Fig. 2,4) is partly a consequence of logging history, given that close to 30% of the region not converted to exclusive human use is covered by forest ≥80 years-old. The predominance of young forest has the potential to reduce adjacency between older, high-value forest and savannah habitat with rich and diverse native bird communities. However, young forest patches may also provide relatively low-cost opportunities to link high-value patches where acquisition costs can be offset by relatively high sequestration rate. Nevertheless, most scenarios preferentially included older stands with more carbon storage, but lower sequestration rates (Fig. S4,S5 in Appendix S2). Several other studies have suggested that carbon credits could be used to advance conservation, particularly on private land to compensate land owners for forgone opportunity costs [18,64,65]. We extended these suggestions by providing a particularly detailed example to demonstrate how land use planners might offset the costs of conservation area design by acquiring habitats that simultaneously maximize the diversity of valued vertebrate communities and realize the economic potential of carbon credit sales.

As an alternative to the sale of carbon credits to reduce conservation costs, land conservation and protection on private lands, accomplished through relationship building and alternative tax plans, could be an option [7,66,67]. But, one of the biggest challenges to realizing a theoretical approach of implementing private land conservation agreements on high value biodiversity landscapes is the need to work with landowners willing to put conservation agreements on their land [7]. In Canada, and British Columbia in particular, there are some incentives in place such as tax credits ([9], e.g. Natural Area Protection Tax Exemption Program), but none of the currently implemented compensation schemes would compensate land owners for the lost opportunity costs of developing their land or using it in other revenue-generating manners such as via agriculture or forestry, as proposed elsewhere [68–70]. Currently, a private conservation agreement approach depends largely on individuals wanting to create a legacy and see their property protected into the future [71]. It was beyond the scope of our study to address issues related to a landowners motivation to participate in conservation [7,72], nor did we want to speculate on the use of tax structure shifts.

Conservation targets

A key goal of our work was to demonstrate novel routes to protecting high-value, Old Forest and Savannah bird communities at landscape scales in western North America. However, the amount of habitat needed to achieve those goals remains uncertain. Policy-driven targets for biodiversity conservation place goals for terrestrial habitat conservation at 17% by 2020 [73], but recent reviews suggest much higher targets (25–75%; [38]). We used a range of targets to explore their influence on reserve design, carbon value and the conservation of Old Forest and Savannah ecosystems, but we found that carbon contributed proportionally less to acquisition costs as targets increased in all scenarios (Fig.4) because higher targets required the acquisition of more expensive parcels. Thus, although total carbon generally increased linearly with conservation target, acquisition costs increased exponentially, causing a decline in relative benefit (Fig.4). However, even for the largest targets (50%) in α and β-diversity scenarios, carbon values reduced acquisition cost by 9.6 and 10.5%, respectively ($235, 227 M; Fig. 4) at $12.5 per Ton.

A number of uncertainties in our study also have the potential to limit its interpretation. First, actual purchase costs may differ from assessed or predicted values [7,74]. Second, it may not be feasible to protect the areas offering the highest conservation value and least cost, particularly if regional representation or the augmentation of existing conservation areas is emphasized [75]. Third, although our results were robust over a range of carbon values, carbon markets remain unpredictable. Nevertheless, carbon markets are of substantial size, the European Union Emissions Trading System for example included 2.1 billion metric tons in 2011 [76]. In 2013 China, the largest national source of greenhouse gases (19.1% of total emissions), introduced pilot emission trading schemes [77,78], joining a growing number of countries with national emission trading schemes [16]. Voluntary carbon markets that are currently the biggest market place for forest carbon offset projects in countries like Canada had a market volume of $572 M in 2011 [76]. Assuming that carbon markets develop further, our results demonstrate that carbon value has the potential to substantially reduce land acquisition costs in human-dominated landscapes, particularly in the Georgia Basin of western North America, where diverse Old Forest and Savannah bird [19] and plant [79] communities still persist in relatively isolated, mature forest and woodland habitats.

Supporting Information

Appendix S1 Further details on our bird score modelling approach as well as residual spatial autocorrelation test results and community scores for individual bird species. We further present maps of the Old Forest, Savannah and β-scores.

Appendix S2 Summary results of carbon storage and carbon sequestration potential modelling and a technical report on the FORECAST modelling approach used.

Appendix S3 The R code we developed for Marxan calibration that can be used in combination with both the 32 and 64bit versions of Marxan. Function results can be used for Marxan post processing in R.

Appendix S4 The expert elicitation introduction document that expert birders used as guidelines for their input.

Acknowledgments

R. Butler, R. & R. Cannings, I. Criuckshank, K. Ferguson, S. Hannon, J. Kimm, A. Nightingale, S. Roias, A. Rousseau, R. Shortinghuis and A. Stewart offered expert advice. We also thank many private land owners, Parks Canada and BC Parks for facilitating access and are grateful for the helpful comments of V. LeMay, M. Boyce and two anonymous reviewers.

Author Contributions

Conceived and designed the experiments: RS TM PA. Performed the experiments: RS PA. Analyzed the data: RS. Contributed reagents/materials/analysis tools: RS PA. Wrote the paper: RS TM PA.

References

1. Ehrlich PR, Pringle RM (2008) Where does biodiversity go from here? A grim business-as-usual forecast and a hopeful portfolio of partial solutions. Proc Natl Acad Sci 105: 11579–11586. doi:10.1073/pnas.0801911105.
2. Bayon R, Jenkins M (2010) The business of biodiversity. Nature 466: 184–185. doi:10.1038/466184a.
3. Butchart SHM, Walpole M, Collen B, van Strien A, Scharlemann JPW, et al. (2010) Global biodiversity: indicators of recent declines. Science (80-) 328: 1164–1168. doi:10.1126/science.1187512.
4. Estes JA, Terborgh J, Brashares JS, Power ME, Berger J, et al. (2011) Trophic downgrading of planet Earth. Science (80-) 333: 301–306. doi:10.1126/science.1205106.
5. Naidoo R, Balmford A, Ferraro PJ, Polasky S, Ricketts TH, et al. (2006) Integrating economic costs into conservation planning. Trends Ecol Evol 21: 681–687. doi:10.1016/j.tree.2006.10.003.
6. Wunder S (2007) The efficiency of payments for environmental services in tropical conservation. Conserv Biol 21: 48–58.
7. Knight AT, Grantham HS, Smith RJ, McGregor GK, Possingham HP, et al. (2011) Land managers' willingness-to-sell defines conservation opportunity for protected area expansion. Biol Conserv 144: 2623–2630.
8. Armsworth PR, Sanchirico JN (2008) The effectiveness of buying easements as a conservation strategy. Conserv Lett 1: 182–189. Available: http://doi.wiley.com/10.1111/j.1755-263X.2008.00028.x.
9. Parker DP (2004) Land trusts and the choice to conserve land with full ownership or conservation easements. Nat Resour J 44: 483–518. Available: http://lawlibrary.unm.edu/nrj/44/2/08_parker_trusts.pdf.
10. Wilson KA, Underwood EC, Morrison SA, Klausmeyer KR, Murdoch WW, et al. (2007) Conserving biodiversity efficiently: what to do, where, and when. PLoS Biol 5: e223. Available: http://www.ncbi.nlm.nih.gov/pubmed/17713985.
11. Rissman AR, Butsic V (2011) Land trust defense and enforcement of conserved areas. Conserv Lett 4: 31–37.
12. Rissman AR (2013) Rethinking property rights: comparative analysis of conservation easements for wildlife conservation. Environ Conserv 40: 222–230.
13. Venter O, Laurance WF, Iwamura T, Wilson KA, Fuller RA, et al. (2009) Harnessing carbon payments to protect biodiversity. Science (80-) 326: 1368.
14. Agrawal A, Nepstad D, Chhatre A (2011) Reducing emissions from deforestation and forest degradation. Annu Rev Environ Resour 36: 373–396.
15. Phelps J, Webb EL, Adams WM (2012) Biodiversity co-benefits of policies to reduce forest-carbon emissions. Nat Clim Chang 2: 497–503.
16. Climate Commission (2013) The Critical Decade: Global Action Building On Climate Change (full report as pdf). Available: http://apo.org.au/node/33831.
17. Chan KMA, Hoshizaki L, Klinkenberg B (2011) Ecosystem Services in Conservation Planning: Targeted Benefits vs. Co-Benefits or Costs? PLoS One 6: e24378.
18. Douglass LL, Possingham HP, Carwardine J, Klein CJ, Roxburgh SH, et al. (2011) The Effect of Carbon Credits on Savanna Land Management and Priorities for Biodiversity Conservation. PLoS One 6: e23843.
19. Schuster R, Arcese P (2013) Using bird species community occurrence to prioritize forests for old growth restoration. Ecography (Cop) 36: 499–507. doi:10.1111/j.1600-0587.2012.07681.x.
20. Pregitzer KS, Euskirchen ES (2004) Carbon cycling and storage in world forests: biome patterns related to forest age. Glob Chang Biol 10: 2052–2077.
21. Luyssaert S, Schulze E-D, Borner A, Knohl A, Hessenmoller D, et al. (2008) Old-growth forests as global carbon sinks. Nature 455: 213–215.
22. Marsh CJ, Lewis OT, Said I, Ewers RM (2010) Community-level diversity modelling of birds and butterflies on Anjouan, Comoro Islands. Biol Conserv 143: 1364–1374.
23. Mokany K, Harwood TD, Overton JM, Barker GM, Ferrier S (2011) Combining α- and β-diversity models to fill gaps in our knowledge of biodiversity. Ecol Lett 14: 1043–1051. doi:10.1111/j.1461-0248.2011.01675.x.
24. Arponen A, Moilanen A, Ferrier S (2008) A successful community-level strategy for conservation prioritization. J Appl Ecol 45: 1436–1445.
25. Meidinger D, Pojar J (1991) Ecosystems of British Columbia. Victoria, BC: British Columbia Ministry of Forests.
26. Austin MA, Buffett DA, Nicolson DJ, Scudder GGE, Stevens V (2008) Taking Nature's Pulse: The Status of Biodiversity in British Columbia. Victoria, BC: Biodiversity BC. Available: www.biodiversitybc.org.
27. MES (2008) Terrestrial Ecosystem Mapping of the Coastal Douglas-Fir Biogeoclimatic Zone. Madrone Environmental Services LTD., Duncan, BC. Available: http://a100.gov.bc.ca/pub/acat/public/viewReport.do?reportId = 15273.
28. Lea T (2006) Historical Garry oak ecosystems of Vancouver Island, British Columbia, pre-European contact to the present. Davidsonia 17: 34–50.
29. Mosseler A, Thompson I, Pendrel BA (2003) Overview of old-growth forests in Canada from a science perspective. Environ Rev 11: S1–S7. doi:10.1139/a03-018.
30. Turner NJ (2014) Ancient Pathways, Ancestral Knowledge: Ethnobotany and Ecological Wisdom of Indigenous Peoples of Northwestern North America. 2 vol. Montreal: McGill Queen's University Press.
31. MacDougall AS, Beckwith BR, Maslovat CY (2004) Defining conservation strategies with historical perspectives: a case study from a degraded oak grassland ecosystem. Conserv Biol 18: 455–465.
32. Dunwiddie PW, Bakker JD (2011) The future of restoration and management of prairie-oak ecosystems in the Pacific Northwest. Northwest Sci 85: 83–92.
33. McCune JL, Pellatt MG, Vellend M (2013) Multidisciplinary synthesis of long-term human—ecosystem interactions: A perspective from the Garry oak ecosystem of British Columbia. Biol Conserv 166: 293–300.
34. Pellatt MG, Goring SJ, Bodtker KM, Cannon AJ (2012) Using a down-scaled bioclimate envelope model to determine long-term temporal connectivity of Garry oak (Quercus garryana) habitat in western North America: implications for protected area planning. Environ Manage 49: 802–815.
35. Gonzales EK, Arcese P (2008) Herbivory more limiting than competition on early and established native plants in an invaded meadow. Ecology 89: 3282–3289. doi:10.1890/08-0435.1.
36. Martin TG, Arcese P, Scheerder N (2011) Browsing down our natural heritage: Deer impacts on vegetation structure and songbird populations across an island archipelago. Biol Conserv 144: 459–469.
37. Bennett JR, Vellend M, Lilley PL, Cornwell WK, Arcese P (2012) Abundance, rarity and invasion debt among exotic species in a patchy ecosystem. Biol Invasions 15: 707–716. doi:10.1007/s10530-012-0320-z.
38. Noss RF, Dobson AP, Baldwin R, Beier P, Davis CR, et al. (2012) Bolder Thinking for Conservation. Conserv Biol 26: 1–4.
39. Mackenzie DI, Nichols JD, Lachman GB, Droege SJ, Royle JA, et al. (2002) Estimating site occupancy rates when detection probabilities are less than one. Ecology 83: 2248–2255.
40. Kimmins JP, Mailly D, Seely B (1999) Modelling forest ecosystem net primary production: the hybrid simulation approach used in FORECAST. Ecol Modell 122: 195–224.
41. Ministry of Environment BC (2011) Protocol for the Creation of Forest Carbon Offsets in British Columbia. British Columbia Ministry of Environment, Victoria. Available: http://www.env.gov.bc.ca/cas/mitigation/fcop.html.
42. Seely B, Nelson J, Wells R, Peter B, Meitner M, et al. (2004) The application of a hierarchical, decision-support system to evaluate multi-objective forest management strategies: a case study in northeastern British Columbia, Canada. For Ecol Manage 199: 283–305.
43. Meidinger D, Trowbridge R, Macadam A, Tolkamp C (1998) Field Manual for Describing Terrestrial Ecosystems. Victoria, Canada: Crown Publications Inc.
44. Ando A, Camm J, Polasky S, Solow A (1998) Species Distributions, Land Values, and Efficient Conservation. Science (80-) 279: 2126–2128.

45. Polasky S, Camm JD, Garber-Yonts B (2001) Selecting Biological Reserves Cost-Effectively: An Application to Terrestrial Vertebrate Conservation in Oregon. Land Econ 77: 68–78.

46. Ferraro PJ (2003) Assigning priority to environmental policy interventions in a heterogeneous world. J Policy Anal Manag 22: 27–43.

47. R Development Core Team (2012) R: A language and environment for statistical computing 2.15.2, http://www.r-project.org.

48. Pebesma EJ (2004) Multivariable geostatistics in S: the gstat package. Comput Geosci 30: 683–691.

49. Bivand RS, Pebesma EJ, Gómez-Rubio V (2013) Applied Spatial Data Analysis with R. New York, NY: Springer. doi:10.1111/j.1541-0420.2009.01247_1.x.

50. Anderson K, Weinhold D (2008) Valuing future development rights: the costs of conservation easements. Ecol Econ 68: 437–446.

51. Schilling BJ, Sullivan KP, Duke JM (2013) Do Residual Development Options Increase Preserved Farmland Values? J Agric Resour Econ 38.

52. Nickerson CJ, Lynch L (2001) The effect of farmland preservation programs on farmland prices. Am J Agric Econ 83: 341–351.

53. Ball IR, Possingham HP, Watts ME (2009) Marxan and relatives: Software for spatial conservation prioritisation. In: Moilanen A, Wilson K, Possingham HP, editors. Spatial conservation prioritisation quantitative methods and computational tools. Oxford: Oxford University Press. pp. 185–195. Available: http://www.uq.edu.au/spatialecology/docs/Publications/2009_Ball_etal_MarxanAndRelatives.pdf.

54. ESRI (2012) ArcGIS 10.1 Economic and Social Reserach Institute Inc., Redlands, CA. http://www.esri.com/.

55. Beyer HL (2012) Geospatial Modelling Environment (Version 0.7.2.1). (software). URL: http://www.spatialecology.com/gme.

56. Ardron JA, Possingham HP, Klein CJ, editors (2010) Marxan Good Practices Handbook, Version 2. Victoria, BC, Canada: Pacific Marine Analysis and Research Association.

57. PCT (2013) Pacific Carbon Trust Carbon Offset Portfolio. Available: http://pacificcarbontrust.com/assets/Uploads/Pricing/PCTpricing2009-2012.pdf, date accessed: 2013-08-10. Accessed 10 August 2013.

58. Vellend M, Bjorkman AD, McConchie A (2008) Environmentally biased fragmentation of oak savanna habitat on southeastern Vancouver Island, Canada. Biol Conserv 141: 2576–2584.

59. Wiersma YF, Urban DL (2005) Beta Diversity and Nature Reserve System Design in the Yukon, Canada. Conserv Biol 19: 1262–1272. doi:10.1111/j.1523-1739.2005.00099.x.

60. Fairbanks D (2001) Species and environment representation: selecting reserves for the retention of avian diversity in KwaZulu-Natal, South Africa. Biol Conserv 98: 365–379. doi:10.1016/S0006-3207(00)00179-8.

61. Bradshaw CJA, Bowman DMJS, Bond NR, Murphy BP, Moore AD, et al. (2013) Brave new green world—Consequences of a carbon economy for the conservation of Australian biodiversity. Biol Conserv 161: 71–90.

62. Venter O, Meijaard E, Possingham H, Dennis R, Sheil D, et al. (2009) Carbon payments as a safeguard for threatened tropical mammals. Conserv Lett 2: 123–129.

63. Arcese P, Schuster R, Campbell L, Barber A, Martin TG (2014) Deer density and plant palatability predict shrub cover, richness and aboriginal food value in a North American island archipelago. Divers Distrib in press.

64. Crossman ND, Bryan BA, Summers DM (2011) Carbon Payments and Low-Cost Conservation. Conserv Biol 25: 835–845.

65. Evans MC, Carwardine J, Fensham RJ, Butler DW, Wilson KA, et al. (2014) The economics of biodiverse carbon farming: regrowth management as a viable mechanism for restoring deforested agricultural landscapes. Glob Environ Chang in press.

66. Fishburn IS, Kareiva P, Gaston KJ, Armsworth PR (2009) The Growth of Easements as a Conservation Tool. PLoS One 4: 6. Available: http://www.ncbi.nlm.nih.gov/pubmed/19325711.

67. Gordon A, Langford WT, White MD, Todd JA, Bastin L (2011) Modelling trade offs between public and private conservation policies. Biol Conserv 144: 558–566. Available: http://www.sciencedirect.com/science/article/pii/S0006320710004544.

68. Drechsler M, Wätzold F, Johst K, Bergmann H, Settele J (2007) A model-based approach for designing cost-effective compensation payments for conservation of endangered species in real landscapes. Biol Conserv 140: 174–186.

69. Klimek S, gen Kemmermann A, Steinmann H-H, Freese J, Isselstein J (2008) Rewarding farmers for delivering vascular plant diversity in managed grasslands: A transdisciplinary case-study approach. Biol Conserv 141: 2888–2897.

70. Bunn D, Lubell M, Johnson CK (2013) Reforms could boost conservation banking by landowners. Calif Agric 67.

71. Moon K, Cocklin C (2011) Participation in biodiversity conservation: motivations and barriers of Australian landholders. J Rural Stud 27: 331–342.

72. Knight A, Cowling RM, Difford M, Campbell BM (2010) Mapping Human and Social Dimensions of Conservation Opportunity for the Scheduling of Conservation Action on Private Land. Conserv Biol 24: 1348–1358. Available: http://dx.doi.org/10.1111/j.1523-1739.2010.01494.x.

73. Convention on Biological Diversity (2010) Strategic plan for biodiversity 2011–2020 and the Aichi targets. Secretariat of the Convention on Biological Diversity, Montreal.

74. Carwardine J, Wilson KA, Hajkowicz SA, Smith RJ, Klein CJ, et al. (2010) Conservation Planning when Costs Are Uncertain. Conserv Biol 24: 1529–1537.

75. Pressey RL, Cabeza M, Watts ME, Cowling RM, Wilson KA (2007) Conservation planning in a changing world. Trends Ecol Evol 22: 583–592. doi:10.1016/j.tree.2007.10.001.

76. Newell RG, Pizer WA, Raimi D (2013) Carbon markets 15 years after kyoto: Lessons learned, new challenges. J Econ Perspect 27: 123–146.

77. Lo AY (2013) Carbon trading in a socialist market economy: Can China make a difference? Ecol Econ 87: 72–74.

78. Wang Q (2013) China has the capacity to lead in carbon trading. Nature 493: 273.

79. Bennett JR, Arcese P (2013) Human Influence and Classical Biogeographic Predictors of Rare Species Occurrence. Conserv Biol 27: 417–421.

13

Balancing Energy Budget in a Central-Place Forager: Which Habitat to Select in a Heterogeneous Environment?

Martin Patenaude-Monette[1], Marc Bélisle[2], Jean-François Giroux[1]*

1 Groupe de recherche en écologie comportementale et animale, Département des sciences biologiques, Université du Québec à Montréal, Montréal, Québec, Canada, **2** Département de biologie, Université de Sherbrooke, Sherbrooke, Québec, Canada

Abstract

Foraging animals are influenced by the distribution of food resources and predation risk that both vary in space and time. These constraints likely shape trade-offs involving time, energy, nutrition, and predator avoidance leading to a sequence of locations visited by individuals. According to the marginal-value theorem (MVT), a central-place forager must either increase load size or energy content when foraging farther from their central place. Although such a decision rule has the potential to shape movement and habitat selection patterns, few studies have addressed the mechanisms underlying habitat use at the landscape scale. Our objective was therefore to determine how Ring-billed gulls (*Larus delawarensis*) select their foraging habitats while nesting in a colony located in a heterogeneous landscape. Based on locations obtained by fine-scale GPS tracking, we used resource selection functions (RSFs) and residence time analyses to identify habitats selected by gulls for foraging during the incubation and brood rearing periods. We then combined this information to gull survey data, feeding rates, stomach contents, and calorimetric analyses to assess potential trade-offs. Throughout the breeding season, gulls selected landfills and transhipment sites that provided higher mean energy intake than agricultural lands or riparian habitats. They used landfills located farther from the colony where no deterrence program had been implemented but avoided those located closer where deterrence measures took place. On the other hand, gulls selected intensively cultured lands located relatively close to the colony during incubation. The number of gulls was then greater in fields covered by bare soil and peaked during soil preparation and seed sowing, which greatly increase food availability. Breeding Ring-billed gulls thus select habitats according to both their foraging profitability and distance from their nest while accounting for predation risk. This supports the predictions of the MVT for central-place foraging over large spatial scales.

Editor: Michael Sears, Clemson University, United States of America

Funding: The research was supported by grants to JFG and MB from the Natural Sciences and Engineering Research Council of Canada, the Canadian Wildlife Service, ICI Environnement, Falcon Environmental Services, Chamard et Associés, Waste Management, and BFI Canada. MPM was supported by a scholarship from the Natural Sciences and Engineering Research Council of Canada. The funders had no role in study design, data collection and analysis, decision to publish, or preparation of the manuscript.

Competing Interests: ICI Environnement, Falcon Environmental Services, Chamard et Associe´s, Waste Management, and BFI Canada provided funding towards this study. There are no patents, products in development or marketed products to declare.

* Email: giroux.jean-francois@uqam.ca

Introduction

Animals face time and energy constraints leading to trade-offs in their activity budget, which can also be modulated by factors such as the spatio-temporal distribution of food resources, conspecifics, predation risk, and phenology. How animals respond to these constraints in order to maximize their fitness through foraging behaviour has been the main focus of optimal foraging theory [1], [2]. For instance, the marginal-value theorem (MVT) has been used to predict which resource patch an animal should exploit and how long it should stay before moving to another patch or return to its nest or shelter [3], [4]. Assuming that animals maximize their net energy gain, this model has provided relevant qualitative predictions [5]. However, it has been developed and used for small-scale systems in which animals are assumed to incur few or no travel costs and to be highly informed about their environment [1], [2].

This model may therefore be difficult to apply at the landscape level because of information uncertainty about the environment, which influences learning ability and because of the limited motion and navigation capacity of animals [6], [7], [8]. For example, classical central-place foraging models based on the MVT predict that prey load size should increase with the distance traveled by a forager from its central place [4], [9]. However, a forager moving across the landscape with a large load can incur increased travel costs due to greater energy expenditures or can encounter higher predation risks through increased exposure and reduced manoeuvrability [5]. Therefore, the impact of carrying a heavy load can influence the time and energy budget of a central-place forager in different ways, sometime far from the conclusions of the classical models [10].

The MVT predicts that a foraging path is the outcome of balancing trade-offs between energy expenditures and gains, especially within landscapes where resources are heterogeneously distributed. Although it is difficult to use the MVT to make precise predictions under relaxed assumptions, classical central-place foraging models nevertheless allow to predict that distant patches must provide higher energy "prey" than those found in nearby patches [9]. Hence, the profitability of a given load size may vary for a generalist forager traveling through a heterogeneous landscape. Also, habitats providing low energy food should only be used close to the central place whereas habitats with high-energy food may be exploited near or far from the central place. It remains that travel costs may increase the use of poor quality habitats when individuals must sample and learn the quality of their environment [11], [12]. Moreover, temporal variation in habitat availability and forager condition may alter the pattern of habitat use along a distance gradient [13].

Although assessing the costs and benefits of large spatio-temporal scale movements is difficult, analytical methods based on accurate location data (e.g., GPS) are now available to study movement behaviour. Combining these analytical methods with *in situ* observations of individual foraging strategies, patch quality, and environmental conditions while considering the individuals' characteristics has the capacity to provide insights into the cost-benefit trade-offs associated with foraging movements underlying habitat selection [13], [14]. For instance, resource selection functions (RSF) have been widely used to assess habitat selection. They are based on the comparison of relative habitat use (defined by presence-only data) and availability or on the presence/absence of individuals in habitat patches [15]. RSF are particularly informative if a distinction can be made between actively selected locations, such as foraging patches, and the incidentally selected locations visited during inter-patch movements [16], [17]. Bastille-Rousseau *et al.* [18] have advocated the use of a combination of RSF, residence time analysis, and ground surveys to study resource selection and foraging strategies at the landscape level. Considering the hierarchical aspect of the selection process, the difficulty of defining available habitats with presence-only data can be avoided by building RSF based on the habitats actually visited for foraging vs. those crossed when moving to a patch [19], [20]. Measuring the time spent by an animal within the surroundings of recorded locations (residence time) should allow discriminating between locations occurring within foraging patches and those found along movement paths [16], [21].

We used RSF and residence time analyses from GPS-tracking data, as well as survey data, diet characterization and calorimetric analyses to study the processes that determine habitat use by breeding Ring-billed gulls (*Larus delawarensis*). This species is a colonial central-place forager that feeds opportunistically upon a wide variety of prey items found in both aquatic and terrestrial habitats [22], [23]. We expected that gulls should be more likely to forage in a patch where the amount of habitats providing high-energy food increases and that such a relationship should be more pronounced far from the colony so that gulls reach a threshold of profitability. We also hypothesized that gulls should select habitats with a temporally variable food availability only when those habitats provide high food returns. For instance, agricultural lands and lawns should be selected on rainy days when annelids (earthworms) are more available to gulls [24]. By testing these predictions, our study sheds light on the process of habitat selection by animals from an energy trade-off perspective.

Materials and Methods

Ethics statement

Field methods to capture, mark, and collect Ring-billed gulls were approved by the Institutional Animal Protection Committee of the Université du Québec à Montréal (No. 646). The capture and marking of gulls was conducted under Environment Canada scientific permit to capture and band migratory birds (No. 10546) while the collection of specimens was carried out under Environment Canada scientific research permit (No. SC-23)

Study area

We tracked the movements of Ring-billed gulls breeding on Deslauriers Island located in the St. Lawrence River 3 km downstream from Montreal, QC, Canada (45.717°N, 73.433°W). This colony covered 11.4 ha and supported 48,000 pairs at the time of the study. The surrounding foraging area encompassed approximately 6,000 km^2 and consisted of a mosaic of high and low density urban areas, agricultural lands of intensive (soybean, maize, and small cereals) and extensive cultures (hayfields and pastures), as well as riparian habitats along the River and its tributaries (Fig. 1). Four landfills and two waste material transhipment sites were located in the vicinity of the colony. Landfills attract gulls because of the anthropogenic food they supply but the implementation of deterrence programs may reduce their accessibility [25], [26]. During our study, the St-Thomas (41 km) and Lachute (63 km) landfills, as well as the two transhipment sites (12 and 27 km), had no deterrence program. On the other hand, the Ste-Sophie landfill (37 km) initiated a deterrence program in 2009 that combined pyrotechnics and selective culling. However, the program was limited to weekdays from 07:00 to 15:00 thereby leaving some feeding opportunities for gulls [26]. Lastly, the Terrebonne landfill (8 km) conducted a deterrence program since 1995 that included falconry, distress calls, and pyrotechnics. This program was in operation every day from sunrise to sunset, preventing all but few gulls to use the landfill (Thiériot, E., unpublished data).

Telemetry

Breeding Ring-billed gulls were fitted with 10–16-g GiPSy-2 data loggers (TechnoSmart, Italy) between April and June 2009–2010. The loggers represented (mean ± SD) 2.8±0.5% of the birds' body mass (485±49 g). Most gulls were captured and recaptured with nest traps or dip nets but some had to be recaptured by rifle shooting; carcasses were then kept for further analyses (see *Diet and calorimetric analyses*). Data loggers were attached on the two median rectrices with white TESA tape (no. 4651) and programmed to acquire locations at 4-min intervals. Tracking lasted 1 to 3 days depending on battery life. Half of the birds included in the analyses returned to their nest within 15 min after being released and 81% of them returned within 60 min. Birds that spent more than 1 h away from their nest took on average 4.5±3.9 h to return. Breeding stage upon capture was categorized as incubating or brood rearing. Gulls were sexed with genomic DNA isolation from chest feathers [27].

We recaptured 109 Ring-billed gulls (41 females, 68 males) with loggers that provided reliable data (Table S1). After removing locations within a 300-m buffer zone around the colony (see *Data analyses*), there were only 28 missing locations on a potential of 15,948. The remaining 15,920 locations had a low dilution of precision metric (DOP ≤6) and an estimated precision of ±5 m [28]. A total of 67 gulls were followed during incubation (164 foraging trips) and 42 during the brood rearing period (239 foraging trips).

Gull Surveys

We conducted weekly surveys from April to June alternating between three periods (05:00–10:00, 10:00–15:00, and 15:00–20:00) to determine the proportion of time Ring-billed gulls spent

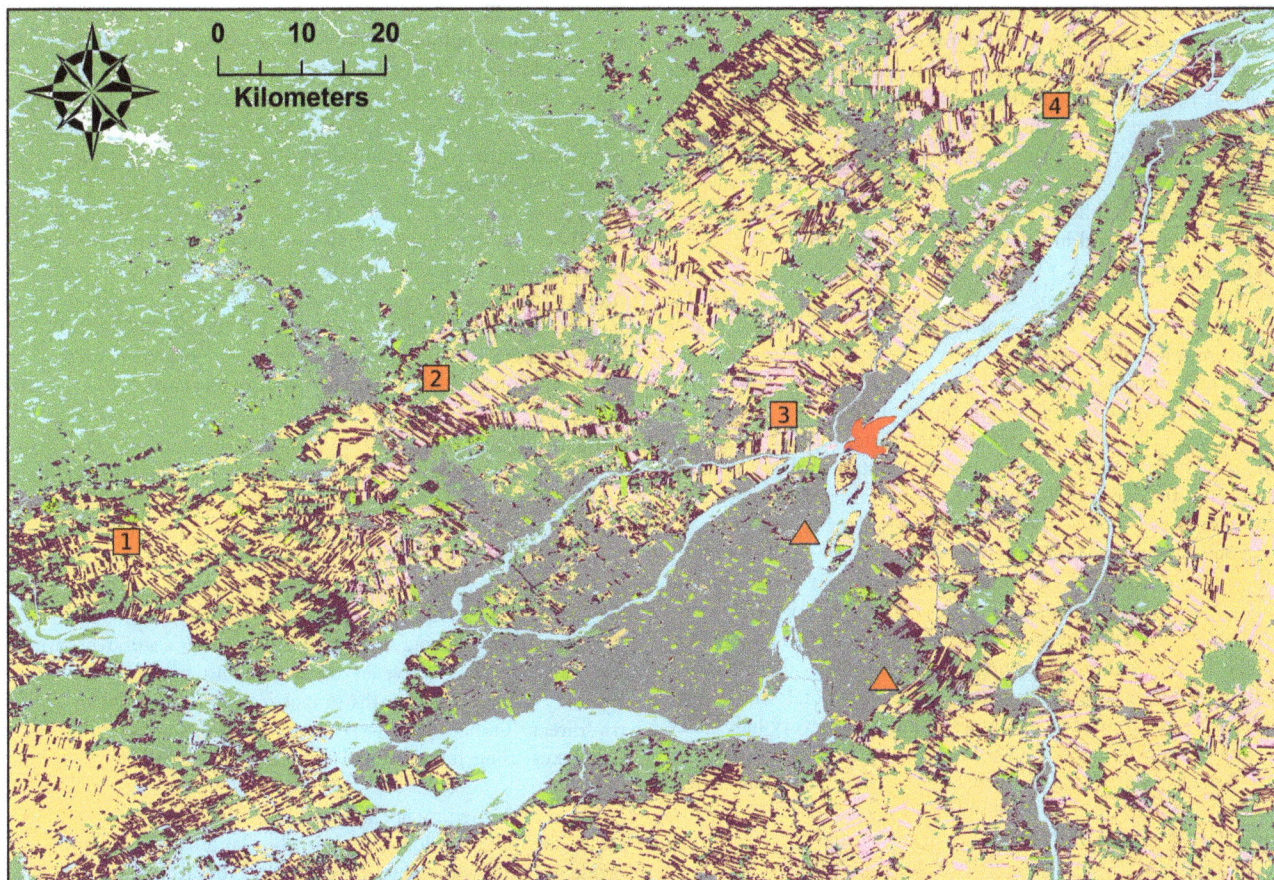

Figure 1. Map of the study area. Land cover types include water (blue), urban areas (gray), intensive cultures (mango), extensive cultures (purple), unidentified cultures (rose), lawns (olive green), and woodlots (dark green). Numbers in squares indicate landfill locations (1- Lachute, 2- Ste-Sophie, 3- Terrebonne, 4- St-Thomas), red triangles indicate transhipment site locations, and the bird pictogram indicates the location of the Deslauriers Island Ring-billed gull colony.

foraging in their main feeding habitats. In agricultural habitats, we surveyed a 50-km roadside transect on each shore of the St. Lawrence River (N= 13 and 21 surveys in 2009 and 2010, respectively). We tallied the number of birds in each flock and performed an instantaneous scan sampling to determine the proportion of birds foraging (head down below the horizontal or

Table 1. Cover percentage of eight habitat types available in the foraging range of 109 Ring-billed gulls breeding on Deslauriers Island established as the minimum convex polygon calculated with all gull locations and mean cover percentage ($\pm$1 SD) in movement (residence time <100 s) and foraging (residence time $\geq$100 s) patches (200-m radius), 2009–2010.

| | % cover | | |
| | Foraging range | Movement patches | Foraging patches |
Habitat type	**(5,565 km^2)**	**(N = 2,599)**	**(N = 4,490)**
Lawns (parks, golf courses, etc.)	1.2	1.8$\pm$8.3	2.1$\pm$9.4
Woodlots	20.6	13.7$\pm$28.8	4.8$\pm$16.0
Urban areas	16.8	27.8$\pm$38.5	23.1$\pm$36.6
Water bodies	5.3	18.7$\pm$34.8	22.4$\pm$37.9
Intensive cultures	39.5	24.3$\pm$34.1	31.4$\pm$38.7
Extensive cultures	11.7	8.4$\pm$16.2	8.0$\pm$15.6
Unidentified cultures	4.1	3.4$\pm$12.3	4.2$\pm$14.3
Landfills/Transhipment sites	0.1	1.3$\pm$11.3[a]	4.2$\pm$20.0[a]

[a]Percent occurrence.

probing into the soil) that we considered as the proportion of time spent foraging [29]. A flock was defined as a group of gulls using the same field type and not separated by more than 200 m from each other. Birds using different field types but closer than 200 m from each other were considered as different flocks. The total number of tractors and their activity (ploughing, harrowing or sowing) was also noted over the entire transect during each survey.

Observations in other habitats were conducted weekly in 2010 at fixed points located in urban ($N=25$ points), suburban ($N=53$ points), and riparian ($N=10$ points) areas on the Montreal Island ($N=16$ surveys) and along the North ($N=18$ surveys) and the South shores ($N=22$ surveys) of the St. Lawrence River. These sites were selected because they were susceptible to be visited by gulls while insuring that observers driving vehicles could stop safely. At each point, gulls using different habitat types (lawns, shores, water, grounds covered with concrete, asphalt or gravel, building roofs, and post lights) were counted and scanned to determine the proportion of birds foraging (erratic flight in emergent insect clouds above waterbodies, feeding on garbage, head down below the horizontal or probing into the soil or water).

Finally, we estimated the proportion of time gulls spent foraging at landfills by conducting 5-h observation periods once a week in 2009 ($N=7$) and five days a week in 2010 ($N=59$) at the Ste-Sophie landfill, again alternating among the three daily periods including periods with and without deterrence. Total bird counts and instantaneous scan sampling were conducted every half hour. The mean daily abundance of gulls was computed for each day as well as the proportion of birds that were actually foraging (flying less than 5 m above the active tipping area, head down below the horizontal or probing into refuse).

Diet and calorimetric analyses

We collected 496 boli from chicks of both sexes during weekly visits to the Deslauriers colony during the rearing period of 2009 and 2010. We selected chicks haphazardly and slightly pressed their proventriculus to make them regurgitate recently swallowed food. Spontaneous regurgitations of adults ($N=13$) captured during banding operations throughout the breeding period were also collected. Samples were frozen until they were analysed. We also kept frozen the carcasses ($N=51$) of adults fitted with data loggers and recaptured by shooting until the content of their oesophagus and proventriculus could be analysed. Similarly, we analysed stomach contents of birds collected by rifle shooting in agricultural lands ($N=69$), riparian areas ($N=54$), and at the Ste-

Sophie landfill ($N=85$). We made sure that birds were actively feeding in these habitats before collecting them. For safety reasons, gulls could not be collected in urban areas. Each food item of a bolus or stomach was separated, identified, dried to constant weight and weighted (± 0.01 g). Food items were grouped into broad categories (e.g., arthropods, annelids, vertebrates, refuse, vegetation, other).

Food availability could not be assessed throughout the 6,000 km^2 of the foraging area to estimate the benefits obtained by gulls when feeding in different habitats. Instead, we relied on the relative area of each habitat and the food quality in these habitats based on energy content of the various food items. We therefore performed duplicate or triplicate calorimetric analyses of each food category using a bomb calorimeter (Parr, model 1108P).

Data analyses

We first created a 300-m buffer zone around Deslauriers Island (colony) to discriminate between foraging trips and short movements to the shore or surrounding shallow water where gulls rest and preen [30]. Our analyses were limited to locations outside this zone. The mean number of foraging trips per day, the mean direct (Euclidean) distance between the colony and the farthest location reached during a foraging trip (whether a stopover or not), the mean distance traveled on a foraging trip and the mean sinuosity of movement paths (traveled distance divided by round trip direct distance, [31]) were compared between breeding stages and sexes using linear mixed models with gull ID as a random factor.

For each foraging trip, we calculated the total amount of time spent at different locations on the landscape by estimating residence time without rediscretization [16]. Residence time was defined as the time spent in a circle of radius r centred on a given location along the foraging path. The circle, with its specific habitat composition and features, could then be viewed as a potential foraging patch. In the absence of precise information regarding the spatio-temporal distribution of resources, the hierarchy of spatial scales at which animals are likely to respond to landscape heterogeneity (i.e., patches, [32]) can only be identified through behaviour [16], [33]. For each trip, we thus computed the coefficient of variation (CV) of residence times for radii ranging between 200 and 2,000 m with 100-m increments. We averaged the CV across paths and plotted them against the circle radii (Fig. S1). The mean CVs of residence time across paths showed a plateau for radii of 200 to 400 m instead of a clear peak.

Table 2. Summary of *a priori* models based on resource selection functions that predict the probability that a breeding Ring-billed gull will forage in a patch (200-m radius) for a 100-s residence time threshold.

Model	Deviance	K	ΔAICc	w_i
H+D+B	7,891	26	0.00	0.813
H+D+R+B	7,886	30	2.94	0.187
H+D	7,940	19	35.06	0.000
H+D+R	7,936	23	38.63	0.000
H+B	8,466	18	558.98	0.000
H+R+B	8,459	22	560.26	0.000
H	8,533	11	611.33	0.000
H+R	8,527	15	613.67	0.000

H: habitat types; D: distance between a location and the colony; R: mean daily rainfall; B: breeding stage (egg incubation vs. chick rearing); K: number of parameters; w_i: Akaike weight.

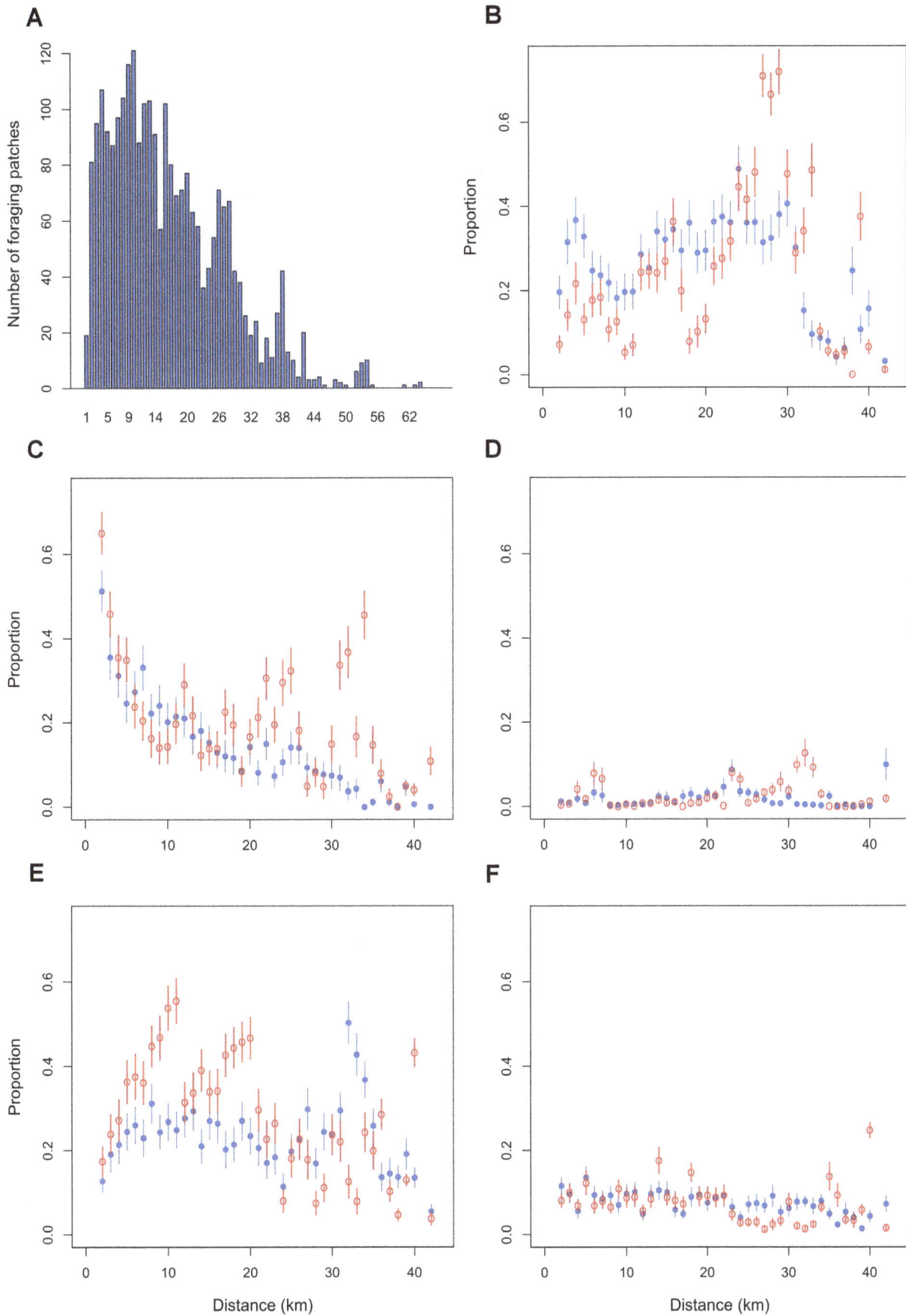

A

B

C

D

E

F

Figure 2. The effect of distance on habitat selection by foraging Ring-billed gulls. Number of foraging patches within 1-km concentric bands from the colony (a). Mean (±1 SD) proportion of urban areas (b), waterbodies (c), lawns (d), intensive cultures (e), and extensive cultures (f) in foraging (blue) and movement patches (red) in relation with the distance from the colony.

There was no significant difference in CVs distribution between males and females. We thus chose a 200-m radius to get a stronger contrast in habitat composition between foraging and movement patches. We finally retained locations distanced by at least two radii to limit spatial autocorrelation.

We calculated the landscape composition within each circle in which residence time was estimated based on a land cover map created in ArcGIS 9.3.1 [34] using both agricultural and topographic data ([35], [36], [37]; planimetric precision <30 m). Landscape composition was defined as the proportion of different habitats including lawns, woodlots, urban areas, and water as well as intensive, extensive and unidentified cultures. Because of their relatively small size, landfills and transhipment centres were noted as presence/absence in each circle. We also measured the distance between each location where a residence time was computed and the nest of the tracked gull. Finally, we calculated the mean daily

rainfall using data from 10 meteorological stations located throughout the entire foraging area [38].

We first described the habitats within the global home range of gulls breeding on Deslauriers Island by estimating the proportion of habitats within the 100% minimum convex polygon drawn using the foraging trip locations of all birds. Next, we built a RSF based on patches visited by gulls on their foraging trips. Considering that a foraging individual must reduce its flying speed and increase its turning rate, we used residence time to discriminate "foraging patches" from "movement patches". We assumed that if a gull spent more than 100 s in a 200-m radius circle, it was actively foraging. Otherwise, we considered that it was moving either between the colony and a foraging patch or between two foraging patches. Gulls observed foraging during surveys typically spent more than 100 s within 200 m from where they were first detected. Moreover, based on the flight speed of

Table 3. Mixed-effects averaged logit resource selection functions quantifying the probability that a breeding Ring-billed gull forage in a patch.

Variable	β	SE	95% CI	
Intercept	−0.612	0.491	−1.575	0.351
Distance*	0.089	0.013	0.063	0.115
Woodlots*	−2.963	0.448	−3.840	−2.086
Lawns	−0.212	0.834	−1.846	1.422
Urban areas*	−2.550	0.521	−3.570	−1.529
Landfills	0.992	0.510	−0.007	1.992
Water*	−1.130	0.516	−2.142	−0.118
Extensive cultures	−1.037	0.619	−2.251	0.177
Intensive cultures	−0.901	0.516	−1.913	0.111
Unidentified cultures	−0.319	0.688	−1.666	1.029
Lawns×Distance	0.017	0.033	−0.049	0.083
Urban areas×Distance*	0.067	0.016	0.036	0.098
Landfill×Distance	0.001	0.017	−0.033	0.035
Water×Distance	0.031	0.016	−0.001	0.063
Extensive cultures×Distance	−0.005	0.021	−0.045	0.035
Intensive cultures×Distance	−0.026	0.015	−0.056	0.004
Unidentified cultures×Distance	−0.036	0.026	−0.087	0.015
Lawns×Incubation	−0.050	0.671	−1.365	1.266
Urban areas×Incubation	0.210	0.244	−0.267	0.688
Landfill×Incubation	0.271	0.446	−0.603	1.145
Water×Incubation	0.205	0.254	−0.294	0.703
Extensive cultures×Incubation	−0.349	0.418	−1.168	0.469
Intensive cultures×Incubation*	1.429	0.237	0.965	1.893
Unidentified cultures×Incubation	0.911	0.474	−0.018	1.839
Lawns×Rainfall*	0.038	0.019	0.001	0.075
Extensive cultures×Rainfall	0.000	0.007	−0.013	0.013
Intensive cultures×Rainfall	−0.002	0.004	−0.009	0.005
Unidentified cultures×Rainfall	−0.003	0.008	−0.019	0.012

Model-averaged coefficients (β), unconditional standard errors (SE), and 95% confidence intervals (CI) are presented. Variables followed by an asterisk are significant (95% CI excluding 0).

Gulls

Tractors

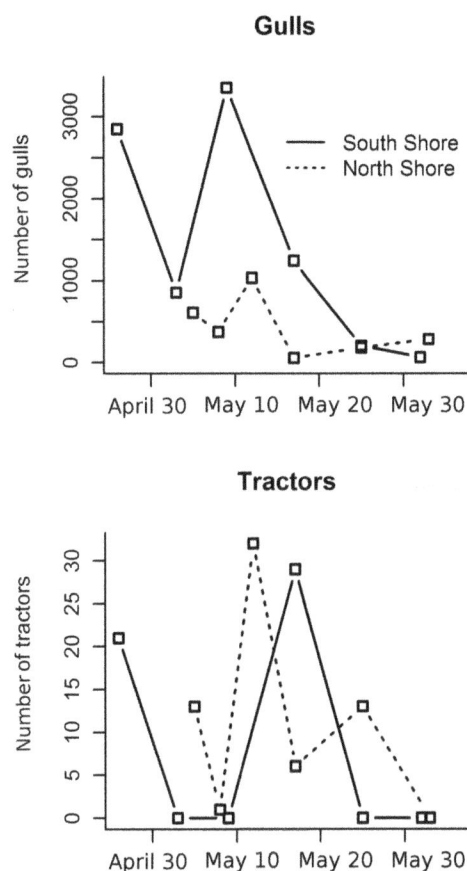

Figure 3. Use of agricultural lands by breeding Ring-billed gulls. Number of gulls and of tractors observed during surveys on the North and South shores of the St. Lawrence River, 2010.

Figure 4. Use of landfills and transhipment sites by Ring-billed gulls. Mean ($\pm$1 SD) proportion of Ring-billed gull locations at landfills and transhipment sites in foraging patches within 1-km concentric bands located at different distances from the colony. All landfills and open transhipment sites were visited by at least one tagged individual. Some sites encompassed more than one band.

Black-headed gulls (*Chroicocephalus ridibundus*) and Lesser Black-backed gulls (*Larus fuscus*), which are respectively slightly smaller and larger than Ring-billed gulls (14.7–15.5 m/s, respectively; [39]), at least 26 s is required for a gull to cross a circle of 200-m radius. The remaining 74 s appears insufficient for a gull to forage significantly in such a circular patch. Although our tracking device did not allow to determine the precise activity of the birds while not moving, we consider justified to assume that gulls were actually foraging in patches where they spend more than 100 s. Indeed, during the breeding period, gulls must brood their eggs or feed their young and must therefore spend as much time as possible on the colony allowing the rest of their time to foraging.

We used mixed effects logistic regressions to quantify the influence of landscape composition on the probability that a gull foraged in a patch along its movement path. Gull ID and foraging trip ID (nested within gull ID) were treated as random factors. The addition of these terms dealt with the hierarchical structure of the data and allowed the estimation of the variability across individuals and foraging trips. Eight different models were built and compared based on the second-order Akaike information criterion (AIC$_c$, [40]). We included the proportion of each habitat type and the occurrence of landfills and transhipment sites in all eight models. We considered the interaction of rainfall with lawns as well as with each type of agricultural cover because annelids are more prevalent under wet conditions [24]. We also included the distance between the location of a gull while foraging and its nest as a proxy for foraging costs and accessibility [41]. We considered

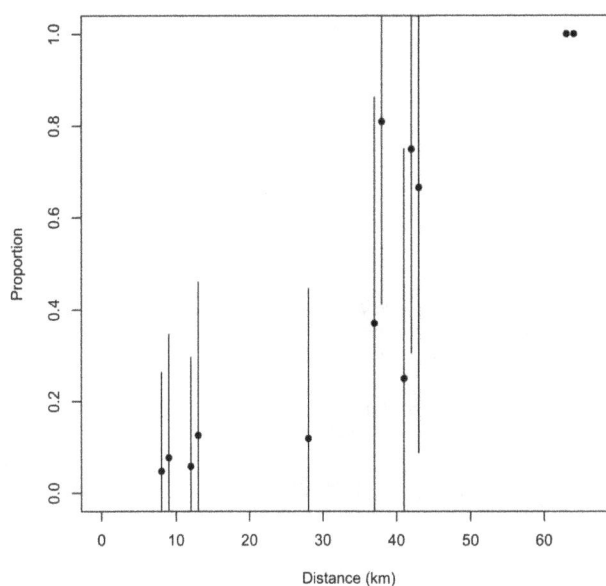

distance both as a main effect and in interaction with the relative amount of each habitat type (except woodlots) as well as with the occurrence of landfills or transhipment sites. We used this approach because we do not know which fitness currency gulls may be maximizing and because the profitability of the different habitats may not scale linearly with distance. Although woodlots are accessible, gulls avoid being under canopy and should thus avoid forest habitats whatever the distance from the colony. We included breeding stage in interactions with each habitat type to take into consideration the gulls' breeding phenology and their associated requirements as well as habitat phenology, particularly for agricultural cover types where farming practices and field conditions vary throughout the season. Finally, we built a second set of eight models, adding the sex of the birds in interaction with each habitat type and the distance of the patch from the colony as males and females differ in size (affecting travel costs and dominance on food patches) and provide different levels of parental care [23]. We fitted mixed effects logistic regressions using the Laplace approximation using the lme4 package (version 0.999375-39; [42]) run in the R statistical environment (version 2.12.2; [43]). AIC$_c$ were computed based on maximum log-likelihoods. Multi-model inference was performed following Burnham and Anderson [41] after testing that there was no problem of collinearity.

The effect of distance on habitat selection may be non-linear partly because central-place foragers often avoid habitats near their central-place by moving further away [44], [45]. Moreover, the relative abundance of different habitats varied with distance from the colony. To overcome this problem, we first draw 1-km wide circular bands up to 67 km from the colony, which corresponds to a few kilometers further than the farthest gull location (see *Results*). For each band with at least five foraging patches, we calculated the mean area covered by each habitat type within patches to estimate habitat use. We also calculated the mean area covered by the different habitats in movement patches

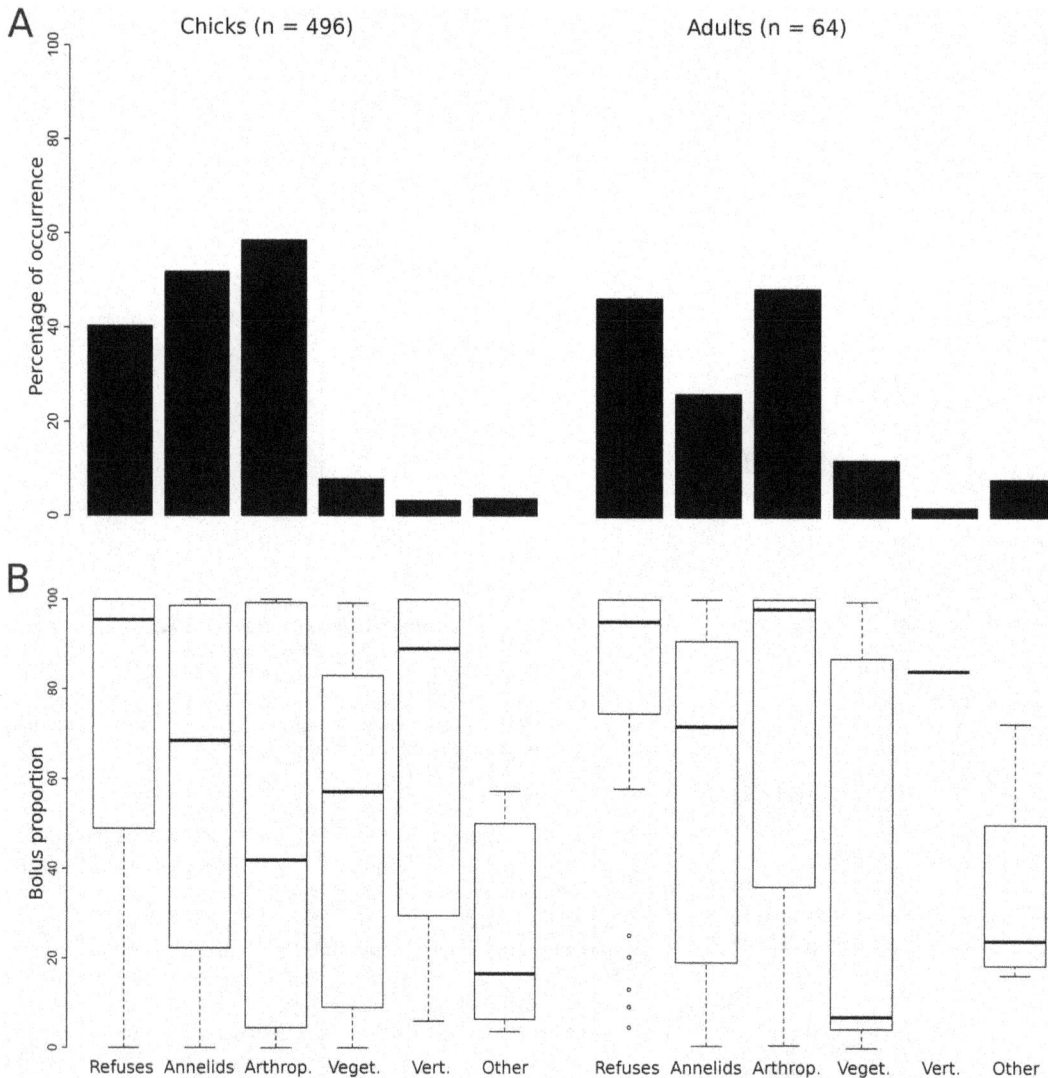

Figure 5. Diet of chicks and breeding adults of Ring-billed gulls. Diet of Ring-billed gull chicks (boli) and breeding adults (boli and stomach contents) at the Deslauriers Island colony, 2009–2010, expressed as the percentage of occurrence of each food category (a) and the proportion based on dry mass of each food category when present (b). Boxplots provide the first (bottom line), second (black midline) and third (top line) quartiles; whiskers extend to observations found up to 1.5 times the interquartile range; observations outside this range are indicated by empty dots.

(≥5 patches) within each circular band. We then plotted these two values for each band and each habitat to explore habitat selection as a function of distance.

The proportion of gulls observed foraging in a flock was considered as the time spent foraging in a given habitat [29]. This proportion was modeled as a two-column matrix, with the first column giving the number of gulls foraging and the second column giving the number of gulls involved in other activities for each flock, using a GLM with a binomial error distribution and logit link function (i.e., a logistic regression) using the stats package run in R. We also assessed whether the abundance of gulls in agricultural lands was related to the total number of tractors encountered along transects, which was considered an index of agricultural field work. This was done using a GLM with a Poisson error distribution and log link function (i.e., a Poisson regression) in R. This model included tractor number, transect location (South or North shore) and their interaction as explanatory variables.

Finally, we computed the proportion of boli containing at least one item of each food category for both chicks and adults. We then calculated the mean relative amount of each food item category when present in a bolus based on dry mass. The energy value of boli (kJ) was calculated for each gull collected at the Ste-Sophie landfill, in agricultural lands and at riparian sites by combining the dry mass of each item found in the stomach and their energy value. We compared the mean energy value of boli across habitats using an ANOVA.

Results

Characteristics of foraging trips

The mean number of foraging trips per day was greater during the rearing period (3.1±1.0 trips/day, ±SD) than during incubation (1.9±0.8 trips/day; $t_{107} = -7.01$, $P<0.001$). The mean direct distance between the colony and the furthest location reached during a foraging trip (whether a stopover or not) was also greater during brood rearing (16.6±12.4 km, maxi-

Table 4. Mean dry mass (g ± SD) and mean energy value (kJ) of nine food items gathered by sub-adult and adult Ring-billed Gulls in three habitat types (N = number of birds with food items present in their stomach).

| Food items | Mean dry mass (g) | | | Energy |
| | Landfills | Agricultural lands | Riparian habitats | |
	(N = 81)	(N = 54)	(N = 22)	(kJ)
Meat	1.84±4.71	-	0.87±2.81	30.1
Bread/rice	1.27±2.62	-	0.47±1.05	19.9
Potatoes/French fries	0.15±0.79	-	0.32±0.89	21.8
Miscellaneous refuse	0.79±2.69	-	-	22.6
Annelids	0.10±0.32	1.99±2.80	-	13.8
Arthropods	0.06±0.24	0.03±0.08	0.09±0.24	22.8
Corn/soybean grains	-	1.44±3.36	-	18.1
Vertebrates	0.01±0.11	-	0.46±0.61	19.4
Miscellaneous	0.10±0.32	0.08±0.35	-	21.3
TOTAL	4.47±6.37	3.55±3.68	2.36±3.39	-

mum = 63.5 km vs. 12.5±9.9 km, maximum = 42.4 km; t_{107} = −2.22, P = 0.03). Furthermore, the mean foraging distance traveled was greater during the rearing period compared to incubation (38.6±29.0 km, maximum = 156 km vs. 30.2±23.8 km, maximum 105 km; t_{107} = −2.55, P = 0.01). However, there was no difference in path sinuosity during the two periods (incubation: 1.2±0.2, maximum = 2.5; rearing: 1.2±0.2, maximum = 2.9; t_{107} = 1.38, P = 0.17). Finally, the mean trip duration was similar throughout the breeding period (incubation: 2.5±2.0 h, maximum = 9.6 h; rearing: 2.3±1.7 h, maximum = 12.4 h; t_{107} = 0.91, P = 0.36) but the trips lasted longer when a landfill was visited (3.5±1.8 h vs. 2.2±1.8 h; t_{107} = 5.95, P<0.0001). No significant effect of sex was found for the trip characteristics (all P>0.21).

Habitat selection

The composition of foraging and movement patches was highly variable (Table 1). Nevertheless, both movement and foraging patches were on average composed of smaller percentages of woodlots and of intensive and extensive cultures than what was found over the whole foraging range. An opposite trend was found for urban areas, waterbodies, landfills, and transhipment sites. While woodlots and urban areas covered a smaller proportion of foraging patches than movement patches, intensive cultures were relatively more important in foraging patches. Landfills and transhipment sites also occurred more often in foraging than movement patches. The distribution of residence time was strongly skewed to the right, with a peak under 100 s and a maximum reaching 19,377 s or 5.4 h (Fig. S2).

Model ranking based on AIC_c remained similar when considering the sex of individuals and its interaction with habitat types or patch distance. Yet, we only show results for models without sex as they performed better with much less parameters ($\Delta AICc$ = 3.04). The best model (w_i = 0.813) included habitat types as well as the distance separating the foraging patch from the colony, the breeding stage and their two-way interactions with habitat types (Table 2). The model that also included rainfall and its interaction with habitat types scored as the second best model (w_i = 0.187), leaving barely any support from the data for the remaining models. Note that the same two models were selected with similar, strong levels of evidence for other residence time thresholds (60, 80, 120, and 140 s) and with patch radii of 200 and 400 m,

underlining the robustness of our results with respect to these two assumptions.

Ring-billed gulls had a greater probability of foraging in patches located farther from the colony (Table 3). The distribution of these patches with respect to their distance from the colony was skewed to the right and showed a noticeable mode at ~10 km notwithstanding habitat types (Fig. 2a). Not surprisingly, gulls strongly avoided foraging in patches that included large amounts of woodlots (Table 3). Patches containing urban areas were significantly avoided close to the colony but increasingly selected further away. In fact, gulls tended to forage in patches with more urban cover compared to movement patches when the birds were between 25 and 35 km from their nest site (Fig. 2b). Because the colony was surrounded by water and gulls foraged little near the colony, there was a significant overall avoidance of this habitat (Table 3). Nevertheless, there was nearly a significant positive interaction between waterbodies and distance. In fact, waterbodies were relatively more important in foraging patches compared to movement patches when the birds were at 12 km or more from the colony (Fig. 2c). As expected, Ring-billed gulls foraged to a greater extent in patches containing lawns on rainy days (Table 3). We observed a greater proportion of lawns in foraging than in movement patches at around 5 km and again between 27 and 35 km (Fig. 2d).

The probability that a gull foraged in a patch increased with the proportion of intensive cultures (i.e., cereal fields) during incubation but tended to decrease during chick rearing (Table 3). More specifically, there were more intensive cultures in foraging than in movement patches up to 23 km from the colony (Fig. 2e). Similarly, the likelihood that gulls foraged in patches with extensive cultures (i.e., hayfields and pastures) tended to decrease with increasing amounts of this habitat (Table 3). The effect of distance on the use of extensive cultures by foraging gulls was not important (Fig. 2f). Gull surveys conducted in agricultural landscapes support the above patterns as the presence of gulls in intensive agricultural lands was related to the occurrence of ploughing, harrowing, and sowing, which all took place during the incubation period (mid-April to mid-May; Fig. 3). Indeed, the number of gulls observed along transects in agricultural lands increased with cultivation activities as indexed by the number of operating tractors seen in the fields (Poisson regression: β ±

SE = 0.064±0.001, $z = 53.9$, $P<0.01$). Of 20,900 gulls counted along transects, 52% were observed on bare soil fields (ploughed or recently sown), 34% on cereal fields with short vegetation (< 10 cm), 8% on stubble cereal fields and the remaining 6% on recently mowed hayfields. Finally, Ring-billed gulls had a greater tendency to forage in patches where a landfill or transhipment site was present (Table 3) and this was especially true as distance from the colony increased (Fig. 4).

Foraging behaviour, diet, and energy

The mean proportion of time that Ring-billed gulls spent foraging varied among habitats (deviance = -1.5×10^4; df = 1351, $P<0.01$). It was higher in agricultural lands (0.54±0.40) than in landfills and transhipment sites (0.17±0.20; $z = -76.9$, $P<0.01$), riparian habitats (0.12±0.25; $z = -78.4$, $P<0.01$), urban areas (0.15±0.32; $z = -53.4$, $P<0.01$) and on lawns (0.43±0.41; $z = -3.1$, $P<0.01$).

The four main food items (i.e., refuse, annelids, arthropods, and vegetation) were found in 40–60% of the boli collected from chicks reared on Deslauriers Island (Fig. 5a). The same items were found in the stomachs and boli of breeding adults, but in lower proportions (25–30%); it was compensated by a greater frequency of vertebrates and miscellaneous items. Yet, vertebrates occurred in less than 10% of the boli/stomachs in both chicks and adults. When refuse items were present, they contributed to a large proportion of the contents based on dry mass, unlike vegetation and miscellaneous items that usually represented a small proportion (Fig. 5b). The importance of annelids and arthropods was much more variable when present. Vertebrates were also quite variable in chick boli, whereas they clearly contributed to a very large proportion of the adult diet when they occurred.

Stomach contents from gulls collected in landfills were largely composed of fat meat typically found in refuse (Table 4). In agricultural lands, stomach contents were composed more or less equally of annelids and grains (soybean and corn). Stomach contents from riparian areas contained edible refuse, wild fishes, and arthropods. By pooling data on the relative importance of each food item and their respective energy content, we found that the mean energy value of stomach contents differed significantly among habitats ($F_{2,144} = 3.51$, $P = 0.03$). It was significantly higher in landfills (112.8±169.8 kJ) than in agricultural lands (55.8±63.6 kJ) and riparian areas (56.5±97.5 kJ), which were not significantly different.

Discussion

By combining analyses of GPS-tracking data and information on the gulls' abundance, diet, and proportion of time spent foraging in different habitats, we found that the distance from the colony and habitat phenology had strong effects on the process of habitat selection by breeding Ring-billed gulls foraging in a heterogeneous environment. For instance, they positively selected areas managed intensively for agriculture at a distance up to about 23 km from the colony but only when fields were being ploughed, harrowed, or sown. Gulls also selected areas where landfills and transhipment sites were present, especially as the distance from the colony increased. The mean energy intake being significantly greater in landfills than in agricultural lands, these results clearly suggest a trade-off by Ring-billed gulls to balance their energy budget. The St. Lawrence River and its tributaries are often used as passageways when flying to and from the insular colony, which resulted in a general avoidance of this habitat as feeding site. Over 12 km, however, gulls may stop along the shores of the rivers and the lakes or feed on emergent insects over water resulting in a selection of this habitat.

Energy trade-offs in selected habitats

The spatial and temporal variation in food availability could not be measured across the 6,000-km^2 study area. Nevertheless, we believe that using energy as an index of food quality and the relative area covered by each habitat allowed us to assess the relative benefits of different habitats. The strong selection for intensive cultures during incubation corresponded to the period when fields were being cultivated and the new cereal shoots were still at a height that allowed the birds to feed without visual obstruction. This seems to be associated with the occurrence of short periods of high food availability. Although it is difficult to differentiate the confounding effects of the breeding stage from the timing of field work and food availability, the positive effect of soil preparation and seed sowing on the abundance of gulls in agricultural lands during the incubation period (vs. brood-rearing) supports the hypothesis that selection for a specific habitat is higher during the peak of food availability. During our surveys, most gulls foraged in bare soil fields as observed for Black-headed gulls [46]. Moreover, half of the gulls' diet in agricultural lands was made of annelids, which are more accessible when tractors are ploughing and harrowing. Sibly and McCleery [24] have shown a positive relationship between the abundance of Herring gulls (Larus argentatus) in agricultural lands and the biomass of earthworms near the ground surface. Yet, the averaged RSFs did not detect an effect of rainfall on the use of agricultural lands despite the positive effect of ground wetness on the availability of annelids and their use by gulls [24]. In agricultural fields, gulls rely on the presence of heavy machinery that cannot work on wet soils. This contrasts with the use of lawns by gulls that was strongly associated with rainfall. Although we could not sample birds using urban areas, the greater availability of annelids on rainy days on lawns and their use by gulls is well established [22]. The other half of the gulls' diet in agricultural lands was made of soybeans and corn, which availability increases when sowing takes place (e.g., seeds accidentally dropped along road and field edges when farmers fill their seeders and seeds sown in superficial ground; M. Patenaude-Monette, pers. obs.). Annelids, soybeans and corn composed a less energy-rich diet than the food gathered by gulls at landfills. Considering that gulls selected the intensive agricultural lands no further than 23 km, we suggest that the profitability of this habitat was limited by the travel costs associated with the distance from the colony and the relatively low energy value of the food.

Gulls selected areas comprising landfills or transhipment sites throughout the breeding season, a period during which food availability at these sites does not vary with time. Although the accessibility (distance from the colony and deterrence program effectiveness) and volume of refuse differed among sites, we could only account for variation in distance from the colony. The selection of landfills was stronger, but also more variable, as the distance increased. Thus, the selection of landfills was probably not constrained by their distance from the colony as was the selection of agricultural lands at the scale of the study area. Nevertheless, its high variability suggests that not all gulls used landfills and transhipment sites. Indeed, landfills and transhipment sites were present in less than 5% of foraging patches of all individuals. Moreover, when refuse food items occurred in boli, they accounted for a much larger proportion of the bolus than any other food items. Furthermore, both the mean bolus mass and the mean energy content of food were much higher in landfills than in any other habitats.

We can hypothesize that gulls incur higher travel costs when foraging in landfills, which are located farther from the colony than agricultural lands [47]. Habitat accessibility is indeed likely to be negatively correlated with the distance separating the foraging site from the nest as travel costs (time, energy) increase with distance [41], [48]. Accordingly, intensively managed agricultural lands may thus provide a profitable net energy gain to foraging gulls despite food items of lower energy value, at least during the incubation period. On the other hand, landfills with their more energy rich food may be valuable foraging sites despite their remoteness and are thereby selected by gulls. The stronger selection observed with increasing distance to the colony (up to 63 km) may result from the fact that the closest sites (<30 km from the colony) included two transhipment sites where refuse is less available than at landfills. Moreover, the Terrebonne landfill that received the largest tonnage of refuse and which is located the closest to the colony has a very effective deterrence program (É. Thiériot, unpublished data).

Time constraints in urban areas

Gulls are known to feed on refuse in commercial and residential areas and on handouts offered by citizens [49]. Nevertheless, we found that breeding Ring-billed Gulls avoided foraging in urban areas located <10 km from the colony, but showed the opposite trend at greater distances. This pattern may result from the profitability of urban areas as foraging sites, which likely depends on the type of development (e.g., residential, commercial, or industrial) and population density. The proportion of time foraging was indeed very low in urban areas where gulls adopted a sit-and-wait strategy to exploit spatially and temporally scattered feeding opportunities (e.g., people handouts and overfilled garbage bins). While the proportion of time foraging was comparable in urban areas and in landfills, foraging opportunities are probably much less predictable in the former habitat. Furthermore, commercial and residential areas of high population densities (i.e., with greater foraging opportunities) were located about 20 km from the colony, which is much further than the closest landfill or agricultural lands. Although urban refuse food may present high energy contents, the time to gather enough refuse is likely too long to make foraging trips to urban areas profitable, particularly during the rearing period when chicks are waiting to be fed at the colony [50], [51]. The situation may nevertheless be different during the post-breeding period when gulls are then actively using urban areas ([52], C. Girault and J.-F. Giroux, unpublished data).

Conclusion

Combining RSF to survey data, diet characterization, and calorimetric analyses allowed us to characterize habitat selection processes of a central-place forager from an energy trade-off perspective. It also shows that other factors such as predation risk associated to deterrence programs at landfills can also play a role in the process of habitat selection at large spatial scales as

suggested through the concept of landscape of fear [53]. This approach was applied to a species that had to move over a large area to find food in a heterogeneous environment where habitat profitability also varied in time. Despite the complexity brought up by travel costs and habitat sampling issues, we were able to show that classical optimal foraging theory can make qualitative predictions applicable at the landscape level. This adds to the few evidences that optimal foraging theory has the potential to be scaled-up to the landscape level as predicted by Lima and Zollner [6]. Moreover, once classical models will have been modified such that their constraints are adapted to large spatio-temporal scales (e.g., [11], [54], [55]), GPS data loggers will allow us to test these models by linking the foraging behaviour of individuals to their breeding performance [14]. Such progress would make significant strides toward understanding the links between movement behaviour, habitat selection, fitness, and population dynamics within heterogeneous landscapes. For instance, this approach could be applied to many gull populations around the world to link their dynamics to food availability through landfill, agriculture, and fishery management.

Supporting Information

Figure S1 Residence times of breeding Ring-billed gulls in relation with patch size. Mean coefficient of variation (CV) of residence times within circular patches of different radii centred on locations obtained by GPS data loggers ($N = 109$ birds).

Figure S2 Frequency distribution of residence times of breeding Ring-billed gulls. Residence times were established for 200-m radius circular patches centred on locations obtained by GPS data loggers ($N = 109$ birds).

Table S1 Characteristics of individual Ring-billed gulls tracked during the study.

Acknowledgments

This research was conducted in partnership with the municipalities of Terrebonne, Repentigny, Laval, Charlemagne, Mascouche, Saint-Hippolyte, Sainte-Sophie, Sainte-Anne-des-Plaines, and Saint-Lin-Laurentides. We thank F. St-Pierre and M. Tremblay for field and laboratory assistance and R. Zamojska for diet and calorimetric analyses. S. Benhamou provided valuable advice about residence time computation while A. Desrochers, P. Peres-Neto, and G. Bastille-Rousseau commented an earlier draft of this manuscript.

Author Contributions

Conceived and designed the experiments: MPM JFG MB. Performed the experiments: MPM JFG. Analyzed the data: MPM MB. Contributed reagents/materials/analysis tools: JFG MB. Wrote the paper: MPM. Provided methodological and editorial comments: JFG MB. Supervised the overall field study: JFG.

References

1. Stephen DW, Krebs JR (1986) Foraging theory. Princeton: Princeton University Press. 247 p.
2. Giraldeau L-A, Caraco T (2000) Social foraging theory. Princeton: Princeton University Press. 376 p.
3. Charnov EL (1976) Optimal foraging, the marginal-value theorem. Theor Popul Biol 9: 129–136.
4. Orians GH, Pearson NE (1979) On the theory of central place foraging. In: Horn DJ, Stairs GR, Mitchell DR, editors. Analysis of ecological systems. Columbus: Ohio State University Press. 155–177.
5. Nonacs P (2001) State dependent behavior and the marginal value theorem. Behav Ecol 12: 71–83.
6. Lima SL, Zollner PA (1996) Towards a behavioral ecology of ecological landscapes. Trends Ecol Evol 11: 131–135.
7. Zollner PA, Lima SL (1999) Search strategies for landscape-level interpatch movements. Ecology 80: 1019–1030.
8. Nathan R, Getz WM, Revilla E, Holyoak M, Kadmon R, et al. (2008) A movement ecology paradigm for unifying organismal movement research. P Natl Acad Sci-Biol 105: 19052–19059.

9. Schoener TW (1979) Generality of the size-distance relation in models of optimal feeding. Am Nat 114: 902–914.

10. Olsson O, Brown JS, Helf KL (2008) A guide to central place effects in foraging. Theor Popul Biol 74: 22–33.

11. Bernstein C, Kacelnik A, Krebs JR (1991) Individual decisions and the distribution of predators in a patchy environment. II. The influence of travel costs and structure of the environment. J Anim Ecol 60: 205–225.

12. Beauchamp G, Bélisle M, Giraldeau L-A (1997) Influence of conspecific attraction on the spatial distribution of learning foragers in a patchy habitat. J Anim Ecol 66: 671–682.

13. Owen-Smith N, Fryxell JM, Merrill EH (2010) Foraging theory upscaled: the behavioural ecology of herbivore movement. Philos T Roy Soc B 365: 2267–2278.

14. Gaillard J-M, Hebblewhite M, Loison A, Fuller M, Powell R, et al. (2010) Habitat-performance relationships: finding the right metric at a given spatial scale. Philos T Roy Soc B 365: 2255–2265.

15. Manly BFJ, McDonald LL, Thomas DL, McDonald TL, Erickson WP (2002) Resource Selection by Animals: Statistical Design and Analysis for Field Studies. Second Edition, London: Kluwer Academic, 221 p.

16. Barraquand F, Benhamou S (2008) Animal movements in heterogeneous landscapes: identifying profitable places and homogeneous movement bouts. Ecology 89: 3336–3348.

17. Beyer HL, Haydon DT, Morales JM, Frair JL, Hebblewhite M, et al. (2010) The interpretation of habitat preference metrics under use-availability designs. Philos T Roy Soc B 365: 2245–2254.

18. Bastille-Rousseau G, Fortin D, Dussault C (2010) Inference from habitat-selection analysis depends on foraging strategies. J Anim Ecol 79: 1157–1163.

19. Fauchald P, Tveraa T (2006) Hierarchical patch dynamics and animal movement pattern. Oecologia 149: 383–395.

20. Freitas C, Kovacs KM, Lydersen C, Ims RA (2008) A novel method for quantifying habitat selection and predicting habitat use. J Appl Ecol 45: 1213–1220.

21. Fauchald P, Tveraa T (2003) Using first-passage time in the analysis of area-restricted search and habitat selection. Ecology 84: 282–288.

22. Brousseau P, Lefebvre J, Giroux J-F (1996) Diet of ring-billed gull chicks in urban and non-urban colonies in Quebec. Colon Waterbird 19: 22–30.

23. Pollet IL, Shutler D, Chardine J, Ryder JP (2012) Ring-billed Gulls (Larus delawarensis). The Birds of North America Online. Ithaca: Cornell Lab of Ornithology. Available: http://bna.birds.cornell.edu.bnaproxy.birds.cornell.edu/bna/species/033doi:10.2173/bna.33. Accessed 5 July 2013.

24. Sibly RM, McCleery RH (1983) The distribution between feeding sites of herring gulls breeding at Walney island, U.K. J Anim Ecol 52: 51–68.

25. Belant JL, Ickes SK, Seamans TW (1998) Importance of landfills to urban-nesting herring and ring-billed gulls. Landscape Urban Plan 43: 11–19.

26. Thiériot E, Molina P, Giroux J-F (2012) Rubber shots not as effective as selective culling in deterring gulls from landfill sites. Appl Anim Behav Sci 142: 109–115.

27. Fridolfsson AK, Ellegren H (1999) A simple and universal method for molecular sexing of non-ratite birds. J Avian Biol 30: 116–121.

28. Frair JL, Fieberg J, Hebblewhite M, Cagnacci F, DeCesare NJ, et al. (2010) Resolving issues of imprecise and habitat-biased locations in ecological analyses using GPS telemetry data. Philos T Roy Soc B 365: 2187–2200.

29. Altmann J (1974) Observational study of behavior: Sampling methods. Behaviour 49: 227–267.

30. Racine F, Giraldeau L-A, Patenaude-Monette M, Giroux J-F (2012). Evidence of social information on food location in a ring-billed gull colony, but the birds do not use it. Anim Behav 84: 175–182.

31. Batschelet E (1981) Circular statistics in biology. London: Academic Press. 371p.

32. Kotliar NB, Wiens JA (1990) Multiple scale of patchiness and patch structure: a hierarchical framework for the study of heterogeneity. Oikos 59: 253–260.

33. Bellier E, Certain G, Planque B, Monestiez P, Bretagnolle V (2010) Modelling habitat selection at multiple scales with multivariate geostatistics: an application to seabirds in open sea. Oikos 119: 988–999.

34. ESRI (2009) ArcGIS version 9.3.1. Redlands, California.

35. FADQ (2010) Bases de données des cultures assurées 2009 et 2010. Direction des ressources informationnelles, la Financière agricole du Québec, Saint-Romuald, Canada. Available: http://www.fadq.qc.ca/geomatique/professionnels_en_geomatique/base_de_donnees_de_cultures_assurees.html. Accessed 5 December 2011.

36. Natural Resources Canada (2009) Land Cover, Circa 2000, Vector. NRCan, Earth Sciences Sector, Centre for Topographic Information, Sherbrooke, Canada. Available: http://geobase.ca/geobase/en/find.do?produit = csc2000v. Accessed 12 January 2012.

37. Natural Resources Canada (2010) CanVec version 1.1. NRCan, Earth Sciences Sector, Centre for Topographic Information, Sherbrooke, Canada. Available: ftp://ftp2.cits.rncan.gc.ca/pub/canvec/. Accessed 12 January 2012.

38. Environment Canada (2010) National Climate Data and Information Archive. Government of Canada website. Available: http://climate.weather.gc.ca. Accessed 10 January 2011.

39. Shamoun-Baranes J, van Loon E (2006) Energetic influence on gull flight strategy selection. J Exp Biol 209: 3489–3498.

40. Burnham KP, Anderson DR (2002) Model selection and multimodel Inference: a practical information-theoretic approach. Second Edition, New-York: Springer-Verlag. 448 p.

41. Matthiopoulos J (2003) The use of space by animals as a function of accessibility and preference. Ecol Model 159: 239–268.

42. Bates D, Maechler M, Bolker B (2011) Lme4: Linear mixed-effects models using S4 classes. R package version 0.999375-39. Available: http://CRAN.R-project.org/package = lme4. Accessed 20 June 2011.

43. R Development Core Team (2011) R: a language and environment for statistical computing. Version 2.12.2, Vienna. http://www.r-project.org/.

44. Gaston AJ, Ydenberg RC, Smith GEJ (2007) Ashmole's halo and population regulation in seabirds. Mar Ornithol 35: 119–126.

45. Elliot KE, Woo KJ, Gaston AJ, Benvenuti S, Dall'Antonia L, et al. (2009) Central-place foraging in an arctic seabird provides evidence for Storer-Ashmole's halo. Auk 126: 613–625.

46. Schwemmer P, Garthe S, Mundry R (2008) Area utilization of gulls in a coastal farmland landscape: habitat mosaic supports niche segregation of opportunistic species. Landscape Ecol 23: 355–367.

47. Wilson RP, Quintana F, Hobson VJ (2012). Construction of energy landscapes can clarify the movement and distribution of foraging animals. P Roy Soc Lond B Bio 279: 975–980.

48. Rosenberg DK, McKelvey KS (1999) Estimation of habitat selection for central place foraging animals. J Wildl Manage 63: 1028–1038.

49. Belant JL (1997) Gulls in urban environments: landscape-level management to reduce conflict. Landscape Urban Plan 38: 245–258.

50. Bukacinska M, Bukacinski D, Spaans AL (1996) Attendance and diet in relation to breeding success in Herring gulls (Larus argentatus). Auk 113: 300–309.

51. Shaffer SA, Costa DP, Weimerskirch H (2003) Foraging effort in relation to the constraints of reproduction in free-ranging albatrosses. Funct Ecol 17: 66–74.

52. Maciusik B, Lenda M, Skorka P (2010) Corridors, local food resources, and climatic conditions affect the utilization of the urban environment by the Black-headed Gull Larus ridibundus in winter. Ecol Res 25: 263–272.

53. Searle KR, Stokes CJ, Gordon IJ (2008) When foraging and fear meet: using foraging hierarchies to inform assessments of landscapes of fear. Behav Ecol 19: 475–482.

54. Amano T, Ushiyama K, Moriguchi S, Fujita G, Higuchi H (2006) Decision-making in group foragers with incomplete information: Test of individual-based model in Geese. Ecol Monogr 76: 601–616.

55. Mueller T, Fagan WF (2008) Search and navigation in dynamic environments – from individual behaviors to population distributions. Oikos 117: 654–664.

14

Stratification of Carbon Fractions and Carbon Management Index in Deep Soil Affected by the Grain-to-Green Program in China

Fazhu Zhao, Gaihe Yang*, Xinhui Han*, Yongzhong Feng, Guangxin Ren

College of Agronomy, Northwest A&F University, Yangling, Shaanxi, China; and The Research Center of Recycle Agricultural Engineering and Technology of Shaanxi Province, Yangling, Shaanxi, China

Abstract

Conversion of slope cropland to perennial vegetation has a significant impact on soil organic carbon (SOC) stock in A horizon. However, the impact on SOC and its fraction stratification is still poorly understood in deep soil in Loess Hilly Region (LHR) of China. Samples were collected from three typical conversion lands, *Robinia psendoacacia* (RP), *Caragana Korshinskii Kom* (CK), and abandoned land (AB), which have been converted from slope croplands (SC) for 30 years in LHR. Contents of SOC, total nitrogen (TN), particulate organic carbon (POC), and labile organic carbon (LOC), and their stratification ratios (SR) and carbon management indexes (CMI) were determined on soil profiles from 0 to 200 cm. Results showed that the SOC, TN, POC and LOC stocks of RP were significantly higher than that of SC in soil layers of 0–10, 10–40, 40–100 and 100–200 cm (P<0.05). Soil layer of 100–200 cm accounted for 27.38–36.62%, 25.10–32.91%, 21.59–31.69% and 21.08–26.83% to SOC, TN, POC and LOC stocks in lands of RP, CK and AB. SR values were >2.0 in most cases of RP, CK and AB. Moreover, CMI values of RP, CK, and AB increased by 11.61–61.53% in soil layer of 100–200 cm compared with SC. Significant positive correlations between SOC stocks and CMI or SR values of both surface soil and deep soil layers indicated that they were suitable indicators for soil quality and carbon changes evaluation. The Grain-to-Green Program (GTGP) had strong influence on improving quantity and activity of SOC pool through all soil layers of converted lands, and deep soil organic carbon should be considered in C cycle induced by GTGP. It was concluded that converting slope croplands to RP forestlands was the most efficient way for sequestering C in LHR soils.

Editor: Raffaella Balestrini, Institute for Plant Protection (IPP), CNR, Italy

Funding: This work was supported by Special Fund for forest-scientific Research in the Public Interest (201304312). The funders had no role in study design, data collection and analysis, decision to publish, or preparation of the manuscript.

Competing Interests: The authors have declared that no competing interests exist.

* E-mail: ygh@nwsuaf.edu.cn (GY); hanxinhui@nwsuaf.edu.cn (XH)

Introduction

Soil organic carbon (SOC) is a dynamic component of the terrestrial system, with internal changes in both vertical and horizontal directions and external exchanges between the atmosphere and the biosphere [1]. SOC storage is estimated at approximately 1500 Pg globally, which is about two and three times the size of carbon pools in the atmosphere and vegetation, respectively [2]. Since carbon uptake and storage is tightly linked to the nitrogen (N) cycle, it is equally important to understand how N pools and fluxes are affected by land use change [3]. Moreover, more than 50% of the total SOC is stored in the subsoil [4]. The proportion of soil organic matter (SOM) stored in the first meter of the world soils below 30 cm depth ranges 46%~63%, except for Podzoluvisols, where 30% of SOC is stored below the depth of 30 cm [4]. A recent study also suggests that in the northern circumpolar permafrost region, at least 61% of the total soil C is stored below 30 cm [5]. Therefore, subsoil C may be even more important in terms of source or sink for CO_2 than topsoil C [6]. Considering the potential role of SOC in atmospheric CO_2 sink, it is important to understand what leads to sequestration of large amounts of SOC in the subsoil or even in deep soil. However, the SOC contents in deep soil layers are not fully understood in LHR of China to date.

As an indicator of soil quality, SOM stratification, which is related to the rate and amount of SOC sequestration [7], is common in many natural ecosystems [8] and managed grasslands and forests [9–10]. Stratification ratio (SR) is defined as the ratio of a soil property at the surface layer to that at a deeper layer. In general, high SR values indicate good soil quality and are usually used to assess agricultural practices [7]. For instance, SR values for SOC at depths of 0–5 cm and 20–40 cm range from 1.1 to 1.5 under traditional tillage (TT) while from 1.6 to 2.6 under conservation tillage (CT) [11]. Little information is available on natural ecosystems and managed shrubs or forests land. Additionally, under semiarid climate, SOC in active fractions, such as particulate organic carbon (POC) and labile organic carbon (LOC) was more sensitive to soil management practices than total SOC [12]. Previous researches have indicated that changing rate of POC and LOC was faster than SOC in whole soil [11], and they could be an early indicator for SOC change in soil [13]. Meanwhile, the carbon management index (CMI), which is derived from the total soil organic C pool and C lability, had been extensively used as a sensitive indicator of SOC variation rate in response to soil management changes [14–15]. Therefore, under semi-arid climate, using SR of total SOC and of different SOC fractions may be useful to reveal how soil management affects soil quality and helpful to understand the mechanism of SOC transformation and cycling in subsoil as well as in deep soil. In LHR of China, soil erosion and desertification are causing a loss of net primary productivity that was estimated as high as 12 kg C ha^{-1}y^{-1} [16]. To counteract soil erosion and other environmental problems, an environmental protection policy was implemented by Chinese central government, which was known as the Grain to Green Program (GTGP). The purpose of GTGP was to convert up to 26.87 million ha low-yield sloped croplands (>25°) into forests, shrubs or grasslands by the end of 2008 [17]. It is the first and the most ambitious "payment–for–ecosystem–services" program in China to date [18]. Although the initial goal of GTGP was to control soil erosion in China, it also plays a significant role in circulation of SOC and total nitrogen (TN). In recent years, a few studies estimated the effects of GTGP on vegetation structure, economic benefits, soil physiochemical properties, and niche characteristics [19–21]. However, SR values of SOC and/or TN and CMI value among different land use types are rarely reported. Especially, information on dynamics of C in deep soil is largely ignored in this region.

This study aimed to: 1) analyze the contents of SOC, TN, POC and LOC and their vertical distributions at the depths of 0–200 cm; 2) assess the stocks of SOC and TN at different soil depths of three land use types; and 3) evaluate the soil quality of different land use types using SR and CMI values as the main assessment parameters.

Materials and Methods

All sites in the watershed we were selected for study was determined through interviews with local farmers (Mr. Yibin Zhang, Soil and Water Conservation Experiment Station, Northwest A&F University, Ansai County, Shaanxi, NW China).We state clearly that no specific permissions were required for the location. We confirm that the location is not privately-owned or protected in any way. We confirm that the field studies do not involve endangered or protected species.

Research area

The study was conducted in the Zhifanggou catchment (36°46′42″–36°46′28″N, 109°13′46″–109°16′03″E), which is located in Ansai county, central LHR (see Fig. 1). Ansai is a typical county characterized by semi-arid climate and hilly loess landscape in the Loess Plateau with an annual average temperature of 8.8°C, and an average annual precipitation of 505 mm. 60% of precipitation occurs between July and September (~300 mm in dry years while >700 mm in wet years). Accumulated temperatures above 0°C and 10°C are 3733°C and 3283°C, respectively. On average, there are about 157 frost-free days and 2415 h sunny time each year. Arable farming mostly occurs on

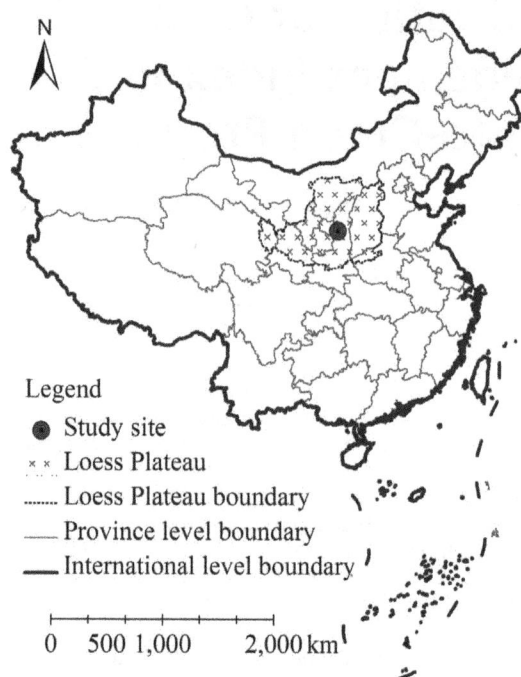

Figure 1. Location of the Loess Plateau and the study site.

sloping lands without irrigation. The loess parent material at the site has an average thickness of approximately 50–80 m and the soil in this region is classified as Calciustepts soil [22]. Sand (2–0.05 mm) and silt (0.05–0.002 mm) account for approximately 29.22% and 63.56% in soil depth of 0–20 cm, respectively. The soil is highly erodible, with an erosion modulus of 10,000–12,000 Mg·km^{-2}·yr^{-1} before the start of restoration efforts [23]. After 30 years vegetation restoration, the area of forest lands significantly increased from 5% to 40% [24].

The Zhifanggou catchment has been an experimental site of the Institute of Soil and Water Conservation, Chinese Academy of Science (CAS) since 1973 [25]. The major agricultural land use type in LHR is slope cropland. Agricultural management in this region, including the major crop types grown, has not been changed significantly since the 1970s. The main crops grown in these sites were millet (Setaria italica) and soybean (Glycine max) rotation, and no irrigation was provided in grown season (depend on rainfall). One crop was grown each year, and fertilizer was applied (mainly manure). After more than 30 years of comprehensive management, the ecological environment of the catchment has been significantly improved [26]. Since late 1970s, slope cropland is replanted with shrubs and woods, mainly Robinia pseudoacacia L. (RP) and Caragana Korshinskii Kom (CK), to control soil erosion (see Table 1). Abandoned cropland was also generated during this period due to its extremely low productivity and long

Table 1. Characteristics of different vegetation types.

Vegetation types	Age	Canopy closure (%)	Litter accumulation (t.ha^{-2})	Undergrowth Vegetation[a]	Species diversity indices
Robinia pseudoacacia L.	30	58	20.5	Lespedeza dahurica - Stipa bungeana	6.5
Caragana Korshinskii Kom	30	50	13.3	Achillea capillaries, Stipa bungeana	3.9

[a]means the main vegetation in forest/shrub land.

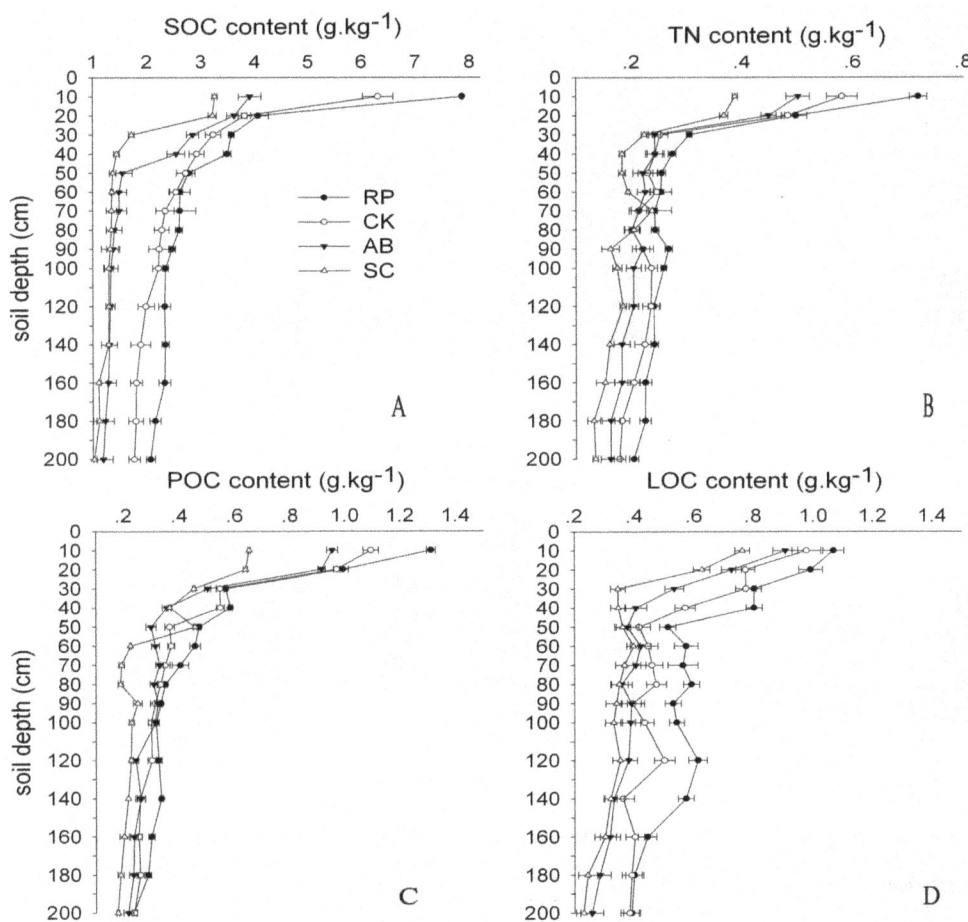

Figure 2. Distribution of soil organic carbon (SOC, A), total nitrogen (TN, B), particulate organic carbon (POC, C), and labile organic carbon (LOC, D) contents of different land used types in soil depth of 0–200 cm. The error bars are the standard errors.

distance from farmers' residences [27–28]. Despite wild grasslands and shrub lands were usually found on steep slopes, these sites were used for firewood collection as well. So the wild vegetation was of limited coverage or even barren for long periods. In 1999, most slope lands were closed for vegetation restoration under the GTGP [29].

Soil sampling

In September 2012, based on land use history, 30 year old *Robinia psendoacacia* (RP), *Caragana Korshinskii Kom* (CK), abandoned land (AB) and slope cropland (SC) in the Zhifanggou catchment were selected. Three 30 m×20 m plots were established for each land use type. All sites were located on the same physiographical units with same slope aspects, same elevation of 1250 m and a spatial distance of 1200 m.

Soil samples were taken at several soil depths using a soil auger (diameter 5 cm) from 10 points within "S" shape at each plot (0–10 cm, 10–20 cm, 20–30 cm, 30–40 cm, 40–50 cm, 50–60 cm, 60–70 cm, 70–80 cm, 80–90 cm, 90–100 cm, 100–120 cm, 120–140 cm, 140–160 cm, 160–180 cm, and 180–200 cm). Then after removing the litter layer, ten soil samples at each depth of each plot were mixed to make one sample. Samples were collected at least 80 cm away from the trees. All samples were sieved through a 2 mm screen, and roots and other debris were removed. Soil samples were air-dried and stored at room temperature for the

determination of soil chemical properties. A ring tube was used to determine the bulk density in each soil depth.

Laboratory analysis

SOC content $(g.kg^{-1})$ and TN content $(g.kg^{-1})$ were determined using $K_2Cr_2O_7$ oxidation method and Kjeldhal method, respectively [30].

To determine POC content, 25 g soil was dispersed with 100 mL of 5 g L^{-1} sodium hexametaphosphate before being. Then, the mixed soil solution was shaken for 1 h at high speed on an end-to-end shaker and screened by a 0.053 mm sieve with several deionized water rinses. The soil remained on the sieve was backwashed into a pre-weighed aluminum box and dried at 60°C for 24 h, then it was grounded for analysis of C [31].

Soil labile organic carbon (LOC) was measured following the method described in Graeme et al. [32]. A 2–6 g air dried soil sample was put into a 50 mL centrifuge tube, and 25 mL of 333 mmolL^{-1} KMnO$_4$ solution was added before being shaken with a rate of 120 rpm for 1 h, and centrifuged for 5 min with a rate of 5,000×g. The upper clear solution was transferred, and diluted by 250 times, and then the absorbance at 565 nm wavelength was determined. The absorbances at values 565 nm with different KMnO$_4$ concentrations were also determined for preparation of standard curve, which was used for the determination of the KMnO$_4$ concentrations. Difference between the

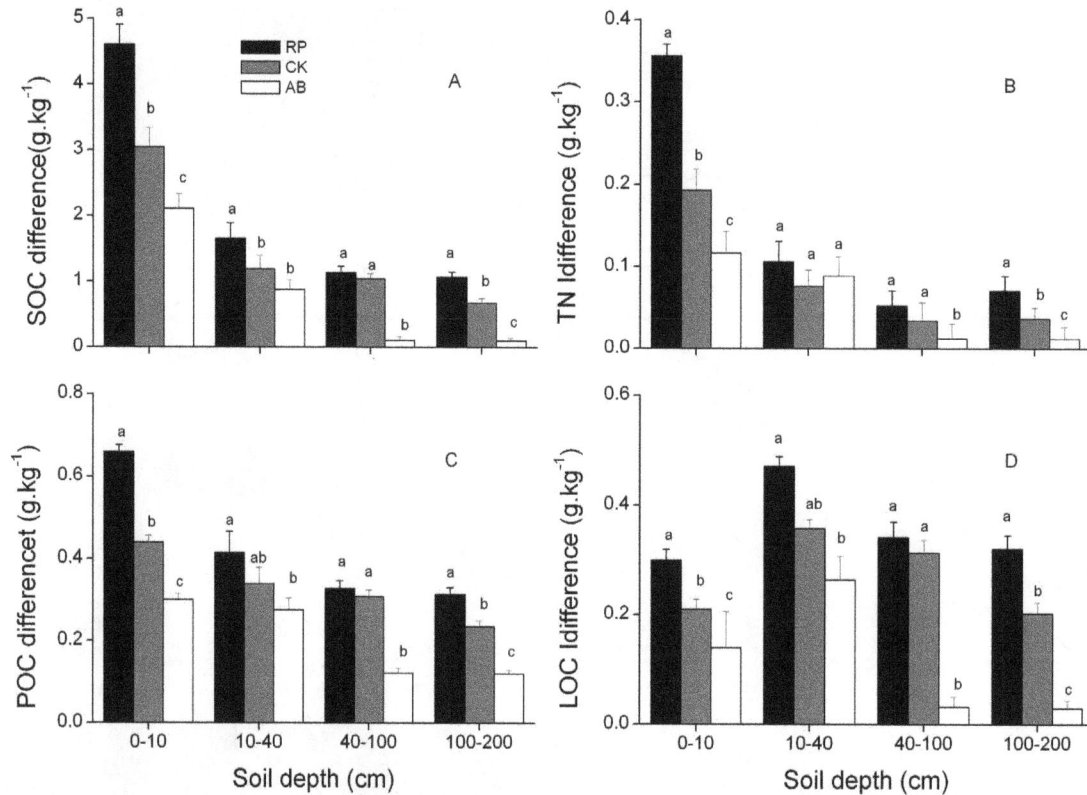

Figure 3. Differences in soil organic carbon (SOC, A), total nitrogen (TN, B), particulate organic carbon (POC, C), labile organic carbon (LOC, D) contents between SC and RP, CK or AB (RP/CK/AB - SC). Error bars are the standard errors. Different lowercase letters indicate significant difference among different land use types within same soil layer (P<0.05). The same for Fig 4

amounts of $KMnO_4$ added and remained was used to calculate labile C concentration in the soil sample.

Calculation of SOC (TN) stocks, SR of SOC (TN, POC, and LOC) and CMI

SOC density (SOCD, TND) represents the total SOC (TN) storage of overall certain sampling depth. SOCD(TND) of different sampling depths were calculated:

$$SOCD(TND) = C_{SOC,TN} \times \rho \times H \times (1 - \delta/100) \times 10^{-1} \quad (1)$$

where SOCD (TND) is the density (Mg·ha^{-1}) of SOC (TN) and $C_{SOC,TN}$ is the content (g·kg^{-1}) of SOC (TN). ρ is the bulk density (g·cm^{-3}), H is the soil horizon thickness (cm), and δ is the fraction (%) of gravels >2 mm in size in soil. Because the soil gravel size of loess in China is mostly below 2 mm, this fraction was assumed to be 0 [33].

SR values (0–10 cm: 10–40 cm, 0–10 cm: 40–100 cm and 0–10 cm: 100–200 cm) were calculated from the contents of SOC, TN, POC and LOC following the method in Franzluebbers (2002).

CMI values were calculated using following procedures:

Firstly, a C pool index (CPI) was calculated:

$$CPI = \frac{sample\ total\ organic\ C\ (g/kg)}{reference\ sample\ total\ C\ (g/kg)} \quad (2)$$

where reference sample is SC soil. Then, a lability index (LI) was calculated:

$$LI = \frac{L\ in\ each\ sampled\ soil}{L\ in\ the\ reference\ soil} \quad (3)$$

where reference soil is SC soil, and L was calculated from the C lability:

$$L = \frac{content\ of\ labile\ C}{content\ of\ non-labile\ C} \quad (4)$$

At last, CMI was calculated:

$$CMI = CPI \times LI \times 100 \quad (5)$$

Statistical analyses

All statistical analyses were carried out with SPSS 17.0. Analysis of variance (ANOVA) and Duncan's Multiple Range Test (DMRT) at 5% level of significance were used to compare the difference in contents and/or stocks of SOC, TN, POC, LOC, CMI, and SR among different land use types or soil depths. A sample linear-regression analysis was used to estimate the relationships between carbon stocks with CMI or SR values.

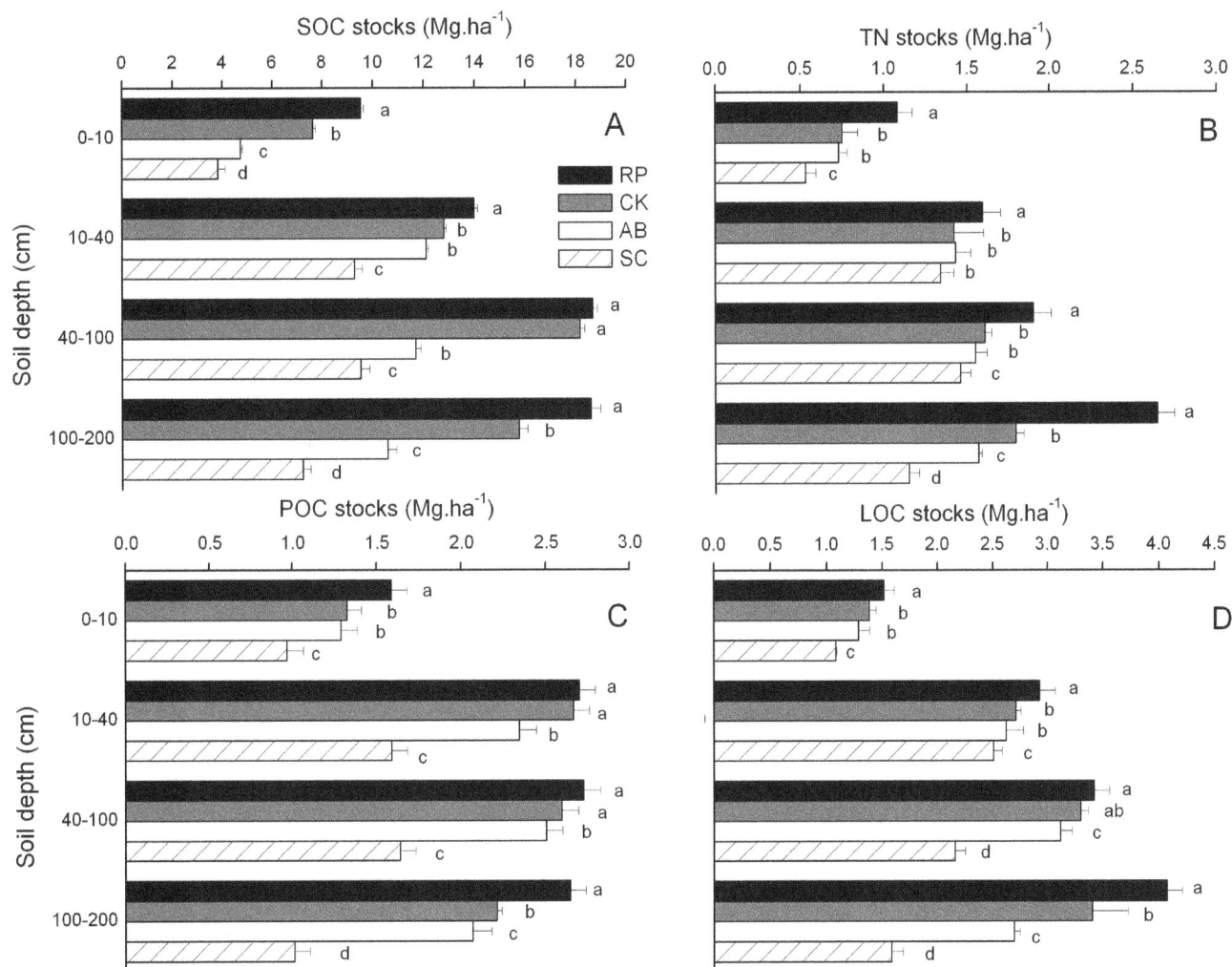

Figure 4. Stocks of soil organic carbon (SOC, A), total nitrogen (TN, B), particulate organic carbon (POC, C), labile organic carbon (LOC, D) of different land use types. The error bars are the standard errors.

Results

Changes in contents of SOC, TN, POC and LOC

The contents of SOC, TN, POC and LOC responded differently as the change of soil depth (Fig. 2). In all land use types, contents of SOC, TN, POC and LOC in top soil (0–10 cm) were 3.26–7.86 g.kg^{-1}, 0.39–0.72 g.kg^{-1}, 0.65–1.31 g.kg^{-1} and 0.76–1.07 g.kg^{-1}, respectively, which were significantly higher than other soil layers (P<0.05). The contents of SOC, TN, POC and LOC decreased significantly in soil depth of 10–40 cm while the decreases trended to be flatter in subsoil (40–100 cm). Additionally, the differences in contents of SOC, TN, POC and LOC in deep subsoil (100–200 cm) were negligible (P<0.05).

The differences in contents of SOC, TN, POC and LOC between three forest/shrub types (RP, CK and AB) and SC are shown in Fig. 3. The differences in SOC, TN, POC and LOC of RP and SC in soil depths of 0–10 cm and 100–200 cm were significantly higher than that between other land use types and SC (P<0.05). The differences in SOC and TN of RP were 33.78% and 45.97% larger than that of CK, and 54.13% and 67.28% larger than that of AB in soil depth of 0–10 cm (P<0.05), while the differences in POC and LOC were 32.8%, 54.0% higher than that of CK, and 23.3% and 45.0% higher than that of AB (P<0.05).

Moreover, the differences in SOC, TN, POC and LOC of RP were 25.05–85.29% higher than that of CK, and 61.78–90.70% higher than that of AB in soil depth of 100–200 cm. Additionally, significant differences in SOC, TN, POC and LOC contents were observed between RP and CK in soil depths of 10–40 cm and 40–100 cm, but there was no difference between CK and AB (P<0.05).

Changes and distribution of SOC, TN, POC and LOC stocks

SOC, TN, POC and LOC stocks of RP, CK and AB were higher than SC in all soil profiles (Fig 4). The SOC, TN, POC and LOC stocks of RP were significantly increased (P<0.05), which were 0.43–5.8 Mg.ha^{-1}, 0.25–4.70 Mg.ha^{-1}, 0.44–9.14 Mg.ha^{-1} and 1.49–11.38 Mg.ha^{-1} higher than that of SC in soil layers of 0–10, 10–40, 40–100 and 100–200 cm, respectively. Moreover, the stocks of SOC, TN, POC and LOC in soil layer of 100–200 cm of RP were higher than that of CK and AB by 15.4–32.1% and 21.8–43.1%, respectively (P<0.05).

The SOC, TN, POC and LOC stocks responded differently as the change of soil depth (Fig. 5). Although the distribution of SOC, TN, POC and LOC stocks in soil depths of 0–10 cm and 10–

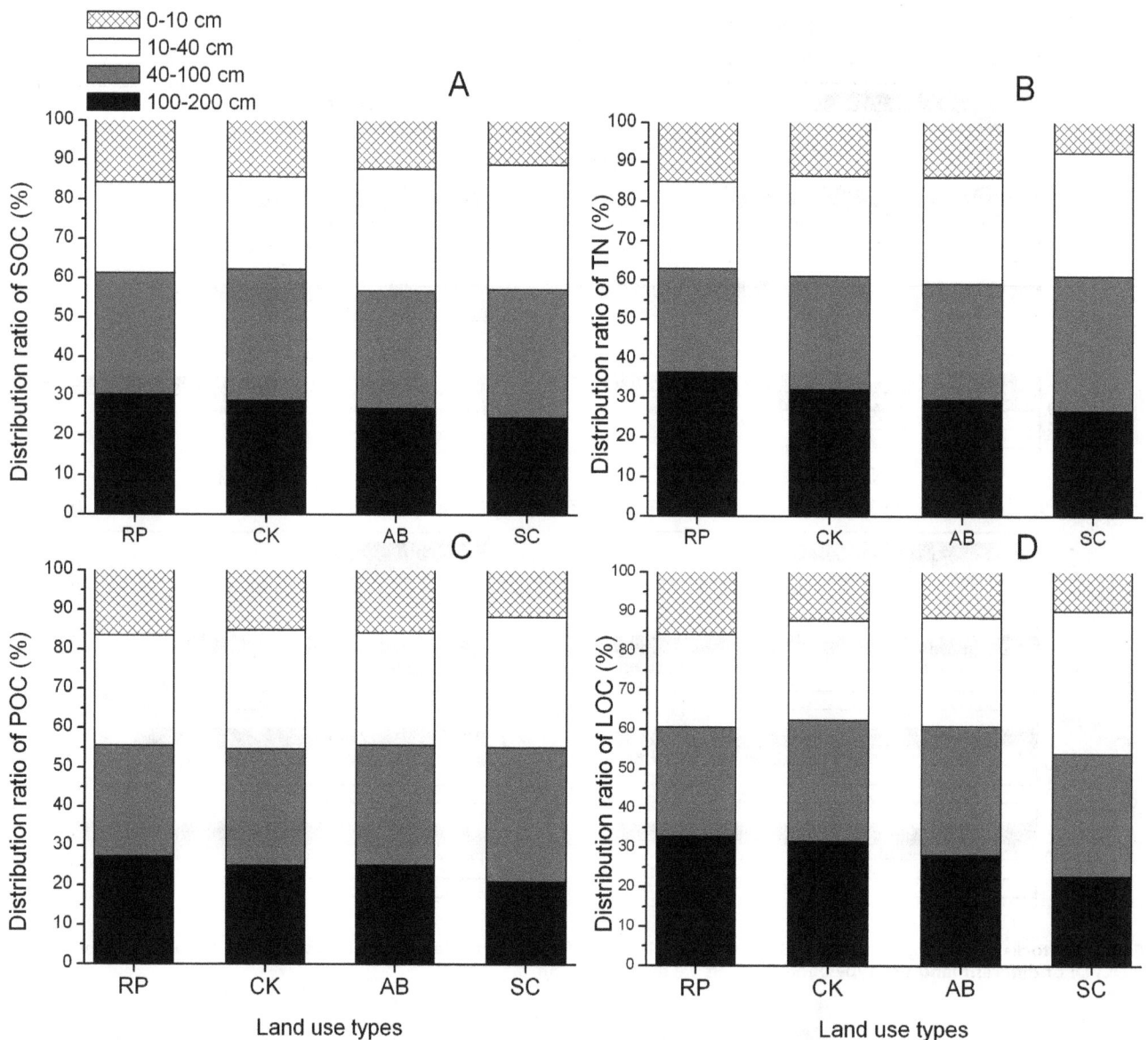

Figure 5. Distribution ratios of soil organic carbon (SOC, A), total nitrogen (TN, B), particulate organic carbon (POC, C), labile organic carbon (LOC, D) in soil depth of 0–200 cm under different land use types.

40 cm accounted for the majority, 26.36–34.06% and 21.08–36.62% were distributed in soil layers of 40–100 cm and 100–200 cm, respectively. Among four land use types, the highest proportion of SOC, TN, POC and LOC stocks were found in RP, while the lowest were in soil depths of SC in 0–10 cm and 100–200 cm of SC. The proportion of SOC, TN, POC and LOC stocks under RP were higher than SC by 4.68%, 7.32%, 4.65% and 5.96% respectively in soil depth of 0–10 cm soil depth, whereas by 5.90%, 9.78%, 6.30% and 10.06% was higher in soil depth of 100–200 cm soil depth respectively.

Change in SR and CMI values

Responses of SR in different land use types to change of soil depth were different (Fig 6). The SR values of SOC, TN and LOC differed significantly among different soil depths (P<0.05), while the SR values of LOC differed only between 0–10:10–40 cm, 0–10:40–100 cm and 0–10:100–200 cm. Among four land use types, the SR values of SOC, TN, POC and LOC of RP were the

highest, but that of SC were the lowest in each soil depth(P<0.05). The SR values of SOC, TN, POC and LOC were in a decreasing order of CK>AB>SC. The SR values differed significantly between CK or AB with SC (P<0.05), while there was no significant difference between CK and AB. Additionally, the ratios of SR values of SOC, TN, POC and LOC in the surface layer (0–10 cm) to that in layer of 10–40 cm were >2.0 in most cases.

The CMI values were significantly affected by land use types. In our study, the CMI values were in a decreasing order of RP>CK>AB>SC in four soil profiles and CMI values were significantly enhanced by RP compared with SC (Fig 7). Averaged CMI values of RP, CK, and AB were 40.60%, 50.54%, 37.81%, and 14.1% higher than that of SC in soil layers of 0–10 cm, 10–40 cm, 40–100 cm and 100–200 cm.

Regression equations to assess CMI/SR values of TN, POC, and LOC (Y) were showed in Table 2. There was a significant

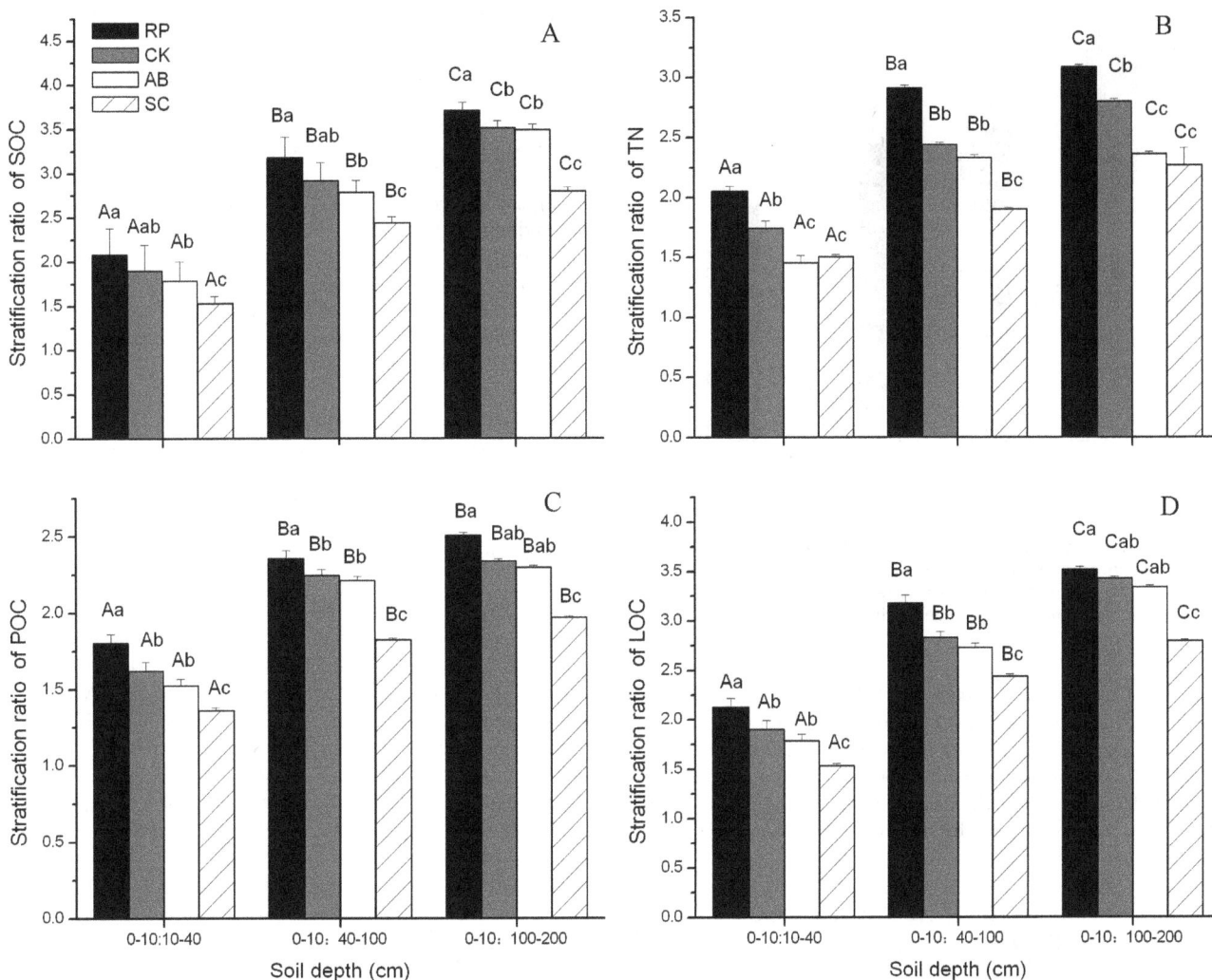

Figure 6. Comparison of stratification ratio of soil organic carbon (SOC, A), total nitrogen (TN, B), particulate organic carbon (POC, C), labile organic carbon (LOC, D) under different land use types. Different uppercase letters indicate significant difference among different soil depths within same land use type while the different lowercase letters indicate significant difference among different land use types within same soil depth. The error bars are the standard errors.

positive correlation between CMI/SR values of TN, POC, and LOC with SOC stocks in surface soil and deep soil.

Discussion

SOC, TN, POC and LOC contents and SOC and TN stocks

Vegetation can greatly influence soil quality, C and N cycling, and regional socioeconomic development [34–35]. It is also reported that converting cropland into land with perennial vegetation would increase the SOC content [36]. Our results showed that land use type and soil depth significantly affected the contents of SOC and TN (Fig. 2). The conclusion that both land use type and soil depth are important factors influencing the soil carbon and nitrogen distribution was consistent with previous studies [35,37]. We also observed that the lowest SOC, TN, POC and LOC contents were found in slope cropland (Fig 4), which essentially agree with a previous study [38], indicating that the conversion of slope cropland to vegetation improves the C and N contents. A possible reason is that the lower residue input into the soil in slope cropland leads to lower SOC and TN contents.

Additionally, our results showed that SOC, TN, POC and LOC contents of RP were greater than that of CK and AB (Fig 4). It infers that the effects of RP on soil C and N play a significant role in land use and ecosystem management. The conclusion was consistent with Qiu et al [39], who reported that RP has potential to improve SOC content in the loessial gully region of the Loess Plateau and the improvements are greater in long-term than middle-term.

Recently it was reported that the depth of sampling is an important factor for the measurement of change in SOC stocks [40], and land use could influence subsoil C pools [41]. We found that SOC, TN, POC and LOC stocks of RP, CK, and AB were higher than SC for different soil profiles, especially in depths of 40–100 cm and 100–200 cm (Fig. 4 and 5). It is demonstrated that converting slope cropland into woodland and shrubland not only affects SOC and TN stocks in surface soil, but also largely influences that in deep soil. The result was consistent with Wang et al [42], who reported that deep layer (50–200 cm) SOC stocks were equivalent to approximately 25% of that in the shallow layer (0–50) in Hilly Loess Plateau. This is mainly due to the fact that

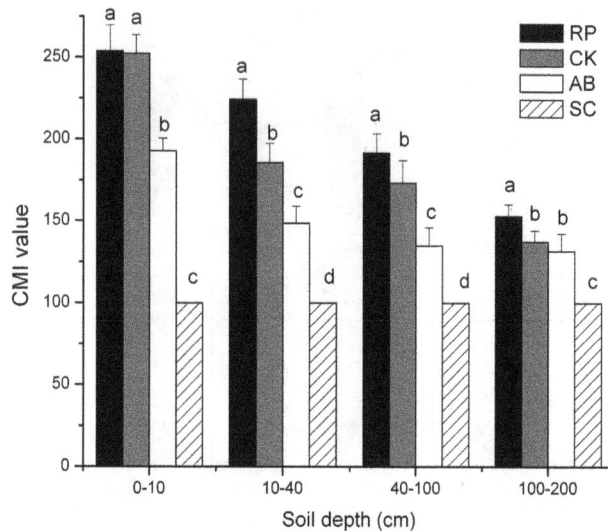

Figure 7. Carbon management index (CMI) values of different land use types at different soil depths. The error bars are the standard errors. Different lowercase letters indicate significant difference among different land use types within same soil depth.

SOC input into subsoil is largely affected by plant roots and root exudates, dissolved organic matter and bioturbation. In addition, most important factors leading to protection of SOC in subsoil include the spatial separation of SOM, microorganisms and extracellular enzyme activity related to the heterogeneity of C input [43]. As a result, stabilized SOC in subsoil is horizontally stratified.

Stratification ratios of SOC, TN, POC, and LOC

According to Franzluebbers [7], SOC SR values >2 in degraded conditions is uncommon, and the SR values of SOC are generally low and seldom reach 2.0. And SR values of soil organic C and N pools with value of >2 would be an indicator that soil quality might be improved [7]. In our study, the most of SR in SOC, TN, POC and LOC was more than 2 after convert slope cropland to forest or shrub land (Fig. 6). This means soil quality was improved in these afforested soils without disturbance. Greater C stratification ratios could be related to the fact that, during soil recovery by re-vegetation or land abandonment, soil was undisturbed thus reducing oxidation and favoring soil C [44]. The result was consistent with Sá et al [45]. Similar results were also reported by Moreno et al [46] and Franzluebbers [7], who reported that stratification of SOC occurs over time when soil tillage and disturbance is stopped and it is usually greater in undisturbed soils than in disturbed soils. In addition, the stratification may increase with time, and SOC, TN, POC and LOC contents are still aggrading but have not reached soil C saturation yet. That is the reason why SR values of SOC, TN, POC and LOC under CK and AB were higher but no significant differences were observed compared with SC (Fig. 6). Sá et al [45] concluded that the SOC pool stabilization may be attained in about 40 years after long-term no-tillage adoption.

Carbon management index

CMI value was calculated to obtain indications of the C dynamics of the system and provide an integrated measure for quantity and quality of SOC [15]. Soils with higher CMI values are considered as better managed [47]. We found that CMI values were significantly enhanced by RP forest compared with CK, AB and SC in both surface soil and subsoil and deep soil (Fig. 7). Soil management under RP plot was more appropriate to improve the SOC status than other land use types. Similar result was reported by Qiu et al [39], who illustrated that RP forest has significantly increased SOC, total nitrogen, ratio of carbon to nitrogen and ratio of carbon to phosphorus compared to other vegetation types. Our result showed that there were significant positive correlations between SOC stocks and CMI/SR in both surface soil and deep soil (Table 2). These findings showed that SR values of SOC, TN,

Table 2. Regression equations among SOC stocks and CMI/SR for different soil layers.

Axis		Soil depth (cm)	Equations	R^2	Significant level
X, CMI/SR[a]	Y, SOC stocks[b]				
CMI	SOC	0–10	Y = 53.5+23.20X	0.91	P = 0.048
		10–40	Y = −146+25.81X	0.97	P = 0.027
		40–100	Y = 24.2+8.60X	0.93	P = 0.023
		100–200	Y = 77.5+4.05X	0.93	P = 0.046
TN	SOC	0–10:10–40	Y = 1.12+0.08X	0.93	P = 0.047
		0–10:40–100	Y = 1.26+0.07X	0.96	P = 0.032
		0–10:100–200	Y = 1.67+0.07X	0.98	P = 0.017
POC	SOC	0–10:10–40	Y = 1.18+0.06X	0.97	P = 0.023
		0–10:40–100	Y = 1.55+0.04X	0.93	P = 0.041
		0–10:100–200	Y = 1.74+0.04X	0.92	P = 0.047
LOC	SOC	0–10:10–40	Y = 1.30+0.08X	0.96	P = 0.031
		0–10:40–100	Y = 1.95+0.05X	0.97	P = 0.025
		0–10:100–200	Y = 2.53+0.05X	0.98	P = 0.013

[a]CMI = carbon management index, SR = stratification ration, TN = SR of total nitrogen, POC = SR of particulate organic carbon, LOC = SR of labile organic carbon.
[b]For the Y-axis, the SOC stocks (0–10 cm, 40–100 cm, and 100–200 cm) were used to analyze correlations between SOC stocks and SR values of TN, POC, and LOC (0–10:10–40, 0–10:40–100, 0–10:100–200).

POM, and LOC, and CMI are suitable indicators for evaluating soil quality and C changes induced by GTGP in surface soil and deep soil.

Conclusion

In this study, the SOC, TN, POC and LOC contents of RP, CK and AB in soil layer of 100–200 cm were higher than SC, especially for RP plot. Although the SOC, TN, POC and LOC stocks in soil layer of 100–200 cm were lower, there was more than 27.38–36.62%, 25.10–32.91%, 21.59–31.69% and 21.08–26.83% of SOC, TN, POC and LOC stocks were distributed in 100–200 cm soil depth under RP, CK and AB. Meanwhile, the SR of SOC, TN, POC and LOC in the surface to lower depth ratio (i.e., 0–10:10–40 cm) was >2.0 in most of case. And SR and as well CMI values were significantly enhanced by RP compared with SC in deep soil (100–200 cm) (P<0.05). Indicating that soil quality

was improved after converting slope land into perennial vegetation, especially under RP plot from surface soil to deep soil. Moreover, there were significant and positive correlations between SOC stocks and CMI or SR of TN, POC, LOC both surface soil and deep soil indicated that the SR and CMI value are suitable indicators for evaluating soil quality and C changes in surface soil as well as in deep soil. We, therefore, propose deep soil organic carbon should be considered in C cycle induced by Grain-to-Green Program (GTGP) and under RP forest is more appropriate strategy to improve the SOC status than other land use types in surface soils and deep soil.

Author Contributions

Conceived and designed the experiments: GY XH. Analyzed the data: FZ. Contributed reagents/materials/analysis tools: YF GR. Wrote the paper: FZ.

References

1. Zhang CS, McGrath D (2004) Geostatistical and GIS analyses on soil organic carbon concentrations in grassland of southeastern Ireland from two different periods. Geoderma 119: 261–275.
2. Jobbágy EG, Jackson RB (2000) The vertical distribution of soil organic carbon and its relation to climate and vegetation. Ecological Applications 10: 423–436.
3. Cole CV, Duxbury J, Freney J, Heinemeyer O, Minami K, et al. (1997) Global estimates of potential mitigation of greenhouse gas emissions by agriculture. Nutrient Cycling in Agroecosystems 49: 221–228.
4. Amundson R (2001) The soil carbon budget in soils. Annual Reviews of Earth and Planetary Sciences 29: 535–562.
5. Guo L, Gifford RM (2002) Soil carbon stock s and land use change: a meta analysis. Global Change Biology 8: 345–360.
6. IPCC (2007) Climate change 2007: the physical Science basis. In: Solomon, S., Qin, D., Manning, M., Chen, Z., et al. (Eds.), Contribution of Working Group I to the Fourth Assessment Report of the Intergovernmental Panel on Climate Change. Cambridge University Press, Cambridge.
7. Franzluebbers AJ (2002) Soil organic matter stratification ratio as an indicator of soil quality. Soil Tillage Research 66: 95–106.
8. Prescott CE, Weetman GF, DeMontigny LE, Preston CM, Keenan RJ (1995) Carbon chemistry and nutrient supply in cedar–hemlock and hemlock –amabilis fir forest floors. In: McFee, W.W., Kelley, J.M. (Eds.), Carbon Forms and Functions in Forest Soils. Soil Sci. Soc. Am., Madison, WI, pp. 377–396.
9. Van Lear DH, Kapeluck PR, Parker MM (1995) Distribution of carbon in a Piedmont soil as affected by loblolly pine management. In: McFee, W.W., Kelley, J.M. (Eds.), Carbon Forms and Functions in Forest Soils. Soil Sci. Soc. Am., Madison, WI, pp. 489–501.
10. Schnabel RR, Franzluebbers AJ, Stout WL, Sanderson MA, Stuedemann JA. (2001) The effects of pasture management practices. In: Follett, R.F., Kimble, J.M., Lal, R. (Eds.), The Potential of US Grazing Lands to Sequester Carbon and Mitigate the Greenhouse Effect. Lewis Publishers, Boca Raton, FL, pp. 291–322.
11. Sa JCM, Lal R (2009) Stratification ratio of soil organic matter pools as an indicator of carbon sequestration in a tillage chronosequence on a Brazilian Oxisol. Soil & Tillage Research 103: 46–56.
12. Haynes RJ (2005) Labile organic matter fractions as central components of the quality of agricultural soils: an overview. Adv. Agron 85: 221–268.
13. Franzluebbers AJ, Arshad MA (1992) Particulate organic carbon content and potential mineralisation as affected by tillage and texture. Soil Science Society of America Journal 61: 1382–1386.
14. Sparling GP (1997) Soil microbial biomass activity and nutrient cycling: an indicator of soil health. In: Pankhurst, C.E., Doube, B.M., Gupta, V.V.S.R. (Eds.), Biological Indicators of Soil Health. CAB International, Wallingford, UK, pp. 97–119.
15. Blair GJ, Lefroy RDB, Lisle L (1995) Soil carbon fractions based on their degree of oxidation and the development of a carbon management index for agricultural systems. Australian Journal of Agricultural Research 46: 1459–1466.
16. Bai ZG, Dent D (2009) Recent land degradation and improvement in China. Ambio 38: 150–156.
17. Jia ZB (2009) Investigation Report on Forestry major problem in 2008. Forestry Press in China. 267–273 (In Chinese)
18. LüY H, Fu BJ, Feng XM, Zeng Y, Liu Y, et al. (2012) A Policy-Driven Large Scale Ecological Restoration: Quantifying Ecosystem Services Changes in the Loess Plateau of China. PLoS ONE 7, e31782. doi:10.1371/journal.pone.0031782.
19. Zhao YT (2010) Analysis on the Necessity and Feasibility of Implementing the Project for Conversion of Cropland to Forest. Ecological Economy 7: 81–83.
20. Wei J, Cheng J, Li W, Liu W (2012) Comparing the Effect of Naturally Restored Forest and Grassland on Carbon Sequestration and Its Vertical Distribution in the Chinese Loess Plateau. PLoS ONE 7(7): e40123. doi:10.1371/journal.pone.0040123

21. Wei XR, Qiu LP, Shao MA, Zhang XC, Gale WJ (2012) The Accumulation of Organic Carbon in Mineral Soils by Afforestation of Abandoned Farmland. PLoS ONE 7(3): e32054. doi:10.1371/journal.pone.0032054
22. Gong ZT, Lei WJ, Chen ZC, Gao YX, Zeng SG, et al. (1999) Chinese Soil Taxonomy. Science Press, Beijing 36–38.
23. Liu G (1999) Soil conservation and sustainable agriculture on the Loess Plateau: challenges and prospective. Ambio 28: 663–668.
24. Xue S, Liu GB, Pan YP, Dai QH, Zhang C, et al. (2009) Evolution of Soil Labile Organic Matter and Carbon Management Index in the Artificial Robinia of Loess Hilly Area. Scientia Agricultura Sinica 4: 1458–1464
25. Jiao JY, Zhang ZG, Bai WJ, Jia YF, Wang N (2012) Assessing the Ecological Success of Restoration by Afforestation on the Chinese Loess Plateau. Restoration Ecology 20: 240–249.
26. Zhang F, Zhang SL, Cheng ZJ, Zhao HY (2007) Time structure and dynamics of the insect communities in bush vegetation restoration areas of Zhifanggou watershed in Loess hilly region. Acta Ecologica Sinica 27: 4555–4562. (in Chinese with English abstract)
27. Chen QB, Wang KQ, Qi S, Sun LD (2003) Soil and water erosion in its relation to slope field productivity in hilly gully areas of the Loess Plateau. Aata Ecologica Sinica 23: 1463–1469.
28. Li FM, Song QH, Jjemba PK, Shi YC (2004) Dynamics of soil microbial biomass C and soil fertility in cropland mulched with plastic film in a semiarid agro-ecosystem. Soil Biology and Biochemistry 36: 1893–1902.
29. Wang Z, Liu GB, Xu MX, Zhang J, Wang Y, et al. (2012) Temporal and spatial variations in soil organic carbon sequestration following revegetation in the hilly Loess Plateau, China. Catena 99: 26–33.
30. Bao SD (2000) Soil and Agricultural Chemistry Analysis. China Agriculture Press, Beijing, China (in Chinese).
31. Cambardella CA, Elliot ET (1992) Particulate soil organic matter changes a grassland cultivation sequence. Soil Science Society of America Journal 56: 777–783.
32. Graeme JB, Rod DBL, Leanne L (1995) Soil carbon fractions based on their degree of oxidation, and the development of a carbon management index for agricultural systems. Australian Journal of Agricultural Research 46:1459–1466
33. Wang YF, Fu BJ, Lu YH, Song CJ, Luan Y (2010) Local-scale spatial variability of soil organic carbon and its stock in the hilly area of the Loess Plateau, China. Qua-ternary Research 73: 70–76.
34. Eaton JM, McGoff NM, Byme KA, Leahy P, Kiely G (2008) Land cover change and soil organic carbon stocks in the Republic of Ireland 1851–2000. Climate Change 91: 317–334.
35. Fu XL, Shao MA, Wei XR, Robertm H (2010) Soil organic carbon and total nitrogen as affected by vegetation types in Northern Loess Plateau of China, Geoderma 155: 31–35.
36. Groenendijk FM, Condron LM, Rijkse WC (2002) Effect of afforestation on organic carbon, nitrogen, and sulfur concentration in New Zealand hill country soils. Geoderma 108: 91–100.
37. Davis M, Nordmeyer A, Henley D, Watt M (2007) Ecosystem carbon accretion 10 years after afforestation of depleted subhumid grassland planted with three densities of Pinus nigra. Global Change Biology 13: 1414–1422
38. Chen LD, Gong J, Fu BJ, Huang ZL, Huang YL, et al. (2007) Effect of land use conversion on soil organic carbon sequestration in the loess hilly area, loess plateau of China. Ecology. Research 22: 641–648.
39. Qiu LP, Zhang XC, Cheng JM, Yin XQ (2010) Effects of black locust (Robinia pseudoacacia) on soil properties in the loessial gully region of the Loess Plateau, China. Plant Soil 332: 207–217.
40. VandenBygaart AJ, Bremer E, McConkey BG, Ellert BH, Janzen HH, et al. (2010) Impact of Sampling Depth on Differences in Soil Carbon Stocks in Long-Term Agroecosystem Experiments. Soil Science Society of America Journal 75: 226–234

41. Strahm BD, Harpison RB, TeRPy TA, Harpington TB, Adams AB, et al. (2009) Changes in dissolved organic matter with depth suggest the potential for postharvest organic matter retention to increase subsurface soil carbon pools. Forest Ecology Management 258: 2347–2352

42. Wang Z, Liu GB, Xu MM (2010) Effect of revegetation on soil organic carbon concentration in deep soil layers in the hilly Loess Plateau of China. Acta Ecologica Sinica 14: 3947–3952 (in Chinese with English abstract)

43. Rumpel C, Kögel-Knabner I (2011) Deep soil organic matter-a key but poorly understood component of terrestrial C cycle. Plant Soil 338: 143–158

44. Fayez R (2012) Soil properties and C dynamics in abandoned and cultivated farmlands in a semi-arid ecosystem. Plant Soil 351: 161–175.

45. Sá JCM, Cerpi CC, Dick WA, Lal R, Vesnke-Filho SP, et al. (2001) Organic matter dynamics and carbon sequestration rates for a tillage chronosequence in a Brazilian Oxisol. Soil Science Society of America Journal 5: 1486–1499.

46. Moreno F, Murillo JM, Pelegrín F, Girón IF (2006) Long-term impact of conservation tillage on stratification ratio of soil organic carbon and loss of total and active CaCO3. Soil Tillage Research 85:86–93

47. Diekow J, Mielniczuk J, Knicker H, Bayer C, Dick DP, et al. (2005) Carbon and nitrogen stocks in physical fractions of a subtropical Acrisol as influenced by long-term no-till cropping systems and N fertilization. Plant Soil 268: 319–328.

15

Assessment of Bacterial *bph* Gene in Amazonian Dark Earth and Their Adjacent Soils

Maria Julia de Lima Brossi[1]*****, **Lucas William Mendes**[1], **Mariana Gomes Germano**[2], **Amanda Barbosa Lima**[1], **Siu Mui Tsai**[1]

1 Cellular and Molecular Biology Laboratory, Center for Nuclear Energy in Agriculture, University of São Paulo, Piracicaba, SP, Brazil, **2** Brazilian Agricultural Research Corporation, Embrapa Soybean, Londrina, PR, Brazil

Abstract

Amazonian Anthrosols are known to harbour distinct and highly diverse microbial communities. As most of the current assessments of these communities are based on taxonomic profiles, the functional gene structure of these communities, such as those responsible for key steps in the carbon cycle, mostly remain elusive. To gain insights into the diversity of catabolic genes involved in the degradation of hydrocarbons in anthropogenic horizons, we analysed the bacterial *bph* gene community structure, composition and abundance using T-RFLP, 454-pyrosequencing and quantitative PCR essays, respectively. Soil samples were collected in two Brazilian Amazon Dark Earth (ADE) sites and at their corresponding non-anthropogenic adjacent soils (ADJ), under two different land use systems, secondary forest (SF) and manioc cultivation (M). Redundancy analysis of T-RFLP data revealed differences in *bph* gene structure according to both soil type and land use. Chemical properties of ADE soils, such as high organic carbon and organic matter, as well as effective cation exchange capacity and pH, were significantly correlated with the structure of *bph* communities. Also, the taxonomic affiliation of *bph* gene sequences revealed the segregation of community composition according to the soil type. Sequences at ADE sites were mostly affiliated to aromatic hydrocarbon degraders belonging to the genera *Streptomyces*, *Sphingomonas*, *Rhodococcus*, *Mycobacterium*, *Conexibacter* and *Burkholderia*. In both land use sites, shannon's diversity indices based on the *bph* gene data were higher in ADE than ADJ soils. Collectively, our findings provide evidence that specific properties in ADE soils shape the structure and composition of *bph* communities. These results provide a basis for further investigations focusing on the bio-exploration of novel enzymes with potential use in the biotechnology/biodegradation industry.

Editor: Niyaz Ahmed, University of Hyderabad, India

Funding: Funding provided by 'Conselho Nacional de Desenvolvimento Científico e Tecnológico' and the 'Fundação de Amparo à Pesquisa do Estado de São Paulo' (FAPESP/Biota 2011/50914-3). The funders had no role in study design, data collection and analysis, decision to publish, or preparation of the manuscript.

Competing Interests: The authors have declared that no competing interests exist.

* E-mail: majubrossi@gmail.com

Introduction

Amazonian Dark Earth (ADE), locally termed *'Terra Preta de Índio'*, are anthropogenic soil horizons built-up by the Pre-Colombian Indians between 500 and 8,700 years ago. These soil sites were formed by the progressive deposit of materials and organic compounds, such as charcoal, bone, and pottery sheds, which gradually shifted the natural physical and chemical properties of the soil. As a result, relatively infertile Amazon soils were progressively converted into highly fertile spots through processes like increasing the cation exchange capacity and the nutrient content, as well as promoting the stabilization of the soil physical structure [1,2]. Substantial increments of organic material in these sites gradually increased the carbon content, yielding to the formation of soil spots with a high proportion of incompletely combusted biomass (biochar). These spots have been reported to reach up to a 70-fold higher amount of carbon than native soils at adjacent locations (ADJ) [3].

The existence of ADE sites close to their natural ADJ soil locations, which present the same geological history, provides a unique opportunity to investigate the role of biotic and abiotic factors influencing the microbial community assembly and dynamics at these sites. Previous studies revealed that ADE and ADJ sites present differences in microbial community composition, and bacterial diversity has been reported to be higher at ADE sites [4,5,6,7,8]. Most of these studies rely on comparisons between the taxonomic profiles of these communities (i.e., based on the taxonomic bacterial 16S rRNA gene). In this sense, the extent to which the local environment shapes the functional profiles of these communities, and influences their performance, remains mostly elusive [9,10,11,12,13].

Soil is one of the most biodiverse ecosystems on Earth, being able to support communities from multiple trophic levels, which are constantly performing the metabolism of diverse and complex substrates. The extreme spatial and temporal heterogeneity of the soil matrix, paired with the myriad of internal and external feedbacks, are known to determine the structure and function of these communities [14]. Microbes are involved in many ecosystem processes, including biodegradation, decomposition and mineralization, inorganic nutrient cycling, disease causation and suppression, and pollutant removal. Soil disturbances are known to cause shifts in microbial activities, shifting the rate of these processes and triggering impacts on the ecosystem performance [15]. Several environmental factors are known to affect microbial community composition in the soil, including soil temperature, moisture, texture, carbon content, nutrient availability, pH, land use history, seasonality, and the content of incompletely combusted biomass, such as biochar [13,16,17,18,19].

The biochar content is the major physical distinction between ADE and their ADJ soils, which is also known to play an important role in global carbon biogeochemistry [20]. In anthrosols, such as ADE, it is predicted that distinct microbial communities can perform unique processes, such as the retention of high-labile carbon [5]. Despite the unique and specialized capabilities of these soils, functional assessments of their microbial community, particularly those of genes encoding important steps in the carbon cycle, are still scarce. Biodegradation through bacterial activity is one of the most important processes occurring in soils regarding organic matter recycling. This process involves genes acting on key steps in the carbon cycle, for the turnover of more recalcitrant organic carbon, as well as for pollutant degradation in the ecosystem [21]. This process, along with biosynthesis, largely governs the carbon cycle in the environment, which is dependent on microbial enzymatic activities that most often use organic compounds as a primary energy source [22].

The primary step involved in the aerobic microbial degradation of aromatic hydrocarbons is an oxidative attack [23,24], where enzymes named oxygenases are responsible for the insertion of molecular oxygen into aromatic benzene rings [25]. Genes encoding these enzymes have been characterized in *Rhodococcus, Acinetobacter, Pseudomonas, Mycobacterium, Burkholderia, inter alia* [25,26]. The α-subunit of oxigenases is known as the catalytic domain involved in the transfer of electrons to oxygen molecules. Due its DNA sequence conservation, this subunit has been currently used as a target gene for the detection of such enzymes in complex communities [21,27,28,29].

In this study, we evaluated the structure, composition and abundance of the bacterial catabolic gene Biphenyl Dioxygenase (*bph*) involved in aromatic hydrocarbon degradation in Amazonian Dark Earth and their adjacent soil locations. We aimed to determine the role of anthropogenic action in the diversity of the *bph* gene in soil bacterial communities. Understanding the diversity of specific bacterial genes in ADE should led to future studies that investigate the microbial ecology of anthropogenic altered soils, especially in regards to their potential source of novel enzymes. Collectively, this study characterized this catabolic gene occurring in Amazonian Dark Earth, and compared the profiles with those obtained from adjacent sites, as well as under distinct land use systems.

Materials and Methods

Ethic statement

No specific permits were required for the described field studies. The locations are not protected. The field studies did not involve endangered or protected species.

Study sites, sample collection and soil chemical analyses

Studied sites are located at the Caldeirão Experimental Station of Amazon Brazilian Agricultural Research Corporation (Embrapa) in Iranduba County in the Brazilian Central Amazon (03°26'00" S, 60°23'00" W). A detailed description of the soil sampling locations is given by Taketani et al [8]. Briefly, the four sites sampled are composed of two Amazonian Dark Earth (ADE) and their two correspondent adjacent soils (Haplic Acrisol, ADJ). These sites are under a ~35 year-old secondary forest (SF) or under manioc (*Manihot esculenta*) cultivation (M). Hereafter these sites are termed as ADE-SF, ADJ-SF, ADE-M and ADJ-M. Soil samples were taken in triplicate from the topsoil layer (ca. top

10 cm), and the overlaying litter was discarded. Each sample contained approximately 300 g of soil and was transported to the laboratory at 4°C to further processing (<24 h). A portion of the samples were frozen (−20°C) for total DNA extraction while the other portion was kept at 4°C for chemical measurements.

Chemical analyses were performed at Amazon Embrapa (Manaus, Brazil), according to instructions provided by the Embrapa protocol [30]. Briefly, soil samples were analysed in triplicate for pH (H_2O, 1:2.5); H+Al (calcium extractor 0.5 mol L^{-1}, pH 7.0); sum of bases (SB); soil organic matter (SOM); soil organic carbon (SOC; Walkely-Black method); extractable fraction of Al, Ca, and Mg (1 M KCl); extractable fraction of P and K (double acid solution of 0.025 M sulphuric acid and 0.05 M hydrochloric acid Mehlich 1); and effective cation exchange capacity (eCEC).

Total soil DNA extraction and PCR amplifications for T-RFLP

Total DNA was extracted using 0.25 g of soil as an initial material. Extractions were carried out in triplicate for each site, using the PowerSoil DNA isolation kit (MoBio Laboratories, Carlsbad, CA, USA), according to the manufacturer's protocol. DNA quality and quantity were measured spectrophotometrically using NanoDrop 1000 (Thermo Scientific, Waltham, EUA).

For T-RFLP analyses the bacterial 16S rRNA gene was amplified with the primer set 27F - FAM labelled (5' AGA GTT TGA TCC TGG CTC AG 3') and 1492r (5' ACC TTG TTA CGA CTT 3') [31]. The *bph* gene was amplified with the primer set BPHD F1 – FAM labelled (5' TAY ATG GGB GAR GAY CCI GT 3') and BPHD R0 (5' ACC CAG TTY TCI CCR TCG TC 3') [21]. For 16S rRNA gene amplification, PCR reactions were carried out in a volume of 25 μL containing 2.5 μL reaction buffer 10× (Invitrogen, Carslbad, CA, USA), 1.5 μL $MgCl_2$ (50 mM), 1 μL of each primer (5 pmol $μL^{-1}$), 0.2 μL (5 U) of Platinum Taq DNA polymerase (Invitrogen), 0.5 mL of deoxyribonucleotide triphosphate mixture (2.5 mM), 0.25 μL of bovine serum albumin (1 ng mL^{-1}), 1 μL of DNA template (ca. 10 ng) and 18.05 μL of sterilized ultrapure water. Amplifications were performed in the GeneAmp PCR System 9700 thermal cycler (Applied Biosystems, Foster City, CA, USA). Reaction conditions were 94°C for 3 min, followed by 35 cycles of 94°C for 30 s, 59°C for 45 s, and 72°C for 1 min with a final extension step at 72°C for 15 min. For the *bph* gene amplification, PCR reactions were carried out in a volume of 25 μL containing 2.5 μL reaction buffer 10× (Invitrogen, Carslbad, CA, USA), 1.5 μL $MgCl_2$ (50 mM), 1.25 μL of each primer (5 pmol $μL^{-1}$), 0.5 μL (5 U) of Platinum Taq DNA polymerase (Invitrogen), 0.3 μL of deoxyribonucleotide triphosphate mixture (2.5 mM), 1 μL of DNA template (ca. 10 ng) and 16.7 μL of sterilized ultrapure water. For *bph* gene, similar PCR cycling conditions were used, except the annealing temperature that was set at 60°C. Negative PCR controls (without DNA template) and positive controls (using the *Escherichia coli* ATCC 25922 DNA for the 16S rRNA gene and the DSM 6899 *Pseudomonas putida* DNA for *bph*) were run in parallel for both amplifications. After the amplifications, 5 μL of obtained products (ca. 60 ng) was digested with the endonuclease *HhaI* (Invitrogen) in 15 μL reaction for 3 h at 37°C. Obtained fragments were further purified using sodium acetate/EDTA precipitation and then mixed with 0.25 μL of the Genescan 500 ROX size standard (Applied Biosystems) and 9.75 μL of deionized formamide. Prior to fragment analysis, samples were denatured at 95°C for 5 min and chilled on ice. Analysis of terminal restriction fragment (T-RF) sizes and quantities was performed on an ABI PRISM 3100 genetic analyzer (Applied Biosystems).

T-RFLP profiles were analysed using PeakScanner v1.0 software (Applied Biosystems, Foster City, CA, USA). Terminal restriction fragments (T-RFs) of less than 25 bp were excluded prior to the analysis. The total values of T-RFs for each soil sample were pulled together to construct a Venn's diagram showing shared T-RFs among samples. The relative abundance of a single T-RF was calculated as percent fluorescence intensity relative to total fluorescence intensity of the peaks [32]. Data from individual samples were combined to soil chemical parameters and subjected to multivariate analysis using Canoco 4.5 (Biometris, Wageningen,The Netherlands) and Primer6 (PrimerE, Ivybridge, United Kingdom). All matrices were initially analysed using de-trended correspondence analysis (DCA) to evaluate the length of the gradient of the species distribution; this analysis indicated linearly distributed data (length of gradient <3), revealing that the best-fit mathematical model for the data was the redundancy analysis (RDA). Forward selection (FS) and the Monte Carlo permutation test were applied with 1,000 random permutations to verify the significance of soil chemical properties upon the microbial community. In addition to P values for the significance of each soil chemical property, RDA and Monte Carlo permutation test supplied information about the marginal effects of environmental variables, quantifying the amount of variance explained by each factor. We used ANOSIM based on relative abundance of T-RFs to test for statistical differences between samples.

454-Pyrosequencing analyses of the bacterial *bph* gene

A partial region of the *bph* gene was amplified for the 454-pyrosequencing using the primer set BPHD F3 (5′ ACT GGA ART TYG CIG CVG A 3′) and BPHD R0 (5′ ACC CAG TTY TCI CCR TCG TC 3′) [20] containing specific Roche 454-pyrosequencing adaptors and barcodes of 8 bp. The expected fragment size was ca. 520 bp. Three independent amplifications were performed for each sample. The 20 μL PCR mixture contained 1× FastStart High Fidelity Reaction Buffer (Roche Diagnostics, Basel, Switzerland), 1.25 mM of each primer, 150 ng mL^{-1} of bovine serum albumin (New England BioLabs, Ipswich, MA, USA), 0.2 mM of dNTPs, 0.5 mL (2.5 U) of FastStart High Fidelity PCR System Enzyme Blend (Roche Diagnostics) and 4 ng of template DNA. The PCR conditions were optimized using the genomic DNA of *Burkholderia xenovorans* LB400 [33], which carries one of the target dioxygenase genes. Amplifications were performed as follows: 95°C for 3 min, 30 cycles of 95°C for 45 s, 60°C for 45 s and 72°C for 40 s, with the final extension of 72°C for 4 min. Triplicate PCR products containing the expected fragment size were purified using the QIAquick Gel Extraction Kit (Qiagen, Hilden, Germany) and QIAquick PCR Purification Kit (Qiagen). DNA concentrations were determined using the NanoDrop ND-1000 spectrophotometer (NanoDrop Technologies, Wilmington, DE, USA). Purified PCR products were pooled and subjected to pyrosequencing using the FLX sequencing system (454 Life Sciences, Branford, CT, USA).

Raw data was filtered for valid sequences using the FunGene Pipeline Repository (http://fungene.cme.msu.edu/FunGenePipeline/). Quality sequences were translated in the correct frame of aminoacids using the RDP FrameBot tool. The RDP pipeline extracted a set of representative sequences from known *bph* sequences to use as subject sequences with FrameBot. The FrameBot produces an optimal alignment between the query and the subject sequences in the presence of frameshifts. Only the protein pairwise alignment with the best score was reported. Protein sequences passing FrameBot were aligned with HMMER using a model trained on the same set of representative sequences used by FrameBot. The aligned protein sequences were chopped

at position 351 of the reference sequence of *Pseudomonas putida* F1 (YP_001268196). The total valid sequences were rarefied to the smallest number of sequences per sample in order to minimize effects of sampling effort upon analysis.

Distance matrices were constructed using the MOTHUR software [34]. The resulting matrices were used to estimate the number of operational protein families (OPF) (i.e. group of proteins that share a common evolutionary origin) and to estimate richness (i.e. Chao1, Jackknife, and ACE indices) and diversity (i.e. Shannon and Simpson indices). Rarefaction curves were constructed at a cutoff level of 94% of amino acid identity. MOTHUR was also used to perform ∫-Libshuff comparisons between the four studied sites. The Good's coverage estimator was used to calculate the sample coverage using the formula $C = 1-(n_i/N)$, where N is the total number of sequences analysed and n_i is the number of reads that occurs only once among the total number of reads analysed at a cutoff value of 94% of amino acid identity [35]. Unweighted UniFrac distances among communities were estimated using a tree constructed *de novo* using FastTree. One representative sequence per OPF was selected and subjected to taxonomic affiliation by the comparison tool of NCBI Tblastx (GenBank) using Blast2Go [36].

Sequence data generated by 454-pyrosequencing are available at the MG-RAST server (http://metagenomics.anl.gov) under the project 'Diversity of bhp gene in Amazon soils' (ID 8489) and accession numbers 4557319.3 (ADE_SF), 4557318.3 (ADE_M), 4557321.3 (ADJ_SF) and 4557320.3 (ADJ_M).

Quantitative PCR (qPCR) of the bacterial 16S rRNA and *bph* genes

The bacterial 16S rRNA gene was amplified with the primer set U968F (5' AAC GCG AAG AAC CTT AC 3') and R1387 (5' CGG TGT GTA CAA GGC CCG GGA ACG 3') [37], which amplify a fragment of approximately 400 bp. The 520 bp fragment of the *bph* gene was amplified with the primer set BPHD F3 (5′ ACT GGA ART TYG CIG CVG A 3′) and BPHD R0 (5′ ACC CAG TTY TCI CCR TCG TC 3′) [21]. qPCR reactions were performed in 10 μL containing 5 μL of SYBR green PCR master mix (Fermentas, Brazil), 1 μL of each primer (5 pmol μL^{-1}), 1 μL of DNA template (ca. 10 ng) and 2 μL of sterilized ultrapure water. Thermocycling conditions for the 16S rRNA gene were set as follows: 94°C for 10 min; 40 cycles of 94°C for 30 s, 56°C for 30 s and 72°C for 40 s. Amplification specificity was checked by a melting curve and the data collection was performed at every 0.7°C. qPCR reactions for the *bph* gene were performed at similar conditions, except for the annealing temperature set at 60°C. Reactions were performed in a StepOnePlus system (Applied Biosytems). The Cts values (cycle threshold) were used as standers for determining the amount of DNA template in each sample. Standard curves were produced for the 16S rRNA and *bph* genes using specific cloned fragments. Gene fragments were quantified in a spectrophotometer (190 a 840 nm - NanoDrop ND-1000) and diluted (10^7 to 10^3 genes μL^{-1} for the 16S rRNA gene and 10^6 to 10^2 genes μL^{-1} for *bph*) to generate each specific standard curves. The gene copy numbers in different soil samples were expressed as log copy numbers of the gene per gram of soil. Statistical comparisons were performed using one-way ANOVA (Tukey's test).

Results

Variation in soil chemical properties

Soil chemical properties were measured for each individual sample collected in ADE and ADJ sites (for a detailed description see Table S1). Statistical differences were observed using Tukey's test. Overall, soil chemical properties of ADE-SF were chemically similar to ADE-M. Likewise, ADJ-SF chemical properties were also very similar to ADJ-M. As expected, major differences were attributed mostly to soil type rather than the land use history.

Higher soil pH values were observed in ADE rather than ADJ sites. While ADJ soils were very acidic with a pH of 3.53 (ADJ-SF) and 3.74 (ADJ-M), ADE sites were only weakly acidic with a pH of 5.51 (ADE-SF) and 5.41 (ADE-M). Sites at ADE showed lower total and exchangeable Al (H+Al), a phenomenon that is likely directly connected to observed variations in soil pH.

Soil organic carbon (SOC), soil organic matter (SOM) and effective cation exchange capacity (eCEC) were higher in ADE sites. Different land uses did not influence these properties in ADJ soils; however, the same properties showed significantly higher values in the site under secondary forest rather than in manioc cultivation, in ADE soil locations. In detail, SOC, SOM and eCEC values were approximately 30% higher in the ADE-SF site when compared to the ADE-M.

Assessment of community structures based on the bacterial 16S rRNA and *bph* genes

T-RFLP analyses for the bacterial 16S rRNA and *bph* genes were performed for the four sites. The obtained profiles were used to determine the richness of terminal restriction fragments (T-RFs) and to perform the multivariate analyses. A total of 152, 144, 147 and 141 T-RFs were obtained for the analysis of the bacterial 16S rRNA gene in ADE-SF, ADE-M, ADJ-SF and ADJ-M sites, respectively. There were 14 T-RFs detected as dominant throughout all sites, accounting for >50% of the total fluorescence detected for the 16S rRNA gene analyses. Tukey's test ($P>0.05$) indicated no difference between sites in the richness of T-RFs for the obtained profiles of bacterial 16S rRNA gene.

For the *bph* gene analysis no dominant T-RFs were found, possibly due to the high heterogeneity of this gene. There were 90, 78, 73 and 69 T-RFs in ADE-SF, ADE-M, ADJ-SF and ADJ-M sites, respectively. Samples from ADE sites showed statistically higher richness of T-RFs (Tukey's test) than the observed at ADJ sites ($P>0.05$).

The Venn's diagram according to soil type showed that ADE and ADJ soils shared more common T-RFs for the bacterial 16S rRNA rather than for *bph* gene. Also, the number of unique T-RFs was higher in ADE sites for both assessed genes (Figure 1a). Conversely, Venn's diagram combining the four sites showed a core containing 122 T-RFs for the bacterial 16S rRNA gene, while the distribution of T-RFs for the *bph* gene was more site specific, and only 6 T-RFs comprised a common core (Figure 1b).

Clustering analysis of T-RFLP data for the *bph* gene segregated samples according to soil type and land use (Figure 2). This analysis revealed the formation of two main clusters: the first cluster (a) included samples from ADE soils (SF and M) and the second (b) samples from ADJ soils (SF and M). This analysis also revealed that ADE and ADJ sites segregated at 12% of similarity. Concerning the ADE sites, land use systems separated different land uses at 27% of similarity. Conversely, for ADJ sites, land use systems differed at 20% similarity (Figure 2).

Redundancy analysis based on T-RFLP data explained 78.1% of the variation in the first two axes, thus confirming the segregation of sites primary according to soil type, and further in relation to different land use types (Figure 3). Replicates within each soil site were very consistent, evidenced by the formation of concise clusters. We also observed that different soil types also correlated differently to measured chemical parameters. More precisely, the *bph* community structure from ADE-SF correlated

mostly with pH, eCEC, SOM and SOC, while sites at ADJ-FS presented a significant correlation to H+Al.

ANOSIM analysis indicated statistical differences between the two soil types and land use systems (Table 1). R-values revealed a clear segregation of *bph* gene structures (R>0.75) in ADJ soils, while the 16S rRNA gene differed to a lower extent across sites (R<0.2).

Diversity of the bacterial *bph* gene across soil types and land uses

To access the composition of *bph* gene, samples were sequenced using 454-pyrosequencing. A total of 7,710 reads matched the barcodes, of which 6,877 reads passed the initial filtering (89.2%) and 5,965 (86.7%) were effectively translated into amino acid sequences using the FrameBot. A total of 4,690 valid amino acid sequences were further rarified to the depth of 750 sequences per sample (the minimum in a single sample) for comparative analysis.

The diversity indices for the *bph* gene (Table 2) revealed that ADE sites (SF and M) ($H' = 4.24$ and 4.05, $L' = 0.024$ and 0.028, respectively) were more diverse than ADJ sites (SF and M) ($H' = 3.19$ and 3.17, $L' = 0.098$ and 0.107, respectively). These indices also showed that sites under SF were more diverse than sites under M. Richness estimators (i.e. Chao1, ACE and Jackknife) also revealed ADE sites (SF and M) (Chao 1 = 238 and 229, ACE = 248 and 286, Jackknife = 271 and 255, respectively) to present higher values than ADJ (Chao 1 = 151 and 127, ACE = 210 and 166, Jackknife = 170 and 130, respectively), with SF sites being also higher than sites under M.

ADE sites presented a total of 159 and 129 OPFs for SF and M sites, respectively. Conversely, these values for ADJ were lower (95 and 90, for SF and M, respectively). These sites also presented a different number of singletons (number of unique reads per OPF): 66, 57, 43 and 36 for ADE-SF, ADE-M, ADJ-SF and ADJ-M, respectively.

Statistical differences among sites for the composition of the *bph* gene were confirmed by ∫-Libshuff ($P<0.001$). Venn's diagrams highlight the number of shared OPFs among samples (Figure 4). The number of shared OPFs between ADE soils under different land uses systems (SF and M) was 84 (41% of the total OPFs presented in ADE sites). Conversely, the number of shared OPFs between ADJ soils under different land uses was 55 (42% of the total OPFs in ADJ sites). Soils under SF presented higher numbers of unique OPFs for both soils (ADE and ADJ) (75 and 40 unique OPFs for SF sites and 45 and 35 unique OPFs for ADE sites, respectively).

The estimation of Good's coverage revealed higher values for ADJ sites (0.86 for ADJ-SF and 0.87 for ADJ-M) than for ADE sites (0.79 for ADE-SF and 0.80 for ADE-M), suggesting a highest number of unique sequences in ADE sites. Rarefaction curves (Figure S1) indicate that ADE (SF and M sites) presented a more diverse community than ADJ (SF and M). Sites under SF (ADE and ADJ) were also comparatively more diverse than sites under M (ADE and ADJ). For all comparative analysis the sampling effort did not covered the richness of *bph* gene. The exception was observed for samples from ADJ sites, where a trend towards a "plateau" was observed.

The Principal Coordinate Analysis (PCoA) based on Unweighted UniFrac distances revealed distinct patterns in phylogenetic community composition (Figure 5). The first axis explained 59.59% of the data variation, and this axis separated samples according to soil type. The second axis explained 22.63% of the data variation, and this axis segregated samples according to land use system.

(a)

16S rRNA

bph

(b)

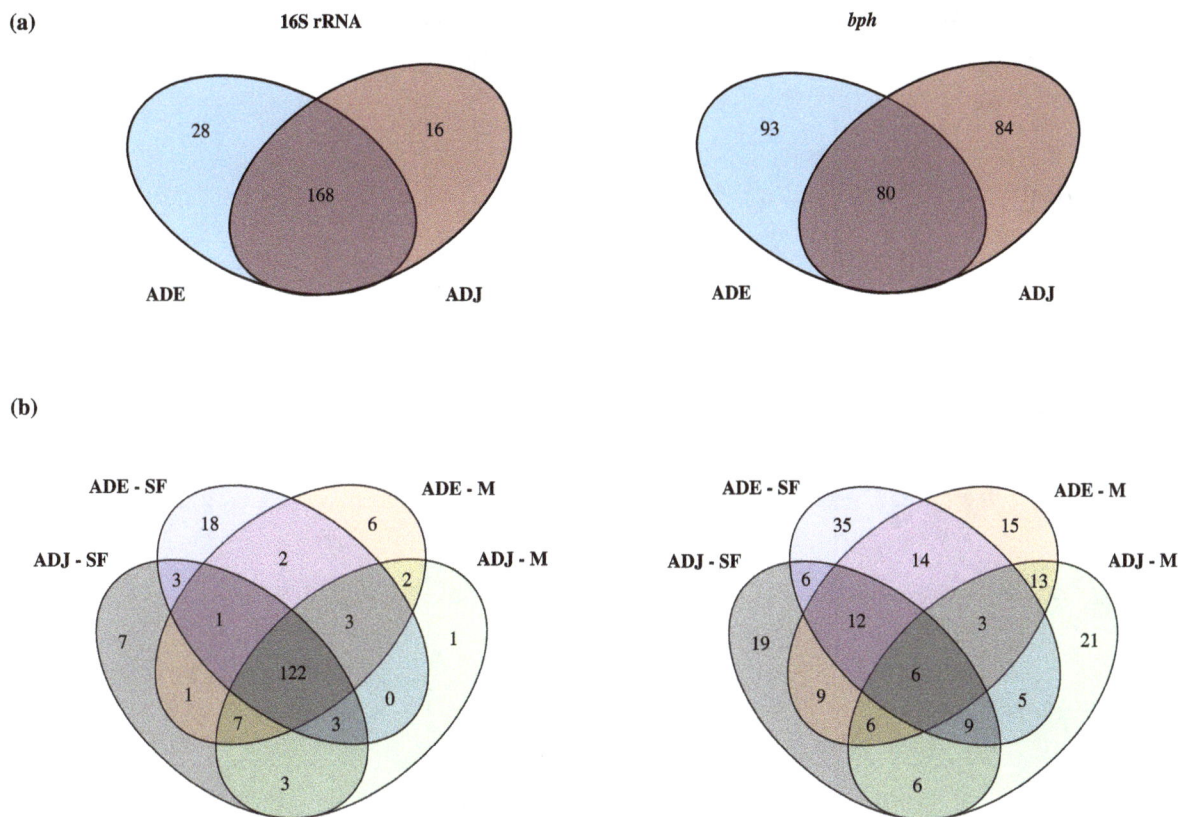

Figure 1. Venn's diagram of T-RFs for 16S rRNA and bph **genes according to (a) soil types and (b) soil types and land uses.** ADE = Amazon Dark Earth; ADJ = Adjacent soils; SF = Secondary Forest; M = Manioc cultivation.

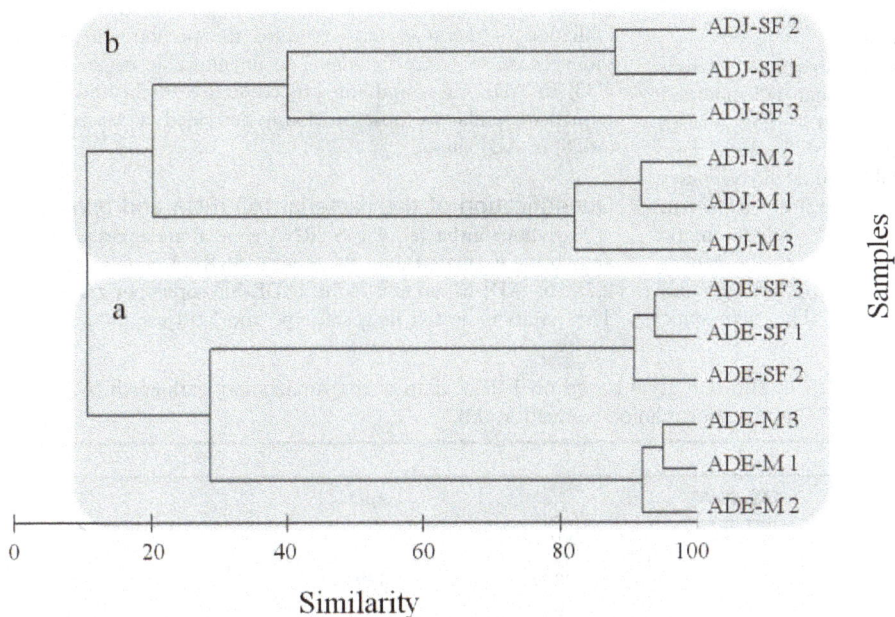

Figure 2. Clustering analysis of T-RFLP data based on Bray-Curtis similarity for the bph **gene.** 'a' and 'b' indicate the segregation patterns according to soil type. ADE = Amazon Dark Earth; ADJ = Adjacent soils; SF = Secondary Forest; M= Manioc cultivation.

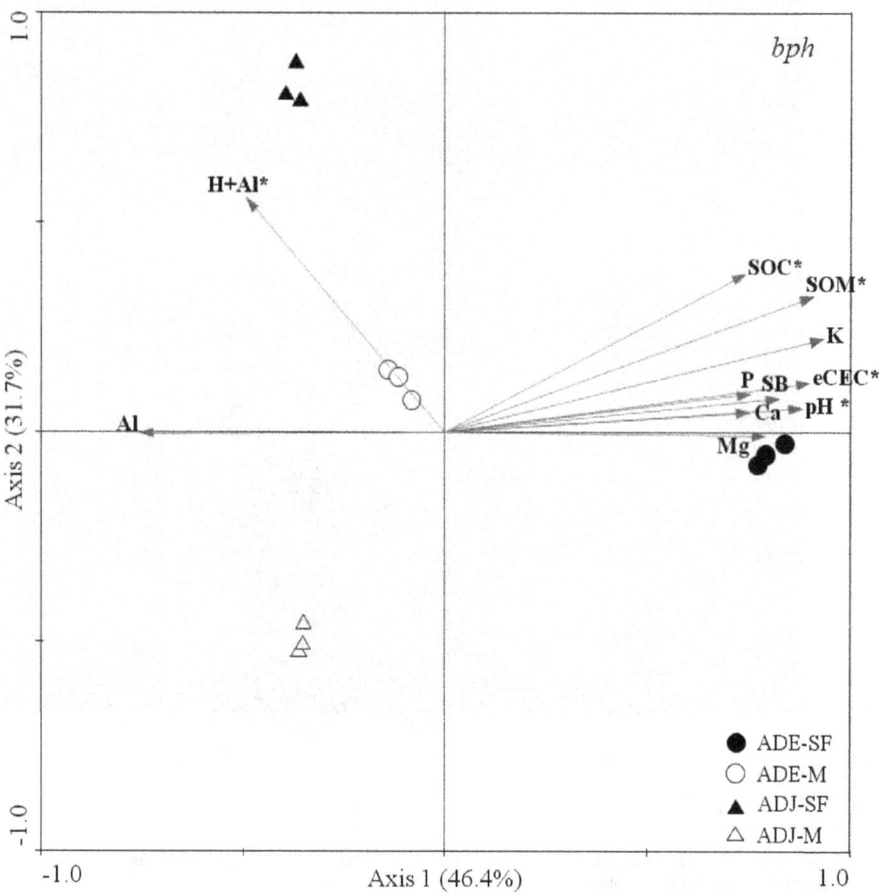

Figure 3. Redundancy analysis (RDA) based on T-RFLP data obtained for the *bph* gene, and soil properties, at the four studied sites. Arrows indicate correlation between the chemical parameters and community structure of samples. The significance of correlations was evaluated via Monte Carlo permutation test and it is indicated as follows: * $p < 0.05$. ADE = Amazon Dark Earth; ADJ = Adjacent soils; SF = Secondary Forest; M = Manioc cultivation.

Taxonomic composition of the *bph* gene

Each obtained OPF were further compared to sequences from the GenBank database for taxonomic assignment. All analyzed sequences matched translated proteins described as dioxygenases or putative dioxygenases, with E-values $< 10^{-3}$. The most abundant differences in dioxygenases (established as dioxygenases at least ten-fold higher in one site than another) are shown in Figure 6. Sequence matches were associated with aromatic hydrocarbon degradation genes belonging mostly to the genera *Streptomyces, Sphingomonas, Rhodococcus, Mycobacterium, Conexibacter* and *Burkholderia*, and uncultured bacterial clones. The taxonomic

affiliation of the reads also revealed the predominance of the dioxygenase sequence belonging to unculturable organisms (rdh cf33) in ADE sites, and the predominance of the dioxygenase sequences similar to those previously described at Australian soil (od16) in ADJ sites.

Quantification of the bacterial 16S rRNA and *bph* genes

Variations in bacterial 16S rRNA gene abundances were clear among the evaluated sites. Bacterial 16S rRNA gene ranged from 2.7×10^{7} (ADJ-M) up to 9.7×10^{7} (ADE-SF) copies per gram of soil. The results indicated that, soil type and land use influenced the

Table 1. ANOSIM test for the bacterial 16S rRNA and *bph* gene based on T-RFLP data of the Amazonian Dark Earth (ADE) and Adjacent soil (ADJ) under secondary forest (SF) and under manioc cultivation (M).

	16S rRNA[a]		*bph*[a]	
	R values	p values	R values	p values
ADE × ADJ	0.17	<0.001	1.00	<0.001
ADE-SF × ADE-CULT	0.11	<0.001	1.00	<0.001
ADJ-SF × ADJ-CULT	0.03	<0.001	1.00	<0.001

[a]Samples were compared using T-RF peak height as a measure of abundance.

Table 2. Comparison of diversity indices and richness estimators for the *bph* gene.

Site	OPFs [a]	Richness			Diversity		ESC [b]
		Chao 1	ACE	Jackknife	Shannon (H')	Simpson (L')	
ADE-SF	159	238	248	271	4.24	0.024	0,79
ADE-M	129	229	286	255	4.05	0.028	0,80
ADJ-SF	95	151	210	170	3.19	0.098	0,86
ADJ-M	90	127	166	130	3.17	0.107	0,87

[a]The operational protein family (OPFs), richness estimators (ACE, Chao1 and Jackknife), diversity indices (Shannon and Simpson) and [b]estimated sample coverage were calculated at a cutoff value of 94% of sequence identity. ADE = Amazon Dark Earth soils; ADJ = Adjacent soils; SF = Secondary forest; M= Manioc cultivation.

abundance of bacteria in the analyzed samples, and higher values were found at ADE-SF site (Figure 7a).

The abundance of the *bph* gene ranged from 1.6×10^6 (ADJ-M) to 2.9×10^6 (ADE SF) copies per gram of soil. Conversely to the data obtained for the total bacteria (i.e. 16S rRNA gene data), the abundance of the *bph* gene was also higher at ADE-SF site (Figure 7b).

Discussion

The aim of this study was to assess bacterial hydrocarbon degrading genes in anthropogenic sites from Brazilian Amazon comparatively to their adjacent locations, under two different land use systems. Recent studies have described the high fertility of ADE soils when compared to adjacent soils in the same area (ADJ), mostly because of their increased pH, higher cation exchange capacity, nutrient content and incompletely combusted biomass [3,6,8,28,38]. Although it is well-known that the taxonomic composition of bacterial communities is strongly influenced by pH [39,40], the pH variation in our data indicate that this is also a strong predictor of the composition and diversity of the bacterial *bph* gene. Variations in soil pH, together with eCEC, SOC and SOM, collectively accounted for 78.1% of the total variation explained in the RDA plot based on T-RFLP data. Higher P values in ADE sites are likely to be an effect of pH, seeing that low acidity soils are known to increase the P solubility [28]. Also, the historical formation of ADE sites (constantly amended with bones and vegetation burning activities) is an intrinsic characteristic of this system, which could possibly explain the high content of phosphorus. There is also a direct relationship between soil P content and eCEC values in ADE sites mainly because the P adsorption decreases due to the formation of complex compounds between P and the organic matter present in the upper layer of the soil[41]. The values of eCEC in our samples were mostly correlated to the organic matter concentration, which was 2-fold higher in ADE than ADJ sites. These findings are likely explained by the higher amount of biochar found in these anthropogenic sites. Biochar is known for retaining soil nutrients due to its specific surface and negative charge density per unit of surface area [20,42,43]. In short, the collective chemical properties intrinsic of ADE sites may play an important role in their high levels of nutrient availability and, ultimately, their fertility. In this context, we hypothesize that such fertility has caused the ADE soils to harbor a higher microbial diversity and functionality than their adjacent soils. Agricultural practices can also alter soil properties, mostly by interfering on the biochemistry of the organic matter available in the system [44]. Our data revealed such an influence in the measurements of SOC, SOM and eCEC, which were significantly higher in ADE sites under secondary forest than under manioc cultivation system. Other chemical properties such as pH, SB and amount of P, Mg and H+Al, did not statistically differ between different land uses in ADE and ADJ sites.

Soil chemical data was also used in regression analyses to understand patterns in the microbial community structure of the *bph* gene. Microbial community structure can be defined by patterns of species abundance and population within a given community [45], which are mostly regulated by the ability of the microorganisms to interact among them and with local conditions [46]. In this study, T-RFLP results did not reveal significant differences among the richness of T-RFs between ADE and ADJ samples for the total bacterial community (16S rRNA). However, ANOSIM revealed significant differences in total bacterial community structure between ADJ and ADE soils types (R = 0.17) and between land uses (R = 0.11 for ADE-SF versus

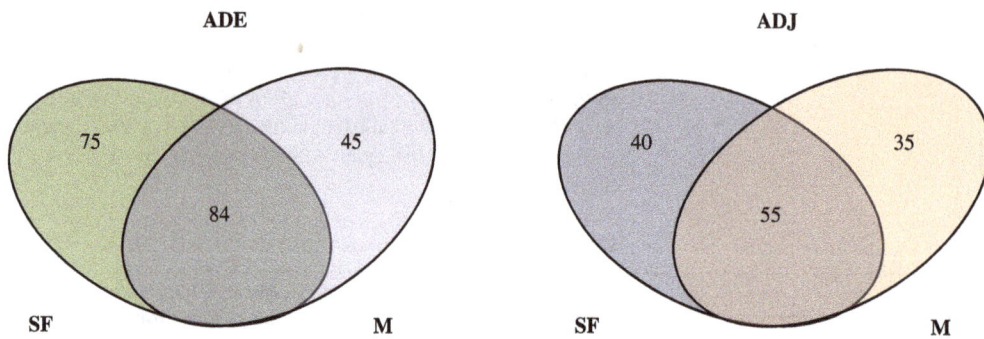

Figure 4. Venn's diagram of *bph* data belonging to operational protein families (OPFs) for different soil types under different land uses. ADE = Amazon Dark Earth; ADJ = Adjacent soils; SF = Secondary Forest; M = Manioc cultivation. Sequences were grouped into OPFs based on sequence identity of 94%.

ADE-M; and R = 0.03 for ADJ-SF versus ADJ-M; Table 1). Lima [47], who similarly assessed the variation of ADE-SF, ADE-M, ADJ-SF and ADJ-M sites, and Cannavan [48], who investigated changes across different ADE sites, also found differences in 16S rRNA community structure between ADE and ADJ sites. Collectively, these studies support the idea that archaeological sites with a long history of anthropogenic activity influence the inhabiting microbiota of their respective soils.

Analysis of the community patterns for the bacterial *bph* gene revealed higher richness of T-RFs in ADE than in ADJ sites. Our results show that *bph* community changes are clear among sites (Figure 2), which can be observed by changes in the abundance of specific T-RFs across sites. In the same way, ANOSIM revealed significant differences in *bph* gene community structure between ADJ and ADE soils types (R = 1.00) and between land uses (R = 1.00 for ADE-SF versus ADE-M; and R = 1.00 for ADJ-SF versus ADJ-M; Table 1). We suspect, despite all variables present in our system, that the high presence of biochar, which chemically

is formed by an aromatic polycyclic structure, could be a factor influencing the observed differences. Táncsics et al [49], by analyzing the structure community of catechol 2,3-dioxygenase genes in aromatic hydrocarbon contaminated environments by T-RFLP, observed that T-RFLP chromatograms obtained from contaminated samples had entirely different T-RFs compared to the control non-contaminated sample.

Redundancy analysis revealed *bph* gene structure to be correlated with different soil parameters (Figure 4). Briefly, ADE-SF correlated with pH, eCEC, SOM and SOC. ADJ-SF showed a significant correlation with H+Al. Several studies have suggested that variations in soil pH and properties related to soil acidity (e.g., K, Al, base saturation) are stronger predictors of the richness and diversity of inhabiting microbial communities [6,18,50,51,52,53]. We extend this concept by advocating that other factors also might play a role in structuring the *bph* gene communities. For instance, the quantity and/or quality of soil organic matter (SOM) and its fractions are likely to regulate

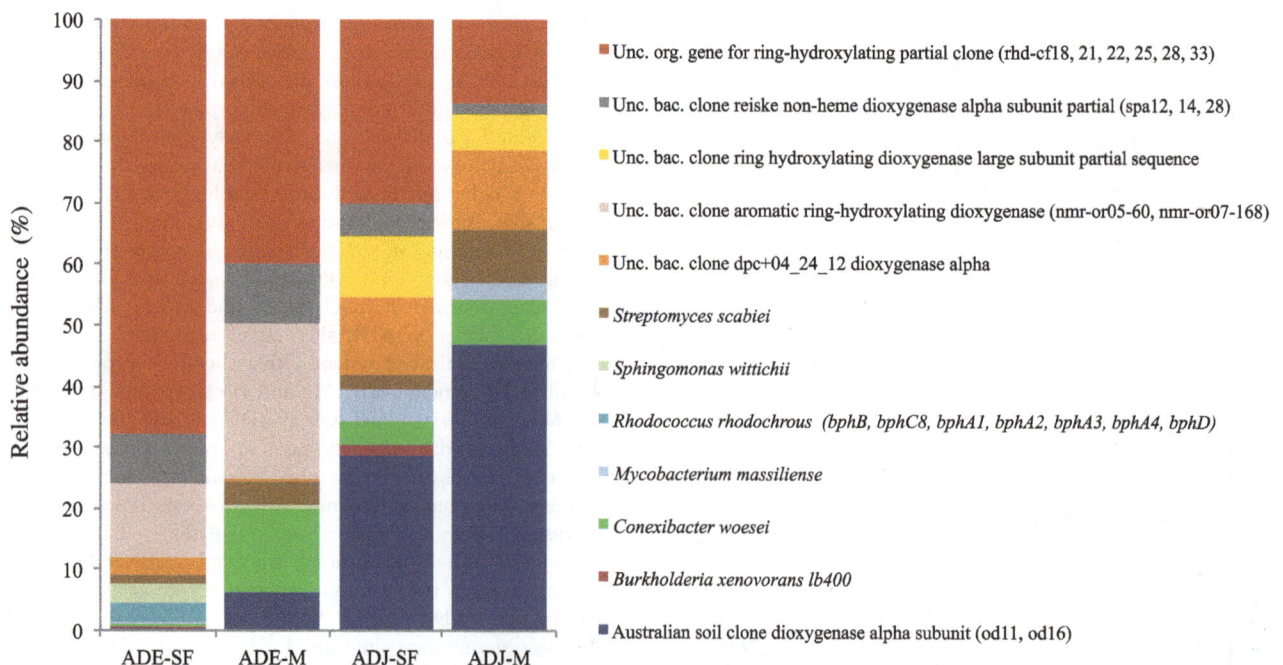

Figure 5. Bar charts representing the taxonomic affiliation of *bph* gene sequences. Sequences were affiliated using the TBlastX tool available in the GenBank database. ADE = Amazon Dark Earth; ADJ = Adjacent soils; SF = Secondary Forest; M = Manioc cultivation.

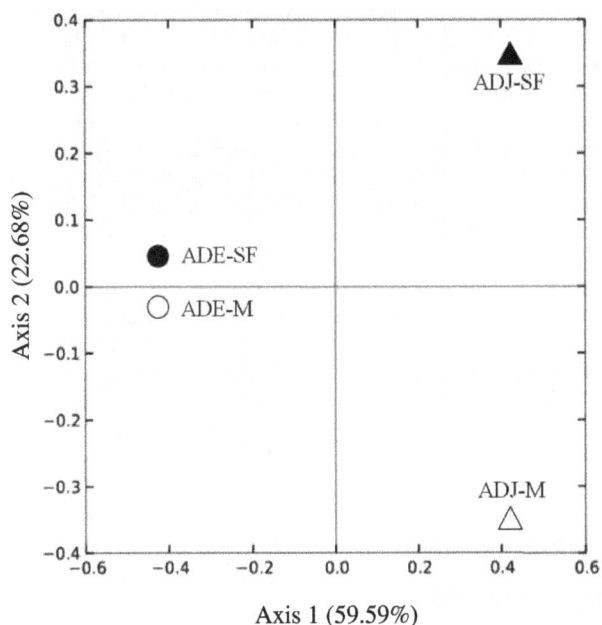

Figure 6. Principal Coordinate Analysis (PCoA) performed with the sequences obtained for the bph gene (based on Unweighted Unifrac distance). ADE = Amazon Dark Earth; ADJ = Adjacent soils; SF = Secondary Forest; M = Manioc cultivation.

microbial community composition and associate function [54]. Since hydrocarbon degradation is one step into the carbon cycle, bph degraders might have an advantage in ADE sites by harnessing energy from elevated levels of SOC and SOM.

Quantification of total bacterial (16S rRNA) and bph gene abundances (Figure 7) revealed higher copy numbers of these

genes in ADE sites, peaking at ADE-SF. These results indicate that soil characteristics of ADE, as well as their land uses has an influence on the abundance of both analyzed genes. Ding et al [55] analyzed the abundance of total bacteria using the16S rRNA gene and the abundance of polycyclic aromatic hydrocarbons ring-hydroxylating dioxygenase (PAH-RHD$_\alpha$) genes by qPCR in two different phenanthrene-contaminated soils (i.e. Luvisol and Cambisol). Their results revealed a significantly higher 16S rRNA gene copy number per gram of soil in both phenanthrene-contaminated soils compared to their controls. PAH-RHD genes were detected only in contaminated soil samples, and the values ranged from 2.0×10^7 (Luvisol phenanthrene-contaminated soil) to 1.7×10^6 (Cambisol phenanthrene-contaminated soil) copies per gram of soil, showing similar values to those found in our study. Similarly, a study about the diversity of naphthalene dioxygenase genes (nahAc) in soil environments from the Maritime Antarctic revealed that the quantities detected in bulk and rhizospheric soils from PAH-affected sites ranged from 6.4×10^4 to 1.7×10^6 nahAc gene copies per gram of soil and presented significantly higher abundance compared with the corresponding counterparts of bulk and rhizospheric soils from non-polluted sites [56].

We used 454-pyrosequencing to investigate the community composition of the bph gene. The use of a high-throughput culture-independent approach enabled an overview of the taxonomic groups occurring in the sampled sites. Taxonomic analyses of 3,000 partial sequences of the bph gene revealed that this gene was most represented by bacteria belonging to the genera *Streptomyces*, *Sphingomonas*, *Rhodococcus*, *Mycobacterium*, *Conexibacter* and *Burkholderia*, in addition to sequences matching uncultured bacteria (Figure 5). Overall, these groups encompass mixed sequences of toluene/biphenyl dioxygenases, such as bphB, bphC8, bphA1, bphA2, bphA3, bphA4 and bphD, and enzymes encoding the alpha subunit of dioxygenases for the degradation of Polycyclic Aromatic Hydrocarbons (PAHs), which are described as environmental widely distributed dioxygenases. Other studies targeting dioxy-

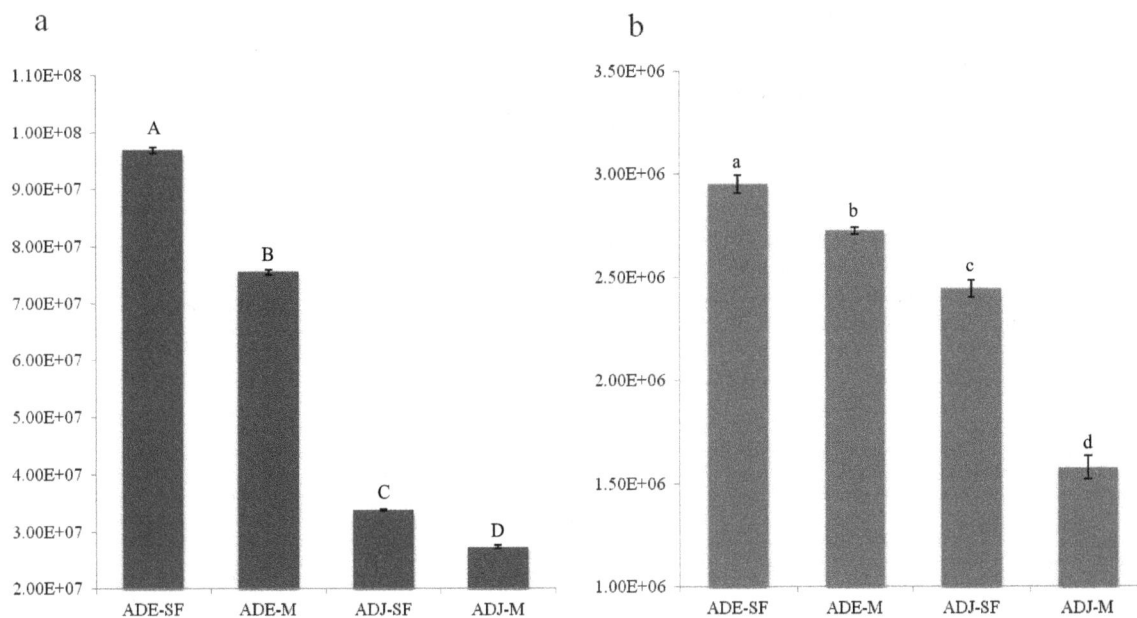

Figure 7. Log gene copy number (y axis) of (a) 16S rRNA and (b) bph genes across the studied sites. Sites are indicated in the x axis. Error bars represent the standard deviation of three independent replicates. ADE = Amazon Dark Earth; ADJ = Adjacent soils; SF = Secondary Forest; M = Manioc cultivation. Different upper case letters refer to differences for the 16S rRNA across sites; while different lower case letters refer to differences for the bph gene (P<0.01, Tukey test).

genases have also reported sequences belonging to *Sphingomonas*, *Rhodococcus*, *Mycobacterium*, *Conexibacter* and *Burkholderia* as major taxonomic groups [21,28,57]. Zhou et al [58] were also able to isolate aromatic hydrocarbon degrading bacteria from mangrove sediments, which were classified as the genera *Mycobacterium* and *Sphingomonas*. Members of these genera have commonly been isolated from diverse sediments and soils [59,60], and they play an important role in hydrocarbon biodegradation [61,62,63]. For instance, Ding et al [55] who studied the diversity of dioxygenases using clone libraries for aromatic ring hydroxylating dioxygenases (ARHD) genes identified gene sequences corresponding to the *phnAc* gene belonging to the *Burkholderia* sp. strain Eh1-1 and PAH-RHD genes of the *Mycobacterium* sp. strain JLS.

Confirming the results of T-RFLP analysis, differences among *bph* communities for ADE and ADJ sites and between SF and M land uses were also observed by differences in their Shannon and Simpson diversity indices (Table 2). Similarly, Germano et al [28] reported a higher diversity of ARHD genes in ADE sites under secondary forest rather than under agricultural cultivation, suggesting that deforestation in these sites has an influence on the diversity of these catabolic genes. We observed that in both soil types, the number of OPF (Table 2) was higher in SF samples, also indicating an influence of the land use system on the richness of this gene. According to Jesus et al [51], the land conversion of tropical forest to agricultural use modifies the size, activity and composition of soil microbial communities. This deeply influences specific bacterial functions, including those acting on organic matter decomposition and nutrient cycling in soils.

The intensive land use by agricultural practices and the conversion of Amazon soil into agricultural areas has been reported to cause significant shifts in the chemical properties of the soil, such as variations in SOC, SOM and eCEC, leading towards a homogenization of the inhabiting bacterial community [64]. However, we hypothesized that, despite an effect of land use, greater differences in community composition would be observed according to soil type. We also expected communities in these different soils to respond differently to agricultural practices. In this context, the literature describes the resilience phenomenon as the ability of the soil to cope with external disturbances and to retain its functional capacity upon the imposition of a stress [65,66,67]. We observed that higher differences in the diversity of *bph* occurred between the different land uses in ADJ sites rather than in ADE sites. Thus, land use type appeared to have a stronger effect on the *bph* community in ADJ sites, maybe due to the higher resilience of the ADE soil against agricultural practices. Principal Coordinate Analysis (PCoA) performed for the *bph* gene also supports these results (Figure 6), revealing that ADE sites (SF and M) clustered closer to each other than observed for ADJ sites (SF and M).

The taxonomic analyses of sequences also revealed that ADE sites harbored distinct *bph* phylogenetic structure from ADJ. These results suggest that the heterogeneity of bacterial *bph* communities could be related to their ability to respond to differences of land use type and soil chemical properties. In a previous study Germano et al [28] compared the phylogenetic structure of *bph* sequences from ADE sites and revealed that most of the protein clusters from these sites group apart from the main well-known dioxygenase groups previously proposed by Kweon et al [27].

In conclusion, we have taken a distinct and highly diverse soil to elucidate the ecological properties and taxonomic affiliation of bacterial communities characterized by the presence of the *bph* gene, which is crucial to the biodegradation of aromatic compounds. These results enable us to understand differences in the structure, abundance and composition of the main active organisms in ADE soils when compared to their adjacent locations. Further studies focusing on the catabolic activities of these communities are needed to enable a collective view of the formation, dynamics and maintenance of functional properties in ADE soils.

Supporting Information

Figure S1   Rarefaction curves of bacterial *bph* gene sequences were grouped into OPF based on distance sequence of 0.06. ADE = Amazonian Dark Earth soils; ADJ = Adjacent soils; SF = Secondary forest; M = Manioc Cultivation.

Table S1   Soil chemical properties of the studied sites. Amazonian Dark Earth (ADE) and Adjacent soil (ADJ). Sites were located under two different land uses: Secondary forest (SF) and under Manioc (*Manihot esculenta*) cultivation (M). Significant differences between sites are followed by different letters (*P*<0.05, Tukey test).

Acknowledgments

We thank Wenceslau Geraldes Teixeira and Western Amazon Embrapa for technical support. We are grateful to ACG Souza, AK Silveira, RS Macedo and TT Souza for helping with sample collection. We thank J Quensen and the Michigan State University Genomics Technology Support Facility team for the 454-pyrosequencing support; and José Elias Gomes and Fábio Duarte for their assistance in molecular analysis. We also thank Cyrus A. Mallon for English revision and helpful comments on the manuscript.

Author Contributions

Conceived and designed the experiments: MJLB MGG ABL SMT. Performed the experiments: MJLB MGG ABL. Analyzed the data: MJLB LWM. Contributed reagents/materials/analysis tools: SMT. Wrote the paper: MJLB SMT.

References

1. Lehmann J, da Silva JP Jr, Steiner C, Nehls T, Zech W, et al. (2003) Nutrient availability and leaching in an archaeological Anthrosol and a Ferralsol of the Central Amazon basin: fertilizer, manure and charcoal amendments. Plant Soil 249: 343–357.

2. Teixeira WG, Martins GC (2003) Soil physical characterization. In: Lehmann J, Kern DC, Glaser B, Woods WI, editors.Amazonian Dark Earths: Origin, properties, management.Dordrecht: Kluwer Academic. pp. 271–286.

3. Glaser B (2007) Prehistorically modified soils of central Amazonia: a model for sustainable agriculture in the twenty-first century. Philos T Roy Soc B 362: 187–196.

4. Kim JS, Sparovek G, Longo RM, De Melo WJ, Crowley D (2007) Bacterial diversity of terra preta and pristine forest soil from the Western Amazon. Soil Biol Biochem 39: 684–690.

5. O'neill B, Grossman J, Tsai SM, Gomes JE, Lehmann J, et al. (2009) Bacterial community composition in Brazilian anthrosols and adjacent soils characterized using culturing and molecular identification. Microbial Ecol 58: 23–35.

6. Taketani RG, Tsai SM (2010) The influence of different land uses on the structure of archaeal communities in Amazonian anthrosols based on 16S rRNA and amoA genes. Microbial Ecol 59: 734–743.

7. Navarrete AA, Cannavan FS, Taketani RG, Tsai SM (2010) A molecular survey of the diversity of microbial communities in different Amazonian agricultural model systems. Diversity 2: 787–809.

8. Taketani RG, Lima AB, Jesus EC, Teixeira WG, Tiedje JM, et al. (2013) Bacterial community composition of anthropogenic biochar and Amazonian anthrosols assessed by 16S rRNA gene 454 pyrosequencing. A van Leeuw J Microb 104: 233–242.

9. Thies J, Suzuki K (2003) Amazonian dark earths: biological measurements. In: Lehmann J, Kern D, Glaser B, Woods W, editors.Amazonian Dark Earths: Origin, Properties, Management.Dordrecht: Kluwer Academic. pp 287–332.

10. Fitter AH, Gilligan CA, Hollingworth K, Kleczkowski A, Twyman RM, et al. (2005) Biodiversity and ecosystem function in soil. Funct Ecol 19: 369–377.

11. Horner-Devine M, Carney K, Bohannan B (2004) An ecological perspective on bacterial biodiversity. Proc R Soc Lond B Biol Sci 271: 113–122.

12. Wawrik B, Kerkhof L, Kukor J, Zylstra G (2005) Effect of different carbon sources on community composition of bacterial enrichments from soil.Appl Environ Microbiol 71: 6776– 6783.

13. Zhou JZ, Xia BC, Treves DS, Wu LY, Marsh TL, et al. (2002) Spatial and resource factors influencing high microbial diversity in soil. Appl Environ Microbiol 68: 326–334.

14. Vogel TM, Simonet P, Jansson JK, Hirsch PR, Tiedje JM, et al. (2009) TerraGenome: a consortium for the sequencing of a soil metagenome. Nat Rev Microbiol 7: 252.

15. Bissett A, Brown MV, Siciliano SD, Thrall PH (2013) Microbial community responses to anthropogenically induced environmental change: towards a systems approach. Ecol Lett 16: 128–139.

16. Buckley DH, Schmidt TM (2001) The structure of microbial communities in soil and the lasting impact of cultivation. Microbial Ecol 42: 11–21.

17. Buckley DH, Schmidt TM (2003) Diversity and dynamics of microbial communities in soils from agro-ecosystems. Environ Microbiol 5: 441–452.

18. Fierer N, Jackson RB (2006) The diversity and biogeography of soil bacterial communities. Proc Natl Acad Sci USA 103: 626– 631.

19. Anderson CR, Condron LM, Clough TJ, Fiers M, Stewart A, et al. (2011) Biochar induced soil microbial community change: Implications for biogeo-chemical cycling of carbon, nitrogen and phosphorus. Pedobiologia 54: 309–220.

20. Liang B, Lehmann J, Sohi SP, Thies JE, O'neill B, et al. (2010) Black carbon affects the cycling of non-black carbon in soil. Org Geochem 41: 206–213.

21. Iwai S, Chai B, Sul WJ, Cole JR, Hashsham SA, et al. (2010) Gene-targeted-metagenomics reveals extensive diversity of aromatic dioxygenase genes in the environment. ISME J 4: 279–285.

22. Wackett LP (2004) Evolution of enzymes for the metabolism of new chemical inputs into the environment. J Biol Chem 279: 41259–41262.

23. Mason JR, Cammack R (1992) The electron-transport proteins of hydroxylating bacterial dioxygenases. Annu Rev Microbiol 46: 277–305.

24. Butler CS, Mason JR (1997) Structure-function analysis of the bacterial aromatic ring-hydroxylating dioxygenases. Adv Microb Physiol 38: 47–84.

25. Hayaishi O (1962) History and scope. In Hayaishi O, editor. Oxygenases. New York: Academic Press.pp. 1–29.

26. Luz AP, Pellizari VH, Whyte LG, Greer C (2004) A survey of indigenous microbial hydrocarbon degradation genes in soils from Antarctica and Brazil. Can J Microbiol 50: 323–333.

27. Kweon O, Kim SJ, Baek S, Chae JC, Adjei MD, et al. (2008) A new classification system for bacterial Rieske non-heme iron aromatic ring-hydroxylating oxygenases. BMC Biochem 9: 11.

28. Germano MG, Cannavan FS, Mendes LW, Lima AB, Teixeira WG, et al. (2012) Functional diversity of bacterial genes associated with aromatic hydrocarbon degradation in anthropogenic dark earth of Amazonia. Pesq Agropec Bras 47: 654–664.

29. Gibson DT, Parales RE (2000) Aromatic hydrocarbon dioxygenases in environmental biotechnology. Curr Opin Biotechnol 11: 236–243.

30. Claessen MEC, Barreto WDO, Paula JL, Duarte MN (1997) Análises químicas para avaliação da fertilidade do solo. Rio de Janeiro: Brazilian Agricultural Research Corporation Press. 212p

31. Amann RI, Ludwig W, Schleifer KH (1995) Phylogenetic identification and in situ detection of individual microbial cells without cultivation. Microbiol Rev 59: 143–169.

32. Culman SW, Gauch HG, Blackwood CB, Thies JE (2008) Analysis of T-RFLP data using analysis of variance and ordination methods: a comparative study. J Microbiol Methods 75: 55–63.

33. Goris J, De Vos P, Caballero-Mellado J, Park JH, Falsen E, et al. (2004) Classification of the PCB and biphenyl-degrading strain LB400 and relatives as Burkholderia xenovorans sp. nov. Int J Syst Evol Microbiol 54: 1677–1681.

34. Schloss PD, Westcott SL, Ryabin T, Hall JR, Hartmann M, et al. (2009) Introducing mothur: open-source, platform-independent, community supported software for describing and comparing microbial communities. Appl Environ Microbiol 75: 7537–7541.

35. Good IJ (1953) The population frequencies of species and the estimation of the population parameters. Biometrika 40: 237–264.

36. Conesa A, Götz S, Garcia-Gomez JM, Terol J, Talon M, et al. (2005) Blast2GO: a universal tool for annotation, visualization and analysis in functional genomics research. Bioinformatics 21: 3674–3676.

37. Heuer H, Krsek M, Baker P, Smalla K, Wellington EMH (1997) Analysis of actinomycete communities by specific amplification of genes encoding 16S rRNA and gelelectrophoresis separation in denaturing gradients. Appl Environ Microbiol 63: 3233–3241.

38. Grossman JM, O'neill BE, Tsai SM, Liang B, Neves E, et al. (2010) Amazonian anthrosols support similar microbial communities that differ distinctly from those extant in adjacent, unmodified soils of the same mineralogy. Microb Ecolol 60: 192–205.

39. Fierer N, Bradford MA, Jackson RB (2007) Toward an ecological classification of soil bacteria. Ecology 88: 1354–1364.

40. Lauber CL, Hamady M, Knight R, Fierer N (2009) Pyrosequencing-based assessment of soil pH as a predictor of soil bacterial community structure at the continental scale. Appl Environ Microbiol 75: 5111–5120.

41. Falcão N, Moreira A, Comenford NB (2009) A fertilidade dos solos de Terra Preta de Índio da Amazônia Central. In: Teixeira WG, Kern DC, Madari BE, Lima HN, Woods W, editors.As Terras Pretas de Índio da Amazônia: sua caracterização e uso deste conhecimento na criação de novas áreas.Manaus: Embrapa Amazônia Ocidental. pp. 189–200.

42. Glaser B, Haumaier L, Guggenberger G, Zech W (2001) The 'Terra Preta' phenomenon: a model for sustainable agriculture in the humid tropics. Naturwissenschaften 88: 37–41.

43. Cunha TJF, Novotny EH, Madari BE, Benites VM, Martin-Neto L, et al. (2009) O Carbono Pirogênico. In: Teixeira WG, Kern DC, Madari BE, Lima HN, Woods W, editors.As Terras Pretas de Índio da Amazônia: sua caracterização e uso deste conhecimento na criação de novas áreas.Manaus: Embrapa Amazônia Ocidental. pp. 264–285.

44. Bünemann EK, Marschner P, Smernik RJ, Conyers M, Mcneill AM (2008) Soil organic phosphorus and microbial community composition as affected by 26 years of different management strategies. Biol Fert Soils 44: 717–726.

45. Ricklefs RE, Miller G (1999) Ecology. New York: W H Freeman. 822p.

46. Bernhard AE, Colbert D, McManus J, Field KG (2005) Microbial community dynamics based on 16S rRNA gene profiles in a Pacific Northwest estuary and its tributaries. FEMS Microbiol Ecol 52: 115–128.

47. Lima AB (2012) Influência da cobertura vegetal nas comunidades de bactérias em Terra Preta de Índio na Amazônia Central brasileira. Piracicaba: University of São Paulo. 116 p.

48. Cannavan FS (2012) A estrutura e composição de comunidades microbianas (Bacteria e Archaea) em fragmentos de carvão pirogênico de Terra Preta de Índio da Amazônia Central. Piracicaba: University of São Paulo. 116 p.

49. Táncsics A, Szabóc I, Bakab E, Szoboszlayc S, Kukolyad J, et al. (2010) Investigation of catechol 2,3-dioxygenase and 16S rRNA gene diversity in hypoxic, petroleum hydrocarbon contaminated groundwater. Syst Appl Microbiol 33: 398–406.

50. Nicol GW, Tscherko D, Chang L, Hammesfahr U, Prosser JI (2006) Crenarchaeal community assembly and microdiversity in developing soils at two sites associated with deglaciation. Environ Microbiol 8: 1382–1393.

51. Jesus EC, Marsh TL, Tiedje JM, Moreira FMS (2009) Changes in land use alter the structure of bacterial communities in Western Amazon soils. ISME J 3: 1004–1011.

52. Nielsen UN, Osler GHR, Campbell CD, Burslem DFRP, van der Wal R (2010) The influence of vegetation type, soil properties and precipitation on the composition of soil mite and microbial communities at the landscape scale. J Biogeogr 37: 1317–1328

53. Wessen E, Hallin S, Philippot L (2010) Differential responses of bacterial and archaeal groups at high taxonomical ranks to soil management. Soil Biol Biochem 42: 1759–1765.

54. Murphy DV, Cookson WR, Braimbridge M, Marschner P, Jones DL, et al. (2011) Relationships between soil organic matter and the soil microbial biomass (size, functional diversity, and community structure) in crop and pasture systems in a semi-arid environment. Soil Res 49: 582–594.

55. Ding GC, Heuer H, Zuhlke S, Spiteller M, Pronk JG, et al. (2010) Soil Type-Dependent Responses to Phenanthrene as Revealed by Determining the Diversity and Abundance of Polycyclic Aromatic Hydrocarbon Ring-Hydrox-ylating Dioxygenase Genes by Using a Novel PCR Detection System. Appl Environ Microbiol 76: 4765–4771.

56. Flocco CG, Gomes NC, Mac CW, Smalla K (2009) Occurrence and diversity of naphthalene dioxygenase genes in soil microbial communities from the maritime Antarctic. Environ Microbiol 11: 700–714.

57. Leigh MB, Pellizari VH, Uhlik O, Sutka R, Rodrigues J, et al. (2007) Biphenyl-utilizing bacteria and their functional genes in a pine root zone contaminated with polychlorinated biphenyls (PCBs). ISME J 1: 134–148.

58. Zhou HW, Guo CL, Wong YS, Tam NFY (2006) Genetic diversityof dioxygenase genes in polycyclic aromatic hydrocarbon-degrading bacteria isolated from mangrove sediments. FEMS Microbiol Lett 262: 148–157.

59. Leys NMEJ, Ryngaert A, Bastiaens L, Verstraete W, Top EM, et al. (2004) Occurrence and phylogenetic diversity of Sphingomonas strains in soils contaminated with polycyclic aromatic hydrocarbons. Appl Environ Microbiol 70: 1944–1955.

60. Miller CD, Hall K, Liang YN, Nieman K, Sorensen D, et al. (2004) Isolation and characterization of polycyclic aromatic hydrocarbon-degrading Mycobac-terium isolates from soil. Microb Ecol 48: 230–238.

61. Khan AA, Wang RF, Cao WW, Doerge DR, Wennerstrom D, et al. (2001) Molecular cloning, nucleotide sequence, and expression of genes encoding a polcyclic aromatic ring dioxygenase fromMycobacterium sp strain PYR-1. Appl Environ Microbiol 67: 3577–3585.

62. Krivobok S, Kuony S, Meyer C, Louwagie M, Willison JC, et al. (2003) Identification of pyrene-induced proteins in Mycobacterium sp strain 6PY1: evidence for two ringhydroxylating dioxygenases. J Bacteriol 185: 3828–3841.

63. Demaneche S, Meyer C, Micoud J, Louwagie M, Willison JC, et al. (2004) Identification and functional analysis of two aromatic-ring-hydroxylating dioxygenases from a Sphingomonas strain that degrades various polycyclic aromatic hydrocarbons. Appl Environ Microbiol 70: 6714–6725.

64. Rodrigues JLM, Pellizari VH, Mueller R, Baek K, Jesus EC, et al. (2013) Conversion of the Amazon rainforest to agriculture results in biotic homogenization of soil bacterial communities. Proc Natl Acad Sci USA 110: 988–993.

65. Arthur E, Schjønning P, Moldrup P, de Jonge LW (2012) Soil resistance and resilience to mechanical stresses for three differently managed sandy loam soils. Geoderma 173–174: 50–60.

66. Gregory AS, Watts CW, Whalley WR, Kuan HL, Griffiths S, et al. (2007) Physical resilience of soil to field compaction and the interactions with plant growth and microbial community structure. Eur J Soil Sci 58: 1221–1232.

67. Schjønning P, Elmholt S, Christensen BT (2004) Soil quality management: concepts and terms. In: Schjønning P, Elmholt S, Christensen BT, editors.Challenges in Modern Agriculture.Wallingford: CABI Publishing. pp. 1–15.

16

Soil Aggregates and Associated Organic Matter under Conventional Tillage, No-Tillage, and Forest Succession after Three Decades

Scott Devine[1], Daniel Markewitz[1]*, Paul Hendrix[2], David Coleman[2]

1 Warnell School of Forestry and Natural Resources, The University of Georgia, Athens, Georgia, United States of America, **2** Odum School of Ecology, The University of Georgia, Athens, Georgia, United States of America

Abstract

Impacts of land use on soil organic C (SOC) are of interest relative to SOC sequestration and soil sustainability. The role of aggregate stability in SOC storage under contrasting land uses has been of particular interest relative to conventional tillage (CT) and no-till (NT) agriculture. This study compares soil structure and SOC fractions at the 30-yr-old Horseshoe Bend Agroecosystem Experiment (HSB). This research is unique in comparing NT and CT with adjacent land concurrently undergoing forest succession (FS) and in sampling to depths (15–28 cm) previously not studied at HSB. A soil moving experiment (SME) was also undertaken to monitor 1-yr changes in SOC and aggregation. After 30 years, enhanced aggregate stability under NT compared to CT was limited to a depth of 5 cm, while enhanced aggregate stability under FS compared to CT occurred to a depth of 28 cm and FS exceeded NT from 5–28 cm. Increases in SOC concentrations generally followed the increases in stability, except that no differences in SOC concentration were observed from 15–28 cm despite greater aggregate stability. Land use differences in SOC were explained equally by differences in particulate organic carbon (POC) and in silt-clay associated fine C. Enhanced structural stability of the SME soil was observed under FS and was linked to an increase of 1 Mg SOC ha^{-1} in 0–5 cm, of which 90% could be attributed to a POC increase. The crushing of macroaggregates in the SME soil also induced a 10% reduction in SOC over 1 yr that occurred under all three land uses from 5–15 cm. The majority of this loss was in the fine C fraction. NT and FS ecosystems had greater aggregation and carbon storage at the soil surface but only FS increased aggregation below the surface, although in the absence of increased carbon storage.

Editor: Mary O'Connor, University of British Columbia, Canada

Funding: This research was funded with internal support for the Warnell School of Forestry and Natural Resources, the Odum School of Ecology, and the University of Georgia. The funders had no role in study design, data collection and analysis, decision to publish, or preparation of the manuscript.

Competing Interests: The authors have declared that no competing interests exist.

* E-mail: dmarke@uga.edu

Introduction

Understanding the mechanisms for soil carbon storage and stabilization relative to land management is increasingly relevant as soil organic carbon (SOC) is known to be a major pool of global C and SOC is critical to sustaining soil productivity. Since physical protection of SOC within stable soil aggregates is considered to be one of the major SOC stabilization mechanisms, [1–3], the effect of land management on aggregate stability is accepted as a key factor in determining SOC levels [4]. There have been many comparisons of soil aggregates under conventional tillage relative to no-tillage agriculture but relatively few that have incorporated other land uses such as abandoned agricultural fields or forest succession, particularly over a decades timescale.

The three-decade agroecosystem experiment at Horseshoe Bend (HSB), Athens, GA, USA, has long focused on SOC differences in side-by-side plots of conventional tillage agriculture (CT) and no-tillage agriculture (NT). This research has demonstrated increased SOC in the 0–5 cm layer of NT [5] and some role for aggregation in SOC protection [6]. More recent work at HSB has incorporated adjacent plots of forest succession (FS), which has demonstrated a similar increase in 0–5 cm SOC relative to CT as

well as an increase in particulate organic carbon (POC) in 5–15 and 15–30 cm subsurface soil layers [7]. Although previous CT-NT comparisons have been reported from HSB this study extends analyses to subsurface soil layers (15–28 cm) previously not studied at HSB for aggregation. These subsurface soils are recognized to affect the carbon balance of CT-NT studies [8]. Furthermore, few studies have compared CT and NT with other land uses but here we simultaneously investigate the role of aggregation in forest succession to evaluate the robust nature of aggregates for SOC stabilization. As such there is an opportunity to understand if physical stabilization through aggregation is functioning similarly in 0–5 cm soils in NT and FS, and if aggregation in subsurface layers of FS is stabilizing POC at rates greater than under agroecosystems.

Under these contrasting land uses (CT-NT-FS) soil structure can both directly and indirectly affect SOC stabilization. Narrow pores can directly exclude extracellular enzymes, microbes, and soil micro-fauna. Pore size also indirectly affects decomposition by regulating fluxes that influence microbial activity, such as oxygen, heat, solutes, and water. This explains how intra-aggregate pores, which are finer (and perhaps more tortuous) than inter-aggregate pores [9], can slow decomposition within aggregates. The abundant research on the relationship between soil aggregation and SOC has produced several contrasting models of how physical protection of SOC occurs.

The aggregate hierarchy model proposes that soil structure arises from aggregation of fine mineral particles into microaggregates (53 μm to 250 μm) continuing to aggregate to macroaggregates (>250 μm) with increasing dependence on more transient, organic binding agents as the scale increases [10]. This model suggests that SOC concentration increases with increasing aggregate size and that cultivation exposes these binding agents to microbial attack with subsequent loss of SOC [11].

Oades [12] later modified this theory hypothesizing that microaggregates formed around POC (53–2000 μm) when enmeshed within stable macroaggregates. Thus, soils lacking mechanical disturbance such as NT or FS result in decreased macroaggregate turnover and higher SOC due to microaggregate formation and stabilization of fine POC (53–250 μm) [13,14]. Loss of SOC in cultivated soils is explained by shorter macroaggregate turnover times and a reduced opportunity for formation of microaggregates around a fine POC core [3].

More recently, it has been hypothesized that increased pore connectivity in soils that are not tilled allows greater movement of dissolved organic carbon (DOC) from larger pores surrounding aggregates into intra-aggregate pore spaces [15]. In tilled soils, disruption of the pore network effectively isolates substantial portions of the intra-aggregate pore space from replenishment of DOC by root exudates, decomposition of POC, or litter leachate laden with DOC. The focus on DOC contrasts with that on POC as the fraction of SOC most susceptible to management and soil structural disruption [16–22].

Given the previous research at HSB the objective of this study was to quantify SOC in various soil aggregate fractions under the different land uses to better understand the role of the aforementioned mechanisms in SOC stabilization [23,24]. It was hypothesized that fractionation of SOC into POC and fine C (presumably adsorbed DOC) offers a means to address differences in the proposed mechanisms of carbon storage. In particular, it was hypothesized that if macro- and micro-aggregate protection of POC is the predominate mechanism increasing SOC then aggregate associated POC contents should be greater in land uses with higher SOC content. Conversely, if micro-scale fluxes of DOC are important to attaining maximum SOC storage capacity, then this should be recovered as fine C (<53 μm) in systems with significantly higher levels of SOC.

In addition to the POC and fine C fractionation, a soil moving experiment (SME) was established to determine the extent to which destructured soil can re-develop macroaggregate stability in the field under the contrasting land uses. This SME better represents land use attributes as opposed to laboratory incubations [5,6]. It was hypothesized that differential recovery rates of macroaggregates in SME soil under the contrasting field conditions would reflect a greater abundance of organic binding agents as indexed by soil microbial biomass (i.e., bacteria and fungi). We hypothesized these rates would be greatest in forests due to an expectation of greater fungal abundance.

Methods

Site history

The Horseshoe Bend (HSB) agroecosystem experiment (33°57′N, 83°23′W) is a long-term comparison of conventional tillage (CT) and no-tillage (NT) treatments begun in 1978. The site is under the management of the Odum School of Ecology at The

University of Georgia so no special permission was required for the current research and no endangered or protected species were involved in or negatively impacted by this research.

Based on a series of aerial photos, HSB had been cleared entirely of trees except for a narrow strip along the Oconee River at least since 1938 [7]. The site was used for pasture and forage production until 1965 when the Institute of Ecology acquired HSB for research [25]. The area underwent secondary succession in 1966 except for the future agroecosystem site, which was tilled to plant a crop of millet for a separate, earlier study [26]. Afterwards, the site went fallow and was used for a prescribed burn study and a N fertilization study [27,28]. In 1978, the study area was divided into equal numbers of CT and NT 0.1 ha plots (n = 4 for each type), and the old-field vegetation was mowed [25].

Site description

Horseshoe Bend is so named because it was established in a bend on the banks of the Oconee River and as such the site is very flat (<1% slope). The soils on this river terrace have previously been identified as clayey, kaolinitic, thermic, Rhodic Kanhapludults [29] but have recently been reclassified as a fine-loamy, kaolinitic, thermic, Typic Kanhapludalf based on new particle-size analysis data in the control section and measurements of base saturation (Table 1) that are >98% in all three treatments at 140 cm (i.e., 125 cm below the top of the kandic horizon). Average rainfall from 1945 to 2003 in Athens was 125 cm with a mean annual temperature of 16.4°C

Experimental design

When the site was originally sub-divided in 1978 the CT-NT treatments were assigned randomly in a completely randomized design [25]. Since the agroecosystem plots were in fallow from 1966, the current NT plots have not been tilled for forty-two years. Tillage has been performed twice annually in most years with moldboard plowing to a depth of 15 cm, followed by disking. In 1981, each agricultural plot was split into different winter crop treatments with winter rye (*Secale cereale*) grown on half of each plot and N-fixing crimson clover (*Trifolium incarnatum*) grown on the other half until 1984 and then again from 1989 to 2007 [25,30]. From 1978–1998, summer crops have included sorghum (*Sorghum bicolor*), soybeans (*Glycine max*), corn (*Zea mays*), and kenaf (*Hibiscus cannabinus*). In 1999 to 2007, a second split-plot was assigned to the agroecosystem experiment for a comparison between summer crops of non-Bt and Bt-cotton (*Gossypium hirsutum* L.) [31]. Soil sampling in the current study sought to estimate an average effect of NT and CT on the measured properties by sampling equally among the winter cover split-plots before compositing; the effects of transgenic cotton on soil C and N were considered negligible, since earlier work showed no difference in decomposition rates between non-Bt and Bt-cotton residues [32].

In 2007, four secondary, hardwood forest plots bordering the tillage experiment were established. These areas represent a third treatment, referred to herein as forest succession (FS). Most likely, these afforested plots were not tilled in 1966, which means they essentially underwent a pasture to forest conversion. All FS plots are on the same terrace as the agroecosystem plots and were clear of forest from 1938 to 1967, as observed in the aerial photographs [7].

Aggregate size fractions and stability

To quantify aggregate stability, four samples were obtained per plot at the end of the summer growing season in October 2007 using a 5 cm diameter corer attached to a slide-hammer to a depth of 28 cm. Each core was divided into 0–5, 5–15, and 15–28 cm

Table 1. Soil chemical attributes (mean ±1SD) for the Horseshoe Bend Agroecosystem Experiment, Athens, GA.

Treatment	Depth	pHs[1]	C	N	P[2]	K	Ca	Mg	ECEC[3]	Bulk Density
	cm		--------g/kg--------		µg/g	--------------------------cmol$_c$/kg--------------------------				g/cm³
Conventional	0–5	5.16±0.04	9.73±0.75	0.75±0.06	4.1±0.6	0.36±0.04	2.13±0.22	0.54±0.03	3.04±0.27	1.45±0.09
Till	5–15	4.90±0.04	6.99±0.58	0.58±0.05	2.2±0.4	0.25±0.03	1.93±0.13	0.49±0.02	2.72±0.18	1.61±0.03
	15–30	5.42±0.08	2.93±0.35	0.27±0.03	1.7±0.1	0.18±0.01	1.92±0.05	0.58±0.03	2.69±0.07	1.68±0.04
	30–50	5.69±0.06	1.84±0.12	0.20±0.02	0.7±0.2	0.12±0.01	1.89±0.07	0.58±0.01	2.58±0.06	1.64±0.03
	50–100	6.00±0.09	1.16±0.05	0.14±0.01	0.4±0.2	0.08±0.01	1.82±0.10	0.41±0.06	2.31±0.15	1.61±0.03
	100–150	6.05±0.06	0.85±0.04	0.10±0.00	0.4±0.1	0.04±0.00	1.87±0.05	0.38±0.07	2.29±0.11	
	150–200	5.98±0.09	0.54±0.02	0.06±0.00	0.4±0.0	0.03±0.01	1.20±0.07	0.33±0.09	1.56±0.12	
No Till	0–5	5.12±0.14	17.93±0.42	1.53±0.04	11.7±2.2	0.34±0.04	4.06±0.35	0.90±0.06	5.32±0.43	1.32±0.02
	5–15	5.09±0.12	6.92±0.31	0.54±0.03	4.3±1.2	0.27±0.05	2.31±0.13	0.57±0.03	3.18±0.17	1.50±0.06
	15–30	5.45±0.04	3.45±0.25	0.30±0.01	1.2±0.2	0.19±0.03	2.00±0.02	0.67±0.04	2.86±0.05	1.55±0.05
	30–50	5.74±0.06	2.14±0.13	0.21±0.01	0.5±0.1	0.14±0.02	1.97±0.10	0.69±0.04	2.80±0.10	1.60±0.05
	50–100	5.87±0.08	1.19±0.08	0.14±0.01	0.4±0.0	0.08±0.01	2.04±0.07	0.43±0.04	2.54±0.07	1.67±0.04
	100–150	5.93±0.05	0.98±0.07	0.11±0.00	0.3±0.0	0.04±0.01	1.99±0.10	0.30±0.07	2.32±0.06	
	150–200	5.93±0.05	0.64±0.06	0.07±0.01	0.3±0.0	0.03±0.00	1.48±0.15	0.27±0.11	1.77±0.19	
Forest	0–5	4.33±0.15	22.4±1.44	1.55±0.11	1.5±0.7	0.23±0.05	2.42±0.51	0.81±0.13	3.97±0.53	1.06±0.05
Succession	5–15	4.04±0.10	8.02±0.34	0.60±0.04	1.0±0.4	0.11±0.03	0.85±0.33	0.38±0.07	2.34±0.20	1.38±0.02
	15–30	4.38±0.06	3.52±0.18	0.30±0.01	0.5±0.1	0.06±0.01	1.34±0.19	0.49±0.02	2.44±0.14	1.65±0.02
	30–50	4.76±0.12	2.11±0.07	0.19±0.01	0.4±0.0	0.04±0.01	1.62±0.20	0.37±0.01	2.23±0.16	1.57±0.06
	50–100	5.19±0.14	1.40±0.21	0.14±0.01	0.4±0.1	0.03±0.00	1.46±0.09	0.37±0.04	1.89±0.08	1.63±0.04
	100–150	5.50±0.16	0.81±0.06	0.08±0.01	0.4±0.1	0.02±0.00	1.04±0.11	0.46±0.11	1.54±0.16	
	150–200	5.30±0.21	0.61±0.07	0.06±0.01	0.3±0.1	0.03±0.00	0.61±0.13	0.55±0.15	1.20±0.26	

N = 4 per treatment type. Soils were collected in 2009.
[1]Salt pH in 0.01 M CaCl$_2$.
[2]Mehlich I extractable P and cations.
[3]Effective cation exchange capacity by sum of cations method.

and then composited by depth. Composite samples were passed through a 10 mm mesh by gently breaking the cores along planes of weakness before air-drying. These air-dried samples were dry-sieved by hand through 6.3, 4, 2, and 1 mm screens to obtain dry aggregate size classes. A separate subsample of the <1 mm dry aggregates was sieved on a 0.25 mm screen. The objective of dry-sieving was to determine the initial aggregate size distribution before wet-sieving [33]. Sub-samples of 50 g were then weighed out by re-combining all the dry aggregate size classes according to their proportional mass for the bulk soil. This ensured that each sub-sample accurately represented the bulk soil dry aggregate size distribution [34]. These 50 g sub-samples were then re-wetted via capillary action followed by wet-sieving with a modified Yoder apparatus [35], which permits complete recovery of all soil from each sample. The method obtains four wet aggregate size classes: (i) >2000 µm; (ii) 250–2000 µm; (iii) 53–250 µm; and (iv) <53 µm. Wet aggregate mean-weighted diameters (MWD) were calculated by multiplying the proportion of soil in each of these four aggregate size classes by the mid-point of the size class. The calculation of the dry MWD used the same size divisions as the wet MWD except no <53 µm fraction was included, since dry sieving material through a 53 µm sieve is not practical. The aggregate stability index was then calculated by dividing the wet MWD by the dry MWD; an index of 1 represents perfect structural stability.

Soil organic C (SOC) fractionation and determination

Soil sub-samples were pulverized in a Spex 8200 ball-mill grinder (Metuchen, NJ) and total SOC and N were determined by dry combustion on a CE Elantech NC 1110 analyzer (Lakewood, NJ). Earlier mineralogical work on subsamples showed no evidence of carbonates so no pre-treatments were performed (Devine, unpublished data, 2008). To obtain particulate organic carbon (POC) and fine-C fractions, a 3–5 g subsample of each aggregate size fraction was dispersed by adding 15–25 mL of 0.5% Na-hexametaphosphate and shaking in 50 mL centrifuge tubes for 16 h on a reciprocal shaker at 200 rpm. The resulting soil slurry was washed through a 53 µm sieve using 0.5 L of DI water to fractionate the soil into: (1) sand+POC and (2) silt+clay+fine-C. A rubber policeman was used to aid in dispersion of stable soil aggregates on top of the 53 µm sieve when necessary. After drying both fractions in a forced-air oven at 60°C, fine-C and POC fractions were pulverized and C and N determined. A mass correction was made for the amount of Na-HMP recovered in the fine fraction prior to estimation of fine-C and N concentrations.

Since the <53 µm fraction does not contain sand, total SOC and POC are calculated on a sand-free basis for each aggregate size fraction in order to make comparisons among all the aggregate size classes [36]:

$$\frac{g\_POC}{kg\_sandfree\_aggregate} = (\frac{g\_POC}{kg\_sand}) * (\frac{kg\_sand}{kg\_silt+clay})$$

$$\frac{g\_SOC}{kg\_sandfree\_aggregate} = (\frac{g\_SOC}{kg\_soil}) * (\frac{kg\_soil}{kg\_silt+clay})$$

Soil moving experiment

A soil moving experiment (SME) was carried out to determine the extent to which destructured soil re-develops aggregate stability among the different land uses. Soil moving or transplant experiments have been carried out to investigate soil processes including C accumulation or N cycling [37,38]. In this study, destructured soil was utilized, as opposed to intact cores, which is similar to the use of sand or sieved soil used in root ingrowth core experiments [39]. First, CT soil from 0–15 cm was collected from all four plots. CT soil was utilized since previous research at HSB has measured this soil to have the least SOC and aggregate stability. Sole use of CT soil, however, also assumes that there is no inherent internal difference between destructured CT, NT, or FS soil and that differences during the experiment would result only from external differences due to land use. After collection soil was dried at 60°C in a forced-air oven, and crushed with a rolling pin to destroy larger macroaggregates and to pass a 1 mm sieve, referred to henceforth as "initial soil." A 10 cm diameter auger was used to remove soil to 20 cm depth from four locations in each CT, NT, and FS plot at the end of July 2007. A PVC pipe of an equal diameter was placed as an insert in each hole to prevent the side walls from collapsing during filling. Each hole was then filled with the initial soil and uniformly tamped with a capped PVC pipe to a bulk density of approximately 1.4 g soil cm^{-3}. Soil packing was tested with success in the laboratory by filling the same PVC pipe insert to the desired bulk density. The PVC insert was removed after packing, and the edge of the SME core marked with flags and aluminum nails. At the end of August 2008, the same 5 cm diameter corer previously used to sample aggregates was used to extract a smaller soil core within the 10 cm diameter SME core. All cores were located (some with the aid of a metal detector), divided into 0–5 and 5–15 cm depth classes, and composited by plot. In one FS plot, only two cores were sampled due to obvious mixing with native soil that had occurred from animal activity; in three other plots (two CT and one FS), only three cores were sampled due to obvious animal activity.

The same procedure as described above was used to measure aggregate stability, total SOC, and POC in the two depth classes. Microbial biomass C was also determined on field moist, replicate subsamples following the 24 hour chloroform fumigation—K_2SO_4 extraction procedure [40]. Extracts were analyzed for non-purgable organic carbon (NPOC) (Shimadzu TOC 5000, Columbia, Maryland, USA). Microbial biomass C (MBC) was estimated following Wu et al. [41]:

$$MBC = (NPOC_{fumigated} - NPOC_{non-fumigated}) * 2.22$$

Statistical analysis

All statistical analyses were completed with SAS 9.3.1 (SAS, Cary, NC). All data were checked for normality using Shapiro-Wilk and Kolmogorov-Smirnoff but no transformations were deemed necessary. Differences in soil properties were tested with a one-way ANOVA by each depth class and aggregate size combination with land use as the main effect (n = 4). Equality of variance among treatment classes was ensured with Levine's test.

The same analysis was performed by each depth class and land use combination with aggregate size as the main effect. For the soil moving experiment where starting values of soil properties were known, changes in total SOC, POC, and fine-C were analyzed using a repeated measures ANOVA by depth class to investigate the effect of time and the land-use x time interaction. When the main effects were significant, post-hoc tests were performed using the LSMEANS statement in PROC GLM to evaluate the significance of mean separations with p-values adjusted by Tukey's honestly significant difference ($\alpha = 0.05$).

Results

Aggregate size distributions and stability

Aggregate stability decreased significantly from FS and NT to CT in the upper 5 cm (Figure 1; see also Table S1). In the 5–28 cm layers FS soil was significantly more stable than both NT and CT (Figure 1). This response can also be seen as: (i) a higher percentage of wet-sieved, large macroaggregates (>2000 μm) in the FS and NT soils from 0–5 cm compared to CT, (ii) more large macroaggregates in FS from 15–28 cm compared to both CT and NT, and (iii) a higher proportion of fines (<53 μm) in 0–15 cm under CT compared to both NT and FS (Figure 2; see also Table S2).

Land use effects on aggregate associated C

There were significant effects of land use on C fraction concentrations in all aggregate size classes from 0–5 cm and the majority of size classes from 5–15 cm. From 0–5 cm, NT and FS aggregate size fractions were significantly elevated for SOC and fine fractions compared to CT. In POC, increasing C from CT>NT>FS was evident in all aggregate sizes but significant at 0.05 for only the >2000 μm (Figure 3; see also Table S3, part 1). From 5–15 cm, there were no significant differences between CT and NT for any of the size class and C fraction combinations, while the two largest aggregate size classes were significantly elevated in FS with respect to NT for SOC and all three classes were greater for POC (Figure 4; see also Table S4, part 1). FS also exceeded CT but only in the small and micro aggregate fractions for POC. There were no significant differences for any of the size class and C fraction combinations from 15–28 cm (Figure 5; see also Table S5, part 1).

Overall, the majority of C at HSB under all land uses is contained in the water-stable macroaggregate fraction to a depth of at least 28 cm (Figure 6; see also Table S6). By summing the C contribution from each aggregate size fraction to calculate a C concentration for each depth class by treatment combination, differences in POC from 0–5 cm explained 34% of the total SOC difference between NT and CT, 43% of the total SOC difference between FS and CT, and 51% of the total SOC difference between FS and NT. From 5–15 cm, differences in POC explained 56% of the difference between FS and CT and 47% of the difference between FS and NT.

Aggregate associated C within a land use

In terms of significant differences among the aggregate size classes within a land use and depth class, the most consistent result was that the <53 μm aggregate fraction had lower total SOC concentrations when expressed on a sand-free basis (Figures 3–5; see also Table S3, part 2, Table S4, part 2, and Table S5, part2). Under FS from 0–15 cm, there were no other significant differences among the size classes >53 μm (Figures 3 and 4) and there were no significant differences in POC among the aggregate size classes for any depth (Figures 3–5).

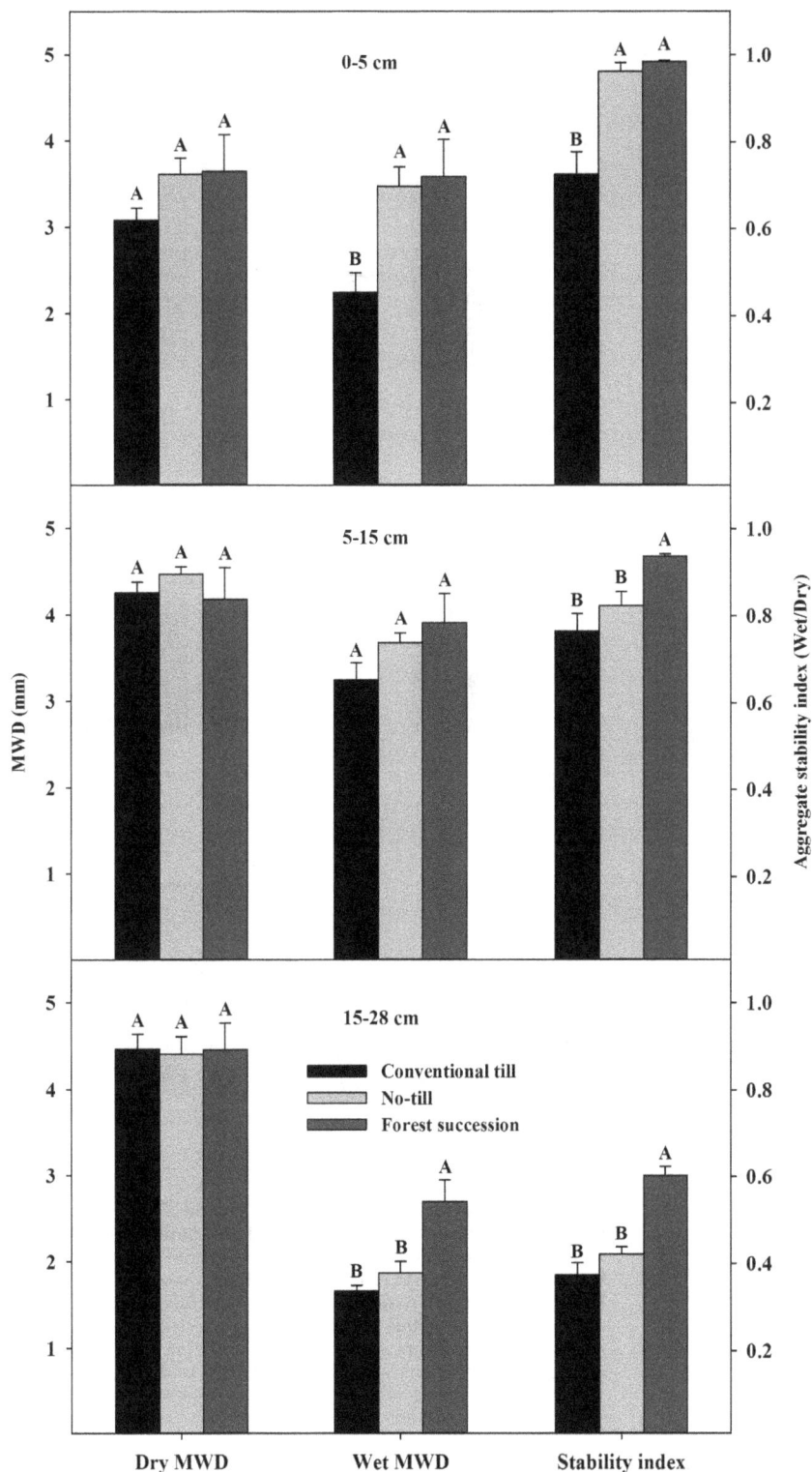

Figure 1. Dry-sieved and wet-sieved aggregates' mean weight diameter (MWD) (mean ± 1 SE) at the Horseshoe Bend Agroecosystem Experiment (n = 4), Athens, GA, USA, October 2007. The aggregate stability index is calculated by dividing the Wet MWD by the Dry MWD. Different letters indicate statistical significance between land uses within a depth class based on Tukey's HSD with a = 0.05. See also Table S1

Under NT, in the upper 15 cm both the large macroaggregates and the <53 μm fractions were SOC depleted relative to the small macroaggregates (250–2000 μm) and microaggregates (Figures 3

and 4). This was similar under CT from 0–5 cm where the small macroaggregates (250–2000 μm) were significantly elevated in total SOC and POC compared to the other size fractions

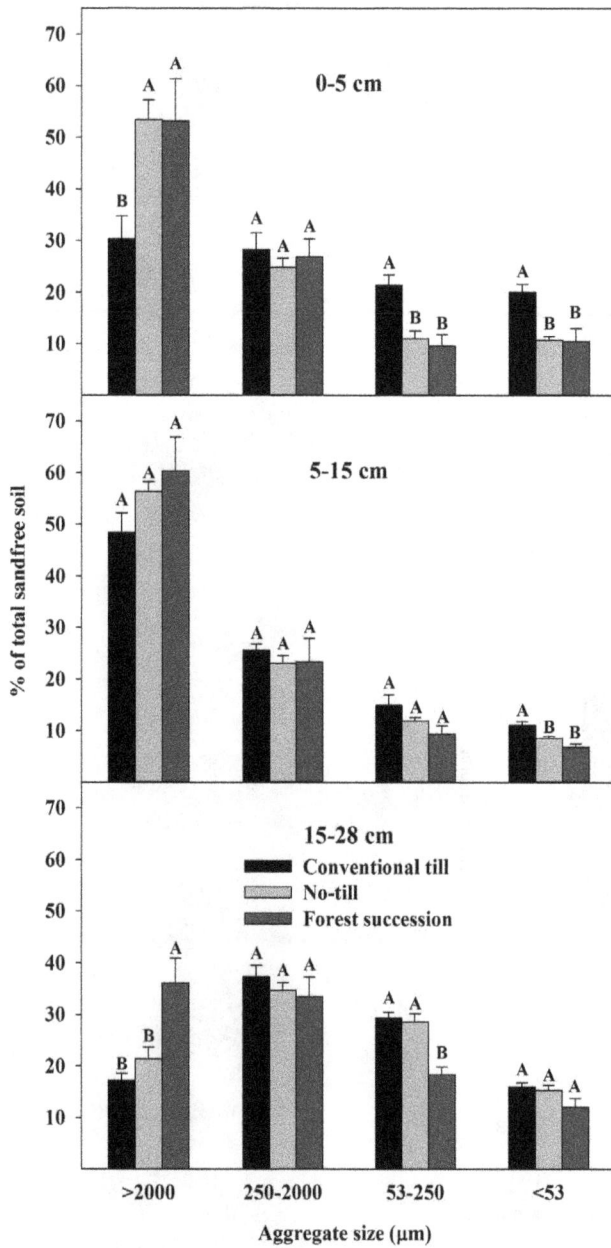

Figure 2. Distribution of water-stable aggregates on a sand-free soil basis (mean ±1 SE) at the Horseshoe Bend Agroecosystem Experiment (n = 4), Athens, GA, USA, October 2007. Different letters within an aggregate size class and depth class represent a significant difference between land uses (α = 0.05). See also Table S2.

Figure 3. Total soil organic C (SOC), fine C (<53 μm), and particulate organic C (POC) (53–2000 μm) concentrations on a sand-free basis from 0–5 cm in four different aggregate size fractions (mean ±1 SE) in the Horseshoe Bend Agroecosystem Experiment (n = 4) Athens, GA, USA, October 2007. Different uppercase letters within an aggregate size class and C fraction represent a significant difference between land uses. Different lowercase letters within a C fraction and a land use represent a significant difference among aggregates size classes (α = 0.05). See also Table S3, part 1 and Table S3, part 2.

(Figure 3). Under CT from 5–15 cm, both the small macroaggregates and microaggregates were elevated in total SOC and POC compared to the large macroaggregates but only the small macroaggregates were significantly elevated compared to the large macroaggregates and only for POC (Figure 4). From 15–28 cm, the large macroaggregates were the most carbon rich under all land uses, having significantly greater total SOC and fine C concentrations compared to the <250 μm size fractions (Figure 5).

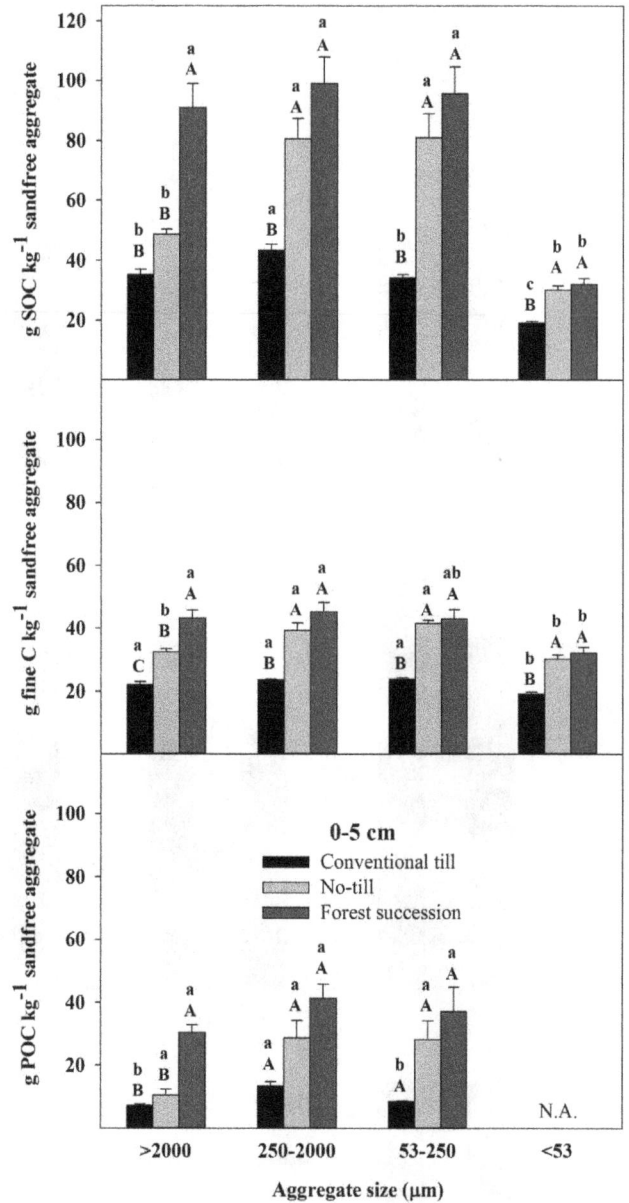

Recovery of POC and fine C

Summation of POC and fine C compared to direct analysis of total SOC revealed that recoveries of POC and fine C were, on average, 86% from 0–5 cm, 88% from 5–15 cm, and 92% from 15–28 cm. A precipitate formed on the glass beakers used to dry the fine C slurry and was more difficult to remove in the beakers containing 0–5 cm samples, suggesting this may have been the source of the <100% recovery of C.

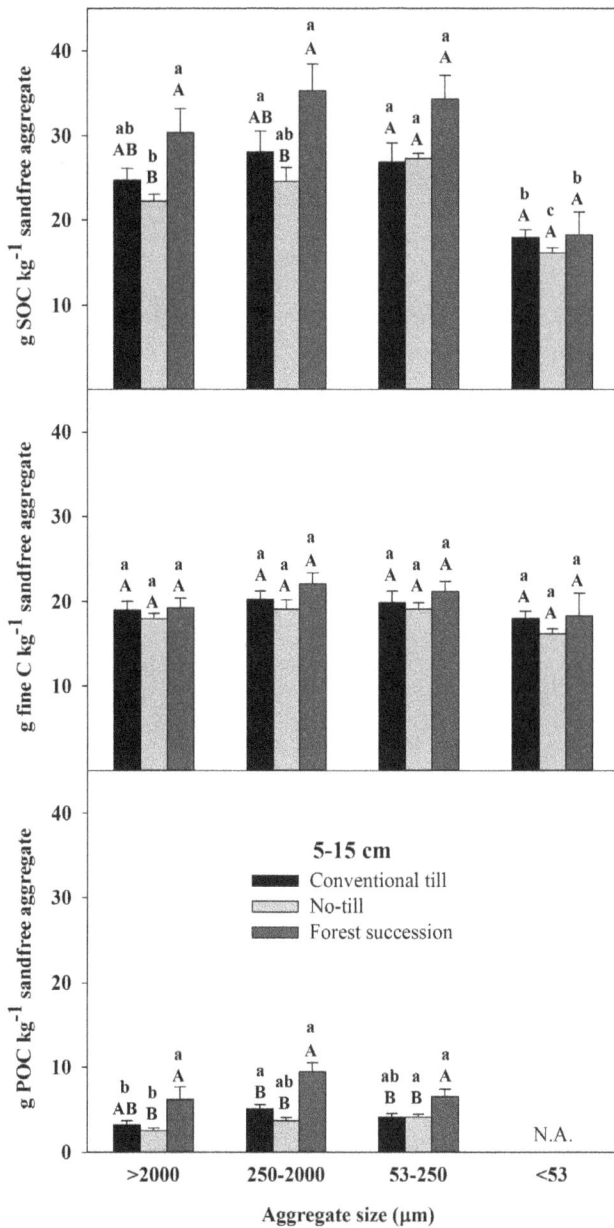

Figure 4. Total soil organic C (SOC), fine C (<53 μm), and particulate organic C (POC) (53–2000 μm) concentrations on a sand-free basis from 5–15 cm in four different aggregate size fractions (mean ±1 SE) in the Horseshoe Bend Agroecosystem Experiment (n = 4) Athens, GA, USA, October 2007. Different uppercase letters within an aggregate size class and C fraction represent a significant difference between land uses. Different lowercase letters within a C fraction and a land use represent a significant difference among aggregates size classes ($\alpha = 0.05$). See also Table S4, part 1 and Table S4, part 2.

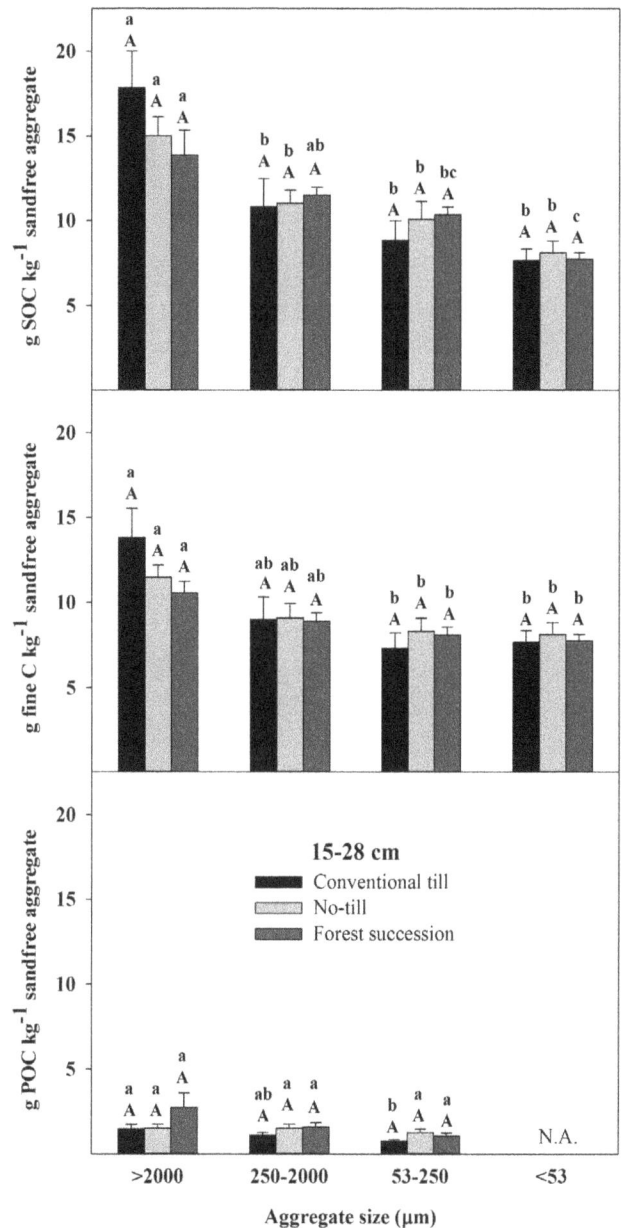

Figure 5. Total soil organic C (SOC), fine C (<53 μm), and particulate organic C (POC) (53–2000 μm) concentrations on a sand-free basis from 15–28 cm in four different aggregate size fractions (mean ±1 SE) in the Horseshoe Bend Agroecosystem Experiment (n = 4) Athens, GA, USA, October 2007. Different uppercase letters within an aggregate size class and C fraction represent a significant difference between land uses. Different lowercase letters within a C fraction and a land use represent a significant difference among aggregates size classes ($\alpha = 0.05$). See also Table S5, part 1 and Table S5, part 2.

Soil moving experiment (SME): Structural recovery, C change, and microbial biomass

The SME soil under FS had a significantly greater stability index compared to both CT and NT at 0–5 and 5–15 cm after one year (Table 2; see also Table S7). There were no significant differences in aggregate stability between CT and NT at either depth in the SME (Table 2). Overall, aggregate stability tended to

be greater from 0–5 cm compared to 5–15 cm under NT and FS. This depth stratification did not occur under CT (Table 2).

Repeated measures analysis showed a significant time x treatment interaction effect from 0–5 cm for total SOC ($p = 0.03$) and POC ($p = 0.0004$) but not fine C ($p = 0.50$), suggesting a significant departure from initial values of SOC and POC was dependent upon land use. The overall effect of time from 0–5 cm was significant for POC ($p = 0.0007$), which increased after one year. The effect of time was not significant

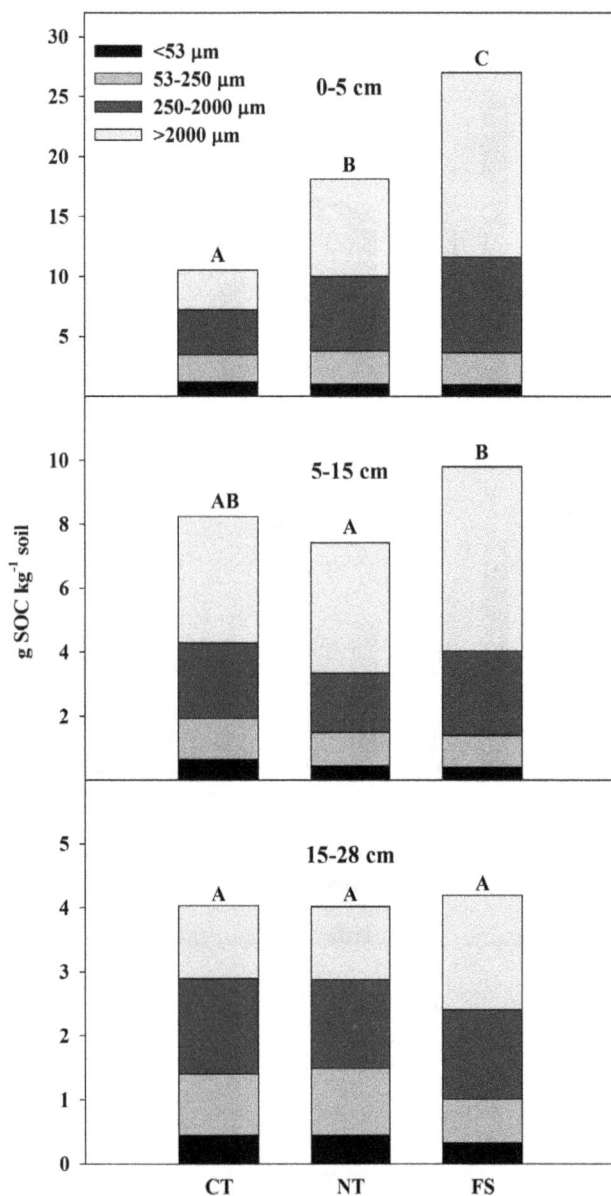

Figure 6. Contributions of four different aggregate size fractions to total soil organic C (SOC) concentrations. Different letters indicate a significant difference between land uses in a depth class ($\alpha = 0.05$). Means of the sums of the four fractions are shown (n = 4) from Horseshoe Bend, Athens, GA, USA, October 2007. See also Table S6.

Table 2. Aggregate stability (mean ±1 SD) at two depths at the end of a soil moving experiment (n = 4).

Depth	Land use	Dry MWD[†]	Wet MWD	Stability index
cm		------------mm----------		Wet/Dry
0–5	Conventional Till	4.3±0.4[a]	1.4±0.2[a]	0.33±0.06[a]
	No-Till	4.5±0.3[a]	1.9±0.3[a]	0.43±0.09[a]
	Forest Succession	3.9±0.3[a]	3.1±0.2[b]	0.79±0.03[b]
5–15	Conventional Till	4.8±0.6[a]	1.2±0.2[a]	0.26±0.08[a]
	No-Till	4.8±0.6[a]	1.0±0.2[a]	0.22±0.05[a]
	Forest Succession	3.6±0.5[b]	2.1±0.3[b]	0.60±0.10[b]

Conventional till soil was initially dried and crushed to pass a 1 mm sieve and then installed under three different land uses for a period of one year at Horseshoe Bend, Athens, GA, USA, July 2007–August 2008.
Different letters indicate statistical significance among treatment means within a depth class based on Tukey's HSD with $\alpha = 0.05$. See also Table S7.
[†]Mean-weighted diameter.

possibly reflecting a lower net loss of POC from 5–15 cm under FS (Table 3). The interaction was not significant for SOC ($p = 0.85$) or fine C ($p = 0.89$), since there was a more uniform decrease in these components from the initial value under all the land uses.

SME microbial biomass carbon (MBC), like aggregate stability and total SOC from 0–5 cm, was elevated in FS relative to the agroecosystem soils, though not significantly ($p > 0.27$; Table 3). Overall, there were no significant differences among the land uses from 0–5 cm in terms of MBC concentrations. CT MBC were significantly elevated compared to NT from 5–15 cm ($p = 0.02$). For all three land uses, MBC tended to be higher from 0–5 cm compared to 5–15 cm (Table 3).

Discussion

Changes in soil structure and SOC after 30 years of different management

Earlier work from HSB reported that the three decades of divergent land use have led to 20% greater SOC content in the FS

Table 3. Total soil organic C (SOC), particulate organic C (POC) and microbial biomass C (MBC) (mean ±1 SD) at two depths at the end of a soil moving experiment (n = 4).

Land use	Depth	SOC	POC	Fine C	MBC
	-cm-	--------------g kg^{-1} soil-------------			--mg kg^{-1}--
Initial	0–5	9.2±0.2	1.8±0.1	7.5±0.1	NA
Conventional till		9.1±0.3[a]	2.0±0.2[a]	7.2±0.2[a]	397±27[a]
No-till		8.7±0.2[a]	1.7±0.1[a]	7.0±0.2[a]	366±30[a]
Forest succession		10.9±1.7[b]	3.3±0.6[b]	7.6±1.2[a]	422±72[a]
Initial	5–15	9.2±0.2	1.8±0.1	7.5±0.1	NA
Conventional till		8.3±0.2[a]	1.5±0.1[a]	6.8±0.2[a]	317±25[a]
No-till		8.3±0.2[a]	1.4±0.1[a]	6.8±0.3[a]	274±13[b]
Forest succession		8.4±0.6[a]	1.7±0.2[a]	6.7±0.5[a]	305±16[ab]

Conventional till soil was initially dried and crushed to pass a 1 mm sieve and then installed under the three different land uses for a period of one year at Horseshoe Bend, Athens, GA, USA, July 2007–August 2008.
Different letters indicate statistical significance between treatment means within a depth class based on Tukey's HSD with $\alpha = 0.05$. See also Table S8.

for SOC ($p = 0.27$) or fine C ($p = 0.35$) for 0–5 cm, which means there was no clear directional change for these components compared to initial values across all land uses (Table 3: see also S8). A test of mean separation showed that total SOC from 0–5 cm was significantly elevated in the FS soil compared to CT ($p = 0.08$) and compared to NT ($p = 0.03$) (Table 3). POC was also significantly elevated from 0–5 cm in the FS soil compared to both CT and NT ($p < 0.002$) (Table 3; see also Table S8). From 5–15 cm, declines in C were observed after one year. The effect of time was significant for SOC ($p < 0.0001$), POC ($p = 0.0004$), and fine C ($p < 0.0001$), meaning a significant decrease in these components from the initial values had occurred (Table 3). The time x treatment interaction was significant for POC at $p = 0.099$

$(61.5\pm4.3$ Mg ha^{-1}) and NT $(60.0\pm4.5$ Mg ha^{-1}) soil profiles (0–2 m) relative to CT $(51.9\pm4.9$ Mg ha^{-1}) [7]. SOC content differences in the 0–5 cm layer accounted for 56% of the 0–2 m mean content difference between FS and CT and 66% of the 0–2 m content difference between NT and CT. These increases in SOC at the surface under FS and NT are linked to enhanced aggregate stability.

The increase in structural stability under NT compared to CT was limited to a depth of 5 cm, while the FS soil demonstrated significantly greater stability compared to CT to a depth of 28 cm and compared to NT from 5–28 cm (Figure 1). Previous work at HSB on aggregate size fractions showed that all aggregates >106 μm from 0–5 cm were significantly more stable in NT compared to CT and from 5–15 cm the large macroaggregates (>2000 μm) were more stable in NT [23]. In the current study, the difference in aggregate stability between NT and FS from 5–15 cm is evident in the aggregate stability metric (Figure 1). The stability metric had lower coefficients of variation within a land-use than did either the dry or wet mean-weighted diameter, indicating this metric may be more sensitive to the effects of land use on soil structure than analysis of aggregate size distributions produced from one methodology (i.e., wet or dry sieving).

Significant differences in SOC were evident among the land uses in different aggregate size fractions. Both NT and FS generally had elevated concentrations in all fractions compared to CT from 0–5 cm with increases in POC and fine C each explaining 40–50% of concentration differences (Figure 3). With respect to our initial hypothesis this relatively equal contribution indicates that both stabilization of POC within microaggregates formed in stable macroaggregates [13,42], and flux of DOC into unsaturated micropores [15] might play a role in soil carbon accumulation. The effect of tillage could disrupt either of these processes and afforestation did not favor either process in the 0–5 cm surface soil.

The FS soil also had elevated concentrations of SOC from 5–15 cm compared to NT for macro- and microaggregates and compared to CT for microaggregates with the increases in POC explaining 35–60% of the concentration differences (Figure 4). For CT and NT there were no differences in SOC concentrations among the aggregate size fractions from 5–15 cm, which has been found previously at HSB [6,23,24,29]. Below 15 cm differences in aggregate stability were evident but significant differences in SOC fractions among the aggregate size fractions were not evident (Figure 5). This demonstrates that enhanced aggregate stability is not always linked to increased stabilization and storage of SOC.

Within the FS soil from 0–15 cm, only the wet-sieved fine fraction (<53 μm) had a significantly lower C concentration than the other fractions within each depth class (Figures 3 and 4), highlighting the importance of POC to increasing total SOC under afforestation. In both agroecosystems, large macroaggregates have significantly lower proportions of POC compared to small macroaggregates at all depths (Table 4; see also Table S9) and most of the difference in SOC concentration at 0–5 cm and 5–15 cm is accounted for by differences in macroaggregate C (Figure 6). As such, most of the C difference in the agroecosystem compared to the FS soil occurs at the macroaggregate level.

In previous research at HSB, Beare et al. [5] observed the highest C concentrations in the large NT microaggregates (106–250 μm) compared to the macroaggregates from 0–5 cm and speculated that this was due to formation of microaggregates around decomposing residues in stable macroaggregates, followed by their release upon macroaggregate breakdown, as originally formulated by Oades [12]. Work that followed this at HSB also found the highest C concentrations in a broader 53–250 μm

microaggregate class from 0–5 cm in NT [6], but this result was not found more recently [24] or in the present work, in which small macroaggregates were found to have similar C concentrations compared to microaggregates (Figure 3). These contrasting results at the same study site may be due to methodological differences in aggregate sieving. Beare et al. [5] used tension-wetted field-moist soil before wet-sieving and Bossuyt et al. [6] used capillary-wetted field moist soil before wet-sieving, whereas Arce Flores and Coleman [24] and the present work air-dried soils for storage and then capillary-wetted soils before wet-sieving. Beare and Bruce [43] clearly demonstrated that aggregate size distributions depend upon both pre-treatment and sieving methodologies.

Finally, aggregate hierarchy theory, which suggests that inter-microaggregate organic matter should increase C concentration of stable macroaggregates compared to microaggregates, is only evident in the 15–28 cm depth class (Figures 3–5). In all the 0–15 cm soils, large macroaggregates (>2000 μm) were relatively depleted in total SOC compared to the small macroaggregates (250–2000 μm) and microaggregates (53–250 μm). The method for isolating water-stable aggregates used repeatedly at HSB may not be forceful enough to isolate large macro-aggregates of higher C concentration in the upper 15 cm. If a more forceful fractionation procedure was used, such as sonication or slaking, then perhaps low C macroaggregates would be dispersed and their particles transferred to smaller fractions. Previous research has also shown that wet-sieving of pre-wetted soil does not support aggregate hierarchy theory, as was done in this study, while slaking dry soil does [11]. Also, aggregate hierarchy has been shown to be less apparent in soils dominated by 1:1 clays with high oxide contents like at HSB compared to soils dominated by 2:1 clays with low oxide contents [44,45].

Soil moving experiment

Significant differences in aggregate stability were evident after moving crushed and sieved (1 mm) soil from the CT plots back to each land use for both the 0–5 and 5–15 cm depth classes after one year (Table 2), suggesting that water-stable macroaggregates can be quickly formed under the different land uses but most quickly under FS, a result consistent with our initial hypothesis.

Similar rapid changes in soil structure were observed in a soil moving experiment in the tropics where larger blocks of soil were transferred from forest to pasture and vice-versa to study the effects of an invasive earthworm species [46]. The effects of different drying-wetting cycles, root densities, and levels of microbial activity are also known to affect soil structural development over time periods of days to a few years [47,48], so the rapid, differential development of aggregate stability under different land uses is consistent with previous and current observations.

Since forest soils are fungal-dominated systems, any observed difference in aggregate stability was hypothesized to result from more rapid binding by fungal hyphae under FS but no significant differences were observed in MBC between land uses (Table 3). The chloroform-incubation method for MBC is related to direct counts of fungal and bacterial cells [49] but the method has failed to efficiently extract fungal biomass from some forest soils [50]. On the other hand, previous work at HSB has shown a sensitivity of the chloroform-incubation method to treatments since both MBC and fungal hyphae lengths significantly decreased after application of fungicide to a mixed meadow that had been recently plowed at HSB [51]. Furthermore, the importance of fungal hyphae in stabilizing aggregates was demonstrated in the agroecosystems by treating sub-plots with fungicides and observing declines in

Table 4. The proportion of particulate organic carbon (POC) within aggregate fractions and whole soil (mean ±1 SD).

Land use	Depth	Large macroaggregates	Small macroaggregates	Microaggregates	Whole soil[1]
	--cm--	------------------------------% POC------------------------------			
CT	0–5	20.2±1.3[a]	31.0±4.2[a]	24.7±0.6[a]	22.6±0.9[a]
NT		21.3±6.3[a]	34.8±7.9[a]	33.7±7.3[a]	26.9±6.6[a]
FS		33.8±5.7[b]	41.8±4.6[a]	38.1±9.8[b]	35.5±5.1[b]
CT	5–15	13.0±2.4[ab]	18.4±1.9[a]	15.3±1.2[ab]	13.8±1.8[a]
NT		11.4±2.6[b]	14.9±2.6[a]	15.0±2.2[a]	12.1±2.3[a]
FS		19.7±6.4[a]	26.9±4.7[b]	18.8±2.0[b]	20.4±5.1[b]
CT	15–28	8.2±0.9[a]	10.4±1.2[a]	8.8±0.8[a]	8.2±0.5[a]
NT		10.1±2.0[a]	13.6±2.8[a]	12.5±1.9[a]	10.8±1.2[a]
FS		20.0±13.6[a]	13.7±3.5[a]	10.3±2.7[a]	15.1±6.7[a]

Horseshoe Bend, Athens, GA, USA, October 2007.
Different letters indicate statistical significance between treatment means within a depth class and aggregate class based on Tukey's HSD with $\alpha = 0.05$. See also Table S9.
[1]Calculated by summation of individual fractions.

aggregate size distributions [29]. As such, further efforts will be needed to conclusively demonstrate the role of fungal hyphae in aggregate formation in FS at HSB.

Along with aggregate stability, total SOC also increased significantly from 0–5 cm under FS from 9.2 g SOC kg^{-1} soil to 10.8 g SOC kg^{-1} soil after one year. This increase in SOC could, however, be explained by mixing of a relatively small portion of the *in situ* FS soil SOC (9–22%) with the SME soil. In contrast, however, the significant increase in POC from 1.8 g POC kg^{-1} soil to 3.3 g POC kg^{-1} soil, which explains 90% of the increase in SOC, cannot be explained by *in situ* soil mixing since POC in 0–5 cm *in situ* forest soil accounts for only 35% of total SOC. The observed increase, which is equal to an increased storage of approximately 1 Mg C ha^{-1}, could result partly from leaf litter inputs under FS that are estimated to be 2 Mg C ha^{-1} yr^{-1} (Markewtiz, 2009, unpublished data) or annual fine root production that has been reported to range globally from 0.3–8.2 Mg C ha^{-1} yr^{-1} [52]. Under a simulated no-till experiment with a Montana Hapludoll 14C labeled SOC in aggregates was derived predominantly from fine root litter [42,53].

Finally, decreases in SOC compared to the initial concentration from 5–15 cm under all three land uses showed that C mineralization exceeded new C inputs in this sub-surface layer. This enhanced decomposition was possibly triggered by the loss of macroaggregate structure from crushing the soil before its installation. Interestingly, decreases in POC concentrations explained less than half (11–36%) of the loss in total SOC under all three land uses, suggesting that fine C was more susceptible to decomposition after structural disruption. This reduction occurred even though the soil used in the moving experiment was CT soil in which macroaggregate protection of SOC had already been impacted during 30 years of continuous moldboard plowing.

Conclusions

Changes in aggregate size distributions after 30 years of differing land management were only clearly evident to a depth of 5 cm between conventional tillage (CT) and no-tillage (NT) systems. When adjacent land undergoing forest succession (FS) was added to the comparison, structural changes were evident to 28 cm. Changes in SOC concentration and composition occurred along with changes in structural stability to a depth of 15 cm,

consistent with a reduced capacity for tilled soil to physically protect organic matter from decomposition. Although differences in stability were evident from 15–28 cm, no significant difference in SOC concentration was observed among the land uses, indicating that increased aggregate stability is not always linked to an increase in SOC storage. SOC results lend support to two different models of how SOC becomes physically protected in systems lacking mechanical disturbance, one that highlights the importance of particulate organic carbon (POC) and the other dissolved organic carbon (DOC). Since POC contributed only about half of the differences in SOC between land uses, it is possible that micro-scale fluxes of DOC are also playing an important role in maintaining higher fine-C concentrations in soils with greater pore connectivity, such as the NT and FS soils in this study. The importance of macroaggregate stability for the protection of both POC and fine C was also demonstrated by the destruction of large macroaggregates in a soil moving experiment in which losses of C from 5–15 cm were observed under all land uses after 1 yr. The SME experiment also demonstrated that the recovery of water-stable macroaggregates can be rapid, since significant increases in stability under FS were observed in the experiment to 15 cm. Finally, a gain of 1 Mg C ha^{-1} was detected in the 0–5 cm soil layer under FS during the SME, 90% of which could be accounted for by a gain in POC. The stabilization of this POC may have been enhanced by recovery of water-stable aggregates under FS.

Supporting Information

Table S1 ANOVA results for Figure 1. ANOVA table reports tests of significance among Land Uses (conventional tillage, no tillage, forest succession) by aggregate attribute (Dry mean weighted diameter, Wet mean weighted diameter, Aggregate Stability {wet/dry}) and soil depth (0–5, 5–15, 15–28 cm).

Table S2 ANOVA results for Figure 2. ANOVA table reports tests of significance among Land Uses (conventional tillage, no tillage, forest succession) by aggregate size fraction (>2000, 250–2000, 53–250, and <53 μm) and soil depth (0–5, 5–15, 15–28 cm).

Table S3 1. ANOVA results for Figure 3 in soil depth 0–5 cm for land use. ANOVA table reports tests of significance among land uses (CT, NT, and FS) within a carbon fraction (Soil organic carbon, particulate organic carbon, fine carbon) and size class (>2000, 250–2000, 53–250, and <53 μm). 2. ANOVA results for Figure 3 in soil depth 0–5 cm for aggregate size class. ANOVA table reports tests of significance among size classes (2000, 250–2000, 53–250 and <53 μm) within a carbon fraction and land use. Note for POC the test is only for three size classes (2000, 250–2000, and 53–250 μm).

Table S4 1. ANOVA results for Figure 4 in soil depth 5–15 cm for land use. ANOVA table reports tests of significance among land uses (CT, NT, and FS) within a carbon fraction and size class. 2. ANOVA results for Figure 4 in soil depth 5–15 cm for aggregate size class. ANOVA table reports tests of significance among size classes (2000, 250–2000, 53–250 and <53 μm) within a carbon fraction and land use.

Table S5 1. ANOVA results for Figure 5 in soil depth 15–28 cm for land use. ANOVA table reports tests of significance among land uses (CT, NT, and FS) within a carbon fraction and size class. 2. ANOVA results for Figure 5 in soil depth 15–28 cm for aggregate size class. ANOVA table reports tests of significance among size classes (2000, 250–2000, 53–250 and <53 μm) within a carbon fraction and land use.

Table S6 ANOVA results for Figure 6. ANOVA table reports tests of significance among Land Uses (conventional tillage, no tillage, forest succession) by soil depth (0–5, 5–15, 15–28 cm) for the sum of C contents in all aggregate fractions.

Table S7 ANOVA results for Table 2. ANOVA table reports tests of significance among Land Uses (conventional tillage, no tillage, forest succession) by aggregate attribute (Dry mean weighted diameter, Wet mean weighted diameter, Aggregate Stability {wet/dry}) and soil depth (0–5, 5–15 cm).

Table S8 ANOVA results for Table 3. ANOVA table reports tests of significance among Land Uses (conventional tillage, no tillage, forest succession) by soil carbon component (soil organic carbon, particulate organic carbon, fine carbon, microbial biomass carbon) and soil depth (0–5, 5–15 cm).

Table S9 ANOVA results for Table 4. ANOVA table reports tests of significance among Land Uses (conventional tillage, no tillage, forest succession) by aggregate size fraction (>2000, 250–2000, 53–250, and <53 μm) and soil depth (0–5, 5–15, 15–28 cm).

Acknowledgments

We are indebted to the legacy of researchers who initiated and maintained the agroecosystem experiment at Horseshoe Bend. This includes but is not limited to: Drs. Gary Barrett, Eugene Odum, Dac Crossley, Benjamin Stinner, Peter Groffman, Michael Beare, and Shuijin Hu. Dr. David Radcliffe offered a critical review of the manuscript. Luke Worsham and Patrick Bussell provided valuable laboratory assistance. The UGA Graduate School and UGA Warnell School of Forestry and Natural Resources provided research support.

Author Contributions

Conceived and designed the experiments: SD DM PH DC. Performed the experiments: SD DM. Analyzed the data: SD DM. Contributed reagents/materials/analysis tools: DM PH DC. Wrote the paper: SD DM PH DC. Historical data: PH DC.

References

1. Mikutta R, Kleber M, Torn MS, Jahn R (2006) Stabilization of soil organic matter: association with minerals or chemical recalcitrance? Biogeochemistry 77: 25–56.
2. von Lutzow M, Kogel-Knabner I, Ekschmitt K, Matzner E, Guggenberger G, et al. (2006) Stabilization of organic matter in temperate soils: mechanisms and their relevance under different soil conditions - a review. European Journal of Soil Science 57: 426–445.
3. Six J, Conant RT, Paul EA, Paustian K (2002) Stabilization mechanisms of soil organic matter: Implications for C-saturation of soils. Plant and Soil 241: 155–176.
4. Six J, Elliott ET, Paustian K, Doran JW (1998) Aggregation and soil organic matter accumulation in cultivated and native grassland soils. Soil Science Society of America Journal 62: 1367–1377.
5. Beare MH, Cabrera ML, Hendrix PF, Coleman DC (1994) Aggregate-protected and unprotected organic-matter pools in conventional-tillage and no-tillage soils. Soil Science Society of America Journal 58: 787–795.
6. Bossuyt H, Six J, Hendrix PF (2002) Aggregate-protected carbon in no-tillage and conventional tillage agroecosystems using carbon-14 labeled plant residue. Soil Science Society of America Journal 66: 1965–1973.
7. Devine S, Markewitz D, Hendrix P, Coleman DC (2011) Soil carbon change through two meters during forest succession alongside a 30-yr agroecosystem experiment. Forest Science 57: 36–50.
8. Baker JM, Ochsner TE, Venterea RT, Griffis TJ (2007) Tillage and soil carbon sequestration – what do we really know? Agriculture, Ecosystems and Environment 118: 1–5.
9. Horn R, Taubner H, Wuttke M, Baumgartl T (1994) Soil physical-properties related to soil-structure. Soil & Tillage Research 30: 187–216.
10. Tisdall JM, Oades JM (1982) Organic matter and water-stable aggregates in soils. Journal of Soil Science 33: 141–163.
11. Elliott ET (1986) Aggregate structure and carbon, nitrogen, and phosphorus in native and cultivated soils. Soil Science Society of America Journal 50: 627–633.
12. Oades JM (1984) Soil organic-matter and structural stability -Mechanisms and implications for management. Plant and Soil 76: 319–337.

13. Six J, Elliott ET, Paustian K (2000) Soil macroaggregate turnover and microaggregate formation: a mechanism for C sequestration under no-tillage agriculture. Soil Biology & Biochemistry 32: 2099–2103.
14. Six J, Callewaert P, Lenders S, De Gryze S, Morris SJ, et al. (2002) Measuring and understanding carbon storage in afforested soils by physical fractionation. Soil Science Society of America Journal 66: 1981–1987.
15. Smucker AJM, Park EJ, Dorner J, Horn R (2007) Soil micropore development and contributions to soluble carbon transport within macroaggregates. Vadose Zone Journal 6: 282–290.
16. Chan KY (2001) Soil particulate organic carbon under different land use and management. Soil Use and Management 17: 217–221.
17. Tiessen H, Stewart JWB (1983) Particle-size fractions and their use in studies of soil organic matter: II. cultivation effects on organic-matter composition in size fractions Soil Science Society of America Journal 47: 509–514.
18. Causarano HJ, Franzluebbers AJ, Shaw JN, Reeves DW, Raper RL, et al. (2008) Soil organic carbon fractions and aggregation in the Southern Piedmont and Coastal Plain. Soil Science Society of America Journal 72: 221–230.
19. Franzluebbers AJ, Arshad MA (1997) Particulate organic carbon content and potential mineralization as affected by tillage and texture. Soil Science Society of America Journal 61: 1382–1386.
20. Skjemstad JO, Swift RS, McGowan JA (2006) Comparison of the particulate organic carbon and permanganate oxidation methods for estimating labile soil organic carbon. Australian Journal of Soil Research 44: 255–263.
21. Wander MM, Traina SJ, Stinner BR, Peters SE (1994) Organic and conventional management effects on biologically-active soil organic-matter pools Soil Science Society of America Journal 58: 1130–1139.
22. Cambardella CA, Elliott ET (1992) Particulate soil organic-matter changes across a grassland cultivation sequence. Soil Science Society of America Journal 56: 777–783.
23. Beare MH, Hendrix PF, Coleman DC (1994) Water-stable aggregates and organic-matter fractions in conventional-tillage and no-tillage soils. Soil Science Society of America Journal 58: 777–786.
24. Arce Flores S, Coleman DC (2006) Comparing water-stable aggregate distributions, organic matter fractions, and carbon turnover using ¹³C natural

abundance in conventional and no tillage soils. Master's Thesis. Athens: Institute of Ecology, University of Georgia.

25. Hendrix PF (1997) Long-term patterns of plant production and soil carbon dynamics in a Georgia Piedmont agroecosystem. In: Paul EA, Paustian K, Elliott ET, Cole CV, editors. Soil Organic Matter in Temperate Agroecosystems: Long-Term Experiments in North America. Boca Raton, FL: CRC Press, Inc. pp. 235–245.

26. Barrett GW (1968) Effects of an acute insecticide stress on a semi-enclosed grassland ecosystem. Ecology 49: 1019–1035.

27. Bakelaar RG, Odum EP (1978) Community and population level responses to fertilization in an old-field ecosystem. Ecology 59: 660–665.

28. Odum EP, Pomeroy SE, Dickinson JC III, Hutcheson K (1973) The effect of a late winter litter burn on the composition, productivity, and diversity of a 4-year old fallow-field in Georgia. Proc Annu Tall Timbers Fire Ecol Conf: 399–412.

29. Beare MH, Hus S, Coleman DC, Hendrix PF (1997) Influences of mycelial fungi on soil aggregation and organic matter storage in conventional and no-tillage soils. Applied Soil Ecology 5: 211–219.

30. Coleman DC, Hunter MD, Hendrix PF, Crossley DAJ, Arce-Flores S, et al. (2009) Long-term consequences of biological and biogeochemical changes in the Horseshoe Bend LTREB agroecosystem, Athens, Ga. In: Bohlen P, editor. Agroecosystem management for ecological, economic and social sustainability. New York: Taylor and Francis.

31. Coleman D, Hunter M, Hendrix P, Crossley D, Simmons B, et al. (2006) Long-term consequences of biochemical and biogeochemical changes in the Horseshoe Bend agroecosystem, Athens, GA. European Journal of Soil Biology 42: S79–S84.

32. Lachnicht SL, Hendrix PF, Potter RL, Coleman DC, Crossley DA (2004) Winter decomposition of transgenic cotton residue in conventional-till and no-till systems. Applied Soil Ecology 27: 135–142.

33. Nimmo JR, Perkins KS (2002) Aggregate stability and size distribution. In: Dane JH, Topp GC, editors. Methods of Soil Analysis: Part 4–Physical Methods. Madison, Wisconsin, , USA: Soil Science Society of America, Inc. pp. 317–328.

34. Kemper WD, Chepil WS (1965) Size distribution of aggregates. In: Black CA, editor. Methods of soil analysis. Madison, WI: ASA. pp. 499–510.

35. Yoder RE (1936) A direct method of aggregate analysis of soils and a study of the physical nature of erosion losses. Journal of the American Society of Agronomy 28: 337–351.

36. Elliott ET, Palm CA, Reuss DE, Monz CA (1991) Organic-matter contained in soil aggregates from a tropical chronosequence - correction for sand and light fraction. Agriculture Ecosystems & Environment 34: 443–451.

37. Balser TC, Firestone MK (2005) Linking microbial community composition and soil processes in a California annual grassland and mixed-conifer forest. Biogeochemistry 73: 395–415.

38. Mack MC, D'Antonio CM (2003) Exotic grasses alter controls over soil nitrogen dynamics in a Hawaiian woodland. Ecological Applications 13: 154–166.

39. Neill C (1992) Comparison of soil coring and ingrowth methods for measuring belowground production. Ecology Letters 73: 1918–1921.

40. Vance ED, Brookes PC, Jenkinson DS (1987) An extraction method for measuring soil microbial biomass. Soil Biology & Biochemistry 19: 703–707.

41. Wu J, Joergensen RG, Pommerening B, Chaussod R, Brookes PC (1990) Measurement of soil microbial biomass C by fumigation extraction - an automated procedure Soil Biology & Biochemistry 22: 1167–1169.

42. Gale WJ, Cambradella CA, Bailey TB (2000) Root-derived carbon and the formation and stabilization of aggregates. Soil Science Society American Journal 64: 201–207.

43. Beare MH, Bruce RR (1993) A comparison of methods for measuring water-stable aggregates - implications for determining environmental-effects on soil structure. Geoderma 56: 87–104.

44. Six J, Paustian K, Elliott ET, Combrink C (2000) Soil structure and organic matter: I. Distribution of aggregate-size classes and aggregate-associated carbon. Soil Science Society of America Journal 64: 681–689.

45. Oades JM, Waters AG (1991) Aggregate hierarchy in soils. Australian Journal of Soil Research 29: 815–828.

46. Barros E, Curmi P, Hallaire V, Chauvel A, Lavelle P (2001) The role of macrofauna in the transformation and reversibility of soil structure of an oxisol in the process of forest to pasture conversion. Geoderma 100: 193–213.

47. Horn R, Dexter AR (1989) Dynamics of soil aggregation in an irrigated desert loess. Soil & Tillage Research 13: 253–266.

48. Feeney DS, Crawford JW, Daniell T, Hallett PD, Nunan N, et al. (2006) Three-dimensional microorganization of the soil-root-microbe system. Microbial Ecology 52: 151–158.

49. Jenkinson DS, Powlson DS, Wedderburn RWM (1976) Effects of biocidal treatments on metabolism in soil. 3. Relationship between soil biovolume, measured by optical microscopy, and flush of decomposition caused by fumigation Soil Biology & Biochemistry 8: 189–202.

50. Ingham ER, Griffiths RP, Cromack K, Entry JA (1991) Comparison of direct vs fumigation incubation microbial biomass estimates from ectomycorrhizal mat and non-mat soils. Soil Biology & Biochemistry 23: 465–471.

51. Hu S, Coleman DC, Hendrix PF, Beare MH (1995) Biotic manipulation effects on soil carbohydrates and microbial biomass in a cultivated soil. Soil Biology & Biochemistry 27: 1127–1135.

52. Nadelhoffer KJ, Raich JW (1992) Fine root production estimates and belowground carbon allocation in forest ecosystems. Ecology 73: 1139–1147.

53. Gale WJ, Cambradella CA (2000) Carbon dynamics of surface residue- and root-derived organic matter under simulated no-till. Soil Science Society American Journal 64: 190–195.

17

Informal Urban Green-Space: Comparison of Quantity and Characteristics in Brisbane, Australia and Sapporo, Japan

Christoph D. D. Rupprecht[1,2]*, Jason A. Byrne[1,2]

1 Environmental Futures Research Institute, Griffith University, Nathan, Queensland, Australia, 2 Griffith School of Environment, Griffith University, Griffith University, Queensland, Australia

Abstract

Informal urban green-space (IGS) such as vacant lots, brownfields and street or railway verges is receiving growing attention from urban scholars. Research has shown IGS can provide recreational space for residents and habitat for flora and fauna, yet we know little about the quantity, spatial distribution, vegetation structure or accessibility of IGS. We also lack a commonly accepted definition of IGS and a method that can be used for its rapid quantitative assessment. This paper advances a definition and typology of IGS that has potential for global application. Based on this definition, IGS land use percentage in central Brisbane, Australia and Sapporo, Japan was systematically surveyed in a 10×10 km grid containing 121 sampling sites of 2,500 m^2 per city, drawing on data recorded in the field and aerial photography. Spatial distribution, vegetation structure and accessibility of IGS were also analyzed. We found approximately 6.3% of the surveyed urban area in Brisbane and 4.8% in Sapporo consisted of IGS, a non-significant difference. The street verge IGS type (80.4% of all IGS) dominated in Brisbane, while lots (42.2%) and gaps (19.2%) were the two largest IGS types in Sapporo. IGS was widely distributed throughout both survey areas. Vegetation structure showed higher tree cover in Brisbane, but higher herb cover in Sapporo. In both cities over 80% of IGS was accessible or partly accessible. The amount of IGS we found suggests it could play a more important role than previously assumed for residents' recreation and nature experience as well as for fauna and flora, because it substantially increased the amount of potentially available greenspace in addition to parks and conservation greenspace. We argue that IGS has potential for recreation and conservation, but poses some challenges to urban planning. To address these challenges, we propose some directions for future research.

Editor: Francisco Moreira, Institute of Agronomy, University of Lisbon, Portugal

Funding: This work was supported by scholarships and research funding from Griffith University (http://www.griffith.edu.au). The funders had no role in study design, data collection and analysis, decision to publish, or preparation of the manuscript.

Competing Interests: The authors have declared that no competing interests exist.

* E-mail: christoph.rupprecht@griffithuni.edu.au

Introduction

Dunn et al. argue that global conservation efforts depend on the interest people have in nature conservation, an interest formed largely through experiencing nature within the cities that people inhabit[1]. Informal urban greenspace (IGS) such as vacant lots, brownfields and street or railway verges comprise one part of this urban nature. Research has found that IGS can play a role in exposing city dwellers to nature – as recreational space for residents and an alternative to traditional greenspace (e.g. parks and playing fields)[2–4], and as habitat for flora and fauna[5–7]. But we presently lack knowledge about the estimated total quantity of IGS in our cities – a key issue, because the quantity of space likely has a strong influence on its potential for recreation and conservation. Questions yet to be answered include: what proportion the different types of IGS contribute to the total amount of IGS in a city, and how does IGS quantity differ between cities? We know little about the spatial distribution (within a city or in different geographical settings) of IGS, its vegetation structure, or its potential accessibility, which are again important factors determining its potential for recreation and conservation. Compounding this problem, scholars presently lack a shared or agreed definition of these taken for granted socio-ecological spaces (though we propose such a definition below). Such a definition is necessary to ensure that researchers are talking about the same concept and vital to creating an integrated research agenda. Finally, we lack a reliable, comprehensive rapid assessment method that can be applied in different geographical contexts and is useful for estimating IGS quantity as a first step in urban planning initiatives to 'green' cities. We take up these tasks in this paper.

This paper reports the results of a study that asked the following four research questions: (1) how does the land use proportion of total IGS and individual IGS subtypes differ between urban core areas in two cities? (2) how do the characteristics (distribution, vegetation structure, accessibility) of IGS differ between urban core areas in two cities? (3) does distance from the city center influence IGS quantity, and (4) how accurate is the IGS land use proportion survey method employed for estimating potential IGS quantity? This study contributes new knowledge in two ways. Our study has for the first time examined how much land likely consists of a wide variety of IGS types in an urban core. Second, it represents the first comprehensive examination comparing IGS

Table 1. Informal urban greenspace typology.

IGS	Examples	Description	Management	Form	Substrates
Street verges	Roadside verges, roundabouts, tree rings, informal trails and footpaths	Vegetated area within 5 m from street not in another IGS category; mostly maintained to prevent high and dense vegetation growth other than street trees; public access unrestricted, use restricted.	Regular vegetation removal (>= once per month); governmental and private stewardship	Small: <100 m², linear	Soil, gravel, stone, concrete, asphalt
Lots	Vacant lots, abandoned lots	Vegetated lot presently not used for residential or commercial purposes; if maintained, usually vegetation removed to ground cover; public access and use restricted.	Irregular veg. removal, medium to long removal intervals; private stewardship	Small-medium: <1 ha, block	Soil, gravel, bricks
Gap	Gap between walls or fences	Vegetated area between two walls, fences or at their base; maintenance can be absent or intense; public access and use often restricted.	Irregular veg. removal; variable removal intervals; private stewardship	Small: <100 m², linear	Soil, gravel
Railway	Rail tracks, verges, stations	Vegetated area within 10 m adjacent to railway tracks not in another IGS category; usually herbicide maintenance to prevent vegetation encroachment on tracks; public access and use mostly restricted.	Regular veg. removal (monthly to yearly); corporate or governmental stewardship	Medium-large: >1 ha, linear	Soil, gravel, stone
Brownfields	Landfill, post-use factory grounds, industrial park	Vegetated area presently not used for industrial or commercial purposes; usually no or very infrequent vegetation removal and maintenance; public access and use mostly restricted.	Irregular veg. removal, long removal intervals; corporate and governmental stewardship	Medium-large: >1 ha, block	Soil, gravel, concrete, asphalt
Waterside	Rivers, canals, water reservoir edges	Vegetated area within 10 m of water body not in another IGS category; occasional removal of vegetation to maintain flood protection and structural integrity; public access and use often possible with some restrictions.	Irregular veg. removal, long removal intervals; governmental stewardship	Small-large: > 10 m² to >1 ha, linear	Soil, stone, concrete, bricks
Structural	Walls, fences, roofs, buildings	Overgrown human artifacts; often vertical; occasional removal of vegetation to maintain structural integrity; public access and use mostly restricted.	Irregular veg. removal, medium to long removal intervals; varying stewardship	Small: <100 m², block	Soil, stone, gravel, wood, metal
Microsite	Vegetation in cracks or holes	Vegetation assemblages in cracks, may develop into structural IGS; maintenance can be absent or intense	Irregular veg. removal, variable removal intervals; variable stewardship	Very small: <1 m², point	Deposits, soil, stone, concrete
Power line	Power line rights of way	Vegetated corridor under and within 25 m of power lines not in another IGS category; vegetation removed periodically to prevent high growth; public access and use mostly unrestricted.	Regular veg. removal (less than yearly); utility or governmental stewardship	Medium-large: >1 ha, linear	Soil

quantity and type within the urban core area of two cities, potentially allowing scholars to examine IGS composition and quantity in other geographical settings.

Methods and Data Collection

Informal urban greenspace (IGS) definition and typology

Cities consist of a patchwork of different spaces, from densely built areas to green space such as urban forests or parklands. But besides these exist also more ambiguous, 'liminal' vegetated spaces, that Jorgensen and Tylecote refer to as 'ambivalent landscapes' [8]. This heterogeneous group of vacant lots, railway verges, utility corridors and waterway embankments is often overgrown with spontaneous vegetation [7], and is managed only to a limited extent (e.g., vegetation removal to protect power lines from overgrowth). They share ambiguities in land tenure, conservation status, maintenance regimes, use, regulation and legitimacy [9], and are best characterized as liminal spaces. Even street verges and suburban lawns can be liminal. While they may have been

planted originally, they are oftentimes a mix of intentionally planted and opportunistic species. Their maintenance level is similar to that of backyard gardens, and depends upon many factors, such as feelings of ownership, cultural beliefs, age, and level of neighbor's surveillance [10,11]. The concept of liminality is derived from several disciplines, but is well-established in the urban geography literature [12,13]. It refers to a state of 'betweenness', intermediacy, or ambiguity of being – the 'indeterminacy of loose space', as Franck and Stevens call it [14]. Liminal spaces are 'at the margins', characterized by emergence and flux, fluidity and malleability, and are neither segregated nor uncontained[15].

This liminality presents a challenge for quantitatively surveying such spaces, which we aim to address by proposing a provisional, non-exclusive definition and typology of a form of liminal green spaces we term 'informal urban green space' (IGS). For the purpose of this study, we have defined IGS as an explicitly socio-ecological entity, rather than a solely biological or cultural object. IGS consists of any urban space with a history of strong

Figure 1. Photos of informal greenspace types following typology in Table 1. Street verges: A) Spontaneous herbal vegetation on sidewalk (Sapporo, Japan), B) Unused, highly maintained nature strip with mix of planted and spontaneous vegetation (Brisbane, Australia), C) Spont. herbal vegetation between street and sidewalk (Sapporo). Lots: D) Former residential vacant lot, remains of garden structure still present (Sapporo), E) Long-term vacant lot in residential area (Brisbane), F) Former residential, long-term vacant lot, "no trespassing" sign (Nagoya, Japan). Gap: G) Space with spontaneous herbal vegetation between two buildings, informal storage use (Sapporo), H) Gap with rudimentarily blocked access in front of building (Sapporo), I) Vegetated gap in sealed surface around fence in industrial zone (Brisbane). Railway: J) Annual grass in verge between rail track and street (Sapporo), K) Vegetated cliff next to rail track (Brisbane), L) Vegetated verge and inter-track space (Sapporo). Brownfield: M) Publicly-owned, large vacant tract with grassland and single trees (Sapporo), N) Old city quarter, overgrown former ceramics factory lot (Tokoname, Japan), O) Vegetated area on municipal land for disaster preparation material storage in urban fringe (Sapporo).

anthropogenic disturbance that is covered at least partly with non-remnant, spontaneous vegetation [5–7] and has a history of strong anthropogenic disturbance. It is neither formally recognized by governing institutions or property owners as greenspace designated for agriculture, forestry, gardening, recreation (either as parks or gardens) or for environmental protection (the typical purposes of most greenspace). Nor is the vegetation contained therein managed for any of these purposes. Any use for recreation is typically informal and transitional (e.g. unsanctioned verge gardening).

IGSs differ in their management (e.g. access, vegetation removal, stewardship), land use and site history, their scale and shape, soil characteristics and local urban context. For example, a small brownfield may be similar to a vacant lot in appearance and size, but their different land use history, vegetation removal periods, and urban context distinguish them. We identified nine

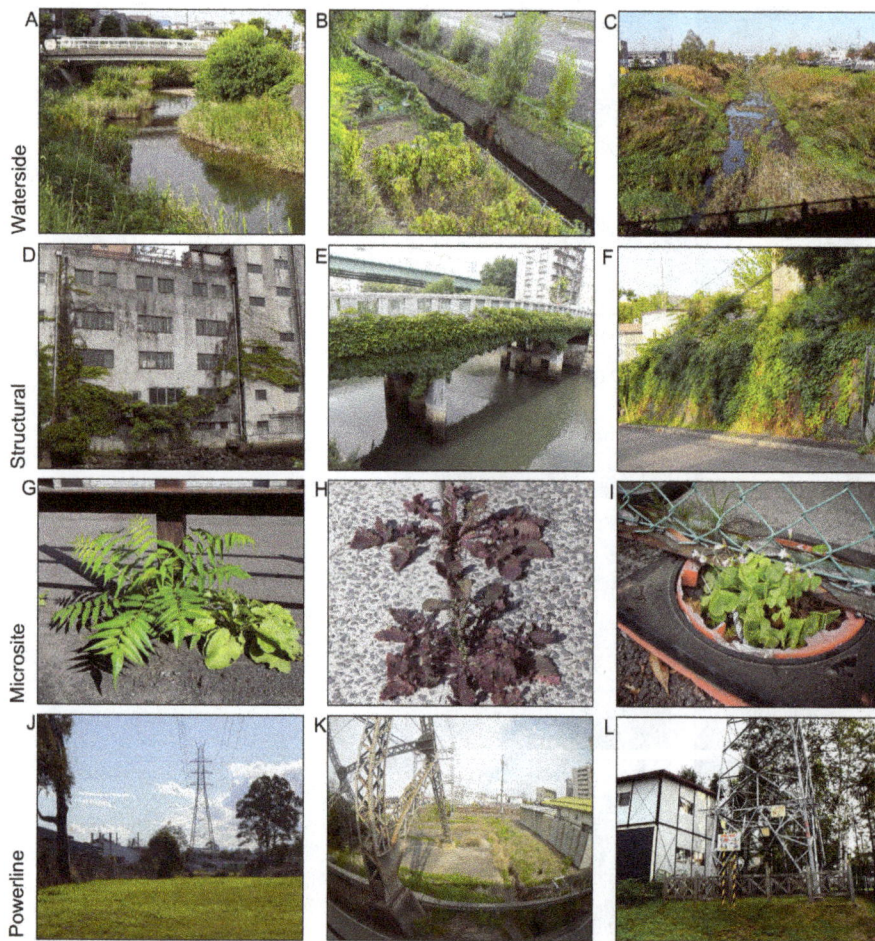

Figure 2. Photos of informal greenspace types following typology in Table 1 (cont.). Waterside: A) Vegetation on soil deposits in concreted river bed (Nagoya), B) Spontaneous vegetation and informal agricultural use of flood-protection stream banks (Sapporo), C) Spontaneously vegetated anthropogenic river banks (Sapporo). Structural: D) Creeping vines on industrial building (Nagoya), E) Overgrown bridge (Nagoya), F) Concrete soil retention wall completely covered in ivy (Sapporo). Microsite: G) Vegetated crack in asphalt on parking lot (Sapporo), H) Vegetation between two sidewalk plates (Brisbane), I) Plant growing out of degraded traffic cone remains (Nagoya). Powerline: J) Powerline reserve in industrial zone (Brisbane), K) Vegetated area around powerline pylon (near Osaka, Japan), L) Vegetated area around powerline pylon (Sapporo).

Table 2. Comparison of cities containing the survey areas.

Characteristics	City of Brisbane (LGA)	Sapporo
Founded	1824, city status 1902	1868, city status 1922
Population	1,089,743 (2011) (2031: 1,27 million)	1,936,189 (2013) (2030: 1,87 million)
Area	1,338 km^2	1,121.12 km^2
Pop. density	814/km^2	1,699/km^2
Peak density	>5,000/km^2	>8,000/km^2
Climate	Humid subtropical (Cfa)	Humid continental (Dfa)
Industry	Tourism, resources, retail, financial services, agriculture hub, education	Tourism, retail, IT, agriculture hub, resources, education
Greenspace	Local parks: 3,290 ha (32 m^2/capita)	Parks: 2,345 ha (12.3 m^2/capita)
	All parks: 11840 ha (115 m^2/capita)	All greenspace: 5,508 ha (28.9 m^2/capita)
Park area planned	40 m^2/capita, minimum 20 m^2/capita	"No greenspace loss, park renovation"

Sources: [46–51].

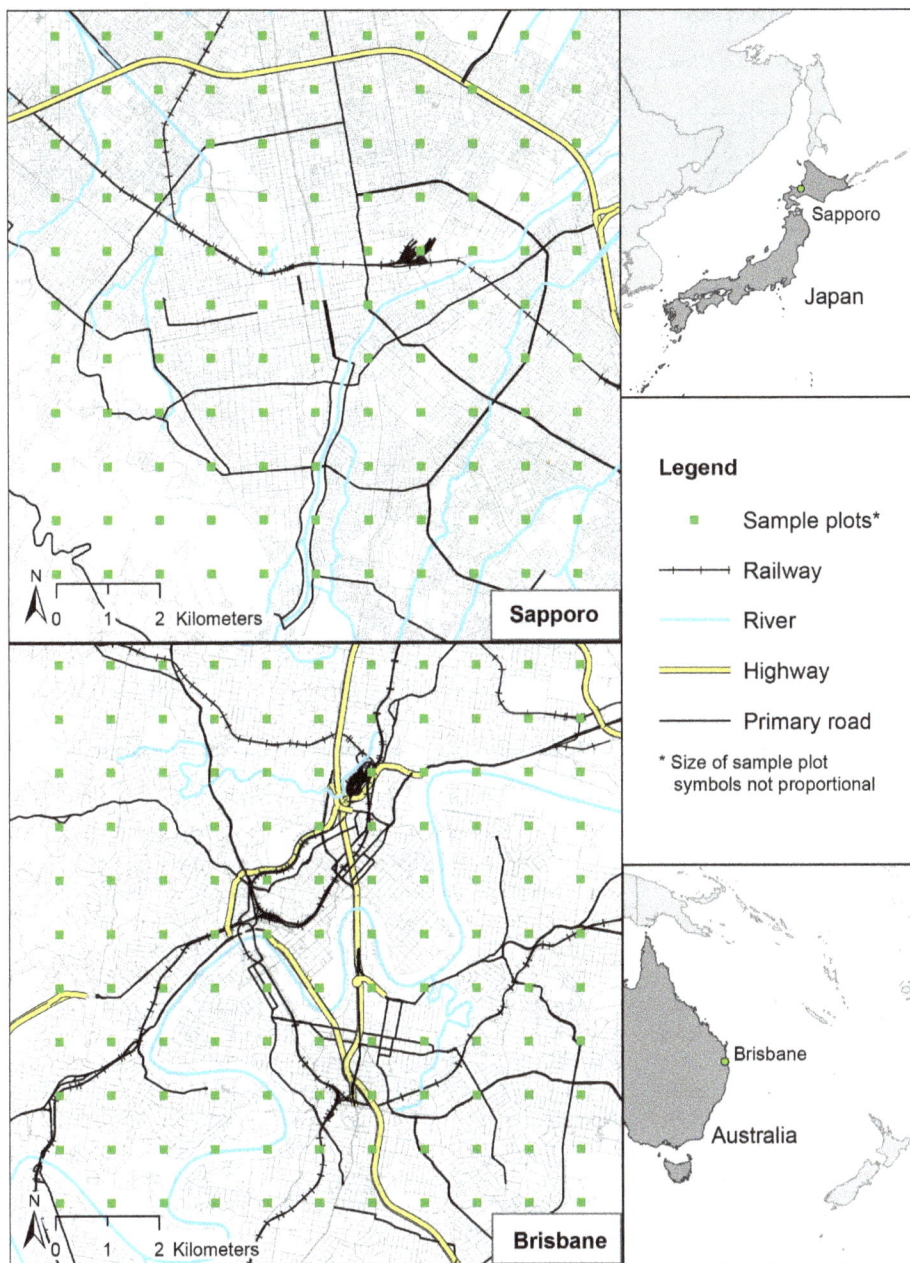

Figure 3. Study locations including sampling sites: Brisbane, Australia (left) and Sapporo, Japan (right).

potentially different subtypes of IGS: street verge, lot, gap, railway, brownfield, waterside, structural, microsite and power line IGS (Table 1, Figure 1, Figure 2). The subtypes are not exclusive, thus an IGS area may be categorized as multiple subtypes (e.g. street verge and gap). Because this typology recognizes the variety of non-traditional greenspace, it provides a better basis to analyze the implications of IGS for planning and conservation than broad terms such as "wasteland" or "derelict land". The distinction between IGS and formal greenspace is not binary, but rather characterized by a gradient of informality: formal recognition as recreational space by the owner provides a criterion to identify a local-government owned vacant lot covered with mowed lawn as IGS, but a low maintenance "wild" private garden as formal greenspace. Secondary-growth urban forests (rather than e.g. small

patches of woody vegetation on a brownfield) represent a borderline case, but in most cases such forests are recognized for silvicultural or recreational value and thus excluded from the definition of IGS used in this article. For the IGS area survey we only recorded IGS larger than one square meter and therefore also excluded microsite IGS.

Study locations

Brisbane (Queensland, Australia) and Sapporo (Hokkaidō, Japan) were chosen as case study cities, because research that examines IGS outside of Europe and the USA is relatively scarce. The two case study cities have similarities and differences that lend them well to comparison (Table 2); they thus provide excellent opportunities for a cross-cultural research. Both cities are relatively

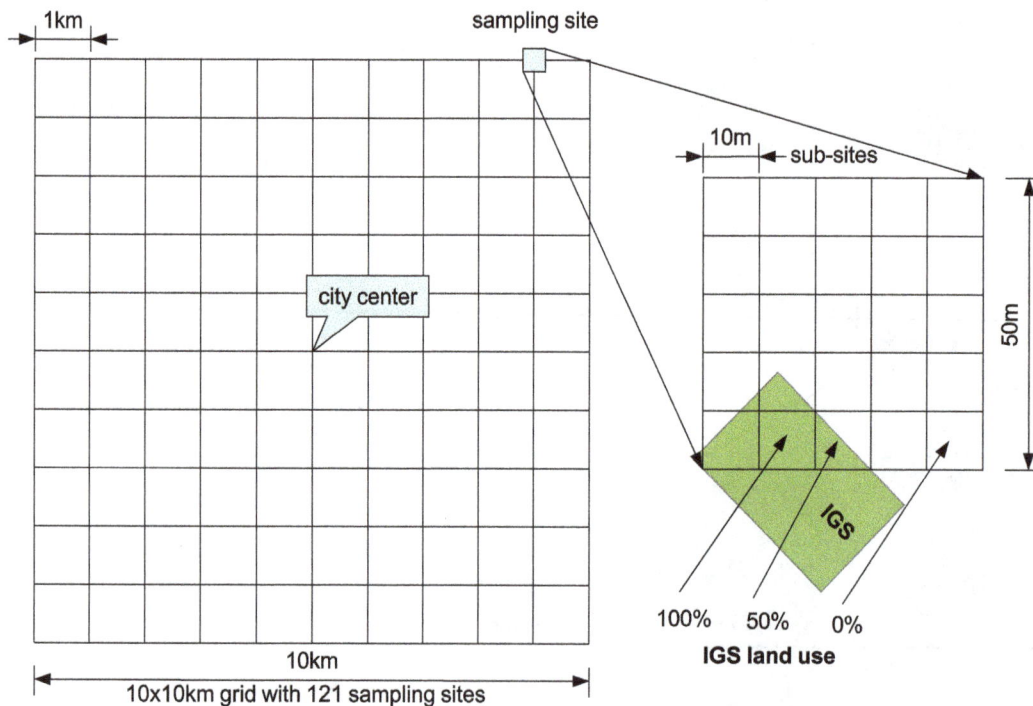

Figure 4. Research design: sampling sites on gridline intersections, with sub-sites and example of IGS percentage calculation.

young (being founded in the 19th century) and they saw most of their growth during the 20th century, especially in the post-second world-war period. Their close geographical size is complemented by a similar urban morphology. Both cities are built around a dense central business district, are situated near to the coast and upland regions, and are intersected by a central river (Figure 3). These similarities contrast with differences in population density, population growth forecasts, and available parks and other greenspaces.

While Sapporo has seen rapid growth throughout the second half of the 20th century and now has a population of about 1.9 million, its population is now stagnating and is predicted to decline in the future. In contrast, Brisbane has a population of around 1 million but is still growing quickly (Table 2). This difference in population development is of particular interest as both expanding cities [16] and shrinking cities [17] have important impacts on urban greenspace provision.

In both cities, formal greenspace consists of networks of over 2,000 public parks, most of them small local parks. Brisbane has 3,290 ha of local parkland (32 m^2/capita), whereas Sapporo has 2,345 ha (12.3 m^2/capita) (Table 2). All parks in Brisbane form an area of 11,840 ha (115 m^2/capita), while all greenspace in Sapporo forms an area of 5,508 ha (28.9 m^2/capita). These areas include forested hillsides in the southwest of both cities, providing recreational benefits to residents and habitat to wildlife.

Research design

To be able to measure the proportion of land use consisting of IGS and compare it between the survey areas in Brisbane and Sapporo, we used a systematic grid sampling design[18,19]. We placed 121 sampling sites of 50 m by 50 m each on the intersecting lines of a 10 km by 10 km grid, centered on the city centers (Figure 4, File S1 Sapporo sampling sites, File S2 Brisbane sampling sites). Surveying only the central area of a city rather

than the whole allowed us to assess a large area despite limited resources, while still covering most of the densely populated areas where access to greenspace may be difficult[16] (Figure 5). This kind of rapid assessment technique can provide an efficient estimate of land use proportions, and can later be followed up with a more detailed, finer resolution assessment if necessary. The General Post Office (Brisbane) and Sapporo City Office (city hall) were chosen as city centers, following common practice in Australia[20,21] and Japan[22,23]. There is no internationally accepted method for determining a city center. There was a one-kilometer distance between any two adjacent sampling sites. Each 2,500 m^2 sampling site was divided into 25 sub-sites of 10 m by 10 m for a total of 3,025 sub-sites to facilitate land use assessment (total surveyed area 302,500 m^2 or 0.299% of the square enclosing all sampling sites (101,002,500 m^2)).

Land use assessment

We used a three-step process to measure the percentage of IGS and other land uses. First, we created a geographic information system (GIS) layer with site locations and projected it on publicly available high-resolution aerial photography data (Google Earth in Brisbane, see http://www.google.com/earth/; Microsoft Bing Maps in Sapporo, see http://www.bing.com/maps/). Second, we surveyed land use type in the field for each sub-site and recorded land use percentage for small land use areas assessable on the ground. This was conducted using a measuring tape and visual estimation for inaccessible site parts (physically or marked with entry-forbidden signs). Sub-sites (25 sub-sites 10×10 m each) for all 121 sampling sites per city were created, as smaller sites allow both easier tape measurement and easier visual estimation. Only land use of one square meter or more was recorded, smaller areas were included in adjacent land use. Land use types, changes in land use since production of the aerial photography, building and land use borders were added to printed field data entry sheets

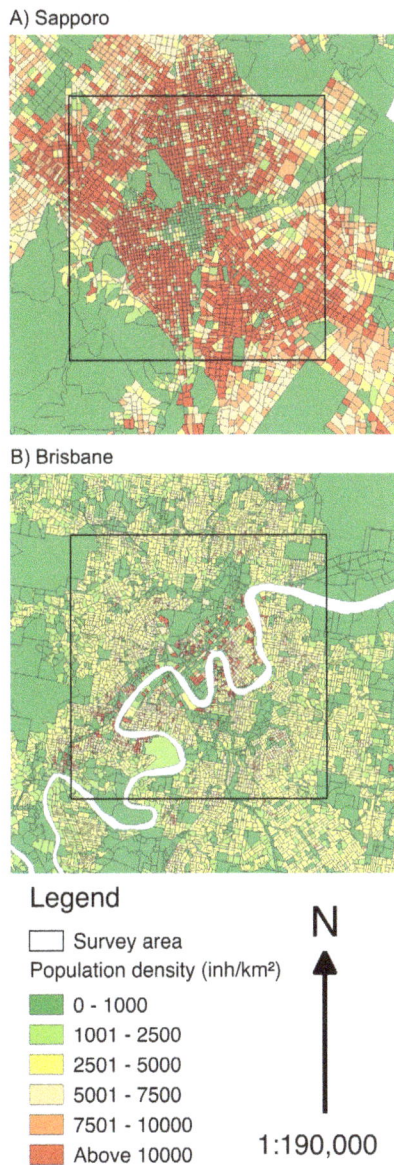

Figure 5. Survey areas and population density of study locations: A) Sapporo, B) Brisbane.

Legend
- Survey area
- Population density (inh/km²)
- 0 - 1000
- 1001 - 2500
- 2501 - 5000
- 5001 - 7500
- 7501 - 10000
- Above 10000

N

1:190,000

method was designed to work without direct site access, and was conducted on publicly accessible land only. Sites or parts of sites located on private, military or access-restricted conservation land were surveyed visually from publicly accessibly land if possible (likewise for vegetation structure and accessibility), or surveyed via aerial photography only (see above). Data collection for this paper did not involve endangered or protected species.

For the final step, we individually estimated percentages on paper in each 10×10 m sub-site for each land use category present in the sub-site. One percent of land use in each of those sub-sites equals one square meter. For complex sites, additional support lines were drawn across the aerial photo, dividing each sub-site into four 5×5 m sites that each represented 25%. Where necessary, these were further divided into 12.5% or 6.25% blocks. To improve the quality of percentage estimates we used non-GIS-compatible high-resolution aerial photography by NearMap (Brisbane, see https://www.nearmap.com/), the photographic collection produced in the field, and Google Street View (for orientation purposes, see https://www.google.com/maps/views/streetview). In one case a sampling site had to be revisited to re-assess a present IGS. Automating this percentage calculation using software was considered but deemed not feasible for the limited number of sampling sites, as the variable quality and nature of the aerial photography (e.g. perspective distortions of higher buildings) used would have required sophisticated software and labor-intensive checks. Future studies of a larger number of sampling sites or cities should, however, consider the use of such software tools.

Vegetation structure assessment

For all IGS types, we visually estimated (with the help of measuring tape) vegetation cover percentage of four different vegetation strata (Figure 6): 1) tree layer cover (all vegetation > 2 m height), 2) bush layer cover (all vegetation between 1 m–2 m height), 3) herb layer cover (all vegetation under 1 m, if vegetation between 30 cm and 1 m height is present), and 4) ground cover (all vegetation under 30 cm, if vegetation between 30 cm and 1 m is not present). Tree layer and herb layer cover are independent, while herb and ground cover are mutually exclusive and thus cannot exceed 100% combined coverage in one IGS. While IGS was defined as vegetated space, ground cover percentage does not have to reach 100% if ground vegetation cover is patchy and includes bare ground (e.g. 10 m² IGS area covered to 50% by herbal layer vegetation, 30% ground layer vegetation, and 20% patchy 1:1 ground layer vegetation/bare ground mix was recorded as 50% herb cover, 40% ground cover). Vegetation cover height of (possibly vertical) structural IGS was measured in a 90° angle from the substrate.

Accessibility assessment

We assessed how accessible IGS areas were on a three-level scale derived from prior research into vacant lot accessibility[26], based on the amount of physical or psychological effort necessary to overcome access barriers. As access barriers we included physical barriers such as fences, walls, chains or barbed wire, as well as symbolic barriers such as signs (e.g. "private ground", "entry forbidden", "no child play", Figure 7). IGS were classified as: accessible, if there were no barriers to access, or very low barriers that required only minimal effort to overcome; partially accessible, if a low fence, a "no entry" sign was present, or space was restricted but not too narrow or high to enter and thus required some effort to overcome the barriers; and not accessible, if a high fence, sign warning of injury or other barriers were present that required considerable effort to overcome them.

containing GIS-layer and aerial photography. Land use was categorized using the IGS typology (Table 1, except microsite IGS) and a customized land use category system (Table S1. Land use category system). This was produced loosely based on the Brisbane land use code system[24,25] and adapted to suit the project after pilot tests. It was further amended in the field if land uses were encountered that could not be properly recorded with the existing categories (e.g. mixed multi-story land use such as bridges or commercial/residential mixed buildings). Extremely rare land uses were filed under the sub-category name "Other" in the category "Other" (category nomenclature followed a category/sub-category system, e.g. "Private greenspace - garden"), recorded and described in a comment field. We documented the site and its surroundings with photographs. Location data was recorded using a handheld GPS device (Trimble Juno ST) at an accessible part of the site edge or at up to 20 m from a site edge. No permission to carry out this study was necessary, as the survey

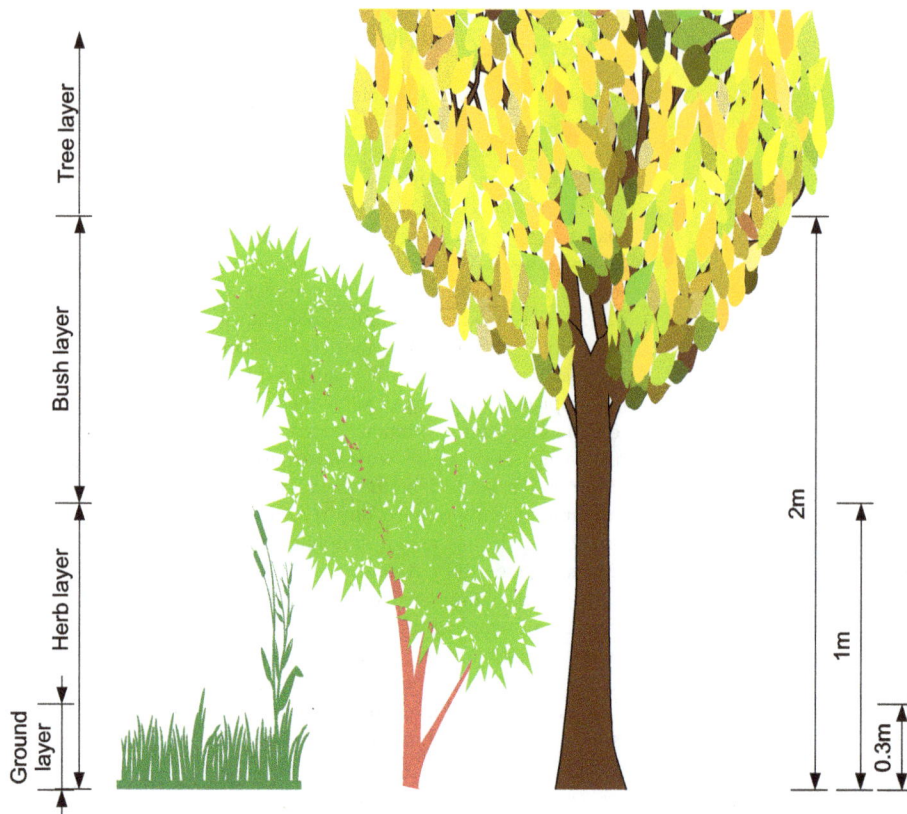

Figure 6. Vegetation structure assessment: tree layer, bush layer, herb layer and ground layer heights.

Data analysis and statistical methods

We used SPSS (v. 21 and 22, OS X) and R (v. 3.02, OS X) to perform descriptive and inferential statistical analyses. Frequency tables were used to describe quantity of IGS, quantity of IGS types, and IGS characteristics (vegetation structure, accessibility). Initial analysis indicated that the sample data was not normally distributed (P-P plots, skewness and kurtosis tests). We therefore used non-parametric tests, namely a Mann-Whitney U test to test for differences in IGS proportion, and a PERMANOVA test with Euclidean distance matrix to test for differences in IGS type proportions between the two survey areas (using R function *adonis*, see http://cc.oulu.fi/~jarioksa/softhelp/vegan/html/adonis. html). All statistical tests were performed on sampling site scale (N = 121 per survey area) with data aggregated from the sub-sites. To test whether distance of the sampling site from the city center were linked with IGS quantity or IGS type quantity, we used a Pearson correlation. A p-value of 0.05 or smaller was interpreted as statistically significant.

Method accuracy assessment

To test how accurate the IGS and land use survey method used in this paper was, we compared the results to land use data from GIS data sets supplied by the local city governments. For this purpose, we first combined the geographic features (e.g. polygons representing residential or green space land use) in the city-supplied data sets (ArcGIS 10, UNION), then removed all features outside the smallest possible square containing all sampling sites (ArcGIS 10, CLIP) to calculate the total land use of the features we wanted to compare. In Brisbane, we compared total combined land use percentages of parks (FGSPK), conservation areas (FGSCN), and sports and recreation areas (FGSSR) from our survey with the total greenspace land use percentage from two Brisbane council greenspace data sets (see Table S1 for land use

Figure 7. Barriers to IGS access. Example photographs: a) IGS inaccessible due to height and missing ladder; b) IGS completely fenced off; c) IGS access restricted by physical (wire) and symbolic barriers (sign).

Table 3. Quantity of IGS and IGS subtypes in Brisbane survey area.

IGS Type	N*	Quantity (m²)	Mean size (m²)	Proportion/area (%)	Proportion/IGS (%)
Lot	32	1,433	44.78	0.47	7.53
Gap	22	117	5.32	0.04	0.61
Street verge	643	15,300	23.79	5.06	80.41
Brownfield	15	967	64.47	0.32	5.08
Waterside	7	125	17.86	0.04	0.66
Waterside/verge	–	–	–	–	–
Structural	38	126	3.32	0.04	0.66
Street verge/gap	–	–	–	–	–
Railway	28	959	34.25	0.32	5.04
Lot/street verge	–	–	–	–	–
Powerline	–	–	–	–	–
Total	785	19,027		6.29	
Extrapolated**		6,353,057		6.29	

*N = number of IGS as recorded in all 3,025 sub-sites.
**Extrapolated to reflect the area of the smallest possible square containing all sampling sites (total square area 101,002,500 m²).

categories). Additionally, we compared total combined residential (all RES categories), garden (PGSGD) and shared greenspace (PGSSG) land use percentages from our survey with the total residential land use percentage of a Brisbane council general land use data set. In Sapporo, we compared park (FGSPK) and sports and recreation area (FGSSR) percentages from our survey with the total non-conservation greenspace land use percentage from a Sapporo City greenspace land use data set. Conservation greenspace was excluded from the comparison in Sapporo because its definition and included greenspace differed substantially between our land use survey and the supplied data set. We then calculated how much the percentages of land use deviate between our land use survey and the supplied GIS data sets.

To check for an accumulation curve and observe the change in land use percentage as sample size increased, we plotted the land use percentage over the number of sites surveyed. Additionally, we plotted the deviation of our land use percentage results from city-supplied datasets (formal greenspace and residential land use in Brisbane, formal greenspace in Sapporo) against the number of sites surveyed. This allowed us to observe what sample size is necessary to achieve a certain level of deviation from the city-supplied datasets.

Results

The surveyed area in Brisbane consisted of 6.3% (19,027 m²) IGS (Table 3), while the surveyed area in Sapporo consisted of IGS to 4.8% (14,559 m²) (Table 4). This difference in IGS proportion was not significant when comparing between the survey areas on site-level (p = .495, N = 242 (121 per survey area),

Table 4. Quantity of IGS and IGS subtypes in Sapporo survey area.

IGS Type	N*	Quantity (m²)	Mean size (m²)	Proportion/area (%)	Proportion/IGS (%)
Lot	159	6144	38.64	2.03	42.20
Gap	386	2796	7.24	0.92	19.20
Street verge	284	2351	8.28	0.78	16.15
Brownfield	22	1458	66.27	0.48	10.01
Waterside	27	1417	52.48	0.47	9.73
Waterside/verge	5	179	35.80	0.06	1.23
Structural	30	93	3.10	0.03	0.64
Street verge/gap	16	68	4.25	0.02	0.47
Railway	7	43	6.14	0.01	0.30
Lot/street verge	1	7	7.00	0.00	0.05
Powerline	2	3	1.50	0.00	0.02
Total	939	14559		4.81	
Extrapolated*		4858220		4.81	

*N = number of IGS as recorded in all 3,025 sub-sites.
*Extrapolated to reflect the area of the smallest possible square containing all sampling sites (total square area 101,002,500 m²).

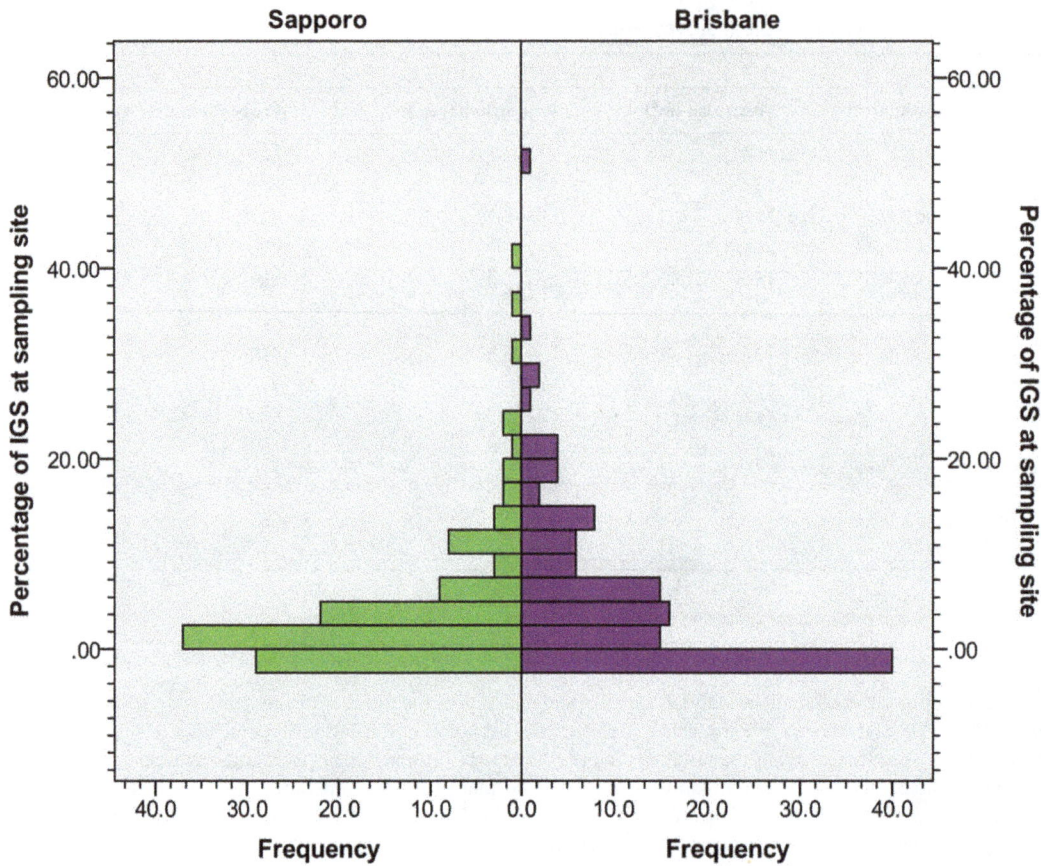

Figure 8. Frequency distribution of IGS land use percentage in sampling sites.

$U = 6953.0$, $z = -.683$; for distribution see Figure 8). In the Brisbane survey area, the mean IGS area per site was 157.25 ($Mdn = 95$, $SD = 214.17$). In the Sapporo survey area, the mean was 120.32 ($Mdn = 41$, $SD = 186.44$). Street verges made up over 80% of IGS in the Brisbane survey area (Table 3), while lots (42.2%) and gaps (19.2%) were the two largest IGS types in the

Sapporo survey area (Table 4). In the Sapporo survey area, IGS consisted of more different IGS types and the proportion of individual IGS types showed IGS was more diverse than in the Brisbane survey area (Table 3, 4).The distribution of IGS types between the two survey areas was significantly different ($p = .0001$, $df = 1$, Pseudo-F $= 19.121$, $MS = 825762$, based on 9999 permu-

Table 5. Comparison of IGS, formal and private greenspace in survey areas.

City	Brisbane survey area		Sapporo survey area	
Greenspace type	Area (m^2)	Area (%)	Area (m^2)	Area (%)
Informal greenspace	19027	6.29	14559	4.81
Parks	16146	5.34	9493	3.14
Sports and recreation	10164	3.36	4423	1.46
Conservation	7641	2.53	32208	10.65
Planted verges	1085	0.36	441	0.15
Total formal GS	35036	11.58	46565	15.39
Gardens	62599	20.69	26193	8.66
Shared greenspace	8434	2.79	5052	1.67
Community land	11592	3.83	13210	4.37
Commercial and industrial	387	0.13	776	0.26
Total private GS	83010	27.44	45231	14.95
Total city greenspace	137073	45.31	106355	35.16

Spatial IGS distribution in Sapporo (top) and Brisbane (bottom) survey areas

Sampling sites*

Percent IGS land use

- 0%
- >0% - 5%
- >5% - 10%
- >10% - 20%
- >20% - 30%
- >30% - 40%
- >40% - 50%
- >50% - 60%

Infrastructure

- Railway
- River
- Highway
- Primary road

0   1   2   4
Kilometers

N

Figure 9. Spatial IGS distribution: percentage of IGS per sampling site in Sapporo (top) and Brisbane (bottom).

tations). In comparison to formal greenspace (parks, sports and recreation, conservation greenspace, and planted verges such as flower-beds) and private greenspace (gardens, shared greenspace, community land, and commercial and industrial greenspace), the area surveyed in Brisbane consisted of more than half as much IGS (6.3%) as formal greenspace (11.6%) and more than one fifth as much IGS than private greenspace (27.4%)(Table 5). In Sapporo, IGS area (4.8%) was almost a third of formal greenspace area (15.4%) and private greenspace (15.0%)(Table 5). Most common non-IGS land use types were small streets (13.3%, INFSS), conservation greenspace (10.7%, FGSCN), and car parks (10.1%, INFCP) in Sapporo, and private gardens (20.7%, PGSGD), small streets (13.6%, INFSS) and residential land use

(12.4%, RESLD) in Brisbane (Table S2. Non-IGS land use in Sapporo and Brisbane).

We found IGS was present in most of the sampling sites in both cities (Figure 9), with obvious exceptions of sites located in areas with other large-scale land use types (e.g. Brisbane river, Mt. Moiwa in the South-West of Sapporo). For vegetation structure, Brisbane IGS had 27.8% mean tree cover, 7.9% mean bush cover, 21.3% mean herb cover, and 72.8% mean ground cover (Table 6). Sapporo IGS had less mean tree cover (6.5%), similar mean bush cover (7.8%), higher herb cover (43.0%) and subsequently lower ground cover (46.0%) (Table 6). For accessibility in Brisbane, 78% of IGS area was accessible, 7% partially accessible and 15% not accessible (Table 7). In Sapporo, the accessible IGS area (68%)

Table 6. Comparison of IGS vegetation structure in survey areas.

City	Brisbane survey area						Sapporo survey area					
IGS Type	N*	Tree (%)	Bush (%)	Herb (%)	Ground (%)	HG (%)**	N*	Tree (%)	Bush (%)	Herb (%)	Ground (%)	HG (%)**
Brownfield	15	0.0	0.0	51.3	34.0	85.3	22	0.0	95.0	100.0	0.0	100.0
Gap	22	0.0	2.3	57.0	21.6	78.6	386	3.2	6.2	45.3	44.4	89.7
Lot	32	23.6	12.5	79.4	11.7	91.1	159	7.6	8.3	36.1	37.7	73.8
Lot/street verge							1	0.0	0.0	100.0	0.0	100.0
Powerline							2	0.0	0.0	0.0	100.0	100.0
Railway	28	1.8	6.6	76.8	4.1	80.9	7	0.0	0.0	75.7	12.9	88.6
Street verge	643	31.7	7.8	10.2	85.3	95.5	284	11.7	2.9	34.4	58.9	93.3
Street verge/gap							16	0.0	0.6	49.4	43.1	92.5
Structural	38	10.5	9.7	73.2	21.3	94.5	30	0.0	6.2	26.3	70.7	97.0
Waterside	7	35.7	21.4	92.9	0.0	92.9	27	0.7	11.9	93.0	7.0	100.0
Waterside/verge							5	64.0	46.0	100.0	0.0	100.0
Total IGS	**785**	**27.8**	**7.9**	**21.3**	**72.8**	**94.0**	**939**	**6.5**	**7.8**	**43.0**	**46.0**	**89.0**

*N = number of IGS as recorded in all 3,025 sub-sites.
**HG = combined percentage of herb and ground cover. Herb and ground cover strata add up to 100% minus ground not covered by vegetation.

Table 7. Comparison of IGS accessibility in survey areas.

Survey area	Accessibility	Lot	Gap	Street verge	Brownfield	Waterside	WS/SV*	Structural	SV/GP*	Railway	LT/SV*	Powerline	Total IGS
Brisbane	Total (N)	32	22	643	15	7	0	38	0	28	0	0	785
	Total (m²)	1433	117	15300	967	125	0	126	0	959	0	0	19027
	Yes (N)	7	3	622	0	7	–	16	–	0	–	–	655
	Yes (N%)	22	14	97	0	100	–	42	–	0	–	–	83
	Yes (m²)	231	10	14433	0	125	–	50	–	0	–	–	14849
	Yes (% of area)	16	9	94	0	100	–	40	–	0	–	–	78
	Partial (N)	12	6	13	0	0	–	8	–	0	–	–	39
	Partial (N%)	38	27	2	0	0	–	21	–	0	–	–	5
	Partial (m²)	661	23	655	0	0	–	28	–	0	–	–	1367
	Partial (% of area)	46	20	4	0	0	–	22	–	0	–	–	7
	No (N)	13	13	8	15	0	–	14	–	28	–	–	91
	No (N%)	41	59	1	100	0	–	37	–	100	–	–	12
	No (m²)	541	84	212	967	0	–	48	–	959	–	–	2811
	No (% of area)	38	72	1	100	0	–	38	–	100	–	–	15
Sapporo	Total (N)	159	386	284	22	27	5	30	16	7	1	2	939
	Total (m²)	6144	2796	2351	1458	1417	179	93	68	43	7	3	14559
	Yes (N)	131	178	265	11	15	0	19	12	0	1	2	634
	Yes (N%)	82	46	93	50	56	0	63	75	0	100	100	68
	Yes (m²)	5032	1154	1800	761	1007	0	73	50	43	7	3	9930
	Yes (% of area)	82	41	77	52	71	0	78	74	100	100	100	68.2
	Partial (N)	17	111	15	11	3	5	7	4	0	0	0	173
	Partial (%)	11	29	5	50	11	100	23	25	0	0	0	18
	Partial (m²)	714	924	441	697	130	179	16	18	0	0	0	3119
	Partial (% of area)	12	33	19	48	9	100	17	26	0	0	0	21.4
	No (N)	11	97	4	0	9	0	4	0	7	0	0	132
	No (%)	7	25	1	0	33	0	13	0	100	0	0	14
	No (m²)	398	718	110	0	280	0	4	0	0	0	0	1510
	No (% of area)	6	26	5	0	20	0	4	0	0	0	0	10.4

*WS/SV = waterside/street verge, SV/GP = street verge/gap. LT/SV = lot/street verge.

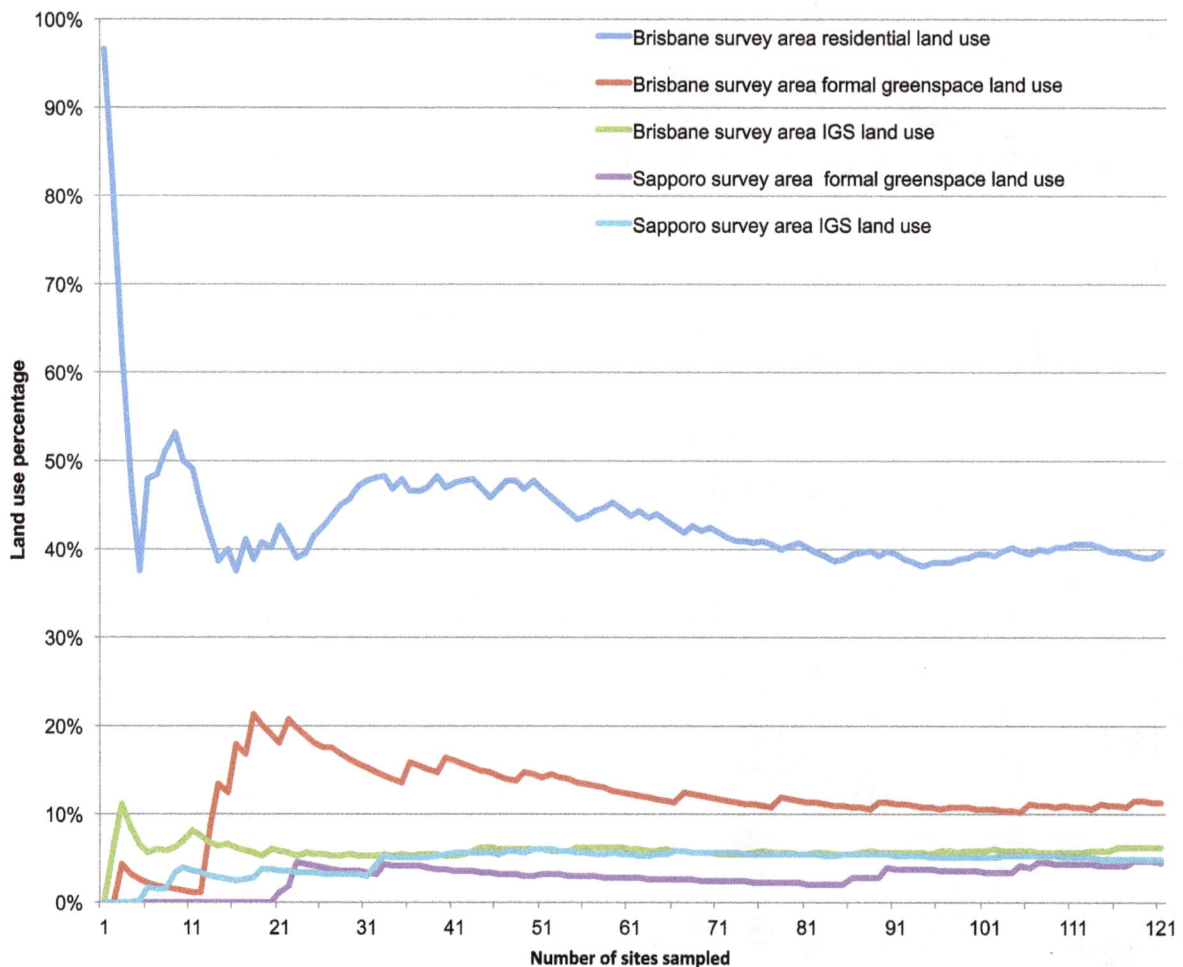

Figure 10. Change in total land use percentage with increasing sample size.

and not accessible area (10%) was smaller, compensated by a larger partially accessible IGS area (21%) (Table 6). Lot, gap and brownfield IGS was more accessible in Sapporo, but street verge and waterside IGS in Brisbane (Table 6). When testing whether the total amount of IGS or the amount of an individual IGS type was correlated with distance to the city center, we found no significant correlations between either total IGS or individual IGS type and distance in either surveyed area.

When measuring the accuracy of our land use survey method, we found the combined percentage of parks, conservation and sports and recreation land use in our survey (11.2%) deviated 8.4% from the combined greenspace land use percentage in Brisbane Council datasets (10.4%). The combined percentage of residential, garden and private shared greenspace in our survey (39.5%) deviated -4.2% from the residential land use percentage in the Brisbane Council dataset (41.3%). In Sapporo, the combined percentage of non-conservation greenspace in our survey (4.8%) deviated −6.9% from the non-conservation greenspace percentage in the Sapporo City dataset (5.1%). As a result, the accuracy of our method was over 90% in both cities when comparing land use percentages of around 5% or more with those of official datasets. A visualization of the change in land use percentage with increasing sample size showed that for a common land use type (residential), good accuracy was reached at a sample size of around 70, while for the rare land use types a sample size of around 90 was

necessary (Figure 10). A deviation from the city-supplied datasets of less than 10% is reached at 60 sampled sites for the residential land use type, but only at 120 sampled sites for the formal greenspace land use types (Figure 11).

Discussion and Conclusions

This study has found similar proportions of IGS in both survey areas. While this could indicate other urban areas may contain a similar percentage of IGS, the conclusions we can draw are limited by the sampling design used. The similarity of the study cases (e.g. age and spatial structures of the cities, size and shape of the survey areas; see Methods) may be partly responsible for the similarity in IGS proportions, so results may vary across survey areas with different characteristics. The survey areas we compared differ in their population density (Figure 5) and cultural context. These two factors seemed to have little influence on the proportion of IGS in the survey areas and its accessibility. However, they may explain the differences in IGS types and vegetation structure. For example, higher population density may influence the amount of land dedicated to street verges through planning policy. The rapid growth Brisbane is experiencing may limit the proportion of lot type IGS, as a high demand for land available for development possibly reduces the time land remains vacant before redevelopment. IGS was also widely, but not equally distributed throughout the survey areas (Figure 9). To better understand the factors

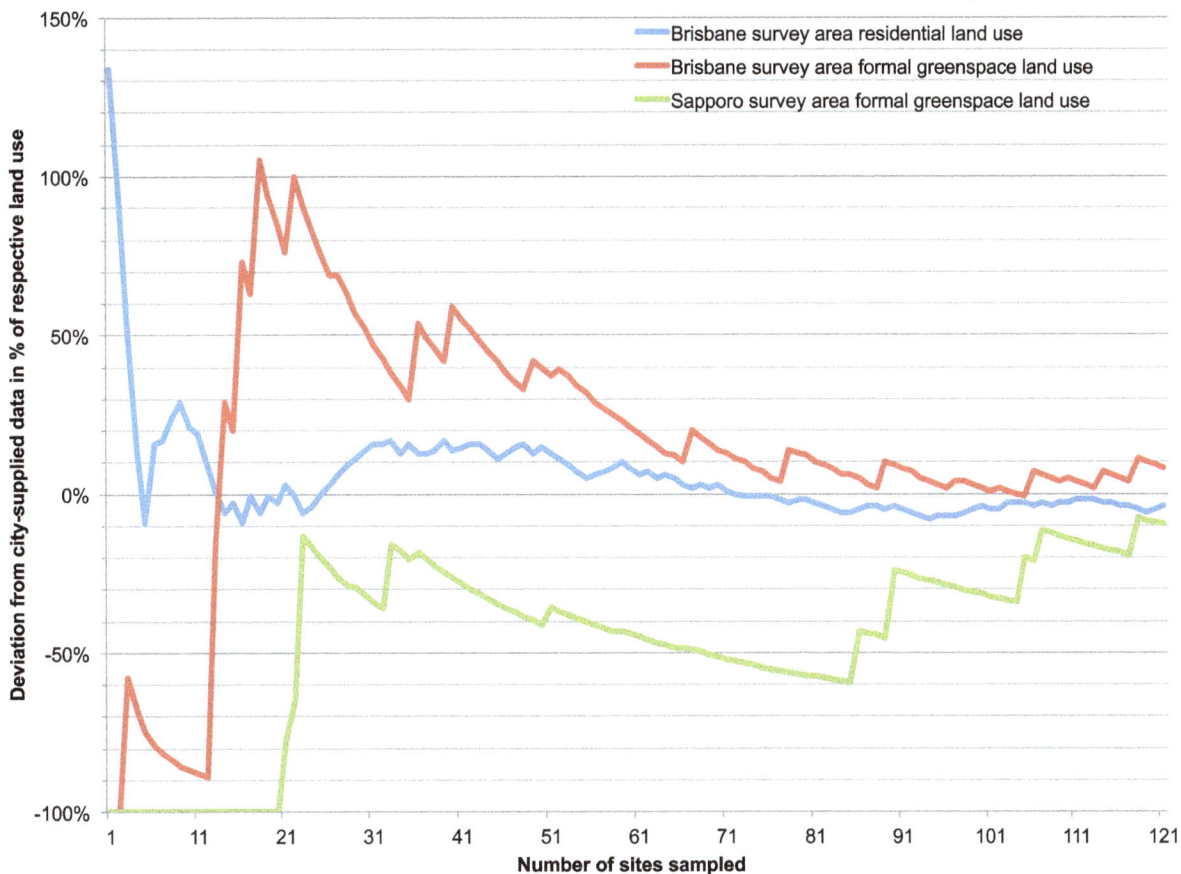

Figure 11. Decrease in deviation of total land use percentage from city-supplied datasets with increasing sample size.

driving IGS occurrence across different cities, future research should seek to compare IGS in multiple locations. Additionally, possible links between IGS occurrence and factors such as other land use types (e.g. formal green space, industrial, residential) or average income around the site could be explored.

The sampling design used in this study has limitations that make it unsuitable for assessing factors of importance to urban conservation, such as the size of individual IGS or the distribution of IGS sizes. Such information is valuable and has been recorded by prior research into roundabouts[27] and vacant lots[26], but these studies require a different sampling design to studies aiming at measuring the percentage of IGS land use in a survey area. However, for research on the size of individual IGS, linear IGS (e.g. railway verges) and microsites represent challenges as they lack clear boundaries.

Most prior research on IGS has pointed out its potential without knowing what proportion of cities consists of IGS. Having this knowledge allows us to identify some potential policy implications that IGS has for recreational use and conservation, by considering its area compared to other greenspace types. IGS accounts for about 14% of total greenspace in the survey areas (Table 5). This suggests IGS may represent an important source of green space, an aspect further emphasized by the fact that the proportion of IGS present in both survey areas was similar. Furthermore, in both survey areas over 80% of IGS was accessible or partly accessible. However, the limitations of the sampling design discussed above also apply here. Additionally, these findings represent just the first necessary step to understand the potential of IGS for recreation

and IGS. Further research is needed to clarify the degree to which IGS is used by residents, whether accessible sites are actually accessed, and which factors may influence residents' perception and appreciation of IGS. These questions are particularly important, because IGS is different from formal greenspace. It lacks common park facilities such as seating or toilets, its informal nature can cause especially adults to perceive it as unsafe or dangerous[28], and the liminal nature of these spaces may limit the degree to which they can be planned – a challenge for urban planners. Yet, being different can also be an advantage. IGS can offer residents an alternative experience to formal green-space[2,29,30], such as opportunities for children to test themselves in a non-controlled environment[31].

The results also have implications for urban conservation. Research has shown IGS plays a role in providing habitat to fauna and flora[32,33] as well as in connecting habitat in and between cities[34,35]. The amount of IGS we found (three times that of conservation greenspace in Brisbane and close to 50% in Sapporo survey areas), its distribution throughout the survey areas, and the relatively complex vegetation structure (Table 6) suggest IGS' role for urban conservation may be more important than previously assumed. Sites with IGS completely surrounded by sites without IGS were rare, suggesting a potential connectivity for species present in IGS (Figure 9). The lack of difference in IGS proportion between the survey areas raises the possibility that a similar percentage of IGS may also be available for conservation in other cities, although the limits of the sampling design discussed above need to be taken into account. The differences in vegetation structure and individual IGS types also imply its actual conservation potential likely depends on its local characteristics. There are concerns about the opportunistic type of species IGS tends to favor[36], and the possible function of IGS as a reservoir or corridor for biological invasions[37]. But research has not always confirmed such connectivity for invasive species[35], and many opportunistic species are adapted to their local environment – they represent what could be called the "de facto native vegetation of the city" [7,38]. As Dunn et al. suggest, the opportunities for nature experience IGS can provide to residents may also be vital for conservation efforts, even beyond urban areas[1].

The potential importance of IGS for urban recreational use and urban conservation has implications for urban and environmental planning. Planners may need to re-think their negative view of 'vacancy' in the urban landscape[39] and acknowledge the benefits of residents' informal creativeness[40]. Expectations for parks, conservation greenspace and private greenspace such as gardens as sole providers of recreational and conservation benefits might need to be reevaluated. Reducing barriers to IGS access[26] or reusing IGS as community gardens[41] are ways to improve IGS utility for residents, but planners should refrain from too much intervention. Some scholars assert that IGS does not necessarily need to be tamed, but should be valued for its informality, ambivalence and special aesthetic [14,29]. In the case of some IGS, freedom of purpose can mean freedom from purpose (e.g. abandoned lots), while other IGS (e.g. utility corridors, railway verges) have purpose but may have to capacity to accommodate additional, informal use. On the other hand, IGS may have negative effects for residents, such as damage to structures caused by spontaneous vegetation [42,43]. The liminality and legally non-public status of many IGS may also present a challenge to planners, as their influence on IGS characteristics such as accessibility is likely limited. It depends not only on the cooperation of the formal space owners, but issues such as liability for injury involve third parties such as insurance companies[44].

The results of our study emphasize more research on IGS is needed to unlock its full potential. Examining how IGS is influenced by its socio-ecological context would be a valuable starting point for future studies. A quantitative examination of how residents use and perceive IGS could provide into the social

aspects. A cross-cultural comparison seems particularly promising, as concepts such as public space may be interpreted differently depending on the cultural context[45]. A possible direction for research on the ecological side could be a comprehensive examination of the value of IGS for biodiversity, either in the form of a systematic literature review providing a synthesis of the many studies focusing on only one or a few IGS types, or in the form of field studies taking into account all IGS types identified in this paper. Finally, replications of the study conducted in this paper in geographic locations around the world would improve our global knowledge of global IGS distribution and provide valuable input for planning policy.

Acknowledgments

We are deeply grateful to Yumi Nakagawa for her invaluable support with data collection and data entry. We thank Jennifer Garden and Jean-Marc Hero for their advice on research design, Alex Lo for helpful comments on the manuscript, Mariola Rafanowicz for assistance in GIS analysis and obtaining data sets, and two anonymous reviewers for their detailed and helpful comments. We also thank the Brisbane City Council and Sapporo City Department of Environment Section Greenspace for providing green space and land use data sets.

Author Contributions

Conceived and designed the experiments: CR JB. Performed the experiments: CR. Analyzed the data: CR. Wrote the paper: CR JB.

References

1. Dunn RR, Gavin MC, Sanchez MC, Solomon JN (2006) The Pigeon Paradox: Dependence of global conservation on urban nature. Conserv Biol 20: 1814–1816. doi:10.1111/j.1523–1739.2006.00533.x.

2. Campo D (2013) The accidental playground. Fordham University Press.

3. Unt A-L, Travlou P, Bell S (2013) Blank Space: Exploring the sublime qualities of urban wilderness at the former fishing harbour in Tallinn, Estonia. Landscape Research: 1–20. doi:10.1080/01426397.2012.742046.

4. Jorgensen A, Keenan R, editors (2012) Urban wildscapes. Routledge.

5. Kowarik I (2011) Novel urban ecosystems, biodiversity, and conservation. Environmental Pollution 159: 1974–1983. doi:10.1016/j.envpol.2011.02.022.

6. Lososová Z, Horsák M, Chytrý M, Cejka T, Danihelka J, et al. (2011) Diversity of Central European urban biota: effects of human-made habitat types on plants and land snails. J Biogeogr 38: 1152–1163. doi:10.1111/j.1365–2699.2011.02475.x.

7. Del Tredici P (2010) Spontaneous urban vegetation: Reflections of change in a globalized world. Nat Cult 5: 299–315. doi:10.3167/nc.2010.050305.

8. Jorgensen A, Tylecote M (2007) Ambivalent landscapes—wilderness in the urban interstices. Landscape Research 32: 443–462. doi:10.1080/01426390701449802.

9. McLain RJ, Hurley PT, Emery MR, Poe MR (2014) Gathering "wild" food in the city: Rethinking the role of foraging in urban ecosystem planning and management. Local Environment 19: 220–240. doi:10.1080/13549839.2013.841659.

10. Head L, Muir P (2006) Suburban life and the boundaries of nature: Resilience and rupture in Australian backyard gardens. Trans Inst Br Geog 31: 505–524. doi:10.1111/j.1475–5661.2006.00228.x.

11. Trigger DS, Head L (2010) Restored nature, familiar culture: Contesting visions for preferred environments in Australian cities. Nat Cult 5: 231–250. doi:10.3167/nc.2010.050302.

12. Howitt R (2001) Frontiers, borders, edges: Liminal challenges to the hegemony of exclusion. Australian Geographical Studies 39: 233–245. doi:10.1111/1467–8470.00142.

13. Davison A (2008) The trouble with nature: Ambivalence in the lives of urban Australian environmentalists. Geoforum 39: 1284–1295. doi:10.1016/j.geoforum.2007.06.011.

14. Franck KA, Stevens Q, editors (2007) Loose space: Possibility and diversity in urban life. Routledge.

15. Moran D (2011) Between outside and inside? Prison visiting rooms as liminal carceral spaces. GeoJournal 78: 339–351. doi:10.1007/s10708-011-9442-6.

16. Byrne J, Sipe N, Searle G (2010) Green around the gills? The challenge of density for urban greenspace planning in SEQ. Australian Planner 47: 162–177. doi:10.1080/07293682.2010.508204.

17. Haase D (2008) Urban ecology of shrinking cities: An unrecognized opportunity? Nat Cult 3: 1–8. doi:10.3167/nc.2008.030101.

18. Royal Geographic Society (n.d.) Royal Geographic Society. Available: http://www.rgs.org. Accessed 2014 May 6.

19. Hirzel A, Guisan A (2002) Which is the optimal sampling strategy for habitat suitability modelling. Ecological Modelling 157: 331–341. doi:10.1016/S0304-3800(02)00203-X.

20. Phinn S, Stanford M, Scarth P, Murray AT, Shyy PT (2002) Monitoring the composition of urban environments based on the vegetation-impervious surface-soil (VIS) model by subpixel analysis techniques. Int J of Remote Sensing 23: 4131–4153.

21. Haby MM, Peat JK, Mellis CM, Anderson SD, Woolcock AJ (1995) An exercise challenge for epidemiological studies of childhood asthma: Validity and repeatability. European Respiratory Journal 8: 729–736.

22. Glickman NJ, Oguri Y (1978) Modeling the urban land market: The case of Japan. Journal of Urban Economics 5: 505–525. doi:10.1016/0094-1190(78)90006-2.

23. Tobita T, Iai S, Kang GC, Konishi Y (2012) Observed damage of wastewater pipelines and estimated manhole uplifts during the 2004 Niigataken Chuetsu, Japan, Earthquake Reston, VA: American Society of Civil Engineers. pp. 1–12. doi:10.1061/41050(357)75.

24. Brisbane City Council (2010) Brisbane council land use codes.

25. Brisbane City Council (2010) City Plan 2000. Brisbane City Council.

26. Hayashi M, Tashiro Y, Kinoshita T (1999) A study on vacant lots enclosed by fences in relation to urbanization. Landscape Research Japan 63: 667–670. doi:10.5632/jila.63.667.

27. Helden AJ, Leather SR (2004) Biodiversity on urban roundabouts—Hemiptera, management and the species-area relationship. Basic and Applied Ecology 5: 367–377. doi:10.1016/j.baae.2004.06.004.

28. Rink D, Herbst H (2011) From wasteland to wilderness - aspects of a new form of urban nature. In: Richter M, Weiland U, editors. Applied urban ecology: A global framework. Chichester, UK: Wiley-Blackwell. pp. 82–92. doi:10.1002/9781444345025.ch7.

29. Foster J (2013) Hiding in plain view: Vacancy and prospect in Paris' Petite Ceinture. Cities. doi:10.1016/j.cities.2013.09.002.

30. Nohl W (1990) Gedankenskizze einer Naturästhetik der Stadt. Landschaft und Stadt 22: 57–67.

31. Mugford K (2012) Nature, nurture; danger, adventure; junkyard, paradise; the role of wildscapes in children's literature. In: Jorgensen A, Keenan R, editors. Urban Wildscapes. Routledge. pp. 80–96.

32. Robinson SL, Lundholm JT (2012) Ecosystem services provided by urban spontaneous vegetation. Urban Ecosystems 15: 545–557. doi:10.1007/s11252-012-0225-8.

33. Meffert PJ, Dziock F (2012) What determines occurrence of threatened bird species on urban wastelands? Biological Conservation 153: 87–96. doi:10.1016/j.biocon.2012.04.018.

34. Francis RA, Hoggart SPG (2012) The flora of urban river wallscapes. River Res Applic 28: 1200–1216. doi:10.1002/rra.1497.

35. Penone C, Machon N, Julliard R, Le Viol I (2012) Do railway edges provide functional connectivity for plant communities in an urban context? Biological Conservation 148: 126–133. doi:10.1016/j.biocon.2012.01.041.

36. Wittig R, Becker U (2010) The spontaneous flora around street trees in cities—A striking example for the worldwide homogenization of the flora of urban habitats. Flora 205: 704–709. doi:10.1016/j.flora.2009.09.001.

37. Pyšek P, Hulme PE (2005) Spatio-temporal dynamics of plant invasions: Linking pattern to process. Ecoscience 12: 302–315. doi:10.2980/i1195-6860-12-3-302.1.

38. Del Tredici P, Pickett STA (2010) Wild urban plants of the Northeast: a field guide. Cornell University Press.

39. Corbin CI (2003) Vacancy and the landscape: Cultural context and design response. Landscape Journal 22: 12–24. doi:10.3368/lj.22.1.12.

40. Jonas MC (2007) Private use of public open space in Tokyo - A study of the hybrid landscape of Tokyo's informal gardens. Journal of Landscape Architecture 2: 18–29.

41. Acres: About Us (n.d.) 596acresorg. Available: http://www.596acres.org/en/about/about-596-acres/. Accessed 13 August 2013.

42. Melander B, Holst N, Grundy AC, Kempenaar C, Riemens MM, et al. (2009) Weed occurrence on pavements in five North European towns. Weed Res 49: 516–525. doi:10.1111/j.1365-3180.2009.00713.x.

43. Lisci M, Monte M, Pacini E (2003) Lichens and higher plants on stone: a review. Int Biodeter Biodegr 51: 1–17. doi:10.1016/S0964-8305(02)00071-9.

44. De Sousa CA (2006) Unearthing the benefits of brownfield to green space projects: An examination of project use and quality of life impacts. Local Environment 11: 577–600. doi:10.1080/13549830600853510.

45. Dimmer C (2012) Re-imagining pubic space. In: Brumann C, Schulz E, editors. Urban Spaces in Japan. Urban Spaces in Japan. pp. 74–105.

46. Byrne J, Sipe N (2010) Green and open space planning for urban consolidation – A review of the literature and best practice. Urban Research Program, Griffith University.

47. Brisbane City Council (2007) CityShape Implementation Strategy. Brisbane City Council.

48. Australian Bureau of Statistics (2013) National Regional Profile: Brisbane (C) (Local Government Area). Australian Bureau of Statistics. Available: http://www.abs.gov.au/ausstats/abs@nrp.nsf/lookup/LGA31000Main+Features12007-2011. Accessed 2013 Oct 7.

49. Sapporo Kankyékyoku Midori No Suishinbu (2011) Sapporoshi midori no kihon keikaku. (Sapporo Basic Green Plan.) City of Sapporo.

50. Sapporo Shiché Seisakujitsu Seisaku Kikakubu (2012) Sapporoshi machizukuri senryaku bishon kanren de-ta. (Sapporo City Town Development Strategic Vision Related Data.) City of Sapporo. Available: http://www.city.sapporo.jp/kikaku/vision/data/newdata.html. Accessed 2013 Oct 20.

51. Sapporo Kankyékyoku Midori No Suishinbu (2012) Sapporoshi no kōen, ryokuchi. (Sapporo City parks and green space.) City of Sapporo. Available: http://www.city.sapporo.jp/ryokuka/shiryo/toukei/documents/1-1.pdf. Accessed 2013 Oct 20.

18

Land Use Compounds Habitat Losses under Projected Climate Change in a Threatened California Ecosystem

Erin Coulter Riordan*, Philip W. Rundel

Department of Ecology and Evolutionary Biology, University of California Los Angeles, Los Angeles, California, United States of America

Abstract

Given the rapidly growing human population in mediterranean-climate systems, land use may pose a more immediate threat to biodiversity than climate change this century, yet few studies address the relative future impacts of both drivers. We assess spatial and temporal patterns of projected 21st century land use and climate change on California sage scrub (CSS), a plant association of considerable diversity and threatened status in the mediterranean-climate California Floristic Province. Using a species distribution modeling approach combined with spatially-explicit land use projections, we model habitat loss for 20 dominant shrub species under unlimited and no dispersal scenarios at two time intervals (early and late century) in two ecoregions in California (Central Coast and South Coast). Overall, projected climate change impacts were highly variable across CSS species and heavily dependent on dispersal assumptions. Projected anthropogenic land use drove greater relative habitat losses compared to projected climate change in many species. This pattern was only significant under assumptions of unlimited dispersal, however, where considerable climate-driven habitat gains offset some concurrent climate-driven habitat losses. Additionally, some of the habitat gained with projected climate change overlapped with projected land use. Most species showed potential northern habitat expansion and southern habitat contraction due to projected climate change, resulting in sharply contrasting patterns of impact between Central and South Coast Ecoregions. In the Central Coast, dispersal could play an important role moderating losses from both climate change and land use. In contrast, high geographic overlap in habitat losses driven by projected climate change and projected land use in the South Coast underscores the potential for compounding negative impacts of both drivers. Limiting habitat conversion may be a broadly beneficial strategy under climate change. We emphasize the importance of addressing both drivers in conservation and resource management planning.

Editor: Ben Bond-Lamberty, DOE Pacific Northwest National Laboratory, United States of America

Funding: This work was supported by the University of California Los Angeles (UCLA) Foundation, Department of Ecology and Evolutionary Biology, UCLA. The funders had no role in study design, data collection and analysis, decision to publish, or preparation of the manuscript.

Competing Interests: The authors have declared that no competing interests exist.

* E-mail: eriordan@ucla.edu

Introduction

The combined impacts of climate change and land use are expected to drive unprecedented rates of environmental change and biodiversity loss this century. Climate is a major driver of species distributions and rising temperatures over the last 100 years have already resulted in significant shifts in species ranges worldwide [1,2]. With mean global surface temperatures expected to rise as high as 4°C by 2100 [3], the persistence of biodiversity may be contingent upon the ability of species to track suitable climatic conditions [4,5]. Considerable attention has focused on predicting potential habitat losses and range shifts of individual species under 21st century climate change, and subsequently forecasting species losses [6,7] yet continued human land use poses a significant, and possibly more immediate threat to species persistence this century [8] through habitat conversion and degradation. The disconnect between relatively high resolution downscaled climate projections and the typically coarse resolution treatment of land use-land cover change projections, has contributed to the exclusion of future land use from assessments of impacts of future environmental change on species and ecosystems, which require local to regional scale analyses [9].

Only recently have studies begun to incorporate projections of both land use and climate change [10,11].

Addressing projected land use in climate change assessments is particularly important for mediterranean-climate regions, where multiple drivers of environmental change are projected to cause some of the highest proportional biodiversity losses worldwide by 2100, chief among which is land use [8]. Harboring nearly 20% of the world's vascular plant species [12], these regions are global biodiversity hotspots [13]. Their dense, rapidly expanding human populations pose a significant threat to biodiversity, as evidenced by increasing numbers of species of conservation concern with growing human population density over the last decade [14]. Projected 21st century climate change poses an additional threat, potentially driving dramatic range shifts and species losses across the biome [15–18], which may be further exasperated by high levels of human land use [19]. Thus, assessing the relative future impacts of both drivers is critical to prioritize conservation planning and effectively protect biodiversity in mediterranean-climate regions under conditions of environmental change.

Focusing on California sage scrub (CSS), a plant association of considerable diversity, endemism, and threatened status in the mediterranean-climate California Floristic Province, we investigate the relative impacts of projected 21st century land use and climate change on CSS habitat suitability and richness. California has the greatest urbanization and population growth of all five mediterranean regions [14]. Nearly 25% of the state's >6,500 native plant taxa have a rare, threatened, or endangered status on federal and/or state agency lists, largely as a result of anthropogenic impacts including habitat degradation and destruction from land use [20]. Furthermore, California's current 37.5 million population is expected to grow to between 43.8 and 147.7 million by 2100 [21]. Substantial shifts in California's climate also pose a considerable threat to the state's biodiversity and ecosystems. General circulation models (GCMs) project an increase in annual mean temperature of 1.35°C to 5.8°C statewide by the end of the century [22], which could drive dramatic range losses for as many as two-thirds of the endemic species of California that comprise over 25% of the state's flora [16].

Within California, CSS coincides with areas of high human impact, occurring primarily in coastal and semiarid interior regions of southern California but also in scattered patches along the central California coast [23–25]. The coastal counties of southern California containing CSS (Ventura, Los Angeles, Orange, and San Diego) also house nearly half (45%) of the state's population, yet only account for seven percent of the state's total land area [26]. Land development combined with habitat conversion to annual grasses driven by anthropogenic practices have reduced CSS to as little as 10% of its original extent in the state [25,27–30], resulting in a large number of associated threatened and sensitive species [31,32]. High land values and development pressure in the region continue to make the conservation of CSS challenging [33]. With rapidly expanding human population in southern California, land use change poses an immediate threat to CSS [34]. In addition, climate change is projected to drive the contraction and replacement of mediterranean-climate conditions with warmer and drier conditions along coastal areas of southern California and northwestern Baja California by 2100 [19,35], areas of high CSS diversity and endemism [23–25,29]. This could cause species losses and changes in CSS diversity patterns as individual species shift in distribution in response to the expansion and contraction of mediterranean climates [36,37]. Furthermore, multiple drivers of environmental change could compound CSS habitat loss [11,38].

Successful CSS conservation this century will likely hinge upon prioritizing management efforts based on the relative impacts of both projected future land use and climate change. Using a species distribution modeling (SDM) framework, we investigate the relative threat of habitat loss from 21st century projected land use and climate change for 20 CSS species. Species distribution models, which define a species range with respect to environmental variables (e.g., climate), provide a means to forecast projected climate change impacts on species and diversity [39–41]. We assess the extent to which habitat loss driven by future land use versus climate change will jeopardize key shrub species and alter species richness patterns. To test if the degree of threat posed by land use and climate change will vary temporally and spatially across CSS, we compare habitat loss impacts at two time intervals (early and late century) in two ecoregions in California (Central Coast and South Coast).

Materials and Methods

Study system and area

California sage scrub is a unique plant association characterized by dominant drought-deciduous shrubs (e.g., *Salvia*) with variable contributions of succulent and evergreen species and a diverse, herbaceous understory. It is primarily distributed in various community compositions along the coast in southern California, USA and northwestern Baja California, Mexico, with scattered patches along the central California coast and the semi-arid interior of southern California [23–25,42]. The southern limit of CSS near El Rosario in Baja California (~30°N) coincides with the southern extent of mediterranean-type climate in North America and the transition to more arid, desert conditions [42]. Within California, CSS is a high conservation priority, providing habitat for over 100 plant and animal species currently considered threatened, endangered, or of special conservation concern [31,32].

At a regional scale, climate strongly influences distributional patterns of CSS species and floristic associations [25,28,43–45]. Initiating new growth after fall rains and growing through the coldest winter months, CSS species are typically limited to lower elevation areas with mild, wet winters and are negatively correlated with minimum winter temperatures [28]. In coastal southern California, CSS is distributed along a gradient of decreasing annual precipitation ranging from 450 to 250 mm [25]. Evapotranspirative stress during summer months, which is related to maximum summer temperature, also appears to be a major factor influencing species distributions [44]. Regionally, CSS occurs across a variety of soils that range broadly in fertility [25,46]. Only at local scales, do topography and soil typically become important in influencing CSS floristic variability [47].

We focused our analysis of the relative future impacts of land use and climate change on individual CSS species in California, where we have spatially explicit projections for both drivers of future habitat loss. We also analyzed geographic patterns of projected land use change and climate change impact on modeled CSS shrub species richness in two California ecoregions most critical to CSS: the Central Coast Ecoregion and South Coast Ecoregion, defined as "Central Western California" and "Southwestern California," respectively, in *The Jepson Manual* [48] (Fig. 1). The Central Coast Ecoregion includes the San Francisco Bay Area at its northern limit, the central California coastline, the inner and other South Coast Ranges, and is bounded by the Santa Ynez Mountains to the south where it borders the South Coast Ecoregion. The South Coast Ecoregion includes the southern California coastline from Point Conception to the U.S.-Mexico border, the Transverse Ranges, and the Peninsular Ranges.

Habitat suitability models

We used a SDM approach to model current and future climatically suitable habitat for 20 dominant CSS shrub species (Table 1). Such climate-based models have been used previously to identify regional patterns of habitat suitability for dominant shrub species and floristic groups in CSS [45]. While edaphic and topographic factors can be important drivers of CSS distribution at local scales (e.g., 50 meters) [47], this is beyond the scope of our analyses, which focus on regional patterns of projected land use and climate change impact on CSS. We modeled current species-climate relationships from herbarium record localities and climate data using Maxent (version 3.3.3) [49], a maximum-entropy modeling algorithm. We chose Maxent for its high performance with spatially biased presence-only data, such as our herbarium record localities [49–51]. The ability of Maxent to calculate probability distributions based on incomplete information is particularly useful for modeling CSS habitat suitability, where current CSS fragmentation patterns may reflect anthropogenic absences rather than environmental limits.

We obtained herbarium record localities collected from 1950 to present from the following herbarium databases: the Consortium of California Herbaria (CCH; http://ucjeps.berkeley.edu/consortium),

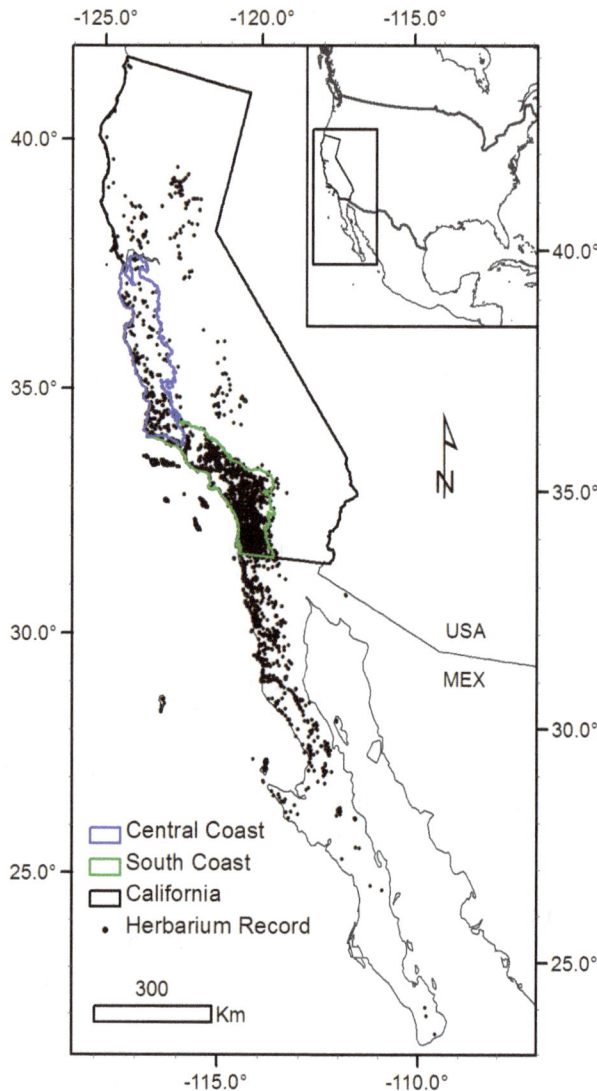

Figure 1. Study Area and Species Localities. Study area and herbarium record locations for the 20 modeled CSS shrub species. Habitat suitability models were created from all herbarium localities, analyses of habitat and land use change were restricted to within California.

the Southwest Environmental Information Network (SEINet; http://swbiodiversity.org/seinet), the Global Biodiversity Informatics Facility (GBIF; http://www.gbif.org), Baja Flora of the San Diego Natural History Museum (http://www.sdnhm.org), and the Red Mundial de Información sobre Biodiversidad (REMIB; http://www.conabio.gob.mx/remib/doctos/remib_esp.html) (Fig. 1). Prior to modeling, we mapped all records to identify and exclude cultivated plants, errors in georeferencing, obvious misidentifications, and duplicate collections.

We obtained current contemporary climate data from Worldclim (http://www.worldclim.org), a set of 19 bioclimatic variables derived from weather station monthly mean temperature and precipitation data collected from 1950 to 2000 and interpolated to 30-arcsec (ca. 1 km) resolution [52]. Finer resolution climate data exist for California [53], but exclude the warmer and drier climatic limit of many CSS species in Baja California. Furthermore, fine scale climate datasets (<1 km resolution) may not be appropriate

for modeling species distributions from herbarium records, as uncertainties in georeferenced locations may limit the accuracy of locality data to coarser resolutions.

We narrowed the 19 bioclimatic variables to those thought to drive regional patterns of CSS distribution (continentality, temperature extremes, water availability, and seasonality in both temperature and precipitation). For highly correlated variable pairs (Pearson correlation coefficient $|r| > 0.80$), we retained the variable with the highest contribution to model performance. Thus, our final seven bioclimatic variables minimized correlations among variables, maximized contribution to model predictions, and represented annual climate trends, seasonality, and extremes relevant to CSS distribution: annual mean temperature (BIO1), temperature seasonality (BIO4), maximum temperature of the warmest month (BIO5), minimum temperature of the coldest month (BIO6), precipitation seasonality (BIO15), precipitation of the warmest quarter (BIO18) and precipitation of the coldest quarter (BIO19).

We selected future climate scenarios representing two possible trajectories of climate change in California under the Intergovernmental Panel on Climate Change Special Report on Emission Scenarios (IPCC-SRES) A1B storyline: (1) a warmer wetter future (CCCMA CGC 3.1) and (2) a warmer drier future (NCAR CCSM 3.0). We obtained 30-arcsec (ca. 1 km) spatial resolution future climate data downscaled from GCM outputs of the IPCC Fourth Assessment Report [3] by the climate change program of the International Center for Tropical Agriculture (CIAT; http://www.ccafs-climate.org/data). We represent 21st century climate at two time steps created from separate 30 year averages, one mid-century (2050s: 2040–2069) and one late-century (2080s: 2070–2100).

We modeled current climate-species relationships in Maxent using all herbarium records to train each species model, then projected these relationships onto future climate layers at two time periods, 2050s and 2080s. For each species model, a constant set of 10,000 background pixels selected randomly over the study area were used as "pseudo absences," the maximum set of iterations was 500, the convergence threshold was set to 10^{-5}, and regularization was set to "auto" allowing Maxent to set the amount of regularization automatically based on our locality and environmental data [54]. We used 10-fold cross-validation to replicate model runs and estimate evaluation statistics for each species. We measured overall model performance using the area under the receiving operator characteristics curve (AUC), which ranges from 0.5 (random prediction) to 1 (maximum prediction) and can be interpreted as the ability of model predictions to discriminate presence sites from random background [49].

We created binary current and future habitat maps (0 = unsuitable, 1 = suitable) from Maxent's logistic output using the maximum sensitivity plus specificity threshold which provides among the most accurate predictions for presence-only data [55,56]. We calculated habitat loss and gain from climate change using two different future dispersal scenarios, a best-case, unlimited dispersal scenario where species can colonize any future suitable habitat, and a worst-case, no dispersal scenario where species cannot disperse to future suitable habitat falling outside of modeled currently suitable habitat. We estimated future CSS species richness by overlaying individual species habitat maps under each climate change and dispersal scenario.

Current and future land use-land cover

We used current (2000) and projected future (2050 and 2080) land use-land cover maps (Fig. S1) consistent with the IPCC-SRES A1B scenario and developed by the United States Geological

Table 1. List of California sage scrub species.

Taxon name	Family	Growth form	N	Mean test AUC (min–max)
Acmispon glaber	Fabaceae	Drought-deciduous subshrub	603	0.927 (0.905–0.942)
Artemisia californica	Asteraceae	Drought-deciduous shrub	242	0.957 (0.934–0.976)
Bahiopsis laciniata	Asteraceae	Drought-deciduous shrub	176	0.968 (0.925–0.986)
Cneoridium dumosum	Rutaceae	Evergreen shrub	117	0.974 (0.957–0.989)
Encelia californica	Asteraceae	Drought-deciduous shrub	204	0.969 (0.955–0.979)
Ericameria ericoides	Asteraceae	Evergreen shrub	85	0.991 (0.985–0.995)
Eriogonum fasciculatum (coastal vars.)	Polygonaceae	Evergreen shrub	665	0.923 (0.913–0.945)
Hazardia squarrosa	Asteraceae	Evergreen shrub	223	0.960 (0.923–0.967)
Hesperoyucca whipplei	Agavaceae	Evergreen shrub (rosette)	226	0.933 (0.880–0.952)
Isocoma menziesii	Asteraceae	Evergreen shrub	313	0.958 (0.948–0.962)
Malosma laurina	Anacardiaceae	Evergreen shrub	218	0.964 (0.949–0.975)
Mimulus aurantiacus	Phrymaceae	Drought-deciduous shrub	793	0.904 (0.890–0.922)
Mirabilis laevis var. crassifolia	Nyctaginaceae	Drought-deciduous subshrub	289	0.944 (0.922–0.952)
Opuntia littoralis	Cactaceae	Succulent shrub	79	0.978 (0.934–0.986)
Rhus integrifolia	Anacardiaceae	Evergreen shrub	196	0.977 (0.971–0.981)
Ribes speciosum	Grossulariaceae	Drought-deciduous shrub	138	0.977 (0.968–0.984)
Salvia apiana	Lamiaceae	Drought-deciduous shrub	305	0.948 (0.939–0.967)
Salvia leucophylla	Lamiaceae	Drought-deciduous shrub	60	0.982 (0.923–0.993)
Salvia mellifera	Lamiaceae	Drought-deciduous shrub	325	0.965 (0.946–0.972)
Xylococcus bicolor	Ericaceae	Evergreen shrub	189	0.977 (0.963–0.986)

Number of herbarium record localities (N) and overall model performance measured as the mean test area under the receiver operating characteristic curve (AUC) score (min–max). Taxonomy follows the second edition *The Jepson Manual: Vascular Plants of California* [48].

Survey (USGS) [9]. These spatially explicit, high resolution (250 m) maps were constructed for 84 ecoregions across the conterminous United States using the FORecasting SCEnarios of future land-use (FORE-SCE) model [57] and integrate both "top-down" drivers of land cover/land use such as demographic change, and local-scale "bottom-up" drivers such as biophysical site conditions [9]. The A1B scenario we considered in our study represents one possible future storyline characterized by rapid economic growth, a global population that peaks in mid-century and declines thereafter, rapid technological innovation, balanced energy sources, and active management of resources [3]. Under this scenario, projected urban growth is high, particularly in coastal areas and near urban centers, large increases in biofuel and food production drive large expansions in agricultural lands, and increased fragmentation of natural land covers is projected in regions with well-developed infrastructure and abundant natural resources [9]. The USGS land use-land cover maps provide five categories of human land use: developed, representing areas of both intensive (e.g., urban and built-up environments, high density housing) and less intensive (e.g., low density housing, parks) uses, cultivated cropland, mechanically disturbed representing areas of forest harvest and clear-cut logging, mining, and hay/pasture. We examined each human land use category individually as well as combined into a single "anthropogenic land use" category. We resampled the 250 m resolution land use-land cover data to match the 1 km resolution of our climate data using the nearest neighbor method in ArcGIS 10.0 (ESRI, Redlands, CA, USA).

Relative projected land use and climate change impacts

We overlaid habitat suitability maps with current and projected land use data to estimate the percent change in habitat under three scenarios, land use only, climate change only, and combined land use and climate change, and two time periods, early century (2000–2050) and late century (2050–2080). We assumed a complete loss of habitat in areas of anthropogenic land use (developed, mechanically disturbed, mining, cultivated croplands, and hay/pasture). All percentages of habitat change were relative to the amount of unconverted current climatically suitable habitat in 2000. In combined land use and climate change scenarios, we calculated the percent overlap in habitat lost or gained due to projected climate change and habitat lost due to projected land use. To investigate the potential impact of both drivers on CSS diversity, we overlaid projected land use change maps with maps of species richness change due to projected climate change. We limited our comparisons of projected land use and climate change impact on individual CSS species to California and CSS species richness to Central Coast and South Coast California Ecoregions. Maxent habitat suitability models, however, were based upon the full range of location data (California and Baja California, Mexico), as using spatially truncated locality data can lead to over-predicting range losses under future climate change [58]. All spatial analyses and model visualizations were performed in ArcMap 10.0 (ESRI, Redlands, CA, USA).

Results

Projected land use-land cover change

Already, anthropogenic surfaces cover 15.6% of the Central Coast and 24.3% of the South Coast (Table S1, Fig. 2). Under the A1B scenario considered in our study, USGS land use-land cover projections show increasing anthropogenic land uses converting natural land covers at rates of 81 km^2yr^{-1} (Central Coast) and

117 km^{2}yr^{-1} (South Coast) early century (2000–2050), slowing to 78 km^{2}yr^{-1} and 75 km^{2}yr^{-1} late century (2050–2080) (Table S1). By 2080, anthropogenic land uses may cover a total of 14,107 km^2, or nearly 30%, of the Central Coast and 19,628 km^2, just over 40%, of the South Coast (Fig. 2). Much of this conversion will be driven by increasing development concentrated near major metropolitan areas such as the San Francisco Bay Area, Los Angeles, and San Diego (Fig. S1). By 2080, development is projected to increase by 4,661 km^2 (Central Coast) and 8,379 km^2 (South Coast) (Table S1). Cultivated croplands are also projected to increase in the Central Coast, covering an additional 1,413 km^2 by 2080. Unlike development, which is projected to slow late century in both ecoregions, the expansion of agricultural land uses (cultivated crops and hay/pasture) is projected to increase at a greater rate late century, though only in the Central Coast. In the South Coast, increasing agricultural land uses will be paired with intensive development of already existing agricultural lands, especially early century, leading to net losses by 2080 (Table 2, Table S1). Much of the projected increase in anthropogenic land uses in the Central and South Coast will be at the expense of grasslands and shrublands (Table 2). The high projected conversion of shrublands is particularly pertinent to CSS, which is included within the shrubland land cover. Projected shrubland decline is greatest in the South Coast with a 5,658 km^2 (26.4%) loss from 2000 to 2080, 4,575 km^2 of which will be due to development, the greatest type of projected land use-land cover change for the ecoregion. In the Central Coast, shrublands are projected to decline by 1,347 km^2 (15.2%) from 2000 to 2080, with development, cultivated croplands, and hay/pasture driving similar degrees of conversion, 423–458 km^2 (Table 2).

Current land use impact on CSS habitat

Overall, Maxent models performed well with a median AUC score of 0.964 (range: 0.904–0.991) for the 20 shrub species in our study, suggesting climate is important in determining regional patterns of CSS habitat suitability (Table 1). Currently, a median 34.7% (range = 19.4–47.5%) of climatically suitable habitat of CSS species has already been converted to anthropogenic land uses (Table S2), with development driving the greatest loss in habitat (median 20.6%, range = 10.4–35.0%), followed by cultivated croplands (median 7.5%, range = 5.6–12.2%), and hay/pasture (median 3.9%, range = 2.1–5.2%). Six species have already lost over 40% of their climatically suitable habitat: *Xylococcus bicolor*, *Rhus integrifolia*, *Encelia californica*, *Opuntia littoralis*, *Bahiopsis laciniata*, and *Cneoridium dumosum* which has the greatest percentage of converted suitable habitat (47.5%). These species are distributed primarily in coastal southern California to northwestern Baja California and are most impacted by land development, which accounts for over 30% of current habitat losses in California. Widely distributed CSS shrub species ranging beyond coastal habitats in central and southern California (*Hesperoyucca whipplei*, *Mimulus aurantiacus*, *Eriogonum fasciculatum*, *Hazardia squarrosa*, *Acmispon glaber*) tend to have a lower degree of total habitat conversion and lower impact of development, which accounts for less than 15% of their current habitat losses (Table S2).

Relative impacts of projected land use and climate change

Habitat loss. Under the A1B scenario we considered, anthropogenic land use alone is projected to drive a median 22.1% loss (range = 12.7–34.8%) of currently unconverted climatically suitable habitat for dominant CSS shrub species by 2050 and an additional median 9.1% loss (5.9–11.0%) from 2050 to 2080 (Table 3). Projected development poses the greatest threat to all species, accounting for a median 16.7% (8.9–30.4%) habitat loss early century (2000–2050), nearly 4 times greater than the combined threats of all other anthropogenic land uses during that period (Table 3). Late century (2050–2080) development poses a

Projected Land Cover-Land Use Change (IPCC-SRES A1B)

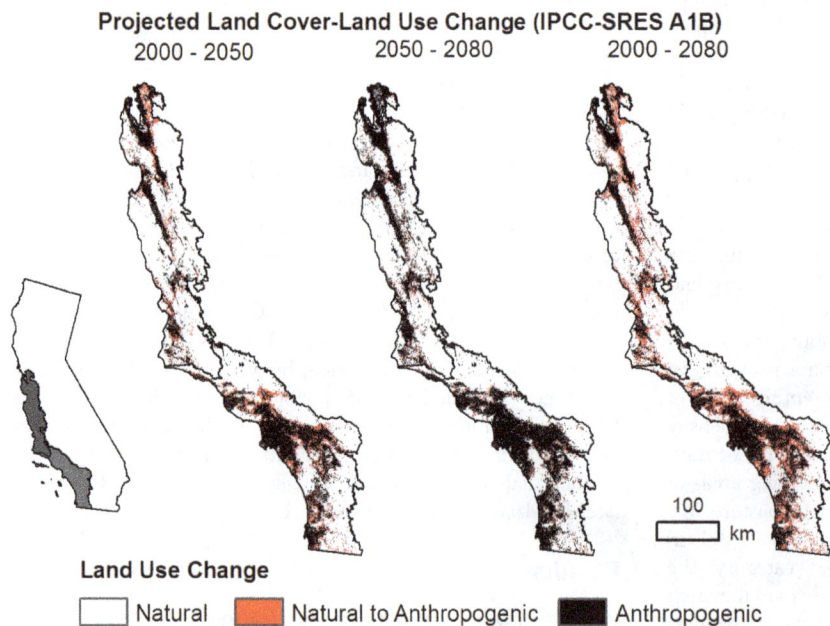

Figure 2. Projected Anthropogenic Land Use Change Maps. Projected 21st century change in anthropogenic land use (2000–2050, 2050–2080, 2000–2080) under the IPCC-SRES A1B scenario. Anthropogenic uses include developed areas, cultivated crops, hay/pasture, mining, and mechanically disturbed (logged) land. Land use-land cover maps were resampled to 1 km resolution from the USGS LandCarbon 250 m resolution land use-land cover maps for the continental United States [9].

Table 2. Primary projected changes in land use-land cover for the Central Coast and South Coast California Ecoregions.

Land Cover	2000–2050		2050–2080		2000–2080	
	Area (km²)	% Ecoregion	Area (km²)	% Ecoregion	Area (km²)	% Ecoregion
Central Coast						
Grassland to developed	2047	4.1	716	1.4	2791	5.6
Grassland to cultivated cropland	786	1.6	762	1.5	1550	3.1
Cultivated cropland to developed	600	1.2	236	0.5	789	1.6
Grassland to hay/pasture	359	0.7	241	0.5	494	1.0
Hay/pasture to developed	285	0.6	128	0.3	478	1.0
Shrubland to cultivated cropland	275	0.6	228	0.5	458	0.9
Shrubland to developed	243	0.5	131	0.3	433	0.9
Shrubland to hay/pasture	222	0.4	225	0.5	423	0.9
South Coast						
Shrubland to developed	3181	6.7	1239	2.6	4575	9.6
Grassland to developed	1763	3.7	359	0.8	2128	4.5
Cultivated cropland to developed	811	1.7	330	0.7	1002	2.1
Shrubland to cultivated cropland	533	1.1	488	1.0	891	1.9
Hay/pasture to developed	382	0.8	110	0.2	464	1.0
Shrubland to hay/pasture	130	0.3	61	0.1	140	0.3
Grassland to cultivated cropland	75	0.2	50	0.1	122	0.3
Evergreen forest to developed	70	0.1	31	0.1	102	0.2

The change in land use-land cover is shown as an area (km²) and as the percent of total land area in each ecoregion. Calculations are based on 1 km resampled USGS LandCarbon 250 m land use-land cover maps [9].

lesser but still notable threat, accounting for a median 5.3% (3.3–8.1) habitat loss, which is 1.8 times greater than the combined median losses of other anthropogenic land uses during that period. Habitat conversion to cultivated cropland and hay/pasture pose relatively low threats, driving median 3.2% (2.0–4.7%) and 2.2% (1.6–3.5%) losses early century and 1.2% (<0.1–0.6%) and 0.8% (0–0.7%) losses late century, respectively. Mining and logging poses negligible threats, accounting for less than 0.1% of total habitat losses. The species with the greatest projected habitat losses continue to be those distributed primarily in coastal southern California and northwestern Baja California, with six species projected to lose 40% or more of unconverted, climatically suitable habitat to anthropogenic land uses by 2080 (Table S2).

In comparison, projected climate change has a much more variable impact on CSS species, potentially driving both losses and gains in climatically suitable habitat (Fig. 3). Overall, we find

patterns of northern habitat expansion and southern habitat contraction with 21st century climate change consistent across warmer wetter and warmer drier climate change trajectories. Assuming scenarios of no dispersal, where species cannot disperse into new climatically suitable habitat, early century climate change alone is projected to drive a median habitat loss of 16.0% (0.2–50.2%) under a warmer wetter future and 14.3% (3.6–43.6%) under a warmer drier future (Table 4). Projected losses increase by a median 9.3% (0.2–24.3%; warmer wetter) and 8.8% (2.0–20.8%; warmer drier) late century, resulting in median cumulative projected habitat losses of 24.6% (0.5–83.0%) and 24.5% (6.8–61.2%), respectively, by the end of the century (2080s). Ten species show cumulative projected climate-driven habitat losses of greater than 40% under at least one climate scenario (Table S3, Table S4). In contrast to the pattern of greater projected land use impact for more narrowly distributed south-coastal species, we find large

Table 3. Summary of loss of CSS species suitable habitat to projected anthropogenic land uses.

Land use type	Median (min–max) percent loss in CSS habitat		
	2000–2050	2050–2080	2000–2080
Development	16.7 (8.9–30.4)	5.3 (3.3–8.1)	22.7 (12.4–39.5)
Cultivated crops	3.2 (2–4.7)	2.2 (1.6–3.5)	4.9 (3.4–7.7)
Hay/pasture	1.2 (0.4–1.8)	0.8 (0.3–1.2)	1.9 (0.6–2.6)
Other	0.1 (<0.1–0.6)	<0.1 (0–0.7)	0.1 (0–0.8)
Total anthropogenic land use	22.1 (12.7–34.8)	9.1 (5.9–11)	31.2 (18.3–45.6)

"Other" land use category includes mining and mechanically disturbed (logging) categories. Table values are the median percent loss (min–max) of modeled current climatically suitable habitat that was unconverted in 2000.

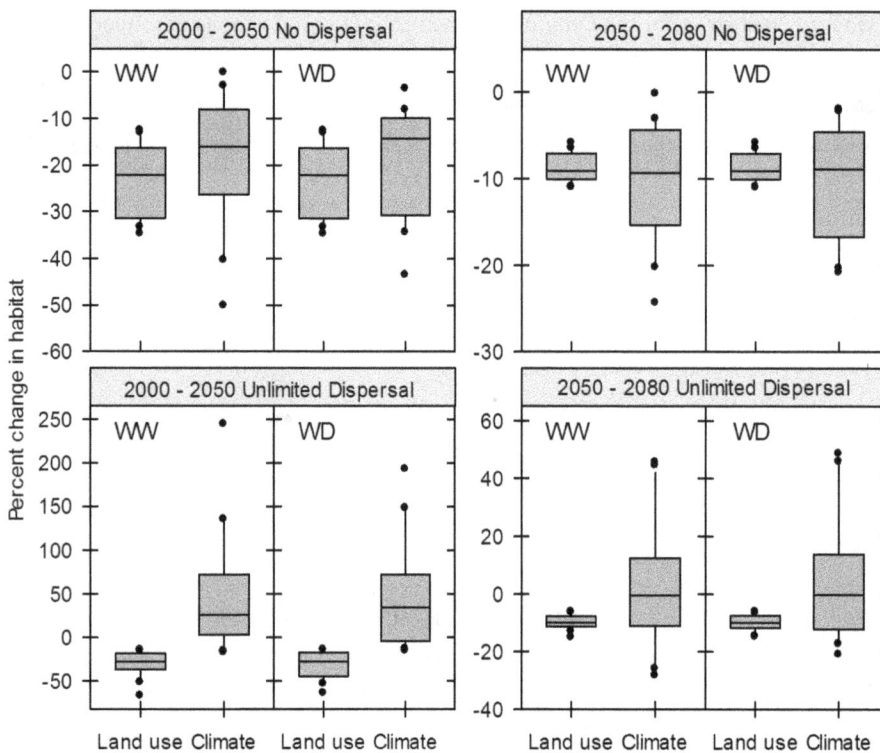

Figure 3. Relative Impacts of Projected Land Use and Climate Change on CSS Species Habitat. Boxplots showing the percent change in CSS species habitat due to projected land use and climate change under the A1B IPCC-SRES scenario. Climate scenarios are abbreviated as WW (warmer wetter CCCMA CGC 3.1) and WD (warmer drier NCAR CCSM 3.0). No dispersal and unlimited dispersal scenarios are shown separately.

habitat losses from projected climate change for both widely distributed (e.g., *E. fasciculatum*) and south-coastal species (e.g., *X. bicolor*).

When impacts due to projected land use are combined with those due to projected climate change under assumptions of no dispersal, overall habitat losses increase to a median 35.1% (23.0–64.6%, warmer wetter) and 37.8% (26.3–51.4%, warmer drier) by the 2050s (Table 4). Late century projected losses increase an additional median 14.7% (9.6–25.4%, warmer wetter) and 16.0% (10.2–24.2%, warmer wetter), leading to median cumulative losses

of ~50%. Only two species, *Artemisia californica* and *Ericameria ericoides*, show projected losses of less than 40% under at least one climate scenario. We also find notable spatial overlap in habitat loss from both drivers (Fig. 4), a median 26.9% (15.8–45.5%, warmer wetter) and 24.5% (13.4–41.5%, warmer drier) of climate-driven losses overlap with projected land use early century, and a median 12.9% (9.7–25.8%, warmer wetter) and 12.5% (6.6–22.5%, warmer drier) overlap late century.

Assuming unlimited dispersal where species can fully expand into all areas of new climatically suitable habitat, climate change is

Table 4. Summary of percent change in suitable CSS species habitat under 21ˢᵗ century land use and climate change scenarios.

Scenario	GCM	Dispersal	Median (min to max) percent change in habitat		
			2000–2050	2050–2080	2000–2080
Land use only	None	N/A	−22.1 (−34.8 to −12.7)	−9.1 (−11 to −5.9)	−31.2 (−45.6 to −18.3)
Climate change only	WW	No	−16 (−50.2 to −0.2)	−9.3 (−24.3 to −0.2)	−24.6 (−83 to −0.5)
		Yes	25.7 (−17.4 to 243.9)	−0.6 (−28.4 to 45.6)	24.5 (−50 to 288.3)
	WD	No	−14.3 (−43.6 to −3.6)	−8.8 (−20.8 to −2)	−24.2 (−61.2 to −6.8)
		Yes	34.5 (−15.3 to 192.4)	−0.2 (−21.1 to 48.6)	35.4 (−32.6 to 240.1)
Climate change and land use	WW	No	−35.1 (−64.6 to −23)	−14.7 (−25.4 to −9.6)	−46.7 (−89.9 to −30.2)
		Yes	11.8 (−43.6 to 176.9)	−7 (−37.5 to 21.8)	1.4 (−62.8 to 193.9)
	WD	No	−37.8 (−51.4 to −26.3)	−16 (−24.2 to −10.2)	−54 (−75.4 to −35.3)
		Yes	9.8 (−25.4 to 130)	−9.4 (−26.1 to 23.3)	1.3 (−43.1 to 151.9)

Table values are the median percent change (min to max) of modeled current climatically suitable habitat that was unconverted in 2000. Climate scenarios (GCM) are abbreviated as WW (warmer wetter; CCCMA CGC 3.1) and WD (warmer drier; NCAR CCSM 3.0).

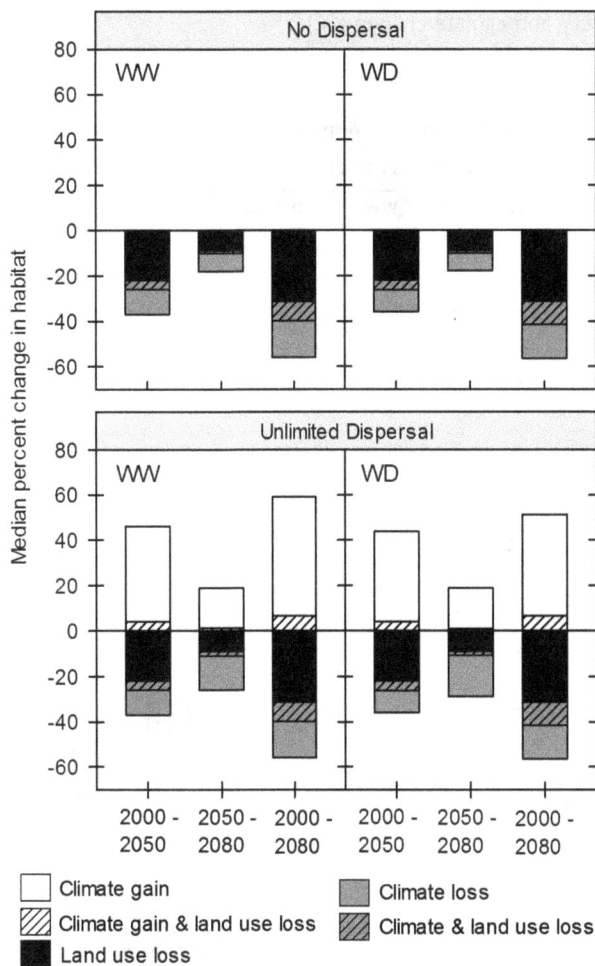

Figure 4. Median Percent Change in CSS Species Habitat from Projected Land Use and Climate Change. Plots show the median percent changes in species habitat due to projected land use and climate change, including percent overlap of impacts, under each climate and dispersal scenario. Climate scenarios are abbreviated as WW (warmer wetter CCCMA CGC 3.1) and WD (warmer drier NCAR CCSM 3.0).

to climate change alone, though the degree of climate change impact remains highly variable among species.

When impacts from projected land use are combined with those from projected climate change under assumptions of unlimited dispersal we find only a slight, cumulative net gain in species habitat by the end of the century, with median projected habitat increases of just 1% between 2000 and the 2080s under both climate scenarios (Table 4). This is due to projected habitat conversion to anthropogenic land uses in both the climatically stable portions of species ranges and areas of projected habitat gain from climate change (Fig. 4). Early century, a median 8.8% (2.3–32.7%, warmer wetter) and 9.1% (2.9–19.1%, warmer drier) of the habitat gained under climate change will be lost to projected anthropogenic land use. Similarly, a median 5.5% (0.8–0.11.6%, warmer wetter) and 4.4% (2.3–7.2%, warmer drier) of the habitat gained late century will be lost to projected anthropogenic land use.

When considered separately, the relative impacts of projected land use and climate do not differ significantly across CSS species during either early or late century under assumptions of no dispersal (all $P>0.20$; two-tailed Wilcoxon Signed Rank Test). While not statistically significant, we do find a pattern of increasing relative threat from climate change during the second part of the century. Early century, projected land use poses a greater threat than projected climate change for 13 species, decreasing to 9–10 species late century, depending upon climate change scenario (Table 5). Under assumptions of unlimited dispersal, habitat gains from projected climate change offset some of the habitat losses also due to climate change, resulting in a greater relative threat of projected land use for all 20 CSS species early century (Table 5). Similar to the no dispersal scenario, the threat posed by projected climate change begins to increase late century with 6–7 species showing greater losses in habitat due to projected climate change relative to losses due to projected land use. Overall, land use is projected to drive greater habitat losses relative to climate change both early and late century under assumptions of unlimited dispersal (all $P<0.04$; two-tailed, paired Wilcoxon Signed Rank test).

Species richness. Under the scenarios considered in our study, patterns of species richness under projected climate change varied geographically across ecoregions (Fig. 5). Currently, modeled CSS shrub species richness is centered in the South Coast Ecoregion in lowland, coastal areas. Under both future climate scenarios, we find considerable declines in modeled species richness due to climate change alone throughout much of the South Coast resulting from the southern contraction of climatically suitable habitat in many CSS species. Under best-case, unlimited dispersal scenarios, only about a quarter (23.3–28.7%) of the land area in the South Coast is projected to experience a net gain in richness due to climate change by the 2080s and 66.4–70.4% of the ecoregion will experience a net loss in richness (Table 6). Losses in species richness due to climate change alone will be greatest under assumptions of no dispersal, with >75% of the ecoregion losing richness. We find a much different outcome in the Central Coast Ecoregion, where northern habitat expansion due to projected climate change for many of our species, assuming unlimited dispersal, results in increased species richness by the end of the century. Assuming unlimited dispersal, over 70% of the Central Coast could experience an increase in species richness due to climate change by the end of the century, with only 12.9–21.8% of the ecoregion experiencing losses in richness (Table 6). Even assuming no dispersal, Central Coast losses in species richness due to climate change are less severe than the South Coast, covering

projected to drive habitat gains for many of CSS species we considered, offsetting some habitat losses (Table S3, Table S4). Projected climate change impacts, however, are highly variable across species (Fig. 3): four species show net habitat gains of 100–200% by the end of the century (2080s) under at least one climate scenario before accounting for projected land use, while nine species show projected net habitat losses under at least one climate scenario (Table S3, Table S4). Under a warmer wetter climate scenario we find a median net habitat gain of 25.7% (−17.4–243.9%) early century and a median net habitat loss of 0.6% (−28.4–45.6%) by late century (Table 4). Under a warmer drier climate scenario, we find a median net habitat gain of 34.5% (−15.3–192.4%) early century and a median net habitat loss of 0.2% (−21.1–48.6%) late century. While nearly all (18) of the CSS species we considered show projected net habitat gains from climate change under at least one climate scenario early century, this number drops to 12 species late century, with only six species showing projected net habitat gains under both climate scenarios (Table 5). Overall, by the end of the century, the majority of species still show a cumulative net increase in suitable habitat due

Table 5. Sensitivity of individual CSS species to projected land use (LU) and climate change (CC).

Species	Net CC gain (unlimited dispersal)				LU>CC (unlimited dispersal)				LU>CC (no dispersal)			
	2000–2050		2050–2080		2000–2050		2050–2080		2000–2050		2050–2080	
	WW	WD	WW	WD	WW	WD	WW	WD	WW	WD	WW	WD
Artemisia californica	X	X	X	X	X	X	X	X	X	X	X	X
Encelia californica	X	X	X	X	X	X	X	X	X	X	X	X
Acmispon glaber	X	X	X	X	X	X	X	X	X	X	X	X
Malosma laurina	X	X	X	X	X	X	X	X	X	X	X	X
Opuntia littoralis	X	X	X	X	X	X	X	X	X	X	X	X
Rhus integrifolia	X	X	X	X	X	X	X	X	X	X	X	X
Isocoma menziesii	X	X		X	X	X	X	X	X	X	X	X
Ericameria ericoides	X	X		X	X	X		X	X	X	X	X
Bahiopsis laciniata	X	X		X	X	X		X	X	X		X
Mirabilis laevis var. crassifolia	X	X	X		X	X	X	X	X	X		
Salvia mellifera	X	X	X		X	X	X	X	X	X		
Ribes speciosum	X	X			X	X	X	X	X	X		
Salvia apiana	X	X			X	X						
Mimulus aurantiacus	X		X		X	X	X					
Cneoridium dumosum		X			X	X	X	X	X	X		X
Xylococcus bicolor		X			X	X		X				
Hesperoyucca whipplei	X				X	X						
Salvia leucophylla	X				X	X	X				X	
Eriogonum fasciculatum					X	X						
Hazardia squarrosa					X	X						

Climate scenarios are abbreviated as WW (warmer wetter; CCCMA CGC 3.1) and WD (warmer drier; NCAR CCSM 3.0). Species with an "X" in the first four columns have projected net habitat gains due to climate change assuming unlimited dispersal. An "X" in the remaining columns indicates a greater threat of habitat loss due to projected land use relative to that due to projected climate change. Species sensitivities are broken into unlimited dispersal and no dispersal assumptions.

just over half (51.8–55.8%) of the ecoregion by the end of the century.

After factoring in projected land use change, we find a disproportionate degree of spatial overlap between areas having losses in species richness due to climate change and areas that are either (1) already converted to anthropogenic land uses or (2) will be converted to anthropogenic land uses (Table 6). This pattern is most pronounced in the South Coast, where high rates of projected land use, particularly development, coincide with considerable habitat contraction driven by climate change for most of the CSS species considered in our study. Both already converted areas and natural areas with projected conversion to anthropogenic land uses in the South Coast have a significantly greater percentage of overlap with areas of projected species richness loss from climate change compared to natural areas in the South Coast without projected anthropogenic conversion (Table 6; all $P<0.0001$ after Bonferroni correction for multiple comparisons; Pearson's Chi-squared test). For example, assuming unlimited dispersal and a warmer wetter future climate in 2050, 82.6% of natural South Coast land with projected anthropogenic conversion ("natural to anthropogenic") will undergo a loss in species richness due to projected climate change by 2080, compared to only 35.9% of South Coast natural lands without projected anthropogenic conversion ("natural"). We also find higher median losses in species richness due to climate change for areas with projected conversion to anthropogenic land uses compared to natural areas without projected conversion, a pattern

that is consistent across time periods, climate scenarios, and dispersal scenarios (Figs. S2, S3).

For the most part, however, this pattern does not hold up for the Central Coast where projected species richness losses and gains due to climate change are similar across categories of projected anthropogenic land use change (Table 6, Figs. S2, S3). Only under a warmer wetter future climate in the 2080s (both dispersal scenarios), do we find a significantly greater percentage of overlap in projected species richness loss due to climate change and projected anthropogenic conversion compared to natural areas without projected anthropogenic conversion (all $P<0.0001$ after Bonferroni correction for multiple comparisons; Pearson's Chi-squared test). Thus, our results indicate that compounding impacts of projected land use and climate change on CSS will be centered primarily in the South Coast Ecoregion under the future scenarios considered in our study.

Discussion

Given the current unprecedented rate of environmental change, successful conservation of biodiversity this century must address the potential impacts of both future land use and climate change on species and ecosystems. The disconnect between relatively high resolution downscaled climate projections and the coarse resolution treatment of land use-land cover change projections, however, has led to the exclusion of future land use from assessments of future environmental change impacts on species and ecosystems,

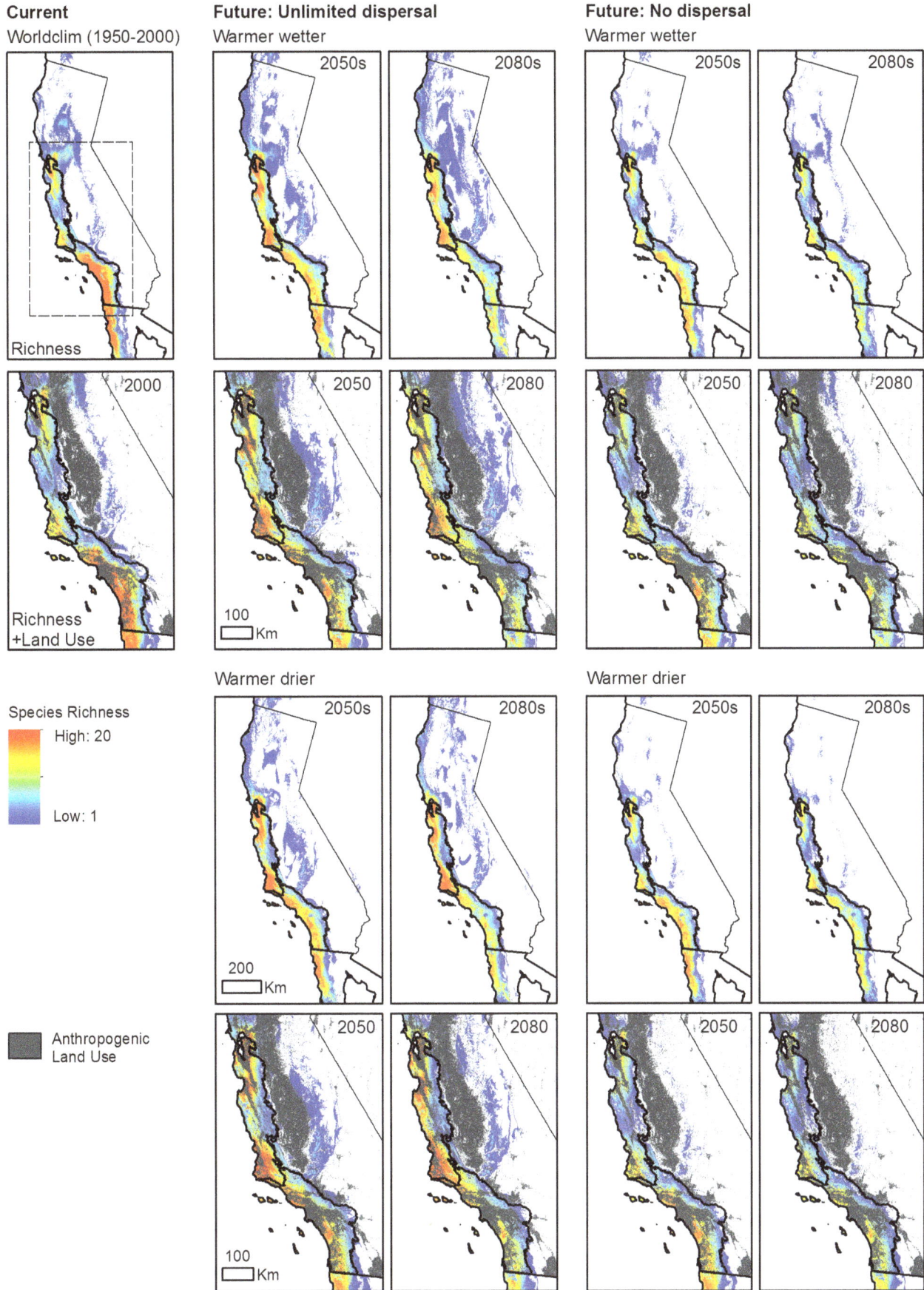

Current
Worldclim (1950-2000)

Future: Unlimited dispersal
Warmer wetter

Future: No dispersal
Warmer wetter

Richness

2000

Richness
+Land Use

2050s 2080s

2050s 2080s

2050 2080

2050 2080

Species Richness
High: 20

Low: 1

Warmer drier

Warmer drier

2050s 2080s

2050s 2080s

100 Km

200 Km

Anthropogenic
Land Use

2050 2080

100 Km

2050 2080

Figure 5. Projected Climate Change and Land Use Impacts on CSS Richness. Current (2000) and projected early century (2050s) and late century (2080s) CSS species richness under two climate change scenarios (top rows: warmer wetter CCCMA CGC 3.1, bottom rows: warmer drier NCAR CCSM 3.0) assuming unlimited dispersal and no dispersal. Anthropogenic land use is shown as gray overlays and includes developed, mining, mechanically disturbed (logging), cultivated croplands, and hay/pasture land uses.

which typically require local to regional scale analyses [9]. By using newly developed, downscaled scenario-based land use projections [9,59], we were able to link future impacts of projected climate change and land use for CSS and examine multiple drivers of environmental change under a consistent future emission storyline (SRES A1B) at a regional scale and larger spatial extent than has previously been possible [34,60].

Our analyses show that during the 21st century, projected land use could pose a threat to CSS that is as large as that posed by projected climate change, if not larger. For many species, projected anthropogenic land use drove greater habitat losses compared to projected climate change, particularly during the first half of the century, which was consistent with the high rates of habitat conversion projected this century in both Central and South Coast Ecoregions. Interestingly, this pattern was only significant under assumptions of unlimited dispersal, where considerable gains in climatically suitable habitat offset some of the concurrent habitat losses from projected climate change. Projected land use within areas of climate-driven habitat gains, however, lowered potential habitat increases in scenarios of combined climate change and land use. We found an increase in the number of species where projected climate change drove greater habitat losses relative to land use during the second half of the century, suggesting that impacts from climate change may rise

through the end of the century. Therefore, mitigating climate change impacts may become increasingly important for CSS management and conservation.

Overall, projected climate change impacts were highly variable across CSS species and heavily dependent on dispersal assumptions, highlighting the importance of dispersal in moderating habitat losses from both land use and climate and suggesting species responses to climate change may be highly individualized. The broad dispersal capacities of many CSS species with small, wind dispersed seeds (e.g., *A. californica, E. californica, E. fasciculatum, M. aurantiacus*) [46], could facilitate a northward expansion of CSS under projected climate change. Such range shifts, however, will also depend upon successful establishment and recruitment of CSS species within the mosaic of chaparral, CSS, and grasslands that currently dominates much of the central coast. These processes will ultimately be driven by both regional (climate) and local factors such as topography, geological substrate, disturbance, and species interactions [24,43,44,46,61–69].

For example, many wind-dispersed CSS species are able to invade areas of chaparral opened by disturbances such as fire [25,65,70]. However, short return intervals of fire, which will also be influenced by projected climate change [71] and urbanization [34], as well as high levels of other anthropogenic disturbances, facilitate conversion of shrublands to exotic grasslands [27,46,72].

Table 6. Percent of land area with projected loss and gain in CSS species richness due to climate change summarized by land use category.

Ecoregion	GCM	Land use	Land area (km^2)		% Land area with richness loss		% Land area with richness gain	
			2050	2080	2050	2080	2050	2080
Central Coast	WW	Natural	37646	35324	1.8 (37.76)	11.3 (47.7)	93.3	82.2
		Natural to Anthropogenic	4124	6448	0.5 (35.0)	14.6 (59.7)	96.3	76.5
		Anthropogenic	7662	7667	0.4 (32.3)	19 (64.4)	96.0	70.7
		Total	*49432*	*49439*	*1.5 (36.7)*	*12.9 (51.8)*	*94.0*	*79.6*
	WD	Natural	37646	35324	13 (43.0)	22 (54.3)	80.2	72.0
		Natural to Anthropogenic	4124	6448	4.1 (32.5)	21.6 (58.4)	91.0	71.1
		Anthropogenic	7662	7667	2.4 (33.6)	21.4 (60.5)	93.5	71.2
		Total	*49432*	*49439*	*10.6 (40.6)*	*21.8 (55.8)*	*83.2*	*71.7*
South Coast	WW	Natural	29864	27606	35.9 (48.1)	47.3 (60.4)	53.8	45.6
		Natural to Anthropogenic	5921	8179	82.6 (87.4)	88.6 (95.6)	12.3	8.3
		Anthropogenic	11527	11538	87.4 (90.8)	96.3 (98.0)	6.9	2.9
		Total	*47312*	*47323*	*54.3 (63.4)*	*66.4 (75.7)*	*37.2*	*28.7*
	WD	Natural	29864	27606	41.3 (53.3)	54.3 (65.4)	44.6	36.4
		Natural to Anthropogenic	5921	8179	80.7 (84.6)	89.3 (93.2)	12.4	7.4
		Anthropogenic	11527	11538	87.4 (89.4)	95.7 (97.5)	8.1	3.2
		Total	*47312*	*47323*	*57.4 (66.1)*	*70.4 (78.0)*	*31.7*	*23.3*

Climate scenarios (GCM) are abbreviated as WW (warmer wetter; CCCMA CGC 3.1) and WD (warmer drier; NCAR CCSM 3.0). Values in parentheses correspond to no dispersal scenarios, all other values correspond to unlimited dispersal scenarios. "Natural" land use corresponds to currently unconverted natural areas that remain unconverted under projected land use change. "Natural to anthropogenic" land use corresponds to currently unconverted areas that will be converted to anthropogenic uses under projected land use change. "Anthropogenic" land use corresponds to currently converted areas that will remain converted under projected land use change. The total area covered by each land use category is reported in Land area (km^2).

Nitrogen deposition from pollution further reinforces this conversion [30,73] and may impede the successful establishment of CSS in new, climatically suitable habitats under climate change. Furthermore, habitat fragmentation poses a formidable barrier to species migration, severely limiting the ability of a species to disperse across a landscape [74]. Thus, the future dynamics of CSS expansion will likely be complex, governed by many factors and processes that are also influenced by anthropogenic change.

Our findings also highlight the potential for future land use and climate change to have compounding negative impacts on CSS, particularly in the South Coast, where we find high geographic overlap in habitat losses driven by projected climate change and projected anthropogenic conversion. The rate of climate change and degree of habitat loss from land use both have thresholds beyond which the probability of population extinction becomes increasingly likely [75,76]. As anthropogenic land use drives habitat loss to a threshold where local populations no longer have sufficient available habitat to persist, concurrent climate change may also surpass a critical rate at which population extinction becomes likely, the position of which may be lowered by habitat loss from land use [76]. Additionally, modeling studies indicate that habitat fragmentation and habitat quality may impact population extinction thresholds such that more habitat is required for population persistence in a fragmented landscape [77]. Fragmented landscapes may also have greater sensitivities to climate change [76].

California sage scrub's coastal distribution in lowland and relatively fertile areas with sizable human populations makes it particularly vulnerable to habitat conversion and fragmentation from development and agriculture [78]. Under a future of increasing land use and climate change, this sensitivity to human land use may result in the loss of CSS species that otherwise may have been able to keep pace with climate change without the additional pressures from land use, especially those species with low colonization ability and/or poor dispersal. Furthermore, our models likely underestimate the impact of projected land use, as we do not address habitat fragmentation, just total habitat loss.

As individual species shift in geographic distribution in response to climate change, patterns of richness and species assemblage could also change dramatically. Under the scenarios considered in our study, we see the potential for increasing richness along the Central Coast resulting from the northern habitat expansion of many CSS species. Though not examined in our study, this could result in novel, no-analog assemblages of species, where future communities and species interactions have no modern-day equivalent. Such community projections under 21st century climate change have been shown for terrestrial breeding bird species in California [36]. In contrast, the South Coast could experience considerable declines in species richness from widespread southern habitat contraction under climate change, particularly in costal San Diego County, currently a region of high CSS floristic diversity and endemism [23–25,29]. Both community reassembly and losses in richness have implications for the numerous sensitive species associated with CSS [31,32].

Ultimately, the response of a species under climate change will be a function of the dynamics at both the leading (expanding) and trailing (low-latitude limit) edges of a species range [5,79], which could have dramatically different processes and mechanisms. Within this context, conservation and management objectives may need to diverge between the Central Coast, where many species may gain climatically suitable habitat, and the South Coast where many species may lose climatically suitable habitat. If 21st century climate change follows a similar trajectory as that examined in our study, the South Coast could represent the trailing edge of many

CSS species ranges. The combined pressures of projected land use and climate change could severely impact local extinction at this trailing edge of CSS species ranges. Meanwhile habitat loss and fragmentation from projected land use in the leading edge of species ranges in the Central Coast could have a large impact on species migrations, severely limiting the ability of species to expand into habitat newly suitable under climate change.

Limiting future habitat conversion and fragmentation from land use is a strategy that could be broadly beneficial, both in preventing further barriers to dispersal so species are more likely to keep pace with climate change, and in maintaining patches of habitat above thresholds where climate change may drive local extinctions. In the South Coast, conservation efforts may need to prioritize the protection of remaining high quality habitat and maintenance of trailing edge populations, which could include developing action plans that mitigate future impacts from both drivers and promote species resilience to climate change. The future persistence of CSS may also hinge upon the successful establishment of species along the Central Coast; however, the role for managers is more complex in this case. How actively should managers facilitate species movements? Does the northern expansion of CSS come at the expense of local natives or other vegetation types? One approach that is likely to benefit a variety of species, no just CSS, is to promote species movements through the protection of strategic migration corridors.

We present a possible future trajectory of change for CSS and our results should be viewed as a hypothesis of how CSS may be impacted by projected land use and climate change. While we provide a direct comparison of two major drivers of future impact for CSS through our use of linked land use and climate change projections, we considered just one (SRES A1B) of many possible future storylines. Additionally, species distribution models are inherently uncertain, from the mechanisms driving species distributions, imperfect modeling methods, to the trajectories of future climate change and the extrapolation of species responses in novel future climates outside the range of contemporary climate used to parameterize models [80]. In using two GCMs, we show a range in the possible trajectory of projected climate change. Different GCMs and future emission scenarios may show different patterns of climate change severity and impact across California. Although uncertainties can also arise from SDM algorithms [81,82], we chose a single algorithm, Maxent, rather than comparing multiple methods, as Maxent has high performance with the spatially biased, presence-only locality data used in our study [50]. Similarly, we use a single model and scenario of projected land use in California. While the land use projections used in our study are perhaps the most thematically and spatially detailed regional dataset available, they represent a first step in the development of tools and models [9], which will undoubtedly continue to be refined.

We were unable to compare or rank the uncertainty arising from each of our model components (climate projections, land use projections, modeling algorithm), which could potentially result in compounded uncertainty. Identifying the greatest sources of uncertainty within combined models is important to inform subsequent management decisions based upon model outputs. Sensitivity analysis may provide a method to compare sources of uncertainty in combined models and has been recently applied to coupled SDM and dynamic population models that incorporate combined impacts of environmental change (climate, land use, disturbance regimes) on species extinction risk [83].

The regional scale and 1 km spatial resolution of our analyses did not include microclimate variability that exists at finer spatial scales [35], which could result in over-predictions of current ranges and future habitat loss due to climate change. Steep microclimatic gradients, such as those due to rugged topography can facilitate species range shifts over shorter distances, making it more likely that a species could keep pace with changing climate. As our modeling framework did not address populations, which are typically dynamic and patchily distributed at local scales, our habitat maps also likely encompass geographically larger and more continuous areas than the current realized distribution of each species. In contrast, our estimates of land use impact are conservative, as we do not consider the additional effects of habitat fragmentation and degradation on CSS persistence and dispersal. Previous papers estimate anthropogenic activities have driven losses of up to 90% of CSS's original extent [25,27–29], considerably higher than our median estimate of 35% current habitat conversion for individual shrub species.

Coupled models can improve predictions by combining dynamic population models and SDMs [84,85], but require detailed species demographic data, which was not available for all 20 species we considered in our study. Additionally, our models do not incorporate dynamic processes that may buffer climate change, such as the capacity of a species for acclimation or adaptation to new environmental conditions. Nevertheless, our findings provide important insight and hypotheses into how 21st century projected land use and climate change may impact CSS species and patterns of species richness. They can best be applied in combination with careful monitoring of CSS and climate change and land use impacts in an adaptive management context.

In conclusion, we emphasize the necessity to include analyses of both projected land use and climate change in conservation and resource management planning. We illustrate the potential for land use and climate change to have compounding negative impacts on CSS, particularly in southern California. We show the potential for the dynamics of CSS to diverge geographically under scenarios of future change, with strikingly different patterns of impact in the Central Coast, which may contain the expanding edge for many species ranges, and the South Coast, which may contain the trailing edge of many species ranges. The persistence and extent of CSS will likely hinge upon the protection of remaining critical habitat in southern California as well as the successful dispersal and establishment of species along the coastal central California. Thus, in the context of future environmental change, conservation objectives and management strategies may need to differ across species ranges and ecoregions.

Supporting Information

Table S1   Current (2000) and projected (2050, 2080) land cover and projected rate of land cover change (km^2 yr^{-1}) by California ecoregion. Total anthropogenic land use is the sum of developed, cultivated cropland, hay/pasture, mining, and mechanically disturbed (logged) land uses. Current and projected land use-land cover data is from USGS LandCarbon [9].

Table S2   Projected conversion of current climatically suitable habitat of CSS species to anthropogenic land uses. The area (km^2) of currently unconverted suitable habitat (Unconvt.) and the percent of total currently suitable habitat already converted to anthropogenic land uses (Convt.) in 2000 are provided in the first two columns. Anthropogenic land uses are abbreviated as developed (D), cultivated crops (C), and hay/pasture (H/P). "Other" includes mechanically disturbed (logging) and mining. All habitat loss values are the percent change in

unconverted current (2000) climatically suitable habitat. Projected land use-land cover data is from USGS LandCarbon [9].

Table S3   Percent change in CSS habitat due to projected land use and climate change under the warmer wetter (CCCMA CGC 3.1) scenario for 2000–2050, 2050–2080, and 2000–2080. Abbreviations: climate change only scenario (CC only) and combined land use and climate change scenario (LU+CC).

Table S4   Percent change in CSS habitat due to projected land use and climate change under the warmer drier (NCAR CCSM 3.0) scenario for 2000–2050, 2050–2080, and 2000–2080. Abbreviations: climate change only scenario (CC only) and combined land use and climate change scenario (LU+CC).

Figure S1   Projected Land use-land cover Maps. USGS historical (2000) and projected (2050, 2080) land use-land cover maps in Central Coast and South Coast Ecoregions of California (back outlines). Projected land use-land cover corresponds to the IPCC-SRES A1B future scenario. All land cover data was resampled to 1 km resolution from the USGS LandCarbon 250 m resolution land-cover-land use maps for the continental United States [9].

Figure S2   Distribution of projected change in CSS richness due to climate change for different land use categories (Fig S2: unlimited dispersal, Fig S3: no dispersal). Figure panels show the distribution (percent of ecoregion land area) of projected change in CSS species richness due to climate change assuming unlimited dispersal (S2) and no dispersal (S3). The top row shows the distribution of projected CSS richness change for the entire ecoregion (Central Coast or South Coast). The next three rows show projected CSS richness change for three categories of projected land use (Natural, Natural to Anthropogenic, and Anthropogenic) within each ecoregion. "Natural" land use corresponds to currently unconverted natural areas that will remain unconverted under projected land use change. "Natural to anthropogenic" land use corresponds to currently unconverted areas that will be converted to anthropogenic uses under projected land use change. "Anthropogenic" land use corresponds to currently converted areas that will remain converted under projected land use change. Gray lines represent early century (2050s) modeled species richness and black lines represent late century (2080s) modeled species richness. The dashed line indicates zero change in species richness.

Figure S3   Distribution of projected change in CSS richness due to climate change for different land use categories (Fig S2: unlimited dispersal, Fig S3: no dispersal). Figure panels show the distribution (percent of ecoregion land area) of projected change in CSS species richness due to climate change assuming unlimited dispersal (S2) and no dispersal (S3). The top row shows the distribution of projected CSS richness change for the entire ecoregion (Central Coast or South Coast). The next three rows show projected CSS richness change for three categories of projected land use (Natural, Natural to Anthropogenic, and Anthropogenic) within each ecoregion. "Natural" land use corresponds to currently unconverted natural areas that will remain unconverted under projected land use change. "Natural to anthropogenic" land use corresponds to currently unconverted areas that will be converted to anthropogenic uses under projected land use change. "Anthropogenic" land use corresponds to currently converted areas that will remain converted under projected land use change. Gray lines represent early century (2050s) modeled species richness and black lines represent late century (2080s) modeled species richness. The dashed line indicates zero change in species richness.

Acknowledgments

We thank Victoria Sork (ULCA), Benjamin Sleeter (USGS), and two anonymous reviewers whose input greatly improved the quality of this manuscript.

Author Contributions

Conceived and designed the experiments: ECR PWR. Performed the experiments: ECR. Analyzed the data: ECR. Wrote the paper: ECR PWR.

References

1. Parmesan C (2006) Ecological and evolutionary responses to recent climate change. Annual Review of Ecology Evolution and Systematics 37: 637–669.
2. Parmesan C, Yohe G (2003) A globally coherent fingerprint of climate change impacts across natural systems. Nature 421: 37–42.
3. Meehl GA, Stocker TF, Collins WD, Friedlingstein P, Gaye AT, et al. (2007) Global climate projections. In: S. Solomon, D. Qin, M. Manning, Z. Chen, M. Marquis et al., editors. Climate Change 2007: The Physical Science Basis Contribution of Working Group I to the Fourth Assessment Report of the Intergovernmental Panel on Climate Change. Cambridge: Cambridge University Press. pp. 747–845.
4. Pearson RG (2006) Climate change and the migration capacity of species. Trends in Ecology & Evolution 21: 111–113.
5. Thuiller W, Albert C, Araujo MB, Berry PM, Cabeza M, et al. (2008) Predicting global change impacts on plant species' distributions: Future challenges. Perspectives in Plant Ecology Evolution and Systematics 9: 137–152.
6. Thomas CD, Cameron A, Green RE, Bakkenes M, Beaumont LJ, et al. (2004) Extinction risk from climate change. Nature 427: 145–148.
7. Thuiller W, Lavorel S, Araujo MB, Sykes MT, Prentice IC (2005) Climate change threats to plant diversity in Europe. Proceedings of the National Academy of Sciences of the United States of America 102: 8245–8250.
8. Sala OE, Chapin FS, Armesto JJ, Berlow E, Bloomfield J, et al. (2000) Biodiversity - Global biodiversity scenarios for the year 2100. Science 287: 1770–1774.
9. Sleeter BM, Sohl TL, Bouchard MA, Reker RR, Soulard CE, et al. (2012) Scenarios of land use and land cover change in the conterminous United States: Utilizing the special report on emission scenarios at ecoregional scales. Global Environmental Change 22: 896–914.
10. Barbet-Massin M, Thuiller W, Jiguet F (2012) The fate of European breeding birds under climate, land-use and dispersal scenarios. Global Change Biology 18: 881–890.
11. Jongsomjit D, Stralberg D, Gardali T, Salas L, Wiens J (2013) Between a rock and a hard place: the impacts of climate change and housing development on breeding birds in California. Landscape Ecology 28: 187–200.
12. Cowling RM, Rundel PW, Lamont BB, Arroyo MK, Arianoutsou M (1996) Plant diversity in Mediterranean-climate regions. Trends in Ecology & Evolution 11: 362–366.
13. Myers N, Mittermeier RA, Mittermeier CG, da Fonseca GAB, Kent J (2000) Biodiversity hotspots for conservation priorities. Nature 403: 853–858.
14. Underwood EC, Viers JH, Klausmeyer KR, Cox RL, Shaw MR (2009) Threats and biodiversity in the mediterranean biome. Diversity and Distributions 15: 188–197.
15. Fitzpatrick MC, Gove AD, Sanders NJ, Dunn RR (2008) Climate change, plant migration, and range collapse in a global biodiversity hotspot: the Banksia (Proteaceae) of Western Australia. Global Change Biology 14: 1337–1352.
16. Loarie SR, Carter BE, Hayhoe K, McMahon S, Moe R, et al. (2008) Climate change and the future of California's endemic flora. PLoS ONE 3: e2502.
17. Midgley GF, Hannah L, Millar D, Rutherford MC, Powrie LW (2002) Assessing the vulnerability of species richness to anthropogenic climate change in a biodiversity hotspot. Global Ecology and Biogeography 11: 445–451.
18. Yates CJ, Elith J, Latimer AM, Le Maitre D, Midgley GF, et al. (2010) Projecting climate change impacts on species distributions in megadiverse South African Cape and Southwest Australian Floristic Regions: Opportunities and challenges. Austral Ecology 35: 374–391.
19. Klausmeyer KR, Shaw MR (2009) Climate change, habitat loss, protected areas and the climate adaptation potential of species in Mediterranean ecosystems worldwide. PLoS ONE 4: e6392.
20. CNPS California Native Plant Society (2012) Inventory of rare and endangered plants (online edition, v8-01a). 2012 ed. Sacramento, CA: California Native Plant Society.
21. Sanstad AH, Johnson H, Goldstein N, Franco G (2011) Projecting long-run socioeconomic and demographic trends in California under the SRES A2 and B1 scenarios. Climatic Change 109: 21–42.
22. Hayhoe K, Cayan D, Field CB, Frumhoff PC, Maurer EP, et al. (2004) Emissions pathways, climate change, and impacts on California. Proceedings of the National Academy of Sciences of the United States of America 101: 12422–12427.
23. Epling C, Lewis H (1942) The centers of distribution of the chaparral and coastal sage associations. American Midland Naturalist 27: 445–462.
24. Westman WE (1983) Xeric mediterranean-type shrubland associations of Alta and Baja California and the community continuum debate. Vegetatio 52: 3–19.
25. Rundel PW (2007) Sage Scrub. In: Barbour MG, Keeler-Wolf T, Schoenherr AA, editors. Terrestrial vegetation of California. 3rd ed. Berkeley: University of California Press. pp. 208–228.
26. CA-DOF State of California Department of Finance (2011) Historical census populations of counties and incorporated cities in California, 1850–2010. Sacramento, CA.
27. Minnich RA, Dezzani RJ (1998) Historical decline of coastal sage scrub in the Riverside-Perris Plain, California. Western Birds 29: 366–391.
28. Taylor RS (2004) A natural history of coastal sage scrub in southern California: Regional floristic patterns and relations to physical geography, how it changes over time, and how well reserves represent its biodiversity. [PhD Dissertation]: University of California, Santa Barbara. 223 p.
29. Westman WE (1981) Diversity relations and succession in Californian coastal sage scrub. Ecology 62: 170–184.
30. Talluto MV, Suding KN (2008) Historical change in coastal sage scrub in southern California, USA in relation to fire frequency and air pollution. Landscape Ecology 23: 803–815.
31. DeSimone SA (1995) California's coastal sage scrub. Fremontia 23: 3:8.
32. O'Leary JF (1990) California coastal sage scrub: General characteristics and considerations for biological conservation. In: Schoenherr AA, editor; Claremont, CA. Southern California Botanists. pp. 24–41.
33. Feldman TD, Jonas AEG (2000) Sage scrub revolution? Property rights, political fragmentation, and conservation planning in Southern California under the federal Endangered Species Act. Annals of the Association of American Geographers 90: 256–292.
34. Syphard AD, Clarke KC, Franklin J (2007) Simulating fire frequency and urban growth in southern California coastal shrublands, USA. Landscape Ecology 22: 431–445.
35. Ackerly DD, Loarie SR, Cornwell WK, Weiss SB, Hamilton H, et al. (2010) The geography of climate change: implications for conservation biogeography. Diversity and Distributions 16: 476–487.
36. Stralberg D, Jongsomjit D, Howell CA, Snyder MA, Alexander JD, et al. (2009) Re-shuffling of species with climate disruption: A no-Analog future for California birds? PLoS ONE 4: e6825.
37. Williams JW, Jackson ST, Kutzbacht JE (2007) Projected distributions of novel and disappearing climates by 2100 AD. Proceedings of the National Academy of Sciences of the United States of America 104: 5738–5742.
38. Syphard AD, Regan HM, Franklin J, Swab RM, Bonebrake TC (2013) Does functional type vulnerability to multiple threats depend on spatial context in Mediterranean-climate regions? Diversity and Distributions 19: 1263–1274.
39. Pearson RG, Dawson TP (2003) Predicting the impacts of climate change on the distribution of species: are bioclimate envelope models useful? Global Ecology and Biogeography 12: 361–371.
40. Guisan A, Thuiller W (2005) Predicting species distribution: offering more than simple habitat models. Ecology Letters 8: 993–1009.
41. Guisan A, Zimmermann NE (2000) Predictive habitat distribution models in ecology. Ecological Modelling 135: 147–186.
42. Shreve F (1936) The transition from desert to chaparral in Baja California. Madroño: 257–264.
43. Kirkpatrick JB, Hutchinson CF (1980) Environmental relationships of Californian coastal sage scrub and some of its component communities and species. Journal of Biogeography 7: 23–38.
44. Westman WE (1981) Factors influencing the distribution of species of Californian coastal sage scrub. Ecology 62: 439–455.
45. Riordan EC, Rundel PW (2009) Modelling the distribution of a threatened habitat: the California sage scrub. Journal of Biogeography 36: 2176–2188.
46. Wells PV (1962) Vegetation in relation to geological substratum and fire in San Luis Obispo Quadrangle, Califonria. Ecological Monographs 32: 79-&.
47. DeSimone SA, Burk JH (1992) Local variation in floristics and distributional factors in Californian coastal sage scrub. Madroño 39: 170–188.
48. Baldwin BG, Goldman DH, Vorobik LA (2012) The Jepson Manual: Vascular Plants of California. Berkeley: University of California Press. 1568 p.
49. Phillips SJ, Anderson RP, Schapire RE (2006) Maximum entropy modeling of species geographic distributions. Ecological Modelling 190: 231–259.
50. Elith J, Graham CH, Anderson RP, Dudik M, Ferrier S, et al. (2006) Novel methods improve prediction of species' distributions from occurrence data. Ecography 29: 129–151.

51. Loiselle BA, Jorgensen PM, Consiglio T, Jimenez I, Blake JG, et al. (2008) Predicting species distributions from herbarium collections: does climate bias in collection sampling influence model outcomes? Journal of Biogeography 35: 105–116.

52. Hijmans RJ, Cameron SE, Parra JL, Jones PG, Jarvis A (2005) Very high resolution interpolated climate surfaces for global land areas. International Journal of Climatology 25: 1965–1978.

53. Flint LE, Flint AL, Thorne JH, Boynton R (2013) Fine-scale hydrologic modeling for regional landscape applications: the California Basin Character-ization Model development and performance. Ecological Processes 2.

54. Phillips SJ, Dudík M (2008) Modeling of species distributions with Maxent: new extensions and a comprehensive evaluation. Ecography 31: 161–175.

55. Liu CR, Berry PM, Dawson TP, Pearson RG (2005) Selecting thresholds of occurrence in the prediction of species distributions. Ecography 28: 385–393.

56. Liu CR, White M, Newell G (2013) Selecting thresholds for the prediction of species occurrence with presence-only data. Journal of Biogeography 40: 778–789.

57. Sohl T, Sayler K (2008) Using the FORE-SCE model to project land-cover change in the southeastern United States. Ecological Modelling 219: 49–65.

58. Barbet-Massin M, Thuiller W, Jiguet F (2010) How much do we overestimate future local extinction rates when restricting the range of occurrence data in climate suitability models? Ecography 33: 878–886.

59. Sohl TL, Sleeter BM, Zhu Z, Sayler KL, Bennett S, et al. (2012) A land-use and land-cover modeling strategy to support a national assessment of carbon stocks and fluxes. Applied Geography 34: 111–124.

60. Syphard AD, Clarke KC, Franklin J, Regan HM, McGinnis M (2011) Forecasts of habitat loss and fragmentation due to urban growth are sensitive to source of input data. Journal of Environmental Management 92: 1882–1893.

61. Parsons DJ (1976) Vegetation structure in mediterranean scrub communities of California and Chile. Journal of Ecology 64: 435–447.

62. Parsons DJ, Moldenke AR (1975) Convergence in vegetation strucutre along analogous climatic gradients in California and Chile. Ecology 56: 950–957.

63. Harrison AT, Small E, Mooney HA (1971) Drought relationships and distribution of two mediterranean-climate California plant communities. Ecology 52: 869–875.

64. Zedler PH, Gautier CR, McMaster GS (1983) Vegetation change in response to extreme events - the effect of a short interval between fires in California chaparral and coastal scrub. Ecology 64: 809–818.

65. Keeley JE, Keeley SC (1984) Postfire recovery of California coastal sage scrub. American Midland Naturalist 111: 105–117.

66. Malanson GP, Westman WE (1991) Modeling interactive effects of climate change, air-pollution, and fire on a California shrubland. Climatic Change 18: 363–376.

67. Callaway RM, Davis FW (1993) Vegetation dynamics, fire, and the physical-environment in coastal central California. Ecology 74: 1567–1578.

68. Franklin J (1995) Predictive vegetation mapping: geographic modelling of biospatial patterns in relation to environmental gradients. Progress in Physical Geography 19: 474–499.

69. Meentemeyer RK, Moody A, Franklin J (2001) Landscape-scale patterns of shrub-species abundance in California chaparral - the role of topographically mediated resource gradients. Plant Ecology 156: 19–41.

70. Keeley JE, Fotheringham CJ, Baer-Keeley M (2005) Determinants of postfire recovery and succession in Mediterranean-climate shrublands of California. Ecological Applications 15: 1515–1534.

71. Westerling AL, Bryant BP, Preisler HK, Holmes TP, Hidalgo HG, et al. (2011) Climate change and growth scenarios for California wildfire. Climatic Change 109: 445–463.

72. Stylinski CD, Allen EB (1999) Lack of native species recovery following severe exotic disturbance in southern Californian shrublands. Journal of Applied Ecology 36: 544–554.

73. Padgett PE, Allen EB, Bytnerowicz A, Minich RA (1999) Changes in soil inorganic nitrogen as related to atmospheric nitrogenous pollutants in southern California. Atmospheric Environment 33: 769–781.

74. Collingham YC, Huntley B (2000) Impacts of habitat fragmentation and patch size upon migration rates. Ecological Applications 10: 131–144.

75. Fahrig L (2001) How much habitat is enough? Biological Conservation 100: 65–74.

76. Travis JMJ (2003) Climate change and habitat destruction: a deadly anthropogenic cocktail. Proceedings of the Royal Society B-Biological Sciences 270: 467–473.

77. Fahrig L (2002) Effect of habitat fragmentation on the extinction threshold: A synthesis. Ecological Applications 12: 346–353.

78. O'Leary JF (1995) Coastal sage scrub: Threats and current status. Fremontia 23: 27–31.

79. Hampe A, Petit RJ (2005) Conserving biodiversity under climate change: the rear edge matters. Ecology Letters 8: 461–467.

80. Elith J, Leathwick JR (2009) Species distribution models: Ecological explanation and prediction across space and time. Annual Review of Ecology Evolution and Systematics 40: 677–697.

81. Buisson L, Thuiller W, Casajus N, Lek S, Grenouillet G (2010) Uncertainty in ensemble forecasting of species distribution. Global Change Biology 16: 1145–1157.

82. Pearson RG, Thuiller W, Araujo MB, Martinez-Meyer E, Brotons L, et al. (2006) Model-based uncertainty in species range prediction. Journal of Biogeography 33: 1704–1711.

83. Conlisk E, Syphard AD, Franklin J, Flint L, Flint A, et al. (2013) Uncertainty in assessing the impacts of global change with coupled dynamic species distribution and population models. Global Change Biology 19: 858–869.

84. Keith DA, Akcakaya HR, Thuiller W, Midgley GF, Pearson RG, et al. (2008) Predicting extinction risks under climate change: coupling stochastic population models with dynamic bioclimatic habitat models. Biology Letters 4: 560–563.

85. Conlisk E, Lawson D, Syphard AD, Franklin J, Flint L, et al. (2012) The roles of dispersal, fecundity, and predation in the population persistence of an oak (Quercus engelmannii) under global change. PloS ONE 7: e36391.

19

Seasonal Blowfly Distribution and Abundance in Fragmented Landscapes. Is It Useful in Forensic Inference about Where a Corpse Has Been Decaying?

Jabi Zabala[1,2], Beatriz Díaz[1,3], Marta I. Saloña-Bordas[1]*

1 Department of Zoology and Animal Cell Biology, Faculty of Science and Technology (UPV/EHU), Bilbao, Spain, **2** Sebero Otxoa, 45, 5 B. 48480 Arrigorriaga, Biscay, Spain, **3** Department of Entomology, Aranzadi Science Society, San Sebastián, Guipúzcoa, Spain

Abstract

Blowflies are insects of forensic interest as they may indicate characteristics of the environment where a body has been laying prior to the discovery. In order to estimate changes in community related to landscape and to assess if blowfly species can be used as indicators of the landscape where a corpse has been decaying, we studied the blowfly community and how it is affected by landscape in a 7,000 km² region during a whole year. Using baited traps deployed monthly we collected 28,507 individuals of 10 calliphorid species, 7 of them well represented and distributed in the study area. Multiple Analysis of Variance found changes in abundance between seasons in the 7 analyzed species, and changes related to land use in 4 of them (*Calliphora vomitoria, Lucilia ampullacea, L. caesar* and *L. illustris*). Generalised Linear Model analyses of abundance of these species compared with landscape descriptors at different scales found only a clear significant relationship between summer abundance of *C. vomitoria* and distance to urban areas and degree of urbanisation. This relationship explained more deviance when considering the landscape composition at larger geographical scales (up to 2,500 m around sampling site). For the other species, no clear relationship between land uses and abundance was found, and therefore observed changes in their abundance patterns could be the result of other variables, probably small changes in temperature. Our results suggest that blowfly community composition cannot be used to infer in what kind of landscape a corpse has decayed, at least in highly fragmented habitats, the only exception being the summer abundance of *C. vomitoria*.

Editor: Peter Shaw, Roehampton university, United Kingdom

Funding: Field work and technicians assistance was supported by a research project funded by the Basque Government and the University of the Basque Country. The funders had no role in study design, data collection and analysis, decision to publish, or preparation of the manuscript.

Competing Interests: The authors have declared that no competing interests exist.

* E-mail: m.salona@ehu.es

Introduction

An adequate knowledge of necrophagous blowfly species ecology and geographical abundance has a direct application in forensic science, as well as in other fields of biomedical disciplines. This is due to their role as indicators for forensic studies, myiasis producers, and pathogen vectors [1,2]. Species of the family Calliphoridae are currently the most commonly used in forensic research [1]. There are hitherto many studies on their ecology that compare abundances among seasons and/or habitat categories to detect forensically meaningful species and to help inference in applied research and criminal cases. However, this knowledge is usually restricted to some areas and the ecology of most species involved is poorly understood, especially with regard to what factors rule their distribution and abundance within the landscape. In this way, for instance, Martínez-Sanchez *et al.* found seasonal differences in the abundance of blowflies in pasture and forest areas in Spain [3], and Baz *et al.* found differences in the distribution and abundance of calliphorid species along altitudinal gradients [4]; Hwang and Tuner found spatio-temporal changes in necrophagous dipterans abundance in London area [5]; Brundage *et al.* analyzed changes in the carrion fly community of central California [1], and found that species composition differed through seasons and environments; and Arnaldos *et al.* in Spain [6], together with Kavazos and Wallman in South Eastern Australia [2], reported seasonal changes in abundance and habitat preferences for some species. All these results are encouraging and suggest the possibility that, in case of need, inference on where a corpse has been could be made on the basis of forensic entomology science.

Despite these findings, results are difficult to extrapolate to other areas for several reasons. One is that other geographical areas may harbor different communities with other forensic indicator species [7,8]. A more important one is that most studies use simple and categorical habitat descriptors (i.e urban, rural, and natural) without taking into consideration internal variability that can hardly be applied in complex landscapes. In fact, most studies compare sampling points in rather homogenous patches embedded in a mixed landscape, but paid little attention to how the mosaicism in landscape matrix might affect to the blowfly community and abundance of species.

Habitat use and selection result from several processes that take place at different scales. Johnson [9] defined four orders of habitat selection, ranging from the selection of a large geographical area to microhabitat selection. Nevertheless, studies on blowflies usually relied on the characteristics of the habitat at the sampling point, despite some blowflies may be dispersed beyond 2 km in a single day [10]. Finally, no study has yet analyzed what environmental variables within the habitat categories influence the abundance of blowfly species, and at what landscape scale.

Most of Western Europe makes a good instance of a complex landscape, with dense human population inhabiting urban areas scattered in a highly fragmented rural landscape of bocage, meadows, orchards, cultures, woods, forest cultures, small towns and hamlets, together with a dense network of transport infrastructures. Although some studies analyzed some gradients [3,4,5], no study has yet evaluated how the forensically significant fly community changes within such a landscape; which species are the most important or useful, and how are they affected by landscape composition at different scales. Therefore, we conducted a field survey of blowflies in Western Europe aimed to determine if there are fly species that have a potential use in forensic science to determine the landscape type in which a corpse has been decomposing. To determine it, we analysed: 1) the blowfly community in the area and test which are the potentially useful blowflies of the area in forensic science, if any, 2) what are the landscape variables influencing their presence, and 3) how do these variables and their importance change in different seasons and scales. We considered that, ideally, in order to be useful for forensic purpose, a blowfly species should: i) be abundant and well distributed in order to have good chances of being found in most cases, at least in a certain type of ecosystem; ii) show differences in abundance among seasons and land uses in order to be indicative of a certain kind of area; iii) its abundance correlated with the natural-urbanisation gradient, with the later explaining an important part of the variation in order to be a good indicator of urban, rural or natural areas, while avoiding biases due to other variables.

Results

A total of 28507 adult calliphorids were captured representing ten different species (Table 1). Only adults were collected and considered in further analyses. *Calliphora vicina* and *C. vomitoria* were the most abundant species, with a total of 9883 and 6530 specimens captured year-round. *Lucilia caesar* and *L. ampullacea* were also abundant, represented with 5607 and 2225 specimens, although no individuals were detected in winter. The other six species where not so abundant and less than 1000 specimens were trapped year-round. However, three of these, *L. illustris*, *L. sericata* and *Chrysomya albiceps*, showed a marked seasonal abundance with a peak in summer, and we retained them for analyses in that season. Finally, three species, *L. cuprina*, *L. richardsi* and *L. silvarum*, were very rare and scarce, with less than 100 specimens collected year-round, therefore we regarded their presence as occasional and did not include them in the analyses.

In general, blowflies were most abundant during Spring (12008 specimens captured; 34.5 specimens per trapping day) and Summer (9755 specimens; 36.8 per trapping day) than during Autumn (1679 specimens; 7.0 per trapping day) or Winter (2365 specimens; 9.2 per trapping day). Similarly, summer was the most diverse season and the one with most species (10 species, Shannon diversity index, $S = 1.62$), followed by Autumn and Spring (9 and 10 species, and $S = 1.17$ and $S = 1.14$ respectively). Winter was the season with lowest abundance of specimens, less species (just three, one of which only represented by two individuals), and smallest diversity values ($S = 0.64$).

Using one year-round data and the three habitat categories assigned in the field, a two way MANOVA was applied to check the influence of land use and/or season on the abundance of the seven tested species (Table 2). The MANOVA found statistically significant differences in the composition of blowfly communities among seasons and land uses, as well as in the interaction of the two categories. Species by species inspection of the results revealed that all the seven species showed significant seasonal variations in their abundance, while only *C. vomitoria*, *L. ampullacea*, *L. caesar* and *L. illustris* showed significant variation related to different Land uses. Finally, only *C. vomitoria*, *L. illustris* and *L. caesar* showed significant variations on abundance related to the interaction of Land use and season. These results suggest that all the seven species can be of forensic interest to assess about season, and four of them to elucidate issues related to Land use. Moreover, *C. vomitoria*, *L. illustris* and *L. caesar* appear to be potentially the most useful ones since their abundance is also related to the interaction of both variables.

In order to assess relationships between species' abundance and land use variables and other landscape descriptors, the degree of correlation among species abundance was firstly analysed (Table 3). No strong negative correlations were found, i. e. adults of no species appear to suppress competitively any other adult specimens. Therefore, we analysed relationships between landscape at different scales and blowfly abundance using Generalized Linear Models.

For this purpose, those of the original 60 sites that were less than 200 metres apart from each other were discarded to avoid spatial pseudoreplication, as well as any traps in the same vegetation patch, those for which we got no information, or only partial information, about land uses in the area (digital maps were only available for the Basque Autonomous Region and some neighbouring areas, but in some instances the sampling points were close to borders of regions with no information). Therefore, only results from 55 sampling sites have been used in these analyses. The degree of correlation of predictor variables at different scales (100, 500 and 2500 m) was analyzed to aid in the interpretation of results (See Table S1). The area covered by forest was strongly correlated to the distance to the nearest dense urban area at the three scales. This is logical, since forested areas tend to be further from urban areas than crops or similar land uses. However, it was kept in the analyses because its relationship with urban was variable (Table S1). Altitude and Y UTM were also strongly correlated at the three scales, as a result of the orography of the region where altitude values grow north to south, from the coast to the Iberian plateau. Both values were kept for the analysis because associated to the Y UTM value there might be other climatic features that might affect fly abundance too (i. e. climatic and vegetation transition from Eurosiberian to Mediterranean). Finally, there were varying degrees of negative correlation among the area covered by different land uses (urban, rural and forest) at different scales (Table S1). This was not unexpected, since the total surface was constant at each scale, and large values of a given land use implied low values of the others.

When performing GLMs, inspection of the dispersion parameters of the models and their relation to the degrees of freedom suggested overdispersion in every case, and, in consequence, quasi-Poisson error structures were used [11]. Results of the GLMs for *Calliphora vomitoria* (Table 4) showed a more or less constant pattern of urban areas avoidance. This pattern is clear at every analysed scale during summer, when *C. vomitoria* was significantly more abundant at points far from urban areas. Moreover, these summer models were fair good and explained in every case more than 65% of the variation. At several scale-season interactions there was also a correlation between abundance and fragmentation, and in seasons other than summer, its abundance was related to

Table 1. Captured califorid species and their seasonal abundance.

Species				
Calliphora vicina	Autumn	Winter	Spring	Summer
Captures	94,75	183,14	462,76	262,89
Positive locations	57	54	58	58
Calliphora vomitoria	Autumn	Winter	Spring	Summer
Captures	14,38	91,63	418,19	93,49
Positive locations	45	36	48	42
Lucilia sericata	Autumn	Winter	Spring	Summer
Captures	3,13	0,23	3,97	89,16
Positive locations	10	2	15	44
Chrysomya albiceps	Autumn	Winter	Spring	Summer
Captures	0,75	0	0,43	42,41
Positive locations	4	0	1	44
Lucilia caesar	Autumn	Winter	Spring	Summer
Captures	51,50	0	99,05	487,47
Positive locations	41	0	41	56
Lucilia ampullacea	Autumn	Winter	Spring	Summer
Captures	41,75	0	45,95	163,61
Positive locations	38	0	39	50
Lucilia illustris	Autumn	Winter	Spring	Summer
Captures	3,25	0	3,10	29,40
Positive locations	8	0	12	39
Lucilia cuprina	Autumn	Winter	Spring	Summer
Captures	0,13	0	1,21	5,54
Positive locations	1	0	4	11
Lucilia richardsi	Autumn	Winter	Spring	Summer
Captures	0,25	0	0,34	0,36
Positive locations	2	0	4	3
Lucilia silvarum	Autumn	Winter	Spring	Summer
Captures	0	0	0,17	0,96
Positive locations	0	0	1	2

Captures indicates the average number of specimens collected per each day the traps were kept active in a given season, and Positive locations the number of different sampling units in which the species was found. For further detail: monthly species abundances in different areas, details on spatial distribution in Saloña et al. [12].

Table 2. Results of the two-way MANOVA examining for effects of Land Use and Season on the abundance of selected species.

Global results	Land use		Season		Land use×Season	
	F value	**p**	**F value**	**p**	**F value**	**p**
	3.452	**0.001**	**10.483**	**0.001**	**2.180**	**0.001**
Results by species	F value	p	F value	p	F value	p
C. vomitoria	**9.988**	**<0.001**	**7.912**	**<0.001**	**3.938**	**<0.001**
C. vicina	0.278	0.758	**29.357**	**<0.001**	1.619	0.143
L. caesar	**5.809**	**<0.003**	**25.272**	**<0.001**	2.612	0.018
L. ampullacea	**4.473**	**<0.013**	**15.885**	**<0.001**	0.741	0.613
L. illustris	**9.389**	**<0.001**	**20.963**	**<0.001**	**5.511**	**<0.001**
L. sericata	0.296	0.7441	**13.671**	**<0.001**	0.284	0.944
C. albiceps	0.593	0.553	**24.933**	**<0.001**	0.522	0.791

Results of the global analysis as well as results of individual species are shown. We show the value of the statistic F and the p value for each case, global and specific, for the effect of Land use, Season and the mixed effect of Land use and Season.

geographical coordinates and landscape descriptors, but the relation was regarded "unclear" because the value of the scale parameter was close to zero (<0.001) (several significant relationships with a plain effect on abundance; Table 4).

The case of *L. caesar* was quite different. In most of the GLMs, its abundance was not related to any variable, whilst in some it was significantly related to geographic coordinates or altitude (Table 5). The only exceptions are the spring abundance at the 100 m scale, significantly related to fragmentation; and the autumn GLM at the 500 m scale, which found a significant relationship with the area covered by rural land uses (Table 5). The later however could be

regarded as unclear as a consequence of the almost flat slope of the line (<0.001). Results for *L. ampullacea* (Table 6) are similar to those for *L. caesar*. Abundance of *L. ampullacea* at the 100 and 500 m scales was significantly related to either altitude or geographic coordinates in two of the three seasons analysed. Contrastingly, this effect disappeared at the 2500 m scale where no significant relationship was found. In both cases *L. caesar* and *L. ampullacea*, the GLMs explained only a small part of the variability in the data, less than 50% in every case but in two of the models. Finally, in the instance of *L. illustris* we only had enough data to conduct analyses for the summer period, when no significant

Table 3. Seasonal correlation among blowfly species.

Spring	C. vicina	L. ampullacea	L. caesar			
C. vomitoria	0.526**	0.346**	0.650**			
C. vicina		0.277*				
L. caesar	0.425*	0.303*				
Summer	**C. vicina**	**L. ampullacea**	**L. caesar**	**Ch. albiceps**	**L. sericata**	**L. illustris**
C. vomitoria	−0.078	0.076	0.291*	0.216	−0.180	−0.088
L. illustris	0.371*	0.514**	0.187	0.394*	0.235	
L. sericata	0.331*	−0.174	−0.127	0.186		
Ch. albiceps	0.247	0.111	0.341*			
L. caesar	−0.094	0.630**				
L. ampullacea	−0.037					
Autumn	**C. vicina**	**L. ampullacea**	**L. caesar**			
C. vomitoria	0.444**	0.142	0.215			
C. vicina		0.318*	0.421*			
L. caesar		0.825**				
Winter	**C. vicina**					
C. vomitoria	0.518**					

We show the Pearson's product-moment correlation coefficient, ranging from −1 (strong negative correlation) to +1 (strong positive correlation). Statistical significance of the correlations is shown with an asterisk (*) when p<0.05, and with two (**) when p<0.001.

Table 4. Results of GLMs analyzing relationships between considered variables at different scales and seasons with abundance of *C. vomitoria*.

Season	Spring			Summer			Autumn			Winter		
Calliphora vomitoria												
Scale	100 m	500 m	2500 m	100 m	500 m	2500 m	100 m	500 m	2500 m	100 m	500 m	2500 m
Forest	−	Unclear(*)	Unclear(*)	+	−	Unclear	−	Unclear	Unclear	−	Unclear(*)	Unclear
Rural	−	Unclear	Unclear(*)	+	Unclear(*)	Unclear	Unclear	Unclear	Unclear	−	Unclear	Unclear
Urban	−	Unclear	Unclear(*)	+	Unclear(*)	Unclear(*)	Unclear	Unclear	Unclear	−	Unclear	Unclear(*)
Altitude	+	−	−	+	+	+	+	+	+	−(*)	−	−
Y UTM	Unclear	Unclear	Unclear	Unclear	Unclear	Unclear	+(*)	Unclear	−Unclear	Unclear	Unclear	Unclear
X UTM	Unclear	Unclear	Unclear(*)	Unclear	Unclear	Unclear	Unclear	Unclear	Unclear	Unclear	Unclear	Unclear
Fragmentation	−	−	+(*)	+	+(*)	+	+	+(*)	+(*)	−	−	+
Dist. to Urban	+	+	+(*)	+	+(*)	+(*)	Unclear	Unclear	Unclear	+	Unclear	+
% Explained dev.	30.6	37.4	51.2	63.6	76.1	70.7	27.6	36.7	52.0	32.1	46.1	31.3

The sense of the relationships is shown with + in case of positive relationships and − for negative relationships (i.e. lower abundance with high values for the variable). When the regression was almost flat (scale parameter value < ±0.001), we considered it unclear. Statistically significant relationships are shown with an asterisk (*), and the deviance explained in each case is shown in bottom row (in percentage). We used n = 55 in the 100 m scale, n = 50 in 500, and n = 36 in 2500 m.

Table 5. Results of GLMs analyzing relationships between considered variables at different scales and seasons with abundance of *L. caesar*.

Season	Spring			Summer			Autumn		
Lucilia caesar									
Scale	100 m	500 m	2500 m	100 m	500 m	2500 m	100 m	500 m	2500 m
Forest	+	Unclear	Unclear	+	Unclear	Unclear	+	Unclear	Unclear
Rural	+	Unclear	Unclear	+	Unclear	Unclear	+	Unclear(*)	Unclear
Urban	+	Unclear	Unclear	Unclear	Unclear	Unclear	+	Unclear	Unclear
Altitude	+	−	+	−	−(*)	−	+	+	+
Y UTM	Unclear	Unclear	Unclear	Unclear	Unclear	Unclear	+(*)	Unclear(*)	Unclear
X UTM	Unclear	Unclear	Unclear	Unclear	Unclear	Unclear	Unclear	Unclear	Unclear
Fragmentation	−(*)	−	+	+	+	+	+	−	+
Dist. to Urban	−	+	Unclear	+	+	Unclear	−	−	Unclear
% Explained dev.	35.1	33.9	30.4	29.4	43.8	23.5	43.7	59.3	58.0

The sense of the relationships is shown with + in case of positive relationships and − for negative relationships (i.e. lower abundance with high values for the variable). When the regression was almost flat (scale parameter value < ±0.001), we considered it unclear. Statistically significant relationships are shown with an asterisk (*), and the deviance explained in each case is shown in bottom row (in percentage). We used n = 55 in the 100 m scale, n = 50 in 500, and n = 36 in 2500 m.

Table 6. Results of GLMs analyzing relationships between considered variables at different scales and seasons with abundance of *L. ampullacea*.

Lucilia ampullacea

Season	Spring			Summer			Autumn		
Scale	100 m	500 m	2500 m	100 m	500 m	2500 m	100 m	500 m	2500 m
Forest	+	Unclear	Unclear	+	Unclear	Unclear	+	Unclear	Unclear
Rural	+	Unclear	Unclear	+	Unclear	Unclear	+	Unclear	Unclear
Urban	+	Unclear	Unclear	+	Unclear	Unclear	+	Unclear	Unclear
Altitude	−(*)	−(*)	−	−	−	+	−	Unclear	−
Y UTM	Unclear	Unclear	Unclear	Unclear	Unclear	Unclear	+(*)	Unclear(*)	Unclear
X UTM	Unclear	Unclear	Unclear	Unclear	Unclear	Unclear	Unclear	Unclear	Unclear
Fragmentation	−	−(*)	+	+	+	+	+	−	+
Dist. to Urban	+	−	+	Unclear	+	Unclear	−	−	Unclear
% Explained dev.	36.4	46.0	36.8	34.0	37.4	44.5	37.1	39.8	49.6

The sense of the relationships is shown with + in case of positive relationships and − for negative relationships (i.e. lower abundance with high values for the variable). When the regression was almost flat (scale parameter value < ±0.001), we considered it unclear. Statistically significant relationships are shown with asterisk (*), and the deviance explained in each case is shown in bottom row (in percentage). We used n = 55 in the 100 m scale, n = 50 in 500, and n = 36 in 2500 m.

relationships were found at any scale with the considered predictor variables (Table 7). In addition, models performed very poor at the three scales in terms of explained variance.

The amount of explained variance increased as wide areas are considered. Generalized linear models at the 100 m scale explained an average of 34.7% (SD = 12.33) of the variance; at 500 m, they explained an average of 42.1% (SD = 16.83), and 42.9% (SD = 15.06) of the variance at the 2500 m scale.

Discussion

Calliphorids were abundant in our study site. Species richness, with seven well represented species and three rare ones, was similar to that of previous studies; for instance, 5 spp. in Salamanca [3]; 7 spp in Madrid [12], 8 spp in Aragón [13], 11 spp in Portugal [14], 7 spp in California [1], 16 in Australia [2], among others. Blowfly abundance was higher in summer and spring than in other seasons, and only *Calliphora* species were abundant throughout the year. This result is in concordance with previous studies that found *Calliphora vicina* and *C. vomitoria* year-round while *Lucilia* species and *Ch. albiceps* were abundant only in summer [3,15]. This is apparently due to the thermophilic nature of the later species [3].

The fact that there were no strong negative correlations between different species (Table 3) suggests that there is no competitive exclusion among adults of different species, or at least the existence of enough resources to allow a spatio-temporal coexistence. On the other hand, strong positive correlations of abundance are likely the result of favourable environmental conditions (temperature, humidity, shelter) for blowflies in the area.

Considering the observed differences in seasonal abundance/ year-round abundance of blowflies, it is not surprising that the Multiple Analysis of Variance found significant changes in the seasonal abundance of all the species. In the same way, the effect of "habitat" or land uses in the abundance of certain species is also a well established fact [1,5]. In our case the MANOVA selected only 4 species whose abundance was significantly related to land use in the sampling point. In the case of *C. vomitoria* (Table 4), the species showed a pattern of urban rejection that reached statistical significance during summer at every scale and in spring at the 2500 m scale. Furthermore, the summer models explained an important amount of the variance (up to 75%). At the 500 m scale, *C. vomitoria* also showed higher abundance in areas with dense forest cover, that is in concordance with an avoidance of urban areas, since the variable "forest" is negatively correlated with "urban" and "rural" (Table S1) in our area, however, the scale parameter of these correlations were very low, suggesting scarce effect on abundance. At the 2500 m scale, we found a tendency for urban rejection in winter and spring, which became unclear in autumn. Anyway, these models explained little variance, suggesting that the key variables ruling the model are not included in the analyses. The spring model produced unclear correlations with every land use, and a significant positive effect of fragmented landscape far from urban areas on abundance of *C. vomitoria*. Interestingly, winter models showed at every scale a negative relationship with altitude, that was significant at the 100 m scale; and autumn, at the 100 m scale, a positive one with Y UTM which in turn is strongly negative related to altitude (Table S1). This suggests that in autumn-winter, *C. vomitoria* is more abundant in the lowlands and in the north of the study area, where altitudes are lower and the coast has a warming effect on environmental temperature. Therefore, our results show that *C. vomitoria* is strongly negative related with urban areas in summer and this

Table 7. Results of GLMs analyzing relationships between considered variables at different scales in summer with abundance of *L. illustris*.

	Lucilia illustris		
Season	**Summer**		
Scale	100 m	500 m	2500 m
Forest	–	Unclear	Unclear
Rural	–	Unclear	Unclear
Urban	–	Unclear	Unclear
Altitude	–	Unclear	+
Y UTM	Unclear	Unclear	Unclear
X UTM	Unclear	Unclear	Unclear
Fragmentation	–	+	+
Dist. to Urban	Unclear	Unclear	–
% Explained dev.	12.5	7.4	23.8

The sense of the relationships is shown with + in case of positive relationships and – for negative relationships (i.e. lower abundance with high values for the variable). When the regression was almost flat (scale parameter value $< \pm 0.001$), we considered it unclear. The deviance explained in each case is shown in bottom row (in percentage). We used n = 55 in the 100 m scale, n = 50 in 500, and n = 36 in 2500 m.

single variable explains an important part of variance of the data, while during the rest of the year this pattern does not hold, and in cold months abundance is related to thermal shelter areas. This effect of warm areas in cold months could partly explain the spring model at the 2500 m scale. In that case, areas around sampling points near the coast are sea water and therefore were not included in the model, so this higher abundance in warmer coastal areas results in an apparent avoidance of all the considered land uses. Nevertheless, Grassberger & Frank stated that *C. vomitoria* outnumber the other blowflies in an experiment done under controlled conditions in central Viena [16]. Taking under consideration the vegetation in their research area (a restricted forested backyard adequately preserved with thick underbrush) [16], we should not ignore the existence of adequate environment conditions for the presence of this species in otherwise "urban conditions". This is in agreement with our findings: that using simple categories as descriptors of complex landscapes can have misleading results. Furthermore, the fact that in our models abundance during summer was related to distance to urban areas and not to land covered by urban or other uses suggests an avoidance pattern. *C. vomitoria* were abundant far from urbanised spots but its abundance was not affected by the amount of land devoted to urban uses. The reason for it remains unknown. On the other hand, other research has shown that the soil type and environmental conditions can affect the development of larvae, and that if the environment of the burial area is not considered researchers can incur in important errors in PMI estimation [17]. For instance, that study [17] found *C. vomitoria* on a 105-day-old corpse as a consequence of cold weather and particular burial conditions. In the instances of *L. ampullacea* and *L. caesar*, a statistically significant relationship was found with altitude and/or geographic coordinates, with three exceptions (Tables 5 and 6). In the same way, all these models explained low percentages of deviance. The pattern emerging from the model was similar for this two species: the abundance was higher in low altitudes and high Y UTM values (lower, northern areas). This altitude-geographic pattern and low explained variance suggest correlation with some not included variables, most likely temperature. Indeed, both species are known to be thermophilous [3], and have been previously reported from localities close to the coast in the study

area and nearby ones [15,18]. Regarding *L. caesar*, there are two season-scale combinations whose results differ from that pattern. On the first hand, the unclear (slightly positive) effect of rural areas on abundance in autumn at the 500 m scale is difficult to explain; a possible explanation could be higher availability of food and/or warmer refuges [5] in cattle rearing areas here defined as rural. In addition, the model also selected altitude as a significantly related variable. On the other hand, the negative effect of fragmentation on abundance in Spring at the 100 m scale suggest that the species is more abundant on simpler landscapes, which again is difficult to explain with the scanty knowledge on the ecology of the species. In the instance of *L. ampullacea* there is also a negative effect of fragmentation on abundance in spring, at the 500 m scale in this case, whose possible explanation again cannot be grasped.

Only during summer months, we captured enough specimens of *L. illustris* to conduct analyses, and GLMs found no significant relationships with any variable at any scale. Therefore, variables ruling abundance of *L. illustris* were not among the variables considered in our models, and not even strongly correlated to them.

Therefore, in spite of the fact that all the species analysed show seasonal statistically significant changes in abundance, three out of seven common species, *L. sericata*, *L. illustris* and *Ch. albiceps* are clear indicators of summer in forensic analyses. *C. vicina* and *C. vomitoria* are common year round with maximum abundances in spring time, whereas *L. caesar* and *L ampullacea* can be found through most of the year, with maximum abundance in summer. Regarding land use, the MANOVA identified 4 species that show different changes in abundance, but, after checking that result with individual anovas, the only species that can be considered as a clear indicator of it is *C. vomitoria* during summer, when it showed statistically significant changes in abundance with distance to urban areas, and the models explained important amounts of the observed variance. The negative effect of urban areas, or its equivalent positive effect of rural and forest areas, seemed to be important in some models of *C. vomitoria* for other seasons too, but they explained little variance. Furthermore, winter models clearly showed a switch in predictor factors for low altitudes in cold months, which are interpreted as an indirect effect of temperature. Concerning the other three species that showed significant changes among land uses in the MANOVA (Table 2), GLM results show that this difference is not related to land use itself but mainly to altitude and geographic coordinates, which somehow are correlated to land uses. This is logic, since variations in climate and temperature influence on soil productivity and, therefore, on main land uses.

There is broad evidence in literature of the thermophilic character of many blowfly species, including the species considered in the present study but the two *Calliphora* species [3,5]. Our results suggest that differences in abundance among land uses detected by the MANOVA truly reflect differences in temperature correlated to it. A correlation with temperature, which was not possible to be measured *in situ*, may also explain the low percentages of explained variance. Furthermore, this could probably be the underlying cause for differences in blowfly communities in different land uses detected by other studies [1,3,5,19]. Another factor correlated to land use that might affect abundance of blowflies, is local availability of carcasses and other food sources. Forested areas probably had more carcasses of small to large wild animals available, whereas in urban areas garbage and small carcasses might be the main resource. Rural areas probably are intermediate between the other two. The total abundance of food and its characteristics in each area type remains an unexplored issue, despite being a probable source of variability in blowfly community species composition and abundance.

Analysis of the amount of deviance explained at different scales suggest that considering land uses within large (>2000 m) distances around sampling points explains part of the variation in the data. This is probably related to the long dispersal capability of blowflies [10], and remarks the need to be careful when relating blowfly communities to a particular sampling point characteristics and/or making inferences on the area where a corpse has been found or decayed, considering only surrounding vegetation and land uses in small areas. The fact that explained deviance grew constantly with distance, suggests that abundance of different blowfly species could be affected by landscape structure at even larger distances than those considered in this study. A note of caution is required here since our results are preliminary, and our study was not primarily designed to test this particular point. Notwithstanding, further research is needed on this point, and can be useful for the interpretation of future forensic cases.

Our results have two important implications. One is that forensic inference should be drawn from local studies. Any inferences extrapolated from other studies must be considered with extreme care because communities vary, as has been found in other areas [1,5,20]. Another is that blowfly community composition cannot be used to infer in what kind of landscape a corpse has decayed, at least in complex and heterogeneous areas like Western Europe and other densely populated areas. The only exception to this point would be the summer abundance of *C. vomitoria* related to the urban-non urban gradient. This species might be of outmost importance in forensic research due to its avoidance of hard urban areas and its widespread distribution too.

In order to be able to relate blowfly abundance to landscape with forensic purposes, further research is needed, especially focused on the reliability of using results from other areas (either similar or nearby) in forensic cases. In addition, research on autoecology (temperature ranges, habitat use and selection, abundance and dispersion, and food preferences) of forensically important blowfly species is necessary to enlighten the interpretation of their presence and abundance in forensic cases.

Materials and Methods

Study Area

The study was conducted in the Basque Country (North of Spain), which is an area of about 7000 km^2 of contrasting landscapes. Three mountain ridges run east-west across the territory, with their north aspects catching the humid winds from the Gulf of Biscay and eliciting rain. The north area, by the Gulf of Biscay, is warm, rainy and rugged with altitudes ranging from 0 to some 1500 m above sea level. In general terms, it has a humid temperate climate without dry season, and an average rainfall of 1200–2000 mm. [21,22]. Valley bottoms are densely populated with villages, industrial areas and hamlets scattered in the landscape of meadows, and woodlots. Slopes are covered with forests and half of the surface is occupied by exotic tree cultures (mostly *Pinus radiata* and *Eucalyptus* spp). Highlands are typically meadows and pastures. The central area is a high plateau (some 600 m a.s.l., with an average rainfall of 750–900 mm.) mainly devoted to crops, with some scattered woods, and ranges with forested slopes. Climate there is Atlantic with a neat Mediterranean influence. The southernmost area lies by the Ebro river valley and it is mainly devoted to vineyards. Forests are scarce, occupying mainly mountain slopes and hilltops, and dominant species are evergreen oaks (*Quercus rotundifolia*) and Scottish pine (*P. sylvestris*) in some areas. Climate there is cold in winter and hot and dry in summer, with rainfall ranging from 500 to 1000 mm per year and taking place mainly in winter-spring. Northern valleys tend to be warmer in winter than southern ones, due to their lower altitudes and the proximity of the ocean.

60 sites were sampled, more or less randomly distributed in the study area (Figure 1). A detailed reference to the specific samples used in this study has been previously described [22]. No natural areas or private farms were invaded as traps were placed in borders and rural pathways inaccessible to pets, cattle or humans, and properly labeled reporting the activity. Therefore, no specific permissions were required for these locations, as the field studies did not involve endangered or protected species [23] and all the activities described abided by spanish regulation and international ethical standards. To ensure that samples represent different uses and landscapes in heterogeneous areas, we placed the traps in pairs, one in an urban sampling point (city or village) and another in a nearby less altered one (rural or forest area). Following common forensic entomological practice we considered three environments: urban, rural, and forest [1–6] Traps were kept active for two-three days depending on the month of the year. The design used for the traps follow the model of double bottle baited with pig kidney [5,22]. Traps consisted of plastic bottles of 1.5 l hung on trees or bushes with the bait placed on the bottom of the bottle, close to a small opening that allowed the entrance of blowflies attracted to the bait. Blowflies accessing the trap were retained on a double funnel made with two upper parts of two bottles of similar size, and later collected by the researchers. Pig kidney was selected as bait because of its similarities with human one and its great attraction power [24,25,26]. It was supplied by official licensed retailers, following Spanish regulations for animal by-products [27], and therefore do not fall under the remit of the Institutional Animal Care and Use Committee (IACUC). Traps were set once every month during a natural year (from July 2007 to June 2008), and every second-third day, we revisited the places and collected traps and samples. In some winter months (December–February) traps were kept active for at least three days to compensate the reduced insect activity and ensure representation of the samples. However, to avoid possible biases as a consequence of this, data were transformed into specimens

captured per trapping day previous to analyses. Captured flies were separated from larvae on the kidney (which were used for development experiments) and killed by introducing them in a freezer. Then, all the imagoes collected were preserved in ethanol 70%, and identified in the laboratory to the species level following different keys [18,28,29,30,31,32]. All analyses discussed were conducted using and considering only adult blowflies.

Data Analysis

Captures were summarized by species and seasons. Following common practice [1,6], 4 seasons were considered: Winter (December, January and February), Spring (March, April and May), Summer (June, July and August), and Autumn (September, October and November). Presence and abundance of species were analyzed in different seasons and different scales. Species represented by less than 100 specimens year-round were discarded from analyses, as well as those species-seasons pairs in which the species was scarce (less than 100 individuals captured) or rare (found in less than 20 locations). This ensures the representativeness, comparability and wider applicability of our results. Seasonal diversity of blowflies was assessed using Shannon diversity index [33].

To analyse relationships between species, landscape and seasons, a Multivariate ANOVA (MANOVA) was conducted with year-round data [1]. In this analysis, landscape was defined into three different categories: urban, rural and forest. Urban were dense villages and highly urbanized areas; rural included small farmlands, meadows, pastures and crops; and, as natural were categorized autochthonous forested areas and least modified ones. Assignment of each location to a category was done in the field.

Analysis Scales

To gain further insight in the relationships between species and habitat features, and how it might be affected by landscape composition, seasonal abundance of species was analyzed against landscape descriptors at three different scales. The smallest scale was the area within 100 meters of the trapping point (i. e. an area of 0.0314 km^2). This scale aimed to represent local features that might influence presence/abundance of blowflies. An intermediate scale of 500 meters around the sampling point (0.79 km^2) was also considered, representing the local landscape. Finally, an area within a 2500 meter radius of the sampling point (19.63 km^2) was considered. This distance was set after the average distance that some blowfly species are known to travel in a day [10]. Therefore, we considered that land uses within that distance could affect the presence and abundance of blowflies in the trapping areas, given the length of the trapping period (1–3 days).

Selection and Characterisation of Independent Variables

To describe landscape composition, 1:10 000 digital cartography with EUNIS (European Nature Information System; http://eunis.eea.europa.eu/) land use categories was used. GPS data of trapping points were uploaded into a Geographic Information System (GIS, gvGIS: http://www.gvsig.org) and three radii of 100, 500 and 2500 m were built around them. In order to minimise spatial pseudoreplication, when areas around sampling points overlapped we discarded one of them. The area covered by each EUNIS class was measured in each case and merged onto three categories: Urban, Rural and Forest. Urban included high and low density built areas, industrial areas, urban parks and gardens, roads, railways and associated infrastructures, together with harbours and other artificial areas. Rural included crops, meadows and any other areas for food production managed, including recently abandoned croplands. Finally, Forest category

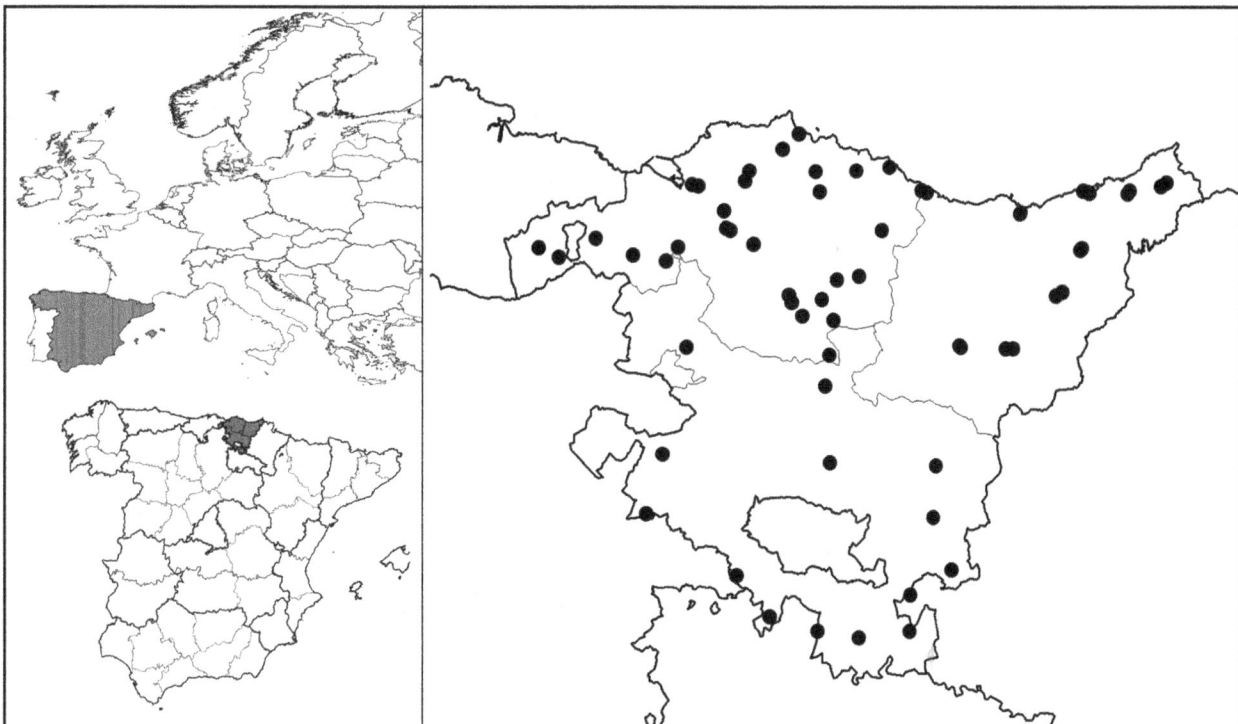

Figure 1. Study area location and sampling point distribution within it.

included all kind of forested lands of any age, including native forest and exotic tree cultures. These 3 categories encompassed more than 90% of the landscape. The remaining 10%, which included mainly sea and other water bodies, as well as rare ecosystems with scarce representation in the territory, was not considered for analysis. In addition, the following variables describing sampling points were also considered: Altitude, referred to the altitude above sea level of the exact point where the trap was placed; Distance to urban area, meaning the distance from the sampling point to the nearest densely built up area; Fragmentation, the number of different land use polygons within the considered area, as an indicator of the degree of mosaicism of the area [34,35]; and X and Y UTM coordinates, to control possible geographic effects. The values for all these variables were calculated for each of the sampling points with the aid of the GIS using available digital cartography and digital model terrains. To detect possible interactions between predictor variables and to consider their effect on results, we built a correlation matrix of the predictor variables using Pearson's product-moment correlation [33].

To enlighten the relationships between seasonal abundance of different species, we first estimated the degree of correlation in the abundance of pairs of species. Strong negative correlations could be interpreted as competitive exclusion and would require to be considered in further analyses, whereas lack of strong negative correlation would allow to species by species analyses. To investigate correlation we used Pearson's product-moment correlation [33]. To analyse relationship between species abundance and landscape descriptors, we used Generalized Linear Models (GLM), which is a generalization of common linear regression that allows for several distribution functions on the response variables [11]. The GLM allows the response variable to be related to the predictor via a link function, and allows the variance to be a function of its predicted value. In our cause we performed GLMs with a Poisson error structure, fit for count data, using a logarithmic link function. We inspected the dispersion parameters of the model and their relation to the degrees of freedom looking for overdispersion [11,36]; and when required, we accounted for it using quasi-poisson error structures [11] using the "stats" package implemented in R [37]. We only evaluated GLMs for species-season combinations for which we captured at least 100 specimens. To assess how the data fits the model we used explained deviance method, which analyses the amount residual deviance of the model against the deviance of the null model [11,36].

All the statistical analyses were conducted using R 2.9 [37], and p values inferior to 0.05 were considered statistically significant in every case.

Acknowledgments

The authors would like to thank the collaboration of Maite Gil Arriortua, Andrea Miguélez Correa and Javier Moneo Pellitero for their technical assistance during field and laboratory work. They also wish to thank, M. Lancia, two anonymous reviewers and academic editor P. Shaw for their critic review and helpful comments of a previous draft.

Author Contributions

Conceived and designed the experiments: MS. Performed the experiments: MS BD. Analyzed the data: JZ BD MS. Contributed reagents/materials/analysis tools: JZ MS. Wrote the paper: JZ BD MS.

References

1. Brundage A, Bros S, Honda JY (2011) Seasonal and habitat abundance and distribution of some forensically important blowflies (Diptera: Calliphoridae) in Central California. Forensic Science International 212: 115–120.
2. Kavazos CRJ, Wallman JF (2012) Community composition of carrion-breeding blowflies (Diptera: Calliphoridae) along an urban gradient in south-eastern Australia. Landscape and Urban Planning 106: 183–190.
3. Martínez-Sánchez A, Rojo S, Marcos-García MA (2000) Annual and spatial activity of dung flies and carrion in a Mediterranean holm-oak pasture ecosystem. Medical and Veterinary Entomology 14: 56–63.
4. Baz A, Cifrián B, Díaz-Aranda LM, Martin-Vega D (2007) The distribution of adult blowflies (Diptera: Calliphoridae) along an altitudinal gradient in Central Spain. Annales de la Societé Entomologique de France, 43: 289–296.
5. Hwang C, Turner BD (2005) Spatial and temporal variability of necrophagous Diptera from urban to rural areas. Medical and Veterinary Entomology 19: 379–391.
6. Arnaldos MI, Romera E, Presa JJ, Luna A, García MD (2004) Studies on seasonal arthropod succession on carrion in southeastern Iberian Peninsula. International Journal of Legal Medicine 118: 197–205.
7. Arnaldos MI, Prado e Castro C, Presa JJ, López-Gallego E, García MD (2006) Importancia de los estudios regionales de fauna sarcosaprófaga. Aplicación a la práctica forense. Ciencia Forense 8: 63–82.
8. Sharanowski BJ, Walker EG, Anderson GS (2008) Insect succession and decomposition patterns on shaded and sunlit carrion in Saskatchewan in three different seasons. Forensic Science International 179: 219–240.
9. Johnson DH (1980) The comparison of usage and availability measurement for evaluating resource preference. Ecology 61: 65–71.
10. Braack LEO, Retief PF (1986) Dispersal, Density and Habitat Preference of the blow-flies Chrysomyia albiceps (Wd) and Chrysomyia marginalis (Wd) (Diptera: Calliphoridae). Onderstepoort. Journal of Veterinary Research 53: 13–18.
11. Crawley MJ (2007) The R book. West Sussex: J. Wiley & Sons, 92 p.
12. García-Rojo AM (2004) Estudio de la sucesión de insectos en cadáveres en Alcalá de Henares (Comunidad Autónoma de Madrid) utilizando cerdos domésticos como modelos animales. Boletín de la Sociedad Entomológica Aragonesa (S.E.A.) 34: 263–269.
13. Castillo Miralbés M (2002) Estudio de la Entomofauna asociada a cadáveres en el Alto Aragón (España). Monografías de la Sociedad Entomológica Aragonesa (S.E.A.), Aragón 6: 96 p.
14. Prado e Castro C, Sousa JP, Arnaldos MI, Gaspar J, García MD (2011) Blowflies (Diptera: Calliphoridae) activity in sun exposed and shaded carrion in Portugal. Annales de la Societé Entomologique de France 47: 128–139.
15. Moneo Pellitero J, Saloña Bordas MI (2007) Califóridos (Diptera: Calliphoridae) de interés forense recogidos en el entorno universitario del campus de Leioa (Vizcaya, España). Boletín de la Sociedad Entomológica Aragonesa (S.E.A.) 40: 479–483.
16. Grassberger M, Frank C (2004) Initial study of Arthropod succession on Pig Carrion in a Central European Urban Habitat. Journal of Medical Entomology 41: 511–523.
17. Turner B, Wiltshire P (1999) Experimental validation of forensic evidence: a study of the decomposition of buried pigs in a heavy clay soil. Forensic Science International 101: 113–22.
18. Peris SV, González Mora D (1991) Los Calliphoridae de España, III Luciliini (Diptera). Boletín de la Real Sociedad Española de Historia Natural 87: 187–207.
19. Nuorteva P (1963) Synanthropy of blowflies (Diptera, Calliphoridae) in Finland. Annales Entomologica Fennica 29: 1–49.
20. Anderson GS, VanLaerhoven SL (1996) Initial studies on insect succession on carrion in Southwestern British Columbia. Journal of Forensic Sciences 41: 617–625.
21. Biurrun I, García-Mijangos I, Loidi J, Campos JA, Herrera M (2011) La vegetación de la Comunidad Autónoma del País Vasco. Leyenda del mapa de series de vegetación a escala 1:50.000. Eusko Jaurlaritza/Gobierno Vasco.
22. Saloña MI, Moneo Pellitero J, Díaz Martín B (2009) Estudio sobre la distribución de Califóridos (Diptera, Calliphoridae) en la Comunidad Autónoma del País Vasco. Boletín de la Asociación Española de Entomología 33: 63–89.
23. Real Decreto 139/2011, de 4 de febrero, para el desarrollo del Listado de Especies Silvestres en Régimen de Protección Especial y del Catálogo Español de Especies Amenazadas. Boletín Oficial del Estado, B.O.E. 46: 20912–20951.
24. Catts EP, Goff ML (1992) Forensic Entomology in Criminal Investigations. Annual Review of Entomology 37: 253–272.

25. Goff ML (1993) Estimation of Postmortem Interval Using Arthropod Development and Successional Patterns. Forensic Science Review 5: 81–94.

26. Schoenly KG, Haskell NH, Mills DK, Bieme-Ndi C, Larsen K, et al. (2006) Using pig carcasses as Model Corpses to Teach Concepts of Forensic Entomology and Ecological Succession. The American Biology Teacher 68: 402–410.

27. Real Decreto 1376/2003, de 7 de noviembre, por el que se establecen las condiciones sanitarias de producción, almacenamiento y comercialización de las carnes frescas y sus derivados en los establecimientos de comercio al por menor. Boletín Oficial del Estado, B.O.E. 273: 40094–40101.

28. Zumpt F (1956) Calliphorinae, in Lindner, E. Die Fliegen der Palaearktischen Region. Schweizerbart, 1: 1–140.

29. Smith KGV (1986) A manual of forensic entomology. London: Trustees of the British Museum (Natural History), 205 p.

30. González Mora D (1989) Los Calliphoridae de España, II: Calliphorini. Eos 65: 39–59.

31. González Mora D, Peris SV (1988) Los Calliphoridae de España, 1: Rhiniinae y Chrysomyinae. Eos 64: 91–139.

32. Rognes K (1991) Blowflies (Diptera, Calliphoridae) of Fennoscandia and Denmark. New York: E.J. Brill/Scandinavian Science Press Ltd., 272 p.

33. Krebs CJ (1999) Ecological methodology. California: Benjamin Cummings, 620 p.

34. Zabala J, Zuberogoitia I, Martínez-Climent JA (2005) Site and landscape features ruling the habitat use and occupancy of the polecat (*Mustela putorius*) in a low density area: a multiscale approach. European Journal of Wildlife Research 51: 157–162.

35. Zabala J, Zuberogoitia I, Martínez-Climent JA, Martínez JE, Azkona A, et al. (2006) Occupancy and abundance of Little Owl *Athene noctua* in an intensively managed forest area in Biscay. Ornis Fennica 83: 97–107.

36. Zuur AF, Ieno EN, Walker NJ, Saveliev AA, Smith GM (2009) Mixed effects models and extension in ecology with R. New York: Springer, 574 p.

37. R Core Team (2013) R: A language and environment for statistical computing. R Foundation for Statistical Computing, Vienna, Austria. URL http://www.R-project.org/.

20

Visibility from Roads Predict the Distribution of Invasive Fishes in Agricultural Ponds

Toshikazu Kizuka[1]*, **Munemitsu Akasaka**[1,2], **Taku Kadoya**[1], **Noriko Takamura**[1]

1 Center for Environmental Biology and Ecosystem Studies, National Institute for Environmental Studies, Tsukuba, Ibaraki, Japan, **2** Faculty of Agriculture, Tokyo University of Agriculture and Technology, Fuchu, Tokyo, Japan

Abstract

Propagule pressure and habitat characteristics are important factors used to predict the distribution of invasive alien species. For species exhibiting strong propagule pressure because of human-mediated introduction of species, indicators of introduction potential must represent the behavioral characteristics of humans. This study examined 64 agricultural ponds to assess the visibility of ponds from surrounding roads and its value as a surrogate of propagule pressure to explain the presence and absence of two invasive fish species. A three-dimensional viewshed analysis using a geographic information system quantified the visual exposure of respective ponds to humans. Binary classification trees were developed as a function of their visibility from roads, as well as five environmental factors: river density, connectivity with upstream dam reservoirs, pond area, chlorophyll *a* concentration, and pond drainage. Traditional indicators of human-mediated introduction (road density and proportion of urban land-use area) were alternatively included for comparison instead of visual exposure. The presence of Bluegill (*Lepomis macrochirus*) was predicted by the ponds' higher visibility from roads and pond connection with upstream dam reservoirs. Results suggest that fish stocking into ponds and their dispersal from upstream sources facilitated species establishment. Largemouth bass (*Micropterus salmoides*) distribution was constrained by chlorophyll *a* concentration, suggesting their lower adaptability to various environments than that of Bluegill. Based on misclassifications from classification trees for Bluegill, pond visual exposure to roads showed greater predictive capability than traditional indicators of human-mediated introduction. Pond visibility is an effective predictor of invasive species distribution. Its wider use might improve management and mitigate further invasion. The visual exposure of recipient ecosystems to humans is important for many invasive species that spread with frequent instances of human-mediated introduction.

Editor: Brock Fenton, University of Western Ontario, Canada

Funding: The present study was supported by the Environment Research and Technology Development Fund (S9) of the Ministry of the Environment, Japan. The funders had no role in study design, data collection and analysis, decision to publish, or preparation of the manuscript.

Competing Interests: The authors have declared that no competing interests exist.

* E-mail: t.kizuka1981@gmail.com

Introduction

Invasive alien species are widely acknowledged as a major threat to the biodiversity of native species. Controlling species invasion is a priority for conservation of native assemblages [1], [2]. Predicting which sites will be susceptible to the introduction and establishment of invasive species and which factors will be associated with their establishment can aid in controlling species invasion [3], [4]. When invasive species have high rates of propagule supply, propagule pressure can play a much more prominent role in predicting their distribution than habitat characteristics do [5].

Propagule pressure of an invasive species is determined by the extent of natural dispersal and human-mediated introduction. When physical barriers limit the extent of natural dispersal, human-mediated movement of a species can strongly affect the propagule supply. The distribution of aquatic organisms is often limited by the extent and availability of natural dispersal. Therefore, human-mediated movement of species is an important structuring influence.

Because of the difficulties in estimating the actual locations, size, and number of instances of human-mediated introduction into recipient areas, various surrogate indicators such as human population [6], [7], roadways [8–10], and urban land use [11–13] have been used to analyze the distributional patterns of invasive plant and animal species. For example, McKinney [14] and Gido et al. [5] found greater numbers of invasive species that had some value to humans (e.g., bait-bucket and sport fish) in highly populated areas at national and state scales in the United States. These studies demonstrated the effectiveness of surrogate indicators of propagule pressure. Their results indicated that the potential accessibility to recipient areas can be a key factor for human-mediated introduction. However, on local scales, for which traditional indicators such as human population, road density, and proportion of urban land-use area often do not show large spatial variation, these indicators might not reflect potential accessibility effectively. In such cases, accessibility probably depends on additional local conditions such as road-use types, travel time and cost, fences, and the visibility of recipient areas. Among them, visibility, the visual exposure of an area to humans, is expected to influence the probability of propagule supply, particularly for species that are introduced intentionally by humans after searching for a recipient area (e.g. a water body for game fishes) in the field. Even if the recipient area is surrounded by high-density roads, accessibility is assumed to be lower in cases where visual exposure to roads is obstructed by rough terrain or blocks of buildings. In such cases, visibility is likely to be more appropriate for evaluating accessibility than road density is, but no examination of their effectiveness has ever been reported in the literature.

Here, we used the visual exposure of a recipient area to roads, its visibility, as a surrogate of propagule pressure to test the effectiveness of visibility for predicting the distribution of invasive species. We also compared this visibility variable with other traditional indicators such as road density and the proportion of urban land-use area (defined as the urban ratio). Visibility has been used widely in archaeological studies [15] and urban landscape planning [16]. The emergence of powerful analytical tools in geographic information system [GIS] software, coupled with computer graphics techniques and increasingly large-scale and higher-resolution terrain data has facilitated the quantification of visibility and has led to its application in various areas of research [17]. Recently, visibility has been applied for ecological studies in relation to habitat selection of wildlife [18], but its application to propagule pressure of invasive species has not been reported.

We used agricultural ponds and two predatory invasive fish species, Bluegill sunfish (*Lepomis macrochirus*) and Largemouth bass (*Micropterus salmoides*), to test the effectiveness of visual exposure to roads. Unlike open terrestrial ecosystems, agricultural ponds are more closed systems; the respective ponds have limited connectivity. For that reason, such systems are appropriate for testing the contribution of human-mediated introduction to propagule pressure. Agricultural ponds and small reservoirs support diverse populations of aquatic animals and plant species [19–22]. They are the most important habitats in terms of both local and regional biodiversity among several aquatic habitat types (i.e., lakes, ponds, ditches, streams, and rivers) in agricultural landscapes [23]. In such shallow water ecosystems, predatory invasive fish species negatively affect macroinvertebrate fauna and indigenous fish species through direct predation and competition [24]. Predicting the distribution of invasive fish species is an important task for biodiversity conservation of pond ecosystems.

Materials and Methods

Study area

This study examined an area of approximately 1000 km² in southwestern Hyogo Prefecture, Japan (Fig. 1). Many agricultural ponds have been created in Hyogo Prefecture (8,395 km²) to irrigate paddy fields. More than 55,000 agricultural ponds, corresponding to about 20% of all ponds in Japan, were recorded in the 1950s [25]. Nevertheless, more than 11,000 of those ponds had been lost by 1997, mainly as a result of urban or residential development [26]. Even where ponds have not been destroyed, their biodiversity has decreased drastically during recent decades [27].

The study area, with elevations of 0–640 m a.s.l., includes mountainous (9% of the study area), hilly (31%), plateau (29%), and lowland (31%) areas (1/200,000 landform classification map provided by the Japanese Ministry of Land, Infrastructure, Transport and Tourism [MLIT], Tokyo, Japan). The study area has a warm temperate climate with an annual mean temperature of 14.4°C (minimum, 3.5°C in January; maximum, 26.4°C in August) and mean annual precipitation of 1198.3 mm (data provided by the Miki Climatological Observatory located within the study area at 145 m a.s.l.). The predominant land uses are woodland (36%), farmland (34%), and urban land (18%) (Fig. 1B; Land utilization segmented mesh data with 100 m meshes; created

in 2009; National Land Information Division, MLIT, Tokyo, Japan). The southern coastal region has been particularly urbanized. The study area, which has an average population density of 1,206 people per square kilometer, is adjacent to the second largest metropolitan area in Japan (Osaka) on the east.

Generally, agricultural ponds are constructed by banking up streams or spring water. For that reason, each pond has an original catchment upstream. The pond water resource depends fundamentally on rainwater that the catchment receives. After the late 1940s, however, massive water resource development projects including dams and canals changed irrigation systems and diversified the ponds' water resources. Now, reservoir and river water are supplied to ponds as necessary via canal networks with main canals of approximately 3 m width and 2 m depth [28]. In the study area, eight agricultural dam reservoirs were constructed during 1940–2000 [29] (Fig. 1C). Some ponds receive reservoir water indirectly from upstream ponds and farmlands in their catchments that are connected directly to the reservoirs' water canals.

Invasive fish species

Bluegill and Largemouth bass, freshwater fish native to North America, are listed among the IUCN 100 worst invasive species in the world. Each has a long and successful invasion history in Japan's lentic environments. Each species is listed as an Invasive Alien Species [IAS] under the Japanese IAS Act adopted in 2005; it is strongly prohibited to transport or stock these living organisms without permission from competent authorities. Negative effects of these invasive fish on lake and pond ecosystems in Japan have been reported from many studies. The abundance of Bluegill and Largemouth bass has negatively affected total richness, endangered species, and functional diversity of various taxa including native fishes, aquatic macrophytes, Odonata, and benthic macroinvertebrates in agricultural ponds [24], [30]. Experimental approaches have revealed direct predation and competition in addition to cascading effects through the chain of bass–crayfish–macrophyte [31]. Bluegill juveniles feed on crustacean zooplankton, which has caused an increase of phytoplankton biomass as a result of reduced grazing pressure [32]. Through alterations of species composition, trophic structure, and ecosystem features, these invasive fishes adversely affect fishery production in many lakes [33].

Bluegill were first introduced to the Freshwater Fish Research Institute of the Fisheries Agency of Japan in 1960 [34]. Some of these offspring were released into the wild, although others were provided to prefectural experimental stations and fishermen for aquaculture as a potential new food resource [35]. Although aquaculture never became established as an industry, their rapid spread became apparent across the country after the 1980s, which was likely related to their stocking by anglers coupled with Largemouth bass as forage. Largemouth bass were first introduced into Japan in 1925 and again in 1972 as a food and game species [36], [37]. Their distribution spread rapidly to other water bodies including agricultural ponds throughout the Japanese Islands after the late 1960s as sport fishing became popular.

Reportedly, Bluegill and Largemouth bass were first introduced, respectively, into the wild in Hyogo Prefecture by 1987 [38] and by 1964 [39]. The population of anglers seemed to decrease after the peak in the late 1990s in the area. Still now, however, agricultural ponds are often used for sport fishing throughout the prefecture. Consequently, human-mediated introduction for sport fishing is assumed to be one factor influencing the present distributions of these invasive species. Anglers generally access agricultural ponds using private vehicles. Therefore, visual

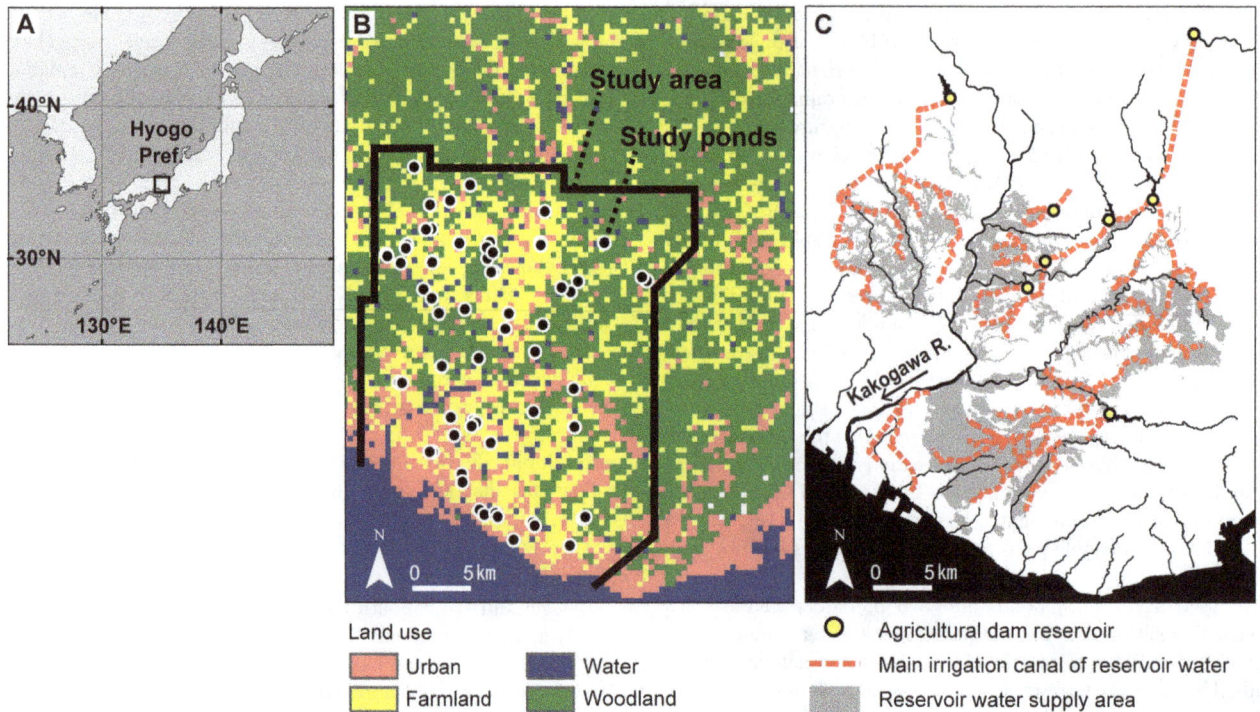

Figure 1. Description of study area. A) Locations of Hyogo Prefecture, Japan, B) study area and ponds with land use, C) agricultural dam reservoirs, irrigation canals, and reservoirs' water supply area. Land use in (B) and irrigation canals in (C) are described respectively using Land utilization segmented mesh data with 100 m meshes (created in 2009; National Land Information Division, National and Regional Policy Bureau, Ministry of Land, Infrastructure, Transport and Tourism, Tokyo, Japan) and the 1/25,000 irrigation canal network maps (provided by Japanese Institute of Irrigation and Drainage, Tokyo, Japan).

detection of the ponds from surrounding roads is regarded as important for access to the ponds. Aside from human-mediated introduction, natural dispersal from reservoirs established upstream and from other ponds via canals occurs. Younger fish, which have less swimming capability, can readily disperse in that manner [40]. Connections with upstream dam reservoirs where the invasive fish reproduce are regarded as another important factor affecting their distribution in agricultural ponds [26]. Additionally, the natural spread of invasive fish might occur via surrounding rivers that connect ponds as water corridors [41].

Largemouth bass are piscivores that can eat fish, insects, crayfish, and zooplankton in agricultural ponds [42]. In contrast, Bluegill are omnivores known to have trophic polymorphism because of their different requirements for efficient resource utilization [43]. In addition to their wide feeding niche, Bluegill show high physiological and behavioral adaptability to changing environments [44], [45].

Selection of study ponds and invasive species survey

The visibility of ponds from roads is affected by landforms and land use [46]. Therefore, we chose 64 ponds based on combinations of landform classifications (lowland, 17 ponds; hilly, 12 ponds; plateau, 35 ponds) and the predominant land use (woodland, 22 ponds; farmland, 31 ponds; urban, 11 ponds) in areas surrounding the ponds. The selected ponds had surface areas of 685 m² to 111,626 m² (mean±SD: 11,052±14,563 m²), with maximum depths of 0.3–6.0 m (2.2±1.3 m). The average pond elevation was 57±33 m a.s.l. Based on interviews with pond managers, we confirmed that the managers and farmers had never stocked their own ponds with largemouth bass or bluegill.

To catch invasive fishes, a Y-shaped fixed net (6.0-m sleeve and 0.7-m open mouth; 4-mm mesh) was set for one night in the littoral zone of each pond during 19 September – 5 October 2006 or 4–11 October 2007. For the same time period, five box-net traps (0.4 m × 0.25 m × 0.4 m with 2-mm mesh) were set for one night along a transect extending from the shallower littoral zone to the deeper limnetic zone in each pond. Furthermore, invasive fishes were sampled using standard D-frame nets (0.35 m open mouth with 2-mm mesh) mainly in littoral zones with aquatic plants (3–12 points of each pond). We counted all individuals of invasive fish species caught in the nets and traps.

We obtained permits for the survey from each pond manager in conjunction with the Agricultural and Environmental Affairs Department, Hyogo Prefecture Government. Surveyed ponds did not involve protected areas and species that required permits for sampling. The sampled invasive alien species were processed in accordance with the Japanese IAS Act.

Environmental characteristics

We assessed eight candidate characteristics that putatively affect the distributions of the invasive fish species. (1) Visibilities of ponds from surrounding roads, (2) connectivity with upstream dam reservoirs, and (3) river densities surrounding ponds were examined as landscape characteristics related to propagule supplies from outside the ponds. For comparison with visibility as a surrogate indicator of human-mediated introduction, (4) the road densities and (5) urban ratio of areas surrounding ponds were assessed. In addition, (6) the surface area, (7) chlorophyll a concentrations (chl.a), and (8) the presence or absence of pond water drainage and drawdown during winter were examined as site characteristics related to the establishment of the invasive fish species. We used surface area as a site characteristic because ponds with greater surface area are assumed to have various microhabitats from shallower littoral zones with aquatic plants to deeper limnetic zones, which provide better opportunities for the establishment of large fishes by supporting spawning and feeding [47]. Furthermore, we used chl.*a* as an indicator of eutrophication. It affected the populations and communities of fish through changes in energy availability, turbidity, and dissolved oxygen concentrations in water [48], [49].

Visibility is an indicator of how much of a target pond can be seen from surrounding roads. Visibility was assessed objectively by generating viewshed maps using ArcGIS 10.0 with the extension 3D Analyst (ESRI, Redlands, USA). They cover the area that is visible from observation points in a three-dimensional digital geographical surface, as represented by Digital Surface Models [DSMs]. We set observation points at 100 m intervals along road polylines derived from the 1/2,500 Digital Map (Spatial Data Framework; published in 2006, Geospatial Information Authority [GSI], Tsukuba, Japan). These observation points were positioned 1.5 m above the surface. Sight lines were drawn with the longest distance of 500 m from each observation point. No constraint was set on the height or direction of sight lines. We used DSMs with 1-m spatial resolution derived using photogrammetry software (SOCET SET v5.5; BAE Systems plc, London, UK) from 1/15,000 aerial photographs taken in 2007. We calculated the number of observation points from which each 1 m^2 parcel was visible (defined as viewshed points) and evaluated the visibility of each pond by aggregating the viewshed points of all parcels located within the polygon of a pond's surface area including the vegetated littoral zone. These pond polygons were delineated based on the orthorectified images of the aerial photographs described above and based on the latest 1/2,500 topographic maps provided by local governments.

We calculated the road densities and urban ratios for buffer areas within 500 m from the edges of pond polygons, which was the same extent as the longest distance of sight lines for visibility analysis. Total lengths of road polylines and total areas of urban land use, for which both data sources were described above, were divided by the buffer area of each pond.

Connectivity between study ponds and upstream dam reservoirs was assessed based on the 1/100,000 reservoir water supply area maps, the 1/25,000 irrigation canal network maps (provided by Japanese Institute of Irrigation and Drainage, Tokyo, Japan), and a list of irrigation facilities. We treated connectivity as a categorical variable in the following statistical analysis: 0, ponds in which reservoir water did not flow; 1, ponds in which reservoir water flows indirectly via upstream farmlands or ponds; 2, ponds in which reservoir water was supplied directly from irrigation canals.

We used 1/25,000 National Land Numerical Information Rivers Data (published in 2012; National Land Information Division, MLIT, Tokyo, Japan) to calculate the river densities surrounding ponds. Graded buffers (5, 10, 25, 50, 100, 250, 500, and 1000 m from the edges of pond polygons) were delineated. Then the total lengths of river lines were divided by each buffer area.

Surface areas of ponds were calculated based on the pond polygons described above. We measured chl.*a* of ponds' water in the summer (end of July in 2006 or 2007) at the approximate center of each pond using a standard method (details in Kadoya et al. [24]). Some pond managers drain pond water to check and repair levees, to improve water quality, and to eradicate invasive fish during winter (November–March), when irrigation water is not required for rice cultivation. Based on interviews with pond managers, we examined whether water was drained periodically or

not at each pond, and treated that information as a binary variable in subsequent analyses.

Statistical analysis

We used a binary classification tree [50], [51] to analyze the presence and absence of invasive fish species in the ponds as a function of six environmental factors: visibility, river density, connectivity with upstream reservoirs, pond area, chl.*a*, and pond drainage. To test the importance of visibility as a surrogate indicator of human-mediated introduction, each of the three indicators (visibility, road density, and urban ratio) was alternatively included as environmental factors of classification tree. Then we compared model performance among these trees for each invasive fish. Splitting variables and the split criteria were determined statistically in each tree. Ten-fold cross-validation was used to obtain estimates of cross-validated relative errors of these trees [52]. These estimates were then shown against tree size. Then the optimal tree was chosen based on the 1–SE rule, which minimizes cross-validated error within one standard error of the minimum [50]. To eliminate the influence of random split of samples in the cross-validation, a series of 50 cross-validations was run for each invasive fish. Then the modal (most likely) single optimal tree was chosen for description. We calculated the overall misclassification, sensitivity (the ability of the model to predict that an invasive fish is present when it is) and specificity (the ability of the model to predict that an invasive fish is not present when it is not) of the optimal tree for all datasets. We used the R statistical environment (R ver. 2.15.0) [53] and the rpart package (rpart ver. 3.1–53) [54] to build models and to evaluate their performance.

Results

Invasive fish species

Bluegill and Largemouth bass were caught respectively in 40 ponds (63% of all studied ponds) and in 12 ponds (19%) (Table 1). Of all ponds, 67% (43 ponds) supported at least one of the invasive fish species. Among them, almost all ponds supported only Bluegill (31 ponds). Three ponds supported only Largemouth bass. Nine ponds supported both species.

Environmental characteristics

The visual exposure of ponds to roads (viewshed points), road density, and the urban ratio varied greatly among ponds, ranging 0–472,752 (median 26,239), 0.14–25.18 (12.21; unit, 10^{-3} m m^{-2}), and 0.00–0.73 (0.12), respectively (Table 2). Scatter plots of these three variables for studied ponds classified by predominant land use in areas surrounding ponds show that ponds surrounded predominantly by paddy fields exhibit moderate road density, but higher visibility than in areas surrounded by urban or woodland land use (Fig. 2A). In urban areas, despite higher road density, visual exposure of ponds was lower than in paddy field areas, probably because sight lines tended to be obscured by a greater number of building blocks surrounding ponds.

Chl.*a* concentration varied greatly among ponds: 1.9–438.8 µg L^{-1} (median 26.2 µg L^{-1}) (Table 2). Over 70% of all ponds exhibited eutrophic (chl.*a*: 8–25 µg L^{-1}) or hypertrophic (≥25 µg L^{-1}) conditions according to the trophic category of lakes [55]; 40% of all ponds had algae blooms. Water from dam reservoirs flowed into about half of the studied ponds either directly (19 ponds) or indirectly (14 ponds). The ponds from which water was drained periodically (27 ponds) were fewer than undrained ponds (37 ponds).

Table 1. Number of ponds in which Bluegill (*Lepomis macrochirus*) or Largemouth bass (*Micropterus salmoides*) were caught, their percentages for total studied ponds, and summaries of total individuals caught in fixed nets, box-net traps, and D-frame nets in 64 agricultural ponds in Hyogo Prefecture, Japan.

Invasive alien fish species	Species presence		Individuals			
	Number of ponds	%	Min.	Max.	Mean	Total
Bluegill (*Lepomis macrochirus*)	40	63	0	246	33.6	2149
Largemouth bass (*Micropterus salmoides*)	12	19	0	407	6.7	428

Classification trees for invasive fish

Based on the optimal tree that incorporated visibility from roads, connectivity with upstream dam reservoirs explained the greatest share of variance in the Bluegill distributions (Fig. 3A). For 82% (27 ponds) of all ponds, reservoir water, supplied directly or indirectly, supported Bluegill. Among ponds without a reservoir water connection (31 ponds), Bluegill tended to be present in ponds with higher visibility (viewshed points $\geq$ 42,669). The tree incorporating visibility showed the lowest misclassification value (17%) with 0.93 for sensitivity and 0.67 for specificity (Table 3), which indicated that visibility is more effective for predicting the Bluegill distribution than traditional indicators such as road density and urban ratio are. An optimal tree using road density instead of visibility revealed that ponds with higher road density ($\geq$ 0.00603 m m^{-2}) were more likely to have Bluegill present (Fig. 3B). This tree had the highest value of sensitivity (0.95), but the lowest value of specificity (0.42), which caused higher misclassification (25%) than the visibility tree did (Table 3). The result suggests overestimation of the presence for Bluegill by the classification tree using road density. However, an optimal tree using the urban ratio had only connectivity with upstream reservoirs as explanatory variables; the urban ratio did not explain the variance in the Bluegill distribution (Fig. 3C). This tree showed the highest misclassification value (30%) among the three models for Bluegill (Table 3).

For Largemouth bass, only chl.*a* presented in all optimal trees, irrespective of surrogate indicators for human-mediated introduction (visibility, road density, and urban ratio) (Fig. 4). Largemouth bass were absent from almost all ponds in which chl.*a* was higher than 38.25 µg L^{-1}, corresponding to the hypertrophic category [55]. No difference of predictive power was found among indicators for Largemouth bass (Table 3). The distribution of Largemouth bass is explainable by an optimal tree model with higher misclassification (39%) than that for Bluegill.

Discussion

Factors affecting invasive fish species distribution

The distribution (presence and absence patterns) of Bluegill, an alien game species in agricultural ponds, was explainable by the surrogate indicators for propagule pressure including the visibility of ponds from roads and the road density surrounding ponds. The result suggests that the probability of human-mediated introduction of the species becomes higher at ponds with higher visibility or denser roads, which concurs with previous findings indicating that the spatial distribution patterns of alien game species are likely to be affected by human-mediated introduction [5], [14].

Previous studies used various indicators related to human populations (e.g. population density [6]), urban land use (e.g. proportional area of urban and industrial land [11]), and roadways (e.g. distance from roads and road use types [9], [10]) as surrogates

of human-mediated introduction to predict the distribution and richness of invasive species. Our study demonstrates the effectiveness of road density in predicting the occurrence of Bluegill, although road density exhibited lower predictive power than pond visibility from roadways. An optimal tree incorporating visibility showed the lowest misclassification value, probably because visibility from roads has best represented behavioral characteristics of human related to the stocking of invasive fish species. In urban areas, although ponds are surrounded by denser road networks, visibility is apparently reduced by the many blocks of buildings that obscure the ponds. Consequently, the classification tree using only road density showed overestimation of the presence for Bluegill.

In contrast to visibility and road density, the urban ratio was not adopted for any classification tree as a predictor in our analyses. Sport anglers are generally known to come not from the immediate proximity of the ponds but from more populated urban areas such as southern coastal regions and metropolitan areas on the eastern edge of the study area. At a local scale with the study area of approximately 30 × 30 km, all ponds within the study area are assumed to be included in the home range for people who live in these populated areas. Therefore urban ratios are not an indicator of the potential accessibility to the ponds. The ratios can not explain the distribution of the invasive fish. Nevertheless, it is clear that urban land use as well as human populations can be strong predictors of species invasions at a larger extent (e.g. country or province scales) as described in many reports of the literature [5], [6], [11]. These results suggest that suitable predictors differ on a spatial scale. The indicators which directly characterize human behaviors related to propagule supply are more effective for local assessments.

For predicting the present distribution of invasive species, the cumulative effect of human-mediated introduction from the initial introduction to the present should be considered. According to the 1/25,000 topographic map (provided by the GSI) and Land utilization segmented mesh data (provided by the MLIT), the landforms and land use surrounding the study ponds did not change considerably after the late 1980s when Bluegill were first introduced into this area. This observation suggests that the visibility of ponds from roads has not changed considerably in the past few decades. Consequently, the present visibility can be regarded as an appropriate indicator of human-mediated introduction from the initial introduction to the present.

Irrespective of visibility, Bluegill tended to be present in ponds to which reservoir water was supplied directly or indirectly. Given that invasive fish species including Bluegill are observed in dam reservoirs in the study area, Bluegill seem to spread spontaneously from upstream established reservoirs via irrigation canals. This finding supports the general notion that dams facilitate biological invasion into freshwater bodies by creating additional habitats with

Table 2. Summary of continuous environmental characteristics for 64 agricultural ponds in Hyogo Prefecture, Japan.

Characteristics	Min.	25th %tile	Median	75th %tile	Max.
Visibility (number of viewshed points)	0	8,996	26,239	89,053	472,752
Road density (10^{-3} m m^{-2})	0.14	7.66	12.21	16.48	25.18
Urban ratio	0.00	0.03	0.12	0.30	0.73
Pond surface area (m^2)	685	4,321	6,897	13,220	111,626
Chlorophyll a (µg L^{-1})	1.9	7.4	26.2	73.6	438.8
River density (10^{-3} m m^{-2})					
5 m buffer	0.00	0.00	0.00	0.00	5.93
10 m buffer	0.00	0.00	0.00	0.00	4.78
25 m buffer	0.00	0.00	0.00	0.00	9.67
50 m buffer	0.00	0.00	0.00	0.00	7.12
100 m buffer	0.00	0.00	0.00	0.00	4.71
250 m buffer	0.00	0.00	0.00	1.13	5.69
500 m buffer	0.00	0.00	0.58	1.20	3.80
1000 m buffer	0.00	0.58	0.83	1.16	1.75

hydrological alterations, fish stocking, and secondary spread to surrounding water bodies [56].

Consequently, variance in Bluegill distributions was explained well solely by propagule supplies from ponds' peripheral areas, including visibility and connectivity with upstream reservoirs. Their distributions were not constrained by any site characteristic, probably because they have high adaptability to environments (see Methods) and readily establish in any pond into which they are introduced. Drainage of pond water did not affect their distributions, probably because they are introduced again even if they are eradicated by pond drying [57], which suggests that pond draining is probably ineffective as a control or eradication technique.

In contrast, the Largemouth bass distribution was explained by a site characteristic: they were absent from hypertrophic ponds. Bonvechio and Bonvechio [58] reported negative correlation between the population of Largemouth bass and chl.a concentration in their native range, which is assumed to result from the lack

of dissolved oxygen [59] with the decomposition of plenty of organic matter and decrease in foraging success with higher turbidity [60]. Largemouth bass are likely to be constrained by site characteristics because of their lower adaptability to environments than that of Bluegill [61].

Generally, because invasive species have a larger niche space and greater environmental tolerance, they spread naturally from introduced sources to non-invaded sites that possess suitable environmental conditions if propagule supplies are sufficient to sustain their populations [62]. Consequently, their distributions in invasion areas tend to be limited by the environmental conditions of sites. The tendency is regarded as strengthening as time passes after their initial introduction, with decreasing relative importance of propagule pressure on their distributions [63]. However, in the present study, propagule pressure dominantly explained the Bluegill distribution despite nearly 20 years since their initial introduction into the study area. This result is attributed to poor connectivity among ponds for their natural dispersal. Dispersion

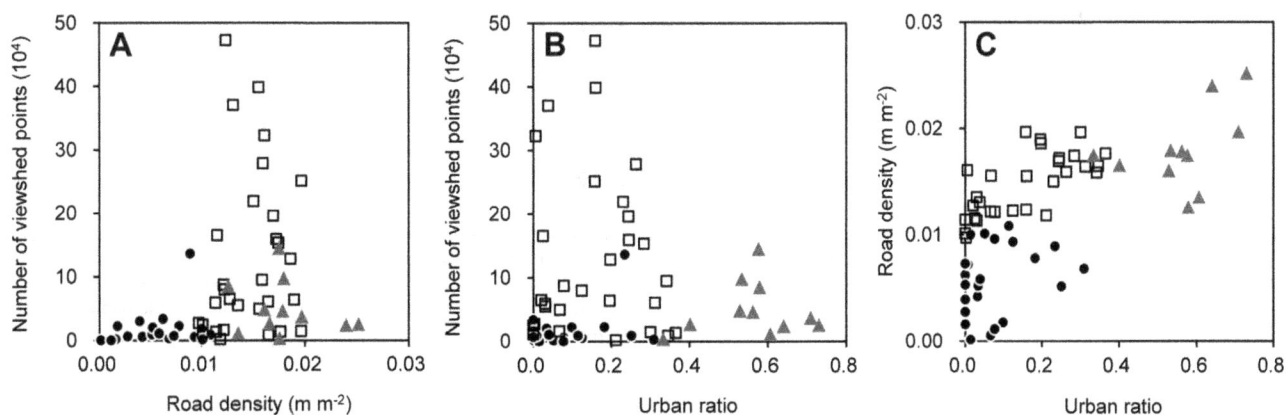

Figure 2. Relation between number of viewshed points, road density and urban ratio. Road density and urban ratio represent the total length of road line for 500 m buffer area (m m^{-2}) and the proportion of urban area for a 500 m buffer area, respectively, for 64 agricultural ponds in Hyogo Prefecture, Japan. Closed circles, open squares, and grayed triangles represent dominant land use types for 500 m buffer areas corresponding respectively to woodland, paddy field, and urban areas.

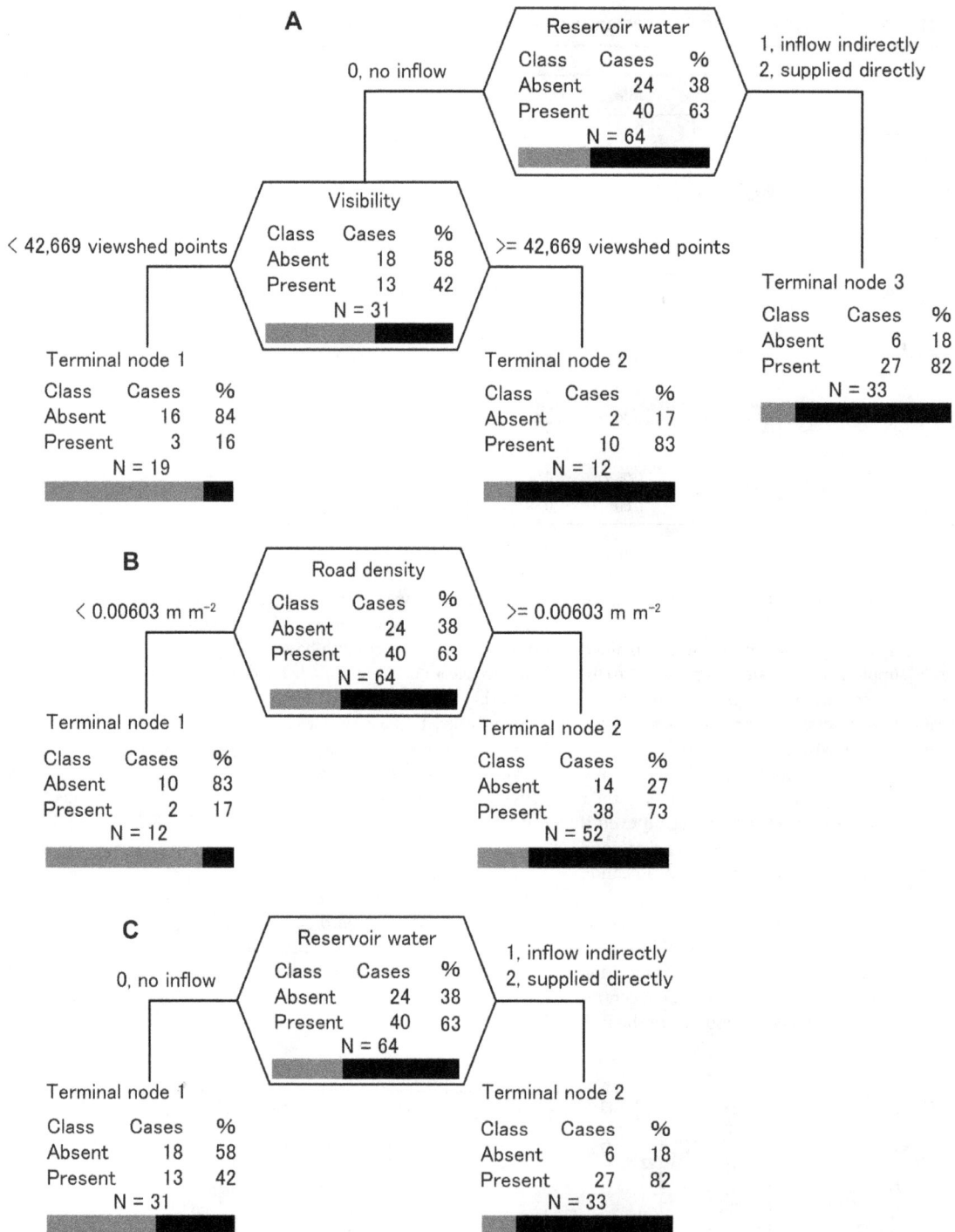

Figure 3. Results of classification tree analysis of the probability of Bluegill (*Lepomis macrochirus*) presence. The optimal tree is shown, which includes (A) visibility, (B) road density, and (C) urban ratio as a surrogate indicator for human-mediated introduction. %, percentage of cases for each class; bars represent the percentage of absent [gray] and present [black]. Although the urban ratio was included among environmental factors, the classification tree procedure omitted the variable from the optimal tree because of its lack of significance (C). The splitting variable name and split criterion are given for each node. The vertical depth of each node is proportional to its improvement value.

corridors are limited to waterways such as rivers, streams, and canals. Invasive fish might only rarely have sufficiently active (fish-vectored) transport to upstream ponds through vertical differences of a few meters between ponds and downstream drainage canals [40]. These results suggest that, for invasive species with limited natural dispersibility, the quantification of propagule pressure might be effective for predicting their distributions even if a long time passes after their initial introduction.

Table 3. Overall misclassification, sensitivity, and specificity of optimal classification trees for the probability of Bluegill (*Lepomis macrochirus*) and Largemouth bass (*Micropterus salmoides*) presence in 64 agricultural ponds in Hyogo Prefecture using visibility, road density, and urban ratio as surrogate indicators for human-mediated introduction.

Invasive alien fish species	Indicators	Overall misclassification (%)	Sensitivity	Specificity
Bluegill (*Lepomis macrochirus*)	Visibility	17	0.93	0.67
	Road density	25	0.95	0.42
	Urban ratio	30	0.68	0.75
Largemouth bass (*Micropterus salmoides*)	Visibility	39	0.92	0.54
	Road density	39	0.92	0.54
	Urban ratio	39	0.92	0.54

Efficient management of invasive species and future studies

The efficient management of invasive species demands knowledge of which factor, propagule pressure or site characteristics, is most important for their distribution [64]. Regarding our studied fish species, the Bluegill distribution depended most strongly on the indicators for propagule pressure including visibility from roads and connectivity with upstream reservoirs. Consequently, inhibiting propagule supplies into ponds is probably efficient for preventing any further invasion. These indicators used in the classification tree can be readily assessed for all agricultural ponds in the study area, which enables prediction of which areas will be susceptible to introduction of the species. Recently, international recognition has been given to the importance of local biodiversity strategies as a means of complementing and supporting national biodiversity actions and of contributing to the implementation of a strategic plan for biodiversity [2]. Our results will be useful for local planning in terms of controlling biological invasions.

The visibility of a site and its contribution to site accessibility and visitation are applicable to various alien animal and plant species that spread with frequent instances of human-mediated introduction. For example, angling is the major driver of fish species introduction worldwide [5], [65]. Incidental introduction of aquatic organisms (e.g. sessile invertebrates and macrophytes) via contaminated boating gear, nets, and other equipment is common in many inland water bodies [6], [66]. Visibility might be useful to detect areas of high invasions risk with higher accessibility by vehicles and recreational watercraft. Additionally, increasing accessibility facilitates species invasion into mountainous areas by residents and tourists [10], [12]. It is possible for visibility to become an efficient predictor for species invasions not only in inland water areas but also in terrestrial ecosystems.

Traditionally, the information associated with potential distribution and relative abundance of invasive species predicted by ecological niche modeling has been widely used for the planning of invasion control efforts [62], [64]. However, the availability is assumed to be lower for species that depend strongly on human-mediated dispersion. In this case, predicting their distributions based on an analysis of human behaviors must be more important, demanding indicators for quantifying the behavioral characteristics affected by their psychographic and sociological conditions.

Biological invasion studies particularly addressing behavioral characteristics of human-mediated introduction are very few. Buchan and Padilla [66] examined the relations between the recreational boater movement and invasive zebra mussel patterns in Wisconsin lakes. Niggemann et al. [67] studied the effect of human behavior between settlements on invasive plant dispersion in Germany, based on questionnaire surveys and statistical data. However, such information is generally lacking, leading to a lack of versatility [6]. Recently, various social information (e.g., population and industrial structure) as well as digital map information (e.g., DSMs, Digital Elevation Models [DEMs], and road networks) has become increasingly available as spatial information with large scale and higher resolution. Furthermore, technological advances in data processing using GIS engender increasing development and application of spatial analysis [68]. With these advancements, it has become possible to quantify human behaviors more directly, objectively, and broadly with spatial indicators including visibility, connectivity and travel costs through transportation networks [69], [70]. Using these indicators, control measures for biological invasions are expected to become more effective.

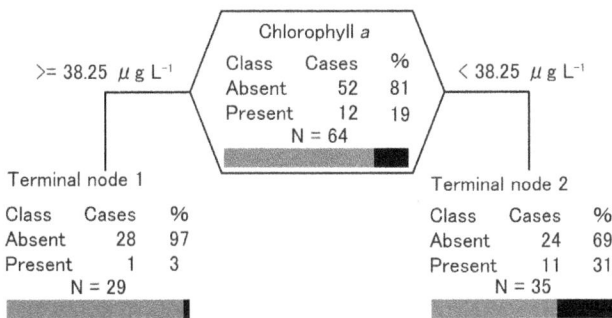

Figure 4. Results of classification tree analysis of the probability of Largemouth bass (*Micropterus salmoides*) presence. %, percentage of cases for each class; bars represent the percentage of absent [gray] and present [black]. Only chlorophyll *a* was used for all optimal trees irrespective of surrogate indicators for human-mediated introduction including visibility, road density, and urban ratio. The split criterion is given for each node.

Acknowledgments

We thank M. Nakagawa (National Institute for Environmental Studies, Tsukuba [NIES]), Y. Oikawa (NIES), and A. Saji (NIES) for field assistance and water analysis. We acknowledge the Agricultural and Environmental Affairs Department, Hyogo Prefecture Government for coordinating field survey. We are also grateful to local land improvement organizations for providing the 1/100,000 reservoir water supply area maps.

Author Contributions

Conceived and designed the experiments: T. Kizuka MA T. Kadoya NT. Performed the experiments: T. Kizuka MA T. Kadoya NT. Analyzed the

data: T. Kizuka MA T. Kadoya NT. Contributed reagents/materials/analysis tools: T. Kizuka MA T. Kadoya NT. Wrote the paper: T. Kizuka MA T. Kadoya NT.

References

1. Mack RN, Simberloff D, Lonsdale WM, Evans H, Clout M, et al. (2000) Biotic invasions: causes, epidemiology, global consequences, and control. Ecol Appl 10: 689–710.
2. CBD (2010) Report of the Tenth Meeting of the Conference of the Parties to the Convention on Biological Diversity. Available: http://www.cbd.int/doc/meetings/cop/cop-10/official/cop-10-27-en.pdf. Accessed 2013 Apr 18.
3. Kolar CS, Lodge DM (2001) Progress in invasion biology: predicting invaders. Trends Ecol Evol 16: 199–204.
4. Lockwood JL, Hoopes MF, Marchetti MP (2007) Invasion Ecology. Malden: Blackwell Publishing Ltd. 304 p.
5. Gido KB, Schaefer JF, Pigg J (2004) Patterns of fish invasions in the Great Plains of North America. Biol Conserv 118: 121–131.
6. Compton TJ, de Winton MD, Leathwick JR, Wadhwa S (2012) Predicting spread of invasive macrophytes in New Zealand lakes using indirect measures of human accessibility. Freshw Biol 57: 938–948.
7. Marini L, Battisti A, Bona E, Federici G, Martini F, et al. (2012) Alien and native plant life-forms respond differently to human and climate pressures. Global Ecol Biogeogr 21: 534–544.
8. Kaufman SD, Snucins Ed, Gunn JM, Selinger W (2009) Impacts of road access on lake trout (Salvelinus namaycush) populations: regional scale effects of overexploitation and the introduction of smallmouth bass (Micropterus dolomieu). Can J Fish Aquat Sci 66: 212–223.
9. Sharma GP, Raghubanshi AS (2009) Plant invasions along roads: a case study from central highlands, India. Environ Monit Assess 157: 191–198.
10. Pollnac F, Seipel T, Repath C, Rew LJ (2012) Plant invasion at landscape and local scales along roadways in the mountainous region of the Greater Yellowstone Ecosystem. Biol Invasions 14: 1753–1763.
11. Chytrý M, Jarošík V, Pyšek P, Hájek O, Knollová I, et al. (2008) Separating habitat invasibility by alien plants. Ecology 89: 1541–1553.
12. Giorgis MA, Tecco PA, Cingolani AM, Renison D, Marcora P, et al. (2011) Factors associated with woody alien species distribution in a newly invaded mountain system of central Argentina. Biol Invasions 13: 1423–1434.
13. Essl F, Mang T, Moser D (2012) Ancient and recent alien species in temperate forests: steady state and time lags. Biol Invasions 14: 1331–1342.
14. McKinney ML (2001) Effects of human population, area, and time on non-native plant and fish diversity in the United States. Biol Conserv 100: 243–322.
15. Fisher P, Farrelly C (1997) Spatial analysis of visible areas from the Bronze Age Cairns of Mull. J Archaeol Sci 24: 581–592.
16. De Montis A, Caschili S (2012) Nuraghes and landscape planning: Coupling viewshed with complex network analysis. Landsc Urban Plan 105: 315–324.
17. Ervin S, Steinitz C (2003) Landscape visibility computation: necessary, but not sufficient. Environ Plan Plan Des 30: 757–766.
18. Alonso JC, Alvarez-Marines JM, Palacin C (2012) Leks in ground-displaying birds: hotspots or safe places? Behav Ecol 23: 491–501.
19. Knutson MG, Richardson WB, Reineke DM, Gray BR, Parmelee JR, et al. (2004) Agricultural ponds support amphibian populations. Ecol Appl 14: 669–684.
20. Oertli B, Biggs J, Cereghino R, Grillas P, Joly P, et al. (2005) Conservation and monitoring of pond biodiversity: introduction. Aquat Conserv Mar Freshw Ecosyst 15: 535–540.
21. Akasaka M, Takamura N, Mitsuhashi H, Kadono Y (2010) Effects of land use on aquatic macrophyte diversity and water quality of ponds. Freshw Biol 55: 909–922.
22. Akasaka M, Takamura N (2011) The relative importance of dispersal and the local environment for species richness in two aquatic plant growth forms. Oikos 120: 38–46.
23. Williams P, Whitfield M, Biggs J, Bray S, Fox G, et al. (2004) Comparative biodiversity of rivers, streams, ditches and ponds in an agricultural landscape in Southern England. Biol Conserv 115: 329–341.
24. Kadoya T, Akasaka M, Aoki T, Takamura N (2011) A proposal of framework to obtain an integrated biodiversity indicator for agricultural ponds incorporating the simultaneous effects of multiple pressures. Ecol Indicat 11: 1396–1402.
25. Uchida K (2003) Irrigation ponds in Japan. Otsu: Kaiseisha Press. 270 p. (in Japanese).
26. Takamura N (2012) Status of biodiversity loss in lakes and ponds in Japan. In: Nakano S, Yahara T, Nakashizuka T, editors. Biodiversity observation network in the Asia–Pacific Region: toward further development of monitoring.Tokyo: Springer. pp. 133–148.
27. Ishii Y, Kadono Y (2003) Changes over 20 years in the macrophytic flora in irrigation ponds in the East-Harima area, Hyogo Prefecture, western Japan. Jpn J Conserv Ecol 8: 25–32. (in Japanese with English abstract).
28. Land Improvement and Consolidation Division, Hyogo Prefecture Government. (1990) Land improvement history in Hyogo. Kobe: Hyogo Prefecture Government. 1570 p.(in Japanese).
29. Japan Dam Foundation (2007) Dam yearbook 2007. Tokyo: Japan Dam Foundation. 1645 p.(in Japanese).
30. Yonekura R, Kita M, Yuma M (2004) Species diversity in native fish community in Japan: comparison between non-invaded and invaded ponds by exotic fish. Ichthyol Res 51: 176–179.
31. Maezono Y, Kobayashi R, Kusahara M, Miyashita T (2005) Direct and indirect effects of exotic bass and bluegill on exotic and native organisms in farm ponds. Ecol Appl 15: 638–650.
32. Fukushima M, Takamura N, Sun L, Nakagawa M, Matsushige K, et al. (1999) Changes in the plankton community following introduction of filter-feeding planktivorous fish. Freshw Biol 42: 719–735.
33. National Federation of Inlandwater Fisheries Cooperatices. (1992) The review of exotic largemouth bass Micropterus salmoides and bluegill Lepomis macrochirus. Tokyo: Fisheries Agency. 221 p.(in Japanese).
34. Maruyama T, Fujii K, Kijima T, Maeda H (1987) Introductory process of foreign new fish species. Tokyo: National Research Institute of Aquaculture, Fisheries Research Agency. 157 p.(in Japanese).
35. Matsuzawa Y, Senou H (2008) Alien fishes of Japan. Tokyo: Bunitsusogo-syuppan. 157 p.(in Japanese).
36. Akaboshi T (1996) Black-bass. Tokyo: Ihatov-syuppan. 189 p.(in Japanese).
37. Senou H (2002) Taxonomy of black bass introduced to Japan. In:Committee for Nature Conservation of the Ichthyological Society of Japan editor. Black bass: its biology and ecosystem effects. Tokyo: Koseisha-koseikaku. pp. 47–59. (in Japanese).
38. Kawamura K, Yonekura R, Katano O, Taniguchi Y, Saitoh K (2006) Origin and dispersal of bluegill sunfish, Lepomis macrochirus, in Japan and Korea. Mol Ecol 15: 613–621.
39. Maruyama T (2002) Bass fishing and administrative actions to it. In: Committee for Nature Conservation of the Ichthyological Society of Japan, editor. Black bass: its biology and ecosystem effects. Tokyo: Koseisha-koseikaku. pp. 99–125. (in Japanese).
40. Miyata H, Kunimoto M, Inoue M (2007) Spatial and temporal patterns in the density of largemouth bass and bluegill in pond associated lotic habitats. Ecol Civ Eng 10: 117–129. (in Japanese with English abstract).
41. Freund JG, Hartman KJ (2005) Largemouth bass habitat interactions among off-channel and main river habitats in an Ohio River navigation pool. J Freshw Ecol 20: 735–742.
42. Takamura K (2007) Performance as a fish predator of largemouth bass [Micropterus salmoides (Lacepède)] invading Japanese freshwaters: a review. Ecol Res 22: 940–946.
43. Yonekura R, Nakai K, Yuma M (2002) Trophic polymorphism in introduced bluegill in Japan. Ecol Res 17: 49–57.
44. Gross MR, Charnov EL (1980) Alternative male life histories in bluegill sunfish. Proceedings of the National Academy of Sciences of the United States of America 77: 6937–6940.
45. Chipps SR, Dunbar JA, Wahl DH (2004) Phenotypic variation and vulnerability to predation in juvenile bluegill sunfish (Lepomis macrochirus). Oecologia 138: 32–38.
46. Sevenant M, Antrop M (2007) Settlement models, land use and visibility in rural landscapes: Two case studies in Greece. Landsc Urban Plan 80: 362–374.
47. Irwin ER, Jackson JR, Noble RL (2002) A reservoir landscape for age-0 largemouth bass. In: Philipp DP, Ridgway MS, editors. Black bass: ecology, conservation, and management. American Fisheries Society Symposium 31. Bethesda: American Fisheries Society. pp. 61–71.
48. Jackson ZJ, Quist MC, Downing JA, Larsceid JG (2010) Common carp (Cyprinus carpio), sport fishes, and water quality: Ecological thresholds in agriculturally eutrophic lakes. Lake Reserv Manag 26: 14–22.
49. Welch EB, Cooke GD, Jones JR, Gendusa TC (2011) DO–Temperature habitat loss due to eutrophication in Tenkiller Reservoir, Oklahoma, USA. Lake Reserv Manag 27: 271–285.
50. Breiman L, Friedman JH, Olshen RA, Stone CG (1984) Classification and regression trees. Boca Raton: CRC Press LLC. 368 p.
51. De'ath G, Fabricius KE (2000) Classification and regression trees: a powerful yet simple technique for ecological data analysis. Ecology 81: 3178–3192.
52. Bourg NA, McShea WJ, Gill DE (2005) Putting a CART before the search: successful habitat prediction for a rare forest herb. Ecology 86: 2793–2804.
53. R Development Core Team (2012) R: A language and environment for statistical computing. R Foundation for Statistical Computing. Available: http://www.R-project.org/. Accessed 2014 Apr 27.
54. Therneau TM, Atkinson EJ (1997) An introduction to recursive partitioning using rpart routines. Mayo Foundation. Available: http://www.mayo.edu/hsr/techrpt/61.pdf. Accessed 18 April 2013.
55. OECD (1982) Eutrophication of waters, monitoring, assessment and control. Paris: OECD. 154 p.
56. Johnson PTJ, Olden JD, Zanden MJV (2008) Dam invaders: impoundments facilitate biological invasions into freshwaters. Front Ecol Environ 6: 357–363.

57. Ushio N, Imada M, Akasaka M, Takamura N (2009) Effects of pond management on the distributions of aquatic invaders in Japanese farm ponds. Jpn J Limnol 70: 261–266. (in Japanese with English abstract).

58. Bonvechio KI, Bonvechio TF (2006) Relationship between habitat and sport fish populations over a 20-year period at West Lake Tohopekaliga, Florida. N Am J Fish Manag 26: 124–133.

59. Hasler CT, Suski CD, Hanson KC, Cooke SJ, Tufts BL (2009) The influence of dissolved oxygen on winter habitat selection by largemouth bass: an integration of field biotelemetry studies and laboratory experiments. Physiol Biochem Zool 82: 143–152.

60. Shoup DE, Wahl DH (2009) The effects of turbidity on prey selection by piscivorous largemouth bass. Trans Am Fish Soc 138: 1018–1027.

61. Moyle PB, Marchetti MP (2006) Predicting invasion success: freshwater fishes in California as a model. BioScience 56: 515–524.

62. MacIsaac HJ, Herborg L-M, Muirhead JR (2007) Modeling biological invasions of inland waters. In: Gherardi F, editor. Biological invaders in inland waters: Profiles, distribution, and threats. Dordrecht: Springer. pp. 347–368.

63. Kondo T, Tsuyuzaki S (2003) Natural regeneration patterns of the introduced larch, *Larix kaempferi* (Pinaceae), on the volcano Mount Koma, northern Japan. Divers Distrib 5: 223–233.

64. Vander Zanden MJ, Olden JD (2008) A management framework for preventing the secondary spread of aquatic invasive species. Can J Fish Aquat Sci 65: 1512–1522.

65. Tricarico E (2012) A review on pathways and drivers of use regarding non-native freshwater fish introductions in the Mediterranean region. Fish Manag Ecol 19: 133–141.

66. Buchan LA, Padilla DK (1999) Estimating the probability of long-distance overland dispersal of invading aquatic species. Ecol Appl 9: 254–265.

67. Niggemann M, Jetzkowitz J, Brunzel S, Wichmann MC, Bialozyt R (2009) Distribution patterns of plants explained by human movement behaviour. Ecol Model 220: 1339–1346.

68. Lorini ML, Paese A, Uezu A (2011) GIS and spatial analysis meet conservation: a promising synergy to address biodiversity issues. Nat & Conservação 9: 129–144.

69. Borgatti SP, Everett MG (2006) A graph-theoretic perspective on centrality. Soc Network 28: 466–484.

70. Drake DAR, Mandrak NE (2010) Least-cost transportation networks predict spatial interaction of invasion vectors. Ecol Appl 20: 2286–2299.

21

Dynamics and Sources of Soil Organic C Following Afforestation of Croplands with Poplar in a Semi-Arid Region in Northeast China

Ya-Lin Hu[1]*, Li-Le Hu[2], De-Hui Zeng[1]*

1 State Key Laboratory of Forest and Soil Ecology, Institute of Applied Ecology, Chinese Academy of Sciences, Shenyang, People's Republic of China, **2** Chinese Research Academy of Environmental Sciences, Beijing, People's Republic of China

Abstract

Afforestation of former croplands has been proposed as a promising way to mitigate rising atmospheric CO_2 concentration in view of the commitment to the Kyoto Protocol. Central to this C sequestration is the dynamics of soil organic C (SOC) storage and stability with the development of afforested plantations. Our previous study showed that SOC storage was not changed after afforestation except for the 0–10 cm layer in a semi-arid region of Keerqin Sandy Lands, northeast China. In this study, soil organic C was further separated into light and heavy fractions using the density fractionation method, and their organic C concentration and ^{13}C signature were analyzed to investigate the turnover of old vs. new SOC in the afforested soils. Surface layer (0–10 cm) soil samples were collected from 14 paired plots of poplar (*Populus* × *xiaozhuanica* W. Y. Hsu & Liang) plantations with different stand basal areas (the sum of the cross-sectional area of all live trees in a stand), ranging from 0.2 to 32.6 m^2 ha^{-1}, and reference maize (*Zea mays* L.) croplands at the same sites as our previous study. Soil Δ_C stocks (Δ_C refers to the difference in SOC content between a poplar plantation and the paired cropland) in bulk soil and light fraction were positively correlated with stand basal area ($R^2 = 0.48$, $p < 0.01$ and $R^2 = 0.40$, $p = 0.02$, respectively), but not for the heavy fraction. SOC_{crop} (SOC derived from crops) contents in the light and heavy fractions in poplar plantations were significantly lower as compared with SOC contents in croplands, but tree-derived C in bulk soil, light and heavy fraction pools increased gradually with increasing stand basal area after afforestation. Our study indicated that cropland afforestation could sequester new C derived from trees into surface mineral soil, but did not enhance the stability of SOC due to a fast turnover of SOC in this semi-arid region.

Editor: Ben Bond-Lamberty, DOE Pacific Northwest National Laboratory, United States of America

Funding: This work was supported by the National Natural Science Foundation of China (No. 31000297) (http://www.nsfc.gov.cn). The funders had no role in study design, data collection and analysis, decision to publish, or preparation of the manuscript.

Competing Interests: The authors have declared that no competing interests exist.

* E-mail: huyl@iae.ac.cn (YH); zengdh@iae.ac.cn (DZ)

Introduction

Afforestation of degraded croplands has occurred globally within the framework of the Kyoto Protocol [1], and has the potential to mitigate the rising atmospheric CO_2 concentration caused by anthropogenic emissions [2,3]. Consequently, changes of soil organic C (SOC) stocks after cropland afforestation have been well studied [3–8], considering the fact that soil has the greatest potential C sink capacity and the longest time period [9–11]. However, most previous studies have suggested that there were large variations in the direction and magnitude of SOC stock changes following afforestation of croplands, which are related to previous land-use, climate, soil texture, tree species, stand age, and management practices [3,4,8]. In addition, the changes of SOC stocks are often not detected by conventional methodologies within a short-time frame for most experiments due to the small changes in soil C when compared with the size of SOC reservoir [6,9]. In order to accurately assess small changes in SOC stocks and stability following cropland afforestation, it is necessary to investigate the dynamics in different SOC pools characterized by different physical and chemical properties, microbial degradability and turnover time [9].

Soil organic matter (SOM) can be divided into discrete fractions with different stability and ecological functions [12,13]. Density fractionation physically separates SOC into light fraction and heavy fraction (LF and HF, respectively), which has been increasingly used to assess SOC dynamics induced by land use change and management practices [8,14–17]. However, soil fractionation and C analysis only provide information on net changes of SOC, but not for soil C balance between the loss of old C and the input of new C. The ^{13}C natural abundance technique offers an elegant approach to quantify the relative contribution of new vs. old SOC, for example, where C3 plants ($\delta^{13}C$ ca. $-28\permil$) grow on soils derived from C4 crops ($\delta^{13}C$ ca. $-12\permil$) [9]. Several studies have elucidated the dynamics in SOC following cropland afforestation using soil fractionation techniques combined with the stable C isotope techniques [9,18–20], but this approach has not been used in semi-arid regions.

The SOC stocks are determined by the balance between the inputs of C derived from litterfall and rhizodeposition and the losses of C mainly through soil organic matter decomposition [1,21]. Therefore, the dynamics of SOC stocks following afforestation are not only correlated with stand age, but also with tree density that influences soil microclimatic conditions and the amount of litterfall [1,22,23]. Stand basal area (BA, the sum of the cross-sectional area of all live trees in a stand) can integrates information of both stand age and tree density, which is more feasible to evaluate the dynamics and sources of SOC following afforestation considering the important roles of trees in SOC inputs from litter and root exudates.

Understanding the changes in SOC stocks and stability following afforestation in semi-arid regions is important principally because of the vast area involved with 2.4 billion ha or ~17.7% of total global land surface area [24], and the different changes in soil C stocks and stability after afforestation compared with humid regions [25,26]. Poplar (*Populus*) species is one of the most widely grown trees on croplands in the Keerqin Sandy Lands, a semi-arid region in northeastern China, in two large scale afforestation programs i.e., the Three-North Shelter Forest Program and the Grain for Green Project. The net changes in above- and below-ground C stocks after afforestation have been well investigated in the Keerqin Sandy Lands [27–29]. In this present study, the objectives are to further evaluate the dynamics and sources of SOC in different soil C pools along BA following afforestation of croplands with poplars in this semi-arid region. We hypothesized that: (1) SOC stocks in bulk soil and light fraction would increase following cropland afforestation due to the enhanced inputs of litterfall with increasing BA, and (2) the sources of SOC would gradually convert from crop-derived sources to tree-derived sources because of the decomposition of old soil C and the accumulation of new soil C. To test these hypotheses, we selected 14 paired stands of poplar plantations with different BAs afforested on croplands and adjacent maize (*Zea mays* L.) croplands as controls in the southeastern region of Keerqin Sandy Lands, and then analyzed SOC concentrations and δ^{13}C values in bulk soil, LF and HF.

Materials and Methods

2.1 Ethics Statement

This study was carried out on collective-owned lands, and the owners of the lands gave us permission to conduct the study on these stands. The field studies did not involve endangered or protected species.

2.2 Site Description and Experimental Design

This study was carried out in the southeastern region of the Keerqin Sandy Lands (42°30′–42°55′N, 122°19′–122°30′E), a typical semi-arid region in northeast China. The climate is temperate continental monsoon. Mean annual temperature is about 5.7°C, ranging from −23.2°C in January to 32.4°C in July (1954–2004). Mean annual precipitation is about 450 mm (ranging from 224 mm to 661 mm during 1954–2004), with more than 60% occurring from June to August, and mean annual potential evaporation ranges from 1300 to 1800 mm with an average length of frost-free season of about 150 days. The soil is a sandy soil with 90.9% sand, 5.0% silt and 4.1% clay, and classified into the Entisol order, Semiaripsamment group (according to the United States Soil Classification System) and developed from sandy parent material through the action of wind [30]. Before croplands are established in this region, the dominant species of the native vegetation include *Agropyron cristatum*, *Arundinella hirta*, *Cleistogenes chinensis*, *Lespedeza davurica* and *Artemisia capillaris* var. *simplex* [31].

Since 1978, a large area of marginal croplands has been afforested with trees in this region under the Three-North Shelter Forest Program and the Grain for Green Project, in order to control windy erosion and desertification. By now, the wind erosion is effectively reduced. In June 2011, we selected 14 poplar (*Populus × xiaozhuanica* W. Y. Hsu & Liang, a hybrid of *P. nigra* var. *italica* and *P. simonii*) plantations (ranging from 2 to 20 years old) afforested on maize croplands and 14 adjacent maize cropland stands as control in Kezuohouqi and Zhangwu counties on the basis of a paired-plot experimental design (Table 1). Most of the paired cropland and plantation stands are conterminous except for several cases, and the distance of sampling point in each paired poplar plantation and cropland is less than 500 m. The topography and soil conditions are similar in each paired cropland and poplar plantation, and the slope of each stand is very gentle and less than 5°. All the poplar plantation stands were planted on maize croplands that had cultivated for at least 20 years before afforestation, and the paired croplands were continually planted to maize. Usually, croplands are fertilized with urea fertilizer each year, while fertilizer is no longer used after afforestation.

2.3 Stand Investigation and Soil Sampling

One 20 × 20 m plot was established in each stand. For poplar plantations, the diameter at breast height (DBH) and tree height were measured for all live trees in each plot. DBH was measured at breast level (1.3 m above ground) using a caliper. For measurement of tree height, we used a long pole to extend vertically to the top of the tree and then measured the length of the pole. The *BA* (m^2 ha^{-1}) was calculated from measurements of the *DBH* (cm) of all trees in a known area (*A*, ha), which is expressed as:

$$BA = \frac{\pi}{4 \times 10000} \times \frac{\sum DBH^2}{A} \qquad (1)$$

Considering the changes of soil organic C stocks were only observed in 0–10 cm layer in our previous study [28] at the same sites, soil samples in the surface layer were only collected in this present study. Four soil cores were sampled randomly using an auger (2.5 cm in diameter) at the surface mineral soil layers (0–10 cm), and thoroughly mixed to form a homogenized sample for each stand (i.e., a total of 28 samples including 14 samples from poplar plantations and 14 samples from croplands). Soil samples were dried at room temperature (20 °C) and then passed through a mesh sieve with a size of 2 mm. Soil bulk density (ρ) was determined at three randomized sampling points in each plot for calculation of SOC content. For measurement of soil bulk density, a metal corer (volume is 100 cm^3) was driven into the soil at the desired depth, and then soil samples were oven dried at 115°C for 24 h and weighed. Soil bulk density was calculated as:

$$\rho = \frac{M}{V} \qquad (2)$$

where *M* is dry mass of soil and *V* is volume of soil (i.e., 100 cm^3).

2.4 Soil Density Fractionation

A soil sample was physically separated into two pools by the modified density fractionation method of Sohi et al. [32]. Briefly, 10 g of air-dried soil (<2 mm) were placed in a centrifuge tube with 40 mL sodium iodide (NaI) solution at a density of 1.7 g cm^{-3}. The tubes were shaken up and down by hand for ten times. The release of light fraction was accelerated by sonication at 58 Watts for 180 s using a sonicator (Bilon96, Bilon Instruments Co., Ltd, China). After sonication was finished, the tubes were

Table 1. Stand location and characteristics.

Polar plantation stand								Cropland stand		
Plot	Location	Elevation (m)	Tree height (m)	DBH[†] (cm)	Density (Trees ha^{-1})	Stand basal area (m^2 ha^{-1})		Plot	Location	Elevation (m)
F1	42°54′11″N, 122°23′30″E	252	3.35	2.92	1100	0.85		C1[‡]	42°54′04″N, 122°23′32″E	255
F2	42°37′06″N, 122°20′54″E	166	4.38	4.62	1075	1.85		C2	42°37′04″N, 122°20′52″E	166
F3	42°54′00″N, 122°23′25″E	252	4.80	6.35	850	2.92		C3	42°53′59″N, 122°23′23″E	251
F4	42°53′04″N, 122°24′08″E	250	5.26	5.73	1275	3.65		C4	42°52′58″N, 122°24′09″E	250
F5	42°54′02″N, 122°25′32″E	247	6.85	7.39	1300	6.11		C5	42°54′00″N, 122°25′34″E	246
F6	42°53′18″N, 122°24′11″E	251	8.77	9.21	1100	8.20		C6	42°53′17″N, 122°24′07″E	252
F7	42°59′25″N, 122°20′25″E	244	8.95	11.58	825	9.24		C7	42°59′26″N, 122°20′20″E	244
F8	42°37′58″N, 122°22′24″E	186	15.80	13.37	750	10.69		C8	42°38′01″N, 122°22′24″E	185
F9	42°54′13″N, 122°23′36″E	252	15.20	15.10	700	13.15		C9	42°54′17″N, 122°23′41″E	251
F10	42°53′59″N, 122°25′10″E	249	18.70	10.54	1575	14.62		C10	42°53′56″N, 122°25′07″E	248
F11	42°55′42″N, 122°24′39″E	246	17.76	13.16	1025	15.69		C11	42°55′42″N, 122°24′37″E	246
F12	42°37′06″N, 122°20′58″E	165	30.72	22.00	625	24.79		C12	42°37′04″N, 122°20′52″E	165
F13	42°54′25″N, 122°23′28″E	250	20.68	17.11	1150	28.51		C13	42°54′23″N, 122°23′28″E	250
F14	42°37′05″N, 122°21′06″E	164	29.87	22.90	775	32.60		C14	42°37′06″N, 122°21′05″E	165

[†]DBH: Diameter at breast height (1.3 m); an average value for all living trees in each plot.
[‡]All croplands were planted to maize.

centrifuged at 6000 rpm for 20 min. The floating material was aspirated together with the NaI solution from the surface of tubes, and then filtered using Whatman GF/A filter papers. This procedure was repeated three times. The material collected on the filter paper (light fraction, LF) and the residue remaining in the centrifuge tube (heavy fraction, HF) were rinsed thoroughly with deionized water and collected. The samples of LF and HF were dried at 60°C for 48 h.

2.5 SOC Concentration and ^{13}C Analysis

The samples of bulk soil, LF and HF were ground to a fine powder with a ball mill and analyzed for SOC concentration and δ^{13}C. SOC concentration was determined using the Walkey and Black $K_2Cr_2O_7$–H_2SO_4 oxidation method [33]. The isotope ratio ^{13}C/^{12}C was determined using isotope ratio mass spectrometer (Finnigan DELTAplusXP, Thermo Fisher Scientific, USA), and the ^{13}C abundance was expressed in delta-units (δ^{13}C, ‰) according to the following equation:

$$\delta^{13}C = [(R_{sam}/R_{std}) - 1] \times 1000 \qquad (3)$$

where R_{sam} is the ^{13}C/^{12}C ratio of soil sample and R_{std} is the ^{13}C/^{12}C ratio of the international Pee Dee formation belemnite carbonate standard (PDB).

2.6 Data Calculation and Statistical Analysis

Soil organic C content of bulk soil, LF and HF was calculated as:

$$SOC_{cont} = c \times \rho \times d \times r \qquad (4)$$

where SOC_{cont} was soil organic C content of bulk soil, LF or HF; c was SOC concentration of bulk soil, LF or HF; ρ was soil bulk density; d was soil depth (i.e. 10 cm); and r was the dry mass ratio of LF or HF to bulk soil.

We estimated the sources of SOC in the poplar plantations based on an isotope mass balance and δ^{13}C values, and the fractional tree-derived SOC (F_{tree-C}) was calculated using a two-component mixing equation [34]:

$$F_{Tree-C}(\%) = \frac{\delta^{13}C_{poplar} - \delta^{13}C_{crop}}{\delta^{13}C_{tree} - \delta^{13}C_{crop}} \times 100\% \qquad (5)$$

where $\delta^{13}C_{poplar}$ and $\delta^{13}C_{crop}$ values are actual measured values in bulk soil, LF or HF in poplar plantation and its paired cropland, respectively. The $\delta^{13}C_{tree}$ value is the measured δ^{13}C of poplar leaf litter (−29.63‰). Subsequently, the mass of tree-derived SOC (SOC_{tree}) and crop-derived SOC (SOC_{crop}) were calculated as:

$$SOC_{tree} = F_{Tree-C} \times SOC_{cont} \qquad (6)$$

$$SOC_{crop} = SOC_{cont} - SOC_{tree} \qquad (7)$$

All statistical analyses were done using the open source statistical software R version 2.14.1. Paired t tests were used to examine the changes of SOC concentration and SOC content in bulk soil, LF and HF between poplar plantations and the paired croplands. To test the dynamics and sources of SOC after afforestation, the relationships of BA and SOC content, and BA and δ^{13}C of bulk soil, LF and HF in poplar plantation stands were tested with a linear regression analysis. The significance level was set at $\alpha = 0.05$ for all the statistical analyses unless otherwise noted.

Results

3.1 SOC in Bulk Soil, Light Fraction and Heavy Fraction

Across all popular plantation stands, the average SOC concentration and SOC content in bulk soil were 64% and 54% higher than that across croplands, respectively (all $p<0.001$)

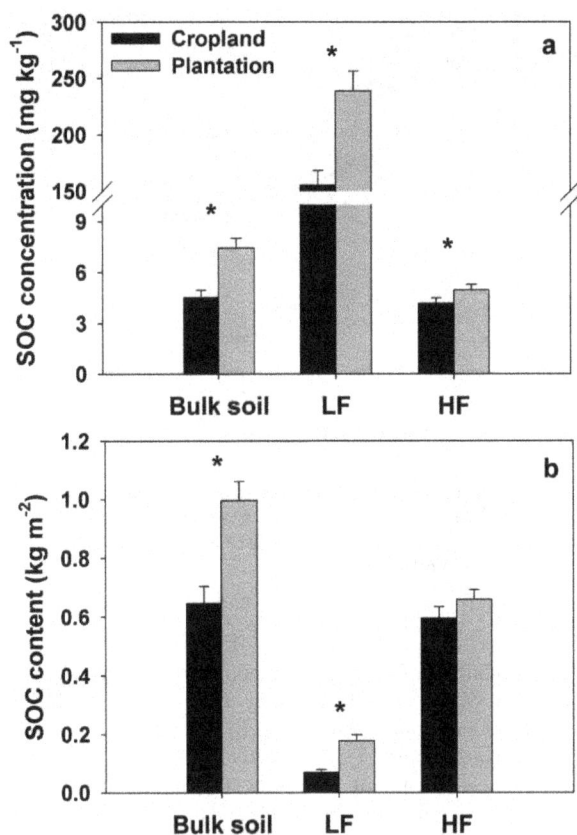

Figure 1. Averages of SOC concentration (a) and content (b) of bulk soil, LF and HF in croplands and poplar plantations. The asterisk above the bars indicates a significant difference between cropland and poplar plantation at α = 0.05 level. The vertical error bars are standard errors of the means (n = 14).

(Fig. 1). SOC concentrations in bulk soil ranged from 4.67 to 12.50 g kg^{-1} in poplar plantations, and from 2.39 to 7.28 g kg^{-1} in croplands. SOC contents had a range from 0.67 to 1.54 kg C m^{-2} in poplar plantations, and a range from 0.38 to 1.06 kg C m^{-2} in cropland stands.

SOC concentrations in LF and HF in poplar plantations (an average of 238 g kg^{-1} in LF and 4.97 g kg^{-1} in HF, respectively) were also significantly higher than that in croplands (an average of 155 g kg^{-1} in LF and 4.19 g kg^{-1} in HF, respectively) (Fig. 1a). SOC was mainly stored in the HF pools in both poplar plantations and croplands. Average values of SOC content in LF and HF were 0.18 and 0.66 kg m^{-2} in poplar plantations, and 0.07 and 0.59 kg m^{-2} in croplands, respectively. SOC content in the LF of poplar plantations was significantly higher than that in croplands (p<0.001), but not for HF (p = 0.12) (Fig. 1b).

3.2 Changes of SOC in Bulk Soil, LF and HF with BA

Soil Δ_C stocks (Δ_C refers to the difference in SOC content between a poplar plantation and the paired cropland) in bulk soil had a linear increase trend with increasing BA (Fig. 2). Similarly, there was a significant positive correlation between soil Δ_C stocks in LF and BA, but not for HF.

3.3 Soil δ^{13}C in Poplar Plantations

Soil δ^{13}C values in bulk soil and LF were negatively correlated with BA (Fig. 3). Soil δ^{13}C in bulk soil was depleted from −21‰

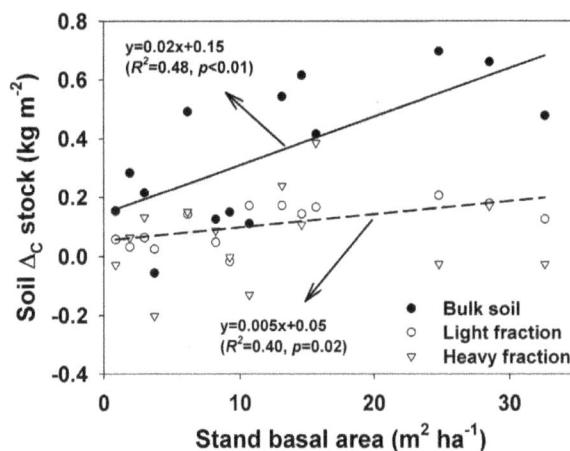

Figure 2. Dynamics of soil Δ_C stocks in bulk soil, LF and HF along stand basal area in poplar plantation stands. Solid line is regression for the bulk soil, and dash line is regression for light fraction.

(BA was 0.85 m^2 ha^{-1}) to −27‰ (BA was 28.51 m^2 ha^{-1}), and soil δ^{13}C in LF was depleted from −22‰ (BA was 0.85 m^2 ha^{-1}) to −30‰ (BA was 28.51 m^2 ha^{-1}). However, the relationship between soil δ^{13}C in HF and BA was not significant (p = 0.18). Soil δ^{13}C in HF (average value of −24‰) were significantly higher than that in LF (an average value of −28‰) in poplar plantations (p<0.001).

SOC$_{crop}$ (SOC derived from crops) content in bulk soil of poplar plantations was slightly lower as compared with SOC content in croplands (p = 0.17), while SOC$_{crop}$ contents in LF and HF were 51% and 27% lower, respectively (all p<0.05) (Fig. 4). SOC$_{tree}$ (SOC derived from poplar trees) contents in bulk soil, LF and HF were all increased significantly with increasing BA (Fig. 5). SOC$_{tree}$ contents in bulk soil ranged from 0.08 kg m^{-2} in the poplar stand (BA was 0.85 m^2 ha^{-1}) to 0.93 kg m^{-2} in the poplar stand (BA was 32.6 m^2 ha^{-1}). The percentages of SOC$_{tree}$ to total SOC content in poplar plantations were on average 40% in bulk soil, 77% in LF, and 33% in HF.

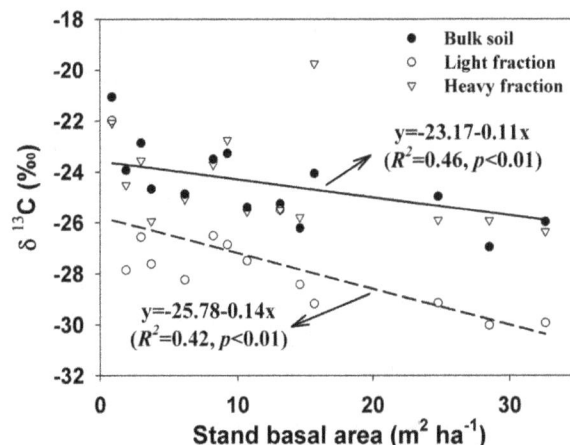

Figure 3. Dynamics of soil δ^{13}C in bulk soil, LF and HF along stand basal area in poplar plantations. Solid line is regression for bulk soil, and dash line is regression for light fraction.

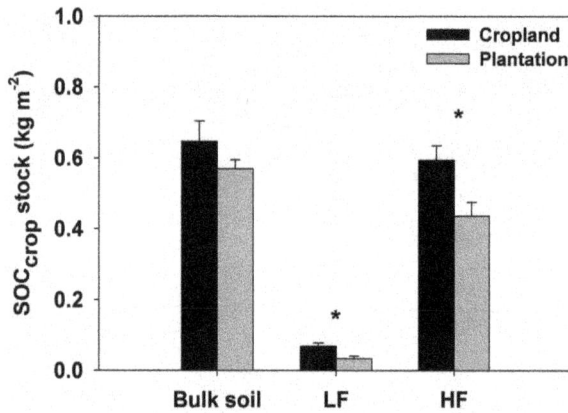

Figure 4. Averages of SOC$_{crop}$ contents in bulk soil, LF and HF in croplands and poplar plantations. The asterisk above the vertical bars indicates a significant difference between cropland and poplar plantation at $\alpha = 0.05$ level. The vertical error bars are standard errors of the means ($n = 14$).

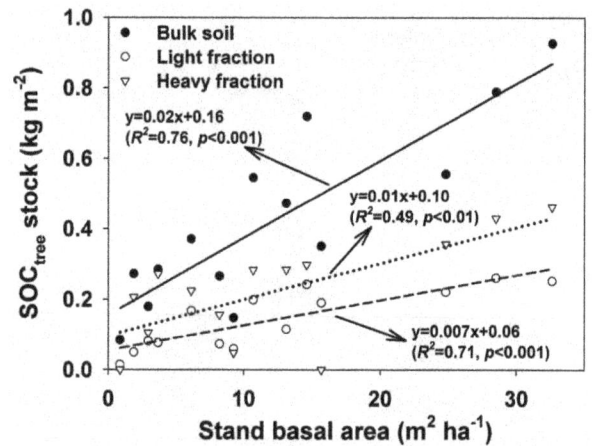

Figure 5. Dynamics of SOC$_{tree}$ contents in bulk soil, LF and HF along stand basal area in poplar plantations. Solid line is regression for bulk soil, and dash line is regression for light fraction, and dot line is regression for heavy fraction.

Discussion

4.1 Changes of SOC in Bulk Soil, LF and HF

In this study, the afforested soil had higher SOC content than croplands (an average of 0.35 kg C m^{-2}) in the surface soils, implying that afforestation with hybrid poplar on croplands could sequester 12.8 Mg CO$_2$ ha^{-1} into surface mineral soils in this semi-arid region. Similarly, Mao et al. [29] observed that SOC sequestration in 1 m depth was 1.9 kg C m^{-2} in a 20-year-old poplar plantation that was afforested on the marginal agricultural land. Li et al. [35] reported that soil sequestered 0.39 kg C m^{-2} in 0–15 cm depth in a 25-year-old Mongolian pine (*Pinus sylvestris* var. *mongolica*) afforested on active sand dune in the Keerqin Sandy Lands. Increased SOC stocks induced by afforestation with poplar trees on croplands were also reported in other regions [36–38]. When land use change from crop to forest, an average increase of 18% in SOC stock was observed, and afforested broadleaved tree species were more effective in sequestering CO$_2$ into soils [3].

Furthermore, this present study showed that soil Δ_C stocks increased with increasing BA, suggesting the potential capacity of storing soil C would enhance with the increase of BA after afforestation. It is consistent with results in the literature: for example, Vesterdal et al. [39] found that SOC contents in 0–5 cm increased with stand age after afforestation with Norway spruce (*Picea abies* (L.) Karst) on former arable lands. Sartori et al. [38] also observeed that SOC stocks in 0–10 cm layer had an increasing trend along a choronosequence of poplar plantations in the Columbia Plateau, Oregon, USA. Mao et al. [29] and Arevalo et al. [36] all found that SOC stocks increased with stand age though there was a loss of SOC in the early stage after poplar plantations were established on croplands. It has been proposed that the increases of soil C inputs from litterfall with increasing stand age or tree density, and the lack of tillage disturbance cause the increase in soil C stocks after cropland afforestation [8,19].

The light fraction of SOM is commonly referred to as plant-like SOM with high C concentration, and the heavy fraction contains more decomposed SOM with lower C concentration [12]. Consistently, we found that SOC concentrations (ranging from 99 to 270 g kg^{-1} in croplands and 138 to 351 g kg^{-1} in poplar plantations, respectively) in the LF were obviously higher than in the HF (ranging from 2.72 to 5.89 g kg^{-1} in croplands and 2.89 to 7.01 g kg^{-1} in poplar plantations, respectively) in this present

study. However, SOC content in the HF accounted for 89% of SOC in croplands and 79% of SOC in poplar plantations (Fig. 1b), indicating that SOC stocks were mainly distributed in the HF in both croplands and poplar plantations. In general, SOC contents in the LF accounted for 17–47% of total SOC contents in the surface mineral soil in a temperate zone [12], and the larger mass of the heavy fraction in soils was also observed by the other studies [8,15–17,40].

Changes of SOC stock in LF are usually more sensitive to land use change and management practices [14–17]. In this study, we observed that SOC concentration and content in the LF were all significantly increased after afforestation (Fig. 1). Similarly, the increases of SOC concentration in the LF following cropland afforestation were also observed by Laik et al. [41]. Furthermore, Li et al. [35] found that afforestation with Mongolian pine on active sand dunes resulted in an increase of light fraction C concentration in Keerqin Sandy Lands. There was a significant positive relationship between soil Δ_C stocks in LF and BA, indicating the gradual increases of SOC sequestration into the LF with increasing BA. It is consistent with the results of Marin-Spiotta et al. [42] who found that the mass and SOC concentrations in the LF increased along a chronosequence of natural reforestation of abandoned tropical pastures. The increases of soil C in the LF could be associated with the enhanced C input with increasing BA [14], considering the significant positive relationship between SOC$_{tree}$ in the LF and BA (Fig. 5).

Soil organic C content in the HF plays an important role in the stability of SOC associated with its slow decomposition due to physical protection. However, we did not observe the increase of SOC stock in HF after afforestation in this semi-arid region. It implies that SOC stability might not be enhanced after afforestation with poplar on croplands. Huang et al. [14] also reported that there were no significant changes in SOC stock in the heavy fraction after afforestation on grasslands. However, Clark et al. [18] observed that the stability of SOC increased when native forests were allowed to invade abandoned agricultural fields in western New England. SOC stability is controlled by soil texture rather than land use management [5]. The enhancement of SOC stability after afforestation more likely occurred in soils with more clay and silt, and under climate conditions with more precipitation and warmer temperature [4,18,26].

4.2 Turnover of SOC

The SOC stocks are determined by the balance between the input of C derived from plant litter and the loss of C mainly through soil organic matter decomposition [1,21]. Our results showed that SOC derived from crops (i.e., old SOC) in the LF and HF in poplar plantations was all significantly lower than that in croplands, though the difference of SOC_{crop} in bulk soil was not significant (Fig. 4), implying that cropland afforestation led to the loss of old SOC. However, the new SOC derived from poplar trees in bulk soil, LF and HF all gradually increased with increasing BA (Fig. 5). Consistently, a net loss of old SOC and a gain of new SOC after cropland or grass afforestation were also observed in several other studies [14,19,20]. Our results imply that the increased C inputs from trees (litter and roots) following afforestation are the major causes enhancing soil C sequestration rather than the inhibition of soil old C decomposition.

Though there was a positive relationship between SOC_{tree} in the HF of the poplar plantations and BA, we could not infer that the long-term stability of SOC derived from trees would be enhanced considering the fast turnover of SOC_{crop} in the HF in poplar plantation stands in this semi-arid region. Galdo et al. [9] and Dondini et al. [19] suggested that afforestation could enhance SOC stability because they did not find the loss of old soil SOC due to the formation of soil aggregates. Furthermore, Paul et al. [5] suggested that the presence of soil C stabilization processes did not necessarily mean that recently incorporated soil C will also be effectively stabilized. In our study, a fast replacement of the old SOC by new SOC might be associated with the weak physical protection in the sandy soil [13], and the frequent drying-rewetting in the semi-arid region [43].

Conclusions

Our results confirm that cropland afforestation with poplars has the potential to sequester C rapidly into soils considering the gradual increases of SOC following afforestation because of the substantial replenishment of old soil C derived from crops by new C derived from trees. However, soil C sequestration was mainly caused by the increase in soil LF, but no significant changes of SOC content were observed in soil passive C pool due to a loss of old soil C in HF in this semi-arid region. It implies that the stability of soil organic C is not enhanced after a short-term afforestation on croplands.

Acknowledgments

We are grateful to Dr. Rong Mao for reviewing this manuscript, and Gui-Gang Lin and Bo Li for their considerable help in field work and laboratory analyses. We also thank two anonymous reviewers and Ben Bond-Lamberty for their helpful comments that substantially improved an earlier version of this manuscript.

Author Contributions

Conceived and designed the experiments: YH DZ. Performed the experiments: YH. Analyzed the data: YH. Contributed reagents/materials/analysis tools: YH LH. Wrote the paper: YH DZ LH.

References

1. Laganière J, Angers DA, Paré D (2010) Carbon accumulation in agricultural soils after afforestation: a meta-analysis. Glob Chang Biol 16: 439–453.
2. Nilsson S, Schopfhauser W (1995) The carbon-sequestration potential of a global afforestation program. Clim Chang 30: 267–293.
3. Guo LB, Gifford RM (2002) Soil carbon stocks and land use change: a meta analysis. Glob Chang Biol 8: 345–360.
4. Paul KI, Polglase PJ, Nyakuengama JG, Khanna PK (2002) Change in soil carbon following afforestation. For Ecol Manage 168: 241–257.
5. Paul S, Flessa H, Veldkamp E, Lpez-Ulloa M (2008) Stabilization of recent soil carbon in the humid tropics following land use changes: evidence from aggregate fractionation and stable isotope analyses. Biogeochemistry 87: 247–263.
6. Morris SJ, Bohm S, Haile-Mariam S, Paul EA (2007) Evaluation of carbon accrual in afforested agricultural soils. Glob Change Biol 13: 1145–1156.
7. Li D, Niu S, Luo Y (2012) Global patterns of the dynamics of soil carbon and nitrogen stocks following afforestation: a meta-analysis. New Phytol 195: 172–181.
8. Tang G, Li K (2013) Tree species controls on soil carbon sequestration and carbon stability following 20 years of afforestation in a valley-type savanna. For Ecol Manage 291: 13–19.
9. Galdo ID, Six J, Peressott A, Cotrufo MF (2003) Assessing the impact of land-use change on soil C sequestration in agricultural soils by means of organic matter fractionation and stable C isotopes. Glob Change Biol 9: 1204–1213.
10. Lal R (2004) Soil carbon sequestration impacts on global climate change and food security. Science 304: 1623–1627.
11. Shi S, Zhang W, Zhang P, Yu Y, Ding F (2013) A synthesis of change in deep soil organic carbon stores with afforestation of agricultural soils. For Ecol Manage 296: 53–63.
12. Christensen BT (1992) Physical fractionation of soil and organic matter in primary particle size and density separates. Adv Soil Sci 20: 1–90.
13. Lützow M, Kögel-Knabner I, Ekschmitt K, Flessa H, Guggenberger G, et al (2007) SOM fractionation methods: relevance to functional pools and to stabilization mechanisms. Soil Biol Biochem 39: 2183–2207.
14. Huang Z, Davis MR, Condron LM, Clinton PW (2011) Soil carbon pools, plant biomarkers and mean carbon residence time after afforestation of grassland with three tree species. Soil Biol Biochem 43: 1341–1349.
15. Janzen HH, Campbell CA, Brandt SA, Lafond GP, Townley-Smith L (1992) Light-fraction organic matter in soils from long-term crop rotations. Soil Sci Soc Am J 56: 1799–1806.
16. John B, Yamashita T, Ludwig B, Flessa H (2005) Storage of organic carbon in aggregate and density fractions of silty soils under different types of land use. Geoderma 128: 63–79.
17. Tan Z, Lal R, Owens L, Izaurralde RC (2007) Distribution of light and heavy fractions of soil organic carbon as related to land use and tillage practice. Soil Till Res 92: 53–59.
18. Clark JD, Planter AF, Johnson AH (2011) Soil organic matter quality in chronosequences of secondary northern hardwood forests in Western New England. Soil Sci Soc Am J 76: 684–693.
19. Dondini M, Groenigen KJ, Galdo ID, Jones MB (2009) Carbon sequestration under Miscanthus: a study of ^{13}C distribution in soil aggregates. Glob Change Biol Bioenergy 1: 321–330.
20. Wei X, Qiu L, Shao M, Zhang X, Gale WJ (2012) The accumulation of organic carbon in mineral soils by afforestation of abandoned farmland. PLoS ONE 7(3): e32054. doi:10.1371/journal.pone.0032054.
21. Arai H, Tokuchi N (2010) Factors contributing to greater soil organic carbon accumulation after afforestation in a Japanese coniferous plantation as determined by stable and radioactive isotopes. Geoderma 157: 243–251.
22. Helmisaari H, Derome J, Nöjd P, Kukkola M (2007) Fine root biomass in relation to site and stand characteristics in Norway spruce and Scots pine stands. Tree Physiol 27: 1493–1504.
23. Litton CM, Ryan MG, Knight DH (2004) Effects of tree density and stand age on carbon allocation patterns in postfire lodgepole pine. Ecol Appl 14: 460–475.
24. Lal R (2004) Carbon sequestration in dryland ecosystems. Environ Manage 33: 528–544.
25. Jackson RB, Banner JL, Jobbágy EG, Pockman WT, Wall DH (2002) Ecosystem carbon loss with woody plant invasion of grasslands. Nature 418: 623–626.
26. Richter DD, Markewitz D, Trumbore SE, Wells CG (1999) Rapid accumulation and turnover of soil carbon in a re-establishing forest. Nature 400: 56–58.
27. Hu YL, Zeng DH, Fan ZP, Chen GS, Zhao Q, et al (2008) Changes in ecosystem carbon stocks following grassland afforestation of semiarid sandy soil in the southeastern Keerqin Sandy Lands, China. J Arid Environ 72: 2193–2200.
28. Hu YL, Zeng DH, Chang SX, Mao R (2013) Dynamics of soil and root C stocks following afforestation of croplands with poplars in a semi-arid region in northeast China. Plant Soil 368: 619–627.
29. Mao R, Zeng DH, Hu YL, Li LJ, Yang D (2010) Soil organic carbon and nitrogen stocks in an age-sequence of poplar stands planted on marginal agricultural land in Northeast China. Plant Soil 332: 277–287.
30. Zhenghu D, Honglang X, Zhibao D, Gang W, Drake S (2007) Morphological, physical and chemical properties of aeolian sandy soils in northern China. J Arid Environ 68: 66–76.
31. Jiao SR (1989) The ecosystem structure and function of sandy-fixation forests in Zhanggutai. Shenyang: Liaoning Science & Technology Press. 9–10 p.
32. Sohi SP, Mahieu N, Arah JRM, Powlson DS, Madari B, et al (2001) A procedure for isolating soil organic matter fractions suitable for modeling. Soil Sci Soc Am J 65: 1121–1128.
33. Nelson DW, Sommers LE (1996) Total carbon, organic carbon and organic matter, in: Sparks DL (Eds.), Methods of soil analysis. Part 3. Chemical methods. Wisconsin: Soil Science Society of America. 961–1010.

34. Hernandez-Ramirez G, Sauer TJ, Cambardella CA, Brandle JR, James DE (2011) Carbon sources and dynamics in afforested and cultivated corn belt soils. Soil Sci Soc Am J 75: 216–225.

35. Li Y, Awada T, Zhou X, Shang W, Chen Y, et al (2012) Mongolian pine plantations enhance soil physico-chemical properties and carbon and nitrogen capacities in semi-arid degraded sandy land in China. Appl Soil Ecol 56: 1–9.

36. Arevalo CBM, Bhatti JS, Chang SX, Sidders D (2011) Land use change effects on ecosystem carbon balance: from agricultural to hybrid poplar plantation. Agric Ecosyst Environ 141: 342–349.

37. Jug A, Makeschin F, Rehfuess KE, Hofmann-Schielle C (1999) Short-rotation plantations of balsam poplars, aspen and willows on former arable land in the Federal Republic of Germany. III. Soil ecological effects. For Ecol Manage 121: 85–99.

38. Sartori F, Lal R, Ebinger MH, Eaton JA (2007) Changes in soil carbon and nutrient pools along a chronosequence of poplar plantations in the Columbia Plateau, Oregon, USA. Agric Ecosyst Environ 122: 325–339.

39. Vesterdal L, Ritter E, Gundersen P (2002) Change in soil organic carbon following afforestation of former arable land. For Ecol Manage 169: 137–147.

40. Bu X, Ruan H, Wang L, Ma W, Ding J, et al (2012) Soil organic matter in density fractions as related to vegetation changes along an altitude gradient in the Wuyi Mountains, southeastern China. Appl Soil Ecol 52: 42–47.

41. Laik R, Kumar K, Das DK, Chaturvedi OP (2009) Labile soil organic matter pools in a calciorthent after 18 years of afforestation by different plantations. Appl Soil Ecol 42: 71–78.

42. Marin-Spiotta E, Silver WL, Swanston CW, Ostertag R (2009) Soil organic matter dynamics during 80 years of reforestation of tropical pastures. Glob Chang Biol 13: 1584–1597.

43. Unger S, Máguas C, Pereira JS, David TS, Werner C (2010) The influence of precipitation pulses on soil respiration-assessing the "Birch effect" by stable carbon isotopes. Soil Biol Biochem 42: 1800–1810.

22

Structure, Composition and Metagenomic Profile of Soil Microbiomes Associated to Agricultural Land Use and Tillage Systems in Argentine Pampas

Belén Carbonetto[1]*, Nicolás Rascovan[1], Roberto Álvarez[2], Alejandro Mentaberry[3], Martin P. Vázquez[1]*

1 Instituto de Agrobiotecnología de Rosario (INDEAR), Predio CCT Rosario, Santa Fe, Argentina, **2** Facultad de Agronomía, Universidad de Buenos Aires, Buenos Aires, Argentina, **3** Departamento de Fisiología y Biología Molecular y Celular, Facultad de Ciencias Exactas y Naturales, Universidad de Buenos Aires, Buenos Aires, Argentina

Abstract

Agriculture is facing a major challenge nowadays: to increase crop production for food and energy while preserving ecosystem functioning and soil quality. Argentine Pampas is one of the main world producers of crops and one of the main adopters of conservation agriculture. Changes in soil chemical and physical properties of Pampas soils due to different tillage systems have been deeply studied. Still, not much evidence has been reported on the effects of agricultural practices on Pampas soil microbiomes. The aim of our study was to investigate the effects of agricultural land use on community structure, composition and metabolic profiles on soil microbiomes of Argentine Pampas. We also compared the effects associated to conventional practices with the effects of no-tillage systems. Our results confirmed the impact on microbiome structure and composition due to agricultural practices. The phyla *Verrucomicrobia*, *Plactomycetes*, *Actinobacteria*, and *Chloroflexi* were more abundant in non cultivated soils while *Gemmatimonadetes*, *Nitrospirae* and WS3 were more abundant in cultivated soils. Effects on metabolic metagenomic profiles were also observed. The relative abundance of genes assigned to transcription, protein modification, nucleotide transport and metabolism, wall and membrane biogenesis and intracellular trafficking and secretion were higher in cultivated fertilized soils than in non cultivated soils. We also observed significant differences in microbiome structure and taxonomic composition between soils under conventional and no-tillage systems. Overall, our results suggest that agronomical land use and the type of tillage system have induced microbiomes to shift their life-history strategies. Microbiomes of cultivated fertilized soils (i.e. higher nutrient amendment) presented tendencies to copiotrophy while microbiomes of non cultivated homogenous soils appeared to have a more oligotrophic life-style. Additionally, we propose that conventional tillage systems may promote copiotrophy more than no-tillage systems by decreasing soil organic matter stability and therefore increasing nutrient availability.

Editor: Kathleen Treseder, UC Irvine, United States of America

Funding: Funding for this work was provided by Agencia Nacional de Promoción Científica y Tecnológica, Argentina-PAE 37164. The funders had no role in study design, data collection and analysis, decision to publish, or preparation of the manuscript.

Competing Interests: The authors have declared that no competing interests exist.

* E-mail: martin.vazquez@indear.com (MPV); belen.carbonetto@indear.com (BC)

Introduction

Agriculture is facing major challenges nowadays. Production will have to double in the next 50 years in order to face growing food demand and bioenergy needs [1,2]. This must be done without increasing environmental threats such as climate change, biodiversity loss and degradation of land and freshwater. Achieving such a goal represents one of the greatest scientific challenges ever. This is in part because of the trade-offs among economic and environmental goals and because of the insufficient knowledge about the biological, biogeochemical and ecological processes that are relevant for sustainable ecosystem functioning [3,4]. Much has been done during the last decade to gain sufficient information on agricultural ecosystem biology, still, more work needs to be done to gain deeper comprehension and to be able to reduce the negative environmental impacts of agriculture [2,5,6]. The main focus should be oriented to soil degradation. Soil fertility, as the capacity to sustain abundant crop production, needs to be preserved. Nowadays soil fertility is maintained by dependence on external inputs; with increasing water contamina-

tion [7]. In this context, the key to understand the behavior of life-supporting elements in soil, such as carbon, nitrogen, and phosphorus lies in the fluxes between their various forms in the environment, which are modulated by biology [8]. Comprehension of soil microorganism dynamics is then essential to understand soil processes that affect fertility. Ecological approaches are being taken into account in soil microbial studies trying to address these questions. These approaches involve diversity and functional analyses of soil communities [9,10]. Scholes & Scholes point out that this complex view is necessary for the comprehension of soil systems and that soil restoration of biological processes is the key to achieving lasting food and environmental security [8].

Argentine Pampas is an important player in this scenario. With a plain area of 50 million ha., nearly 50% of the whole Pampas area is devoted to crop production [11]. Cultivation began in the 19th Century in the central humid portion of the region, in soils of high fertility, and spread in last decades to the south and the semiarid west [12]. Soil degradation (i.e. intense erosion and net loss of nutrients and organic carbon) caused by the use of conventional tillage systems were reported in the Pampas [13–17].

Nowadays between 60 and 80% of production is conducted under conservational no-till practices [18]. Extensive research was done to evaluate the effects of reduced tillage and no-tillage systems on soil physical properties, water content, fertility and crop yields[19]. The main outcome of these analyses points that the adoption of limited tillage systems led to soil improvement, by augmenting organic matter content and soil structure. Still, external fertilization is needed in order to restore nutrient levels and fertility regardless the tillage system employed.

Even though the effects of different tillage practices on soil physical and chemical characteristics have been deeply studied, changes in microbial biodiversity and functioning have been poorly reported in Argentinean Pampas. Most works have studied tillage effects on microbial biomass or specific microbial activities (i.e. utilization of specific substrates, extracellular enzyme production, mineralization, etc.) rather than on full microbiome [13,20,21]. Other studies have focused on the behavior of specific bacterial taxa [22,23]. Reports with an ecological approach (i.e. microbial community analysis) have usually focused on individual effects of land use such as the application of herbicides [24,25]. In these cases, biodiversity variability has been assessed using classical fingerprinting techniques (such as RFLP and DGGE) that lack information about microbial taxonomic identity and only capture the most dominant species in the environment [26,27]. In the last few years, 16S amplicon pyrosequencing has been largely implemented to determine microbial diversity and structure of many different ecosystems worldwide [28,29]. This strategy allows a more exhaustive characterization of community patterns and composition. Moreover, some works have incorporated the use of shotgun metagenomics to study the metabolic potential of soil microbiomes [10,30]. The shotgun approach generates a massive amount of data using random high-trhoughput sequencing of soil isolated DNA. This allows the identification of functional capabilities by gene annotation and the comparison of metabolic profiles between samples. To our knowledge, Figuerola et al. [31] were the only authors studying microbial communities in agronomical soils of Argentine Pampas using high throughput sequencing approaches. They observed differences in microbial community composition of soils under no-tillage systems using 16S pyrosequencing. As a novelty, our efforts focused on assessing the impact of long-term agriculture on Pampas soil microbiomes using both shotgun metagenomics and deep 16S amplicon sequencing approaches. We evaluated the effect of more than a hundred years of agronomical land use on both community features and metabolic profiles of soil microbiomes in comparison with nearby control soils with no agricultural records. We also addressed the differences between the effects of two tillage systems: conventional tillage vs. no-tillage on microbial communities.

Several previous studies of soil microbiomes from different parts of the world showed the effects of agronomical land use on soil microbial communities [10,32–35]. Some of these studies showed differences in trophic strategies between microbial communities related to tillage; and most of them were done in experimental plots. As a novelty, we tested the impact of long term agriculture in soils sampled in production fields in the Argentine Pampas, allowing a deeper insight to the effects of intense land use on soil ecosystems functioning. We confirmed the hypothesis that agronomical practices affected Pampas soil microbiomes by promoting a shift of life-history and trophic strategies. We also showed differences in the effects of contrasting tillage systems (i.e. conventional vs. no- tillage) on community taxonomic and metabolic composition on a long term experiment.

Materials and Methods

Sites description and sampling

Soil samples were taken in production and experimental fields between June and August 2010. To address the effects of agricultural land use on soil microbial communities, three different production farms were sampled in the Rolling Pampas area: "La Estrella", "La Negrita" and "Criadero Klein" (See Rascovan et al and Table S1 for details). Rolling Pampas soils are classified as Typic Argiudolls [36] and mean annual rainfall and temperature were1002 mm and 16.8°C respectively. Two treatments were defined: *cultivated* for production plots, and *no cultivated* for farmhouses parks. Production plots were under cultivation for at least one century under conventional tillage systems, with a mixed rotation of pastures and annual grain crops. During the last 15 years before sampling plots were subjected to continuous crop cultivation under no-tillage systems (i.e. minimal soil disturbance, permanent soil cover, rotations and fertilization). The last crop rotation before sampling was wheat-soybean. Nitrogen and phosphorus fertilizers were applied. Samples were collected one month after soybean harvest. Soil samples were also collected nearby the farmers' houses where no agricultural land use (no tillage nor cultivation) was recorded for the last 30 years except from grass mowing. Parks around farmers' houses are usually considered as undisturbed environments in Argentine Pampas [37]. Soils under no land use were covered with grass and other herbaceous (non-woody) plants common in the region such as *Cirsium sp, Trifolium sp, Micropsis sp, Festuca sp, Dichondra sp, Cyperus sp* and *Taraxacum officinale*. For numerical analyses purposes the three farms are treated as experimental replicates. Four soil samples were taken with an auger from the upper 20 cm soil layer in each farm and treatment. A total of 24 samples were collected in Rolling Pampa soils.

In order to compare effects of contrasting tillage systems, samples were also collected in a 34-year-old experiment located in Balcarce in the Southern Pampas (See Rascovan et al and Table S1 for more details). Samples were taken in experimental plots because no production fields are using conventional tillage for crop production nowadays in the Pampas. Soils in Balcarce are a complex of Typic Argiudolls and Petrocalcic Paleudols and mean annual rainfall and temperatures were 875 mm and 13.8°C respectively. The experiment was carried out in three (175 m^2) experimental plots (n = 3). Treatments were defined as: *no tillage* (NT) and *conventional tillage* (CT). NT plots had minimal soil disturbance and permanent soil cover combined with rotations; which have included pastures and grain crops (soybean, corn, wheat) during the last 16 years. CT plots were managed with moldboard plough. Nitrogen fertilization was performed in NT and CT plots (60 kg N ha-1). Last rotation before sampling was corn-soybean. Two sub-samples were collected from all treatment and replicate plots a month after soybean harvest. A total of 12 samples were collected.

Samples were immediately sent to the lab after collection. Samples used for DNA purification were air dried and sieved through 1 mm mesh to thoroughly homogenize, break aggregates and remove roots and plant detritus, then stored at −80°C. DNA purification and library preparation was previously described in Rascovan. et al.[38].

None of the sampling sites is located in protected areas. Permissions were obtained directly from each farm owner or manager: Alejandro Cattaneo at La Negrita and La Estrella, Roberto Klein at Criadero Klein and Guillermo Studdert at Balcarce experiment.

Soil chemical and physical measurements

Soil organic carbon was determined by wet digestion and organic matter was estimated [39]. Nitrate-nitrogen was analyzed by 2 M KCL extraction and the phenoldisulfonic acid method [40]. Extractable phosphorus was determined by the Bray method [41]. The pH was measured in a soil:water ratio 1:2.5. Salinity was estimated by the determination of electrical conductivity [42]. Texture analysis was performed by the hydrometer method [43] and nitrogen was determined by Kjeldahl method [40].

16S amplicon sequencing and shotgun metagenomic datasets

To analyze the effect of agronomical practices on soil microbiomes, sequence data from the previously reported Pampa dataset [38] was used. In order to evaluate microbial community structure and taxonomic composition, a total of 112,800 high-quality filtered 16S rRNA gene amplicon sequences, obtained from the 42 soil DNA samples (replicates and subsamples were included). DNA shotgun metagenomic data was used to analyze metabolic profiles. Shotgun metagenomic data completed a total of 10,445,170 sequences. In this case sequences were obtained from one subsample per sampling replicated plot.

In brief, libraries were prepared as follows: DNA was isolated from 10 g of soil of each of the 42 soil samples using the Power MaxSoil DNA Isolation Kit following the manufacturer's instructions (MO BIO Laboratories, Inc.). For amplicon libraries the V4 hyper variable region of the 16s rRNA gene was amplified. Duplicated reactions were performed using barcoded bacterial universal primers containing Roche- 454 sequencing A and B adaptors and a nucleotide multiple identifier (MID) to sort samples: 563F: 5′-CGTATCGCCTCCCTCGCGCCATCA-GACGAGTGCGTAYTGGGYDTAAAGNG -3′ (where AC-GAGTGCGT is an example, different MIDs for each sample were used) and 802R (5′-CTATGCGCCTTGCCAGCCCGCT-CAGTACCRGGGTHTCTAATCC, 5′-CTATGCGCCTTGC-CAGCCCGCTCAGTACCAGAGTATCTAATTC, 5′-CTAT-GCGCCTTGCCAGCCCGCTCAGCTACDSRGGTMTCTA-ATC, 5′-CTATGCGCCTTGCCAGCCCGCTCAGTACNVG-GGTATCTAATCC) [44]. All amplicons were cleaned using Ampure DNA capture beads (Agencourt- Beckman Coulter, Inc.) and pooled in equimolar concentrations before sequencing on a Genome Sequencer FLX (454-Roche Applied Sciences) using Titanium Chemistry according to the manufacturer's instructions.

Shotgun metagenomic libraries were prepared by nebulization, followed by tagging with GS-FLX-Titanium Rapid Library MID Adapters Kit (454-Roche Applied Sciences) and sequenced with a Genome Sequencer FLX (454-Roche Applied Sciences) using Titanium Chemistry according to the manufacturer's instructions. Sequencing runs were performed in INDEAR sequencing facility.

All the sequences used in the present study are available in The Sequence Read Archive (SRA) under accession number SRA058523 and SRA056866. See Rascovan et al. [38] for more information.

Amplicon sequence processing, OTU classification and taxonomic assignment

Sequence data were quality controlled and denoised with the amplicon noise.py script of QIIME [45].This script also eliminated chimeras. Sequences obtained from Rolling Pampa soil libraries and Balcarce soil libraries were processed separately. Sequences were clustered into Operational Taxonomic Units (OTUs) using the pick_otus.py script with the Uclust method [46] at 97%

sequence similarity. Rolling Pampas samples yielded 2,591 sequences on average (ranging from 1,455 to 3,991 sequences). Balcarce samples yielded 2,329 reads on average (ranging from 1,211 to 4,755 reads).OTU representative sequences were aligned using PyNast algorithm [47] with QIIME default parameters. Phylogenetic trees containing the aligned sequences were then produced using FastTree [48]. All downstream analyses were determined after each sample was randomly rarefied to 70% the number of reads of the smallest sample (i.e. 1,080 reads for Rolling Pampa libraries and 850 reads for Balcarce libraries). Phylogenetic distances between OTUs were calculated using unweighted and weighted Unifrac [49]. Taxonomic classification of sequences was done with Ribosomal Database Project (RDP) Classifier using Greengenes database using a 50% confidence threshold [50,51].

Microbial community analyses

Unifrac phylogenetic pairwise distances among samples were visualized with principal coordinates analysis (PCoA). Analysis of similarity statistics (ANOSIM) was calculated to test a-priori sampling groups. BIOENV analysis was performed to elucidate which soil properties correlated with community patterns. All calculations were carried out with R packages 'BiodiversityR' and 'Vegan' [52,53]. T- tests were performed with QIIME script otu_category_significance.py, and R scripts in order to elucidate differences in read abundances.

Shotgun metagenomic sequence processing and analysis

SSF files obtained from shotgun sequencing runs were uploaded to the MG-RAST webserver [54] for sequence filter and analyses. Reads more than two standard deviations away from the mean read length were discarded. For dereplication removal MG-RAST used a simple k-mer approach to rapidly identify all 20 character prefix identical sequences. This step is required in order to remove artificial duplicate reads. We obtained an average of 1.28×10^6 filtered reads per sample for Rolling Pampa shotgun libraries, and an average of 304,258 filtered reads per sample for Balcarce libraries.

Filtered high quality sequences were assigned to Cluster of orthologous groups (COG) by the MG-RAST sever pipeline using a similarity-based approach. COGs were assigned with a maximum E value of 10^{-20}, an average alignment of 80 amino acids length and 70% average identity.

Relative abundances were calculated by dividing the number of hits for each COG or COG-category by the total number of filtered reads in each sample. Euclidean distances based on relative abundances were calculated between sample pairs. PCoA visualizations and ANOSIM calculations were performed. All calculations were carried out with R packages 'BiodiversityR' and 'Vegan'. T- tests were performed with QIIME script otu_category-y_significance.py, and R scripts in order to elucidate differences in COG relative abundances between samples.

Results

Microbiome community changes related to agricultural land use

The PCoA visualization revealed clear differences between cultivated and noncultivated soils (ANOSIM R = 0.8406, p ≤ 0.001; Figure 1A).Similar results were obtained when using Bray Curtis distance matrices (Figure S1).

The soil properties (Table S1) that best explained the phylogenetic variation observed in microbial communities were determined using Clarke and Ainsworth's BIOENV analysis. Our results showed that variables that best correlated with community

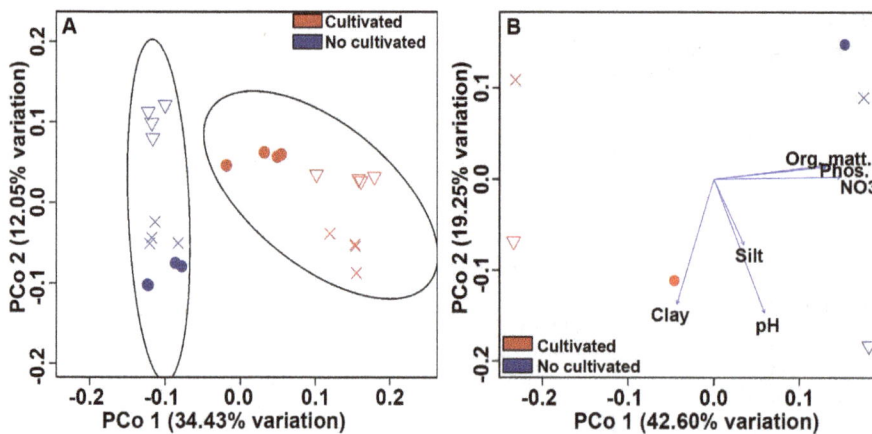

Figure 1. PCoA plots of Pampa production field soil microbiomes based on 97% similarity Weighted Unifrac distance matrices. A) PCoA of cultivated and non-cultivated soil microbiomes. All sub-samples are plotted. Standard error ellipses show 95% confidence areas. B) PCoA biplot of soil properties that best explained variation in community structure. Correlations were calculated using BIOENV on average data of each sampled site (Mantel r = 0.6107, p≤0.05). Circles represent samples from "La Estrella", crosses represent "Criadero Klein" samples and triangles represent "La Negrita" samples.

differences were organic matter, clay and silt content, nitrates, phosphorus and pH (Mantel r = 0.6107, p≤0.05). The PCoA biplot (Figure 1B) showed that organic matter, phosphorus and nitrate levels correlated with the first ordination axis that discriminates between cultivated and non cultivated soils. These three properties were higher in soils under no land use.

Regarding the taxonomic analyses, we observed that members of phyla *Verrucomicrobia*, *Planctomycetes*, *Actinobacteria* and *Chloroflexi* were more abundant in non cultivated soils (p≤0.05) (Figure 2). On the other hand, we found that sequences related to *Gemmatimonadetes*, candidate division WS3 and *Nitrospirae* were enriched in cultivated soils (p≤0.05) (Figure 2). No significant differences were found for *Proteobacteria* (for non of the Clases), *Acidobacteria* and *Bacteroidetes* phyla. The ten mentioned taxa represent on average 95% of total sequences of each sample.

Microbiome metagenomic profile changes related to agricultural land use

We found that cultivated and non cultivated soils also clustered apart when metagenomic functional categories were used for the analysis. The first two components of the PCoA explained over 60% of the variability between samples (Figure 3). Standard deviation ellipses overlapped in the ordination plot, indicating that some features are shared between metagenomes. Still, a positive correlation was observed between metabolic and weighted-Unifrac distance matrices (Mantel r: 0.5036 p≤0.05). The analyses of individual COG categories revealed that the relative abundances of COG categories associated with transcription, protein modification, nucleotide transport and metabolism, wall and membrane biogenesis and intracellular trafficking and secretion were higher in cultivated soils (Figure 4, p≤0.05). A deeper analysis inside COG categories revealed that COGs related to Coenzyme A and acetyl-Coa metabolism, energy storage and starvation or quiescence such as, pantothenate kinase, phospho-transacetylase, and trehalose utilization protein were more abundant in non cultivated than in cultivated soils (Figure S2, p≤0.05). On the other hand, COGs related to rapid regulation systems, tricarboxylic acid cycle and nitrogen assimilation such as urease, citrate synthase, glutamate synthase, fumarate hydratase, S-adenosyl- homocysteine hydrolase, S-adenosyl-methionine synthetase, cobalamin bio-

synthesis protein and ABC transporters, were more abundant in cultivated soils (Figure S2, p≤0.05).

Microbiome community structure and composition related to conventional tillage and no-tillage systems.

To compare the structure of microbiomes under different tillage systems, we collected samples from an experimental field located in Balcarce in the Southern Pampas. The 34-year-old experiment compared two tillage systems: no-tillage (NT) and conventional tillage (CT). Weighted Unifrac analysis showed differences in community structure associated to the tillage system employed

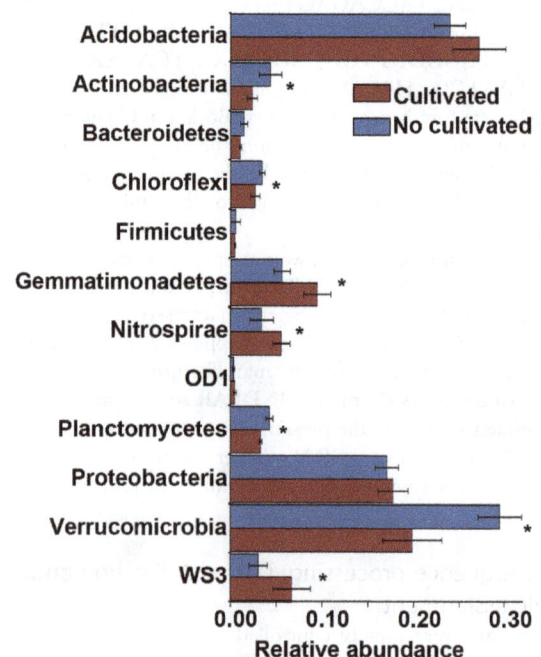

Figure 2. Relative abundances of taxonomic groups in Pampa production field soil microbiomes. Bars represent ± 1 standard error. (*) indicate significant differences (p≤0.05).

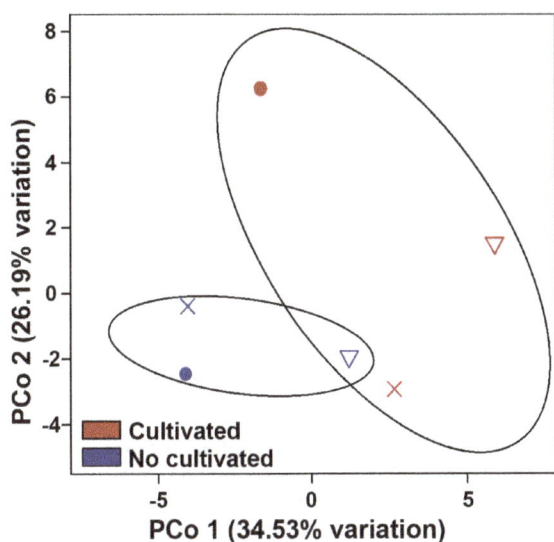

Figure 3. PCoA plot of metagenomic data based on Euclidean distance matrices of COG categories of Pampa production field soil microbiomes. Standard error ellipses show 95% confidence areas. Circles represent samples from "La Estrella", crosses represent "Criadero Klein" samples and triangles represent "La Negrita" samples.

(ANOSIM R = 0.9009, p<0.05, Figure 5A). The first axis explained 31.56% of total variation and separated NT from CT. Additionally, we showed that nitrates were the only soil variable (Table S1) that significantly correlated with community structure (BIOENV analysis, Mantel r = 0.7721, p≤0.01, Figure 5B).

Moreover, microbiomes of CT and NT soils also differed in taxonomic composition. Members of *Acidobacteria, Gemmatimonadetes*, candidate division TM7 and class *Gammaproteobacteria* were more abundant in CT soils, while *Nitrospirae*, candidate divisionWS3 and *Deltaproteobacteria* were more represented in NT soils (Figure 6, p≤0.05).

Microbiome metabolic profiles related to conventional tillage and no- tillage systems

Variation in metagenomic profiles between CT and NT microbiomes was analyzed with PCoA based on Euclidean distance matrices of COG abundances. We could not find significant differences in overall profile metabolic structure between tillage systems (Figure S3). Additionally, we did not find significant correlation between Euclidean metabolic matrices and phylogenetic matrices. However, categories related to intracellular trafficking and secretion, amino acid transport and metabolism, and energy production and conversion were shown to be more abundant in soil under CT (Figure 7).

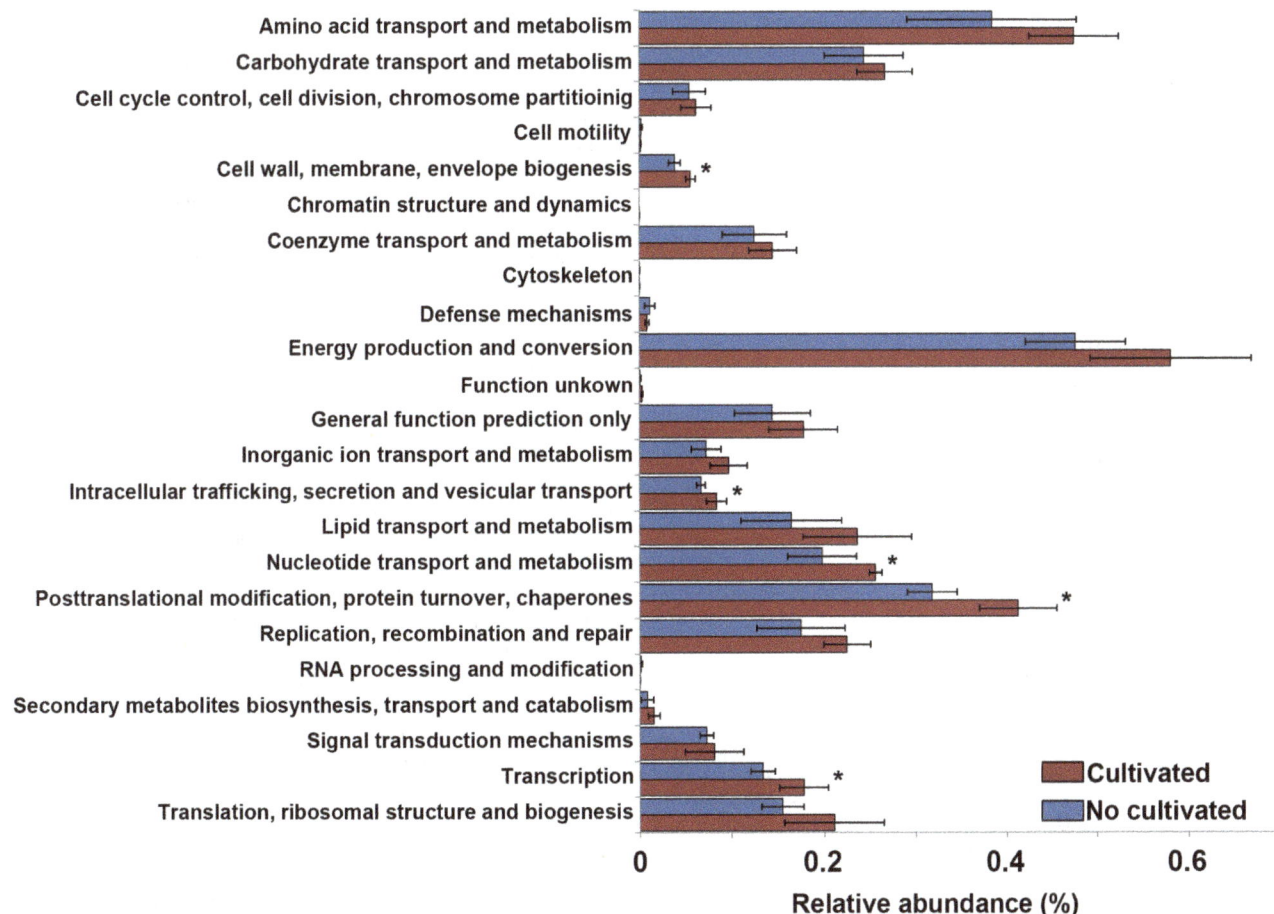

Figure 4. Relative abundances of COG categories in Pampas production field soil microbiomes. Bars represent ±1 standard error. (*) indicate significant differences (p≤0.05).

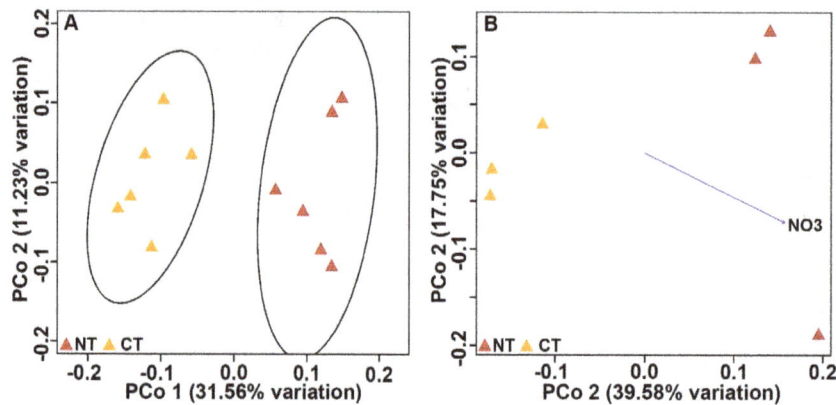

Figure 5. PCoA plots of Balcarce experimental field soil microbiomes based on 97% similarity Weighted Unifrac distance matrices. A) CA: conventional tillage; NT: no-tillage. All sub-samples are plotted. Standard error ellipses show 95% confidence areas. B) PCoA biplot, nitrate was the variable that best explained variation in community structure. Correlations were calculated using BIONEV on average data of each experimental plot (Mantel r = 0.7721, p≤0.01).

Discussion

Much work still needs to be done to get a comprehensive view of the soil microbiomes. Our work is one of the firsts done in the Argentine Pampas at this resolution level, with a combination of metagenomic and phylogentic approaches; and it is aimed to contribute to the comprehension of soil microbiomes function and dynamics. In that context, our results are in agreement with previous works that showed differences in soil microbial community structure and taxonomic composition due to the presence of agricultural land use [9,10,32,33].

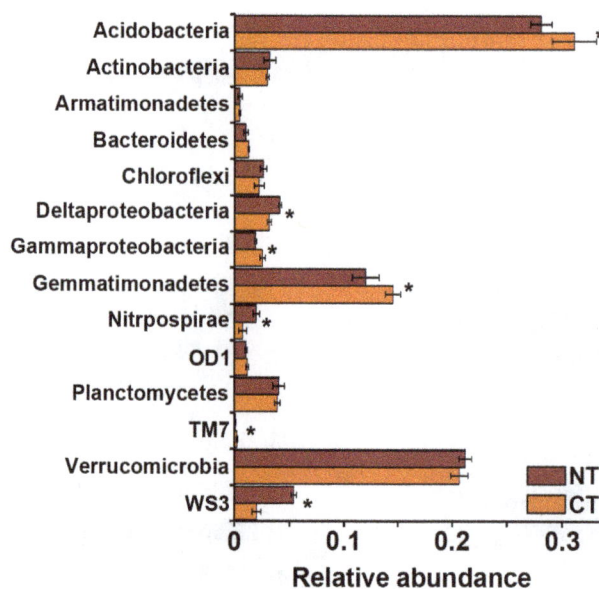

Figure 6. Comparison of relative abundances of taxonomic groups in NT and CT soils. CA: conventional tillage; NT: no-tillage. Bars represent ±1 standard error. (*) indicate significant differences (p≤ 0.05).

Effects of agricultural land use on community composition of soil microbiomes in Argentine Pampas

We observed effects on taxonomic composition at the phylum level. *Verrucomicrobia*, *Plactomycetes*, *Actinobacteria* and *Chloroflexi* were more abundant in soils that were never cultivated; while *Gemmatimonadetes*, *Nitrospirae* and WS3 were more abundant in crop cultivated soils. These results are in agreement with the copiotroph/oligotroph hypothesis [55], that propose that high number of oligotrophic prokaryotes may be found in bogs and soils with high amounts of recalcitrant organic matter [56]. On the other hand, copiotrophic organisms are able to use labile nutrient fractions and to grow at higher rates as a consequence. In our study, non-cultivated soils are considered to be oligotrophic since they present high levels of organic matter highly rich in humic acids [12] and cultivated soils as copiotrophic environments due to fertilization and the seasonal presence of crop residues, which increases organic matter and nitrogen accessibility [57]. Under this assumption, bacteria in non-cultivated soils are expected to be K selected and to present low growth rates and very efficient nutrient uptake systems with higher substrate affinities. In contrast, bacteria in cultivated soils are expected to be r-selected and to have higher rates of activity per biomass unit, higher turnover rates and faster growth rates. Our results showed a trend toward these statements since a reduction in the abundance of taxa with oligotrophic characteristics, such as *Verrucomicrobia* [34], and *Planctomycetes* [58] were detected in cultivated fertilized soils. Moreover, this is in agreement with recent findings that confirm a correlation between *Verrucomicrobia* abundance patterns and conditions of limited nutrient availability in Prairie Soils in the United States [9]. On the other hand, the relative abundance of phylum *Gemmatimonadetes* was increased in fertilized cultivated soils as previously described for nitrogen-fertilized forest soils [59]. Consistently with our results, these authors observed that nitrogen fertilization was related to a higher abundance of *Gemmatimonadetes* and detected no presence of *Verrucomicrobia*. Little is known about *Gemmatimonadetes* ecology and metabolism since only one representative from this phylum has been isolated and characterized [60]. Even though, their presence in environments with a wide range of nutrient concentrations and redox states suggests versatile metabolisms [61].

We can also say that cultivated soils are more heterogeneous environments than non cultivated soils. Crop rotation, periodic fertilization and pesticide application generate temporal and

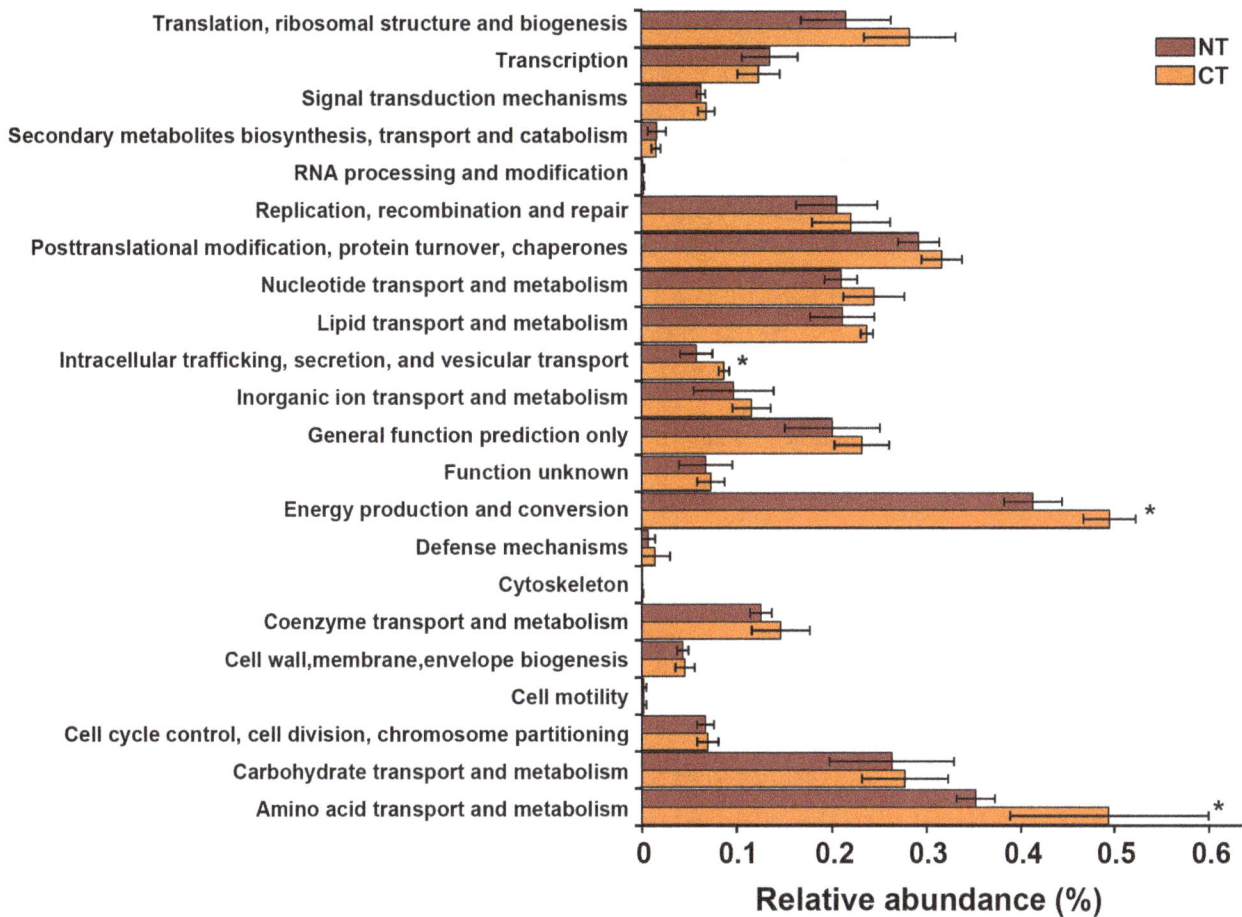

Figure 7. Relative abundances of COG categories in soil microbiomes of tillage systems comparison experiment in Balcarce. CA: conventional tillage; NT: no-tillage. Bars represent ±1 standard erro. (*)indicate significant differences (p≤0.05).

spatial changes in soil chemical properties, and therefore in nutrient availability and approachability for microorganisms. Bacteria dominating in this type of soils should be adapted to heterogeneity. These microorganisms should be able to fine-tune carbon and nitrogen intakes according to their metabolic needs under frequent external changes. This seems to be the case for *Gemmatimonadetes* suggesting a generalist ecological strategy. The high abundance of *Nitrospirae* may also respond to heterogeneous conditions. It has been proposed that *Nitrospirae* lineages occupy different positions on an imaginary scale reaching from K- to r-strategies [62]. We hypothesize that in cultivated soils, *Nitrospirae* K-strategists would be exploiting nitrite in high N microenvironments, while r-strategist would be mining low concentration areas in a nitrogen gradient enhanced by fertilization [63]. This competition would be less fierce in non cultivated soils due to the highest homogeneity and a less marked nitrogen gradient due to the lack of fertilization and more recalcitrant organic matter forms.

Effects of agricultural land use on metabolic profiles of soil microbiomes in Argentine Pampas

Our results from shotgun metagenomic data also indicated a tendency for the microbiomes of cultivated soils towards adaptation to nutrient heterogeneity. The highest relative abundances of sequences assigned to COGs related to transcription, protein modifications, nucleotide transport and metabolism, wall and

membrane biogenesis and intracellular trafficking and secretion in cultivated fertilized soils are consistent with a copiotrophic strategy (i.e. rapid tight metabolism regulation and fast grow rates).Moreover, some of these COG categories were previously shown to be up-represented in copiotrophic marine microorganism genomes [58]. In a deeper look we detected that the relative abundance of sequences assigned to riboswitch regulated genes was higher in cultivated soils (i.e. cobalamin biosynthesis protein, S-adenosyl-methionine synthetase, S-adenosyl- homocysteine hydrolase) [64–66]. These ancient regulators may be playing a central role in a life-style strategy adapted to nutrient heterogeneity since they were described to be the most 'economical' and fast-reacting regulatory systems (no intermediate factors involved) [67]. Moreover, the abundant riboswitch- COGs in cultivated soils were related to synthesis of B vitamins. The levels of these vitamins have already been linked to differences in community composition in marine ecosystems [68]. More studies are needed in order to address this relationship in soil environments; still, our results suggest an important role of B-vitamins in fertilized heterogeneous soils.

Another interesting observation was the highest abundance of glutamate synthase (GOGAT) related COGs in cultivated soils metagenomic profiles. The combined role of GOGAT with glutamine synthetase and glutamate dehydrogenase allows the cells to sense ammonia external levels [69]. The high abundance of this regulation system detected in cultivated soil metagenomes suggests its importance in the detection of N fluxes related with fertilization. Moreover COGs related to the tricarboxylic acid cycle (TCA), such as citrate synthase and succinate dehydrogenase, were more abundant in cultivated soil microbiomes. GOGAT nitrogen assimilation pathway and TCA are related. It has been shown that the concentration of compounds of both pathways changes considerably and rapidly upon nitrogen up shift; in contrast, the concentrations of glycolytic intermediates remains homeostatic [70]. In addition, abundance of urease related COGs were also higher in cultivated soils. The soils sampled in this study have a long history of agricultural land use and have been long fertilized with both type of nitrogen (i.e. ammonia-based and urea-based fertilizers). This kind of environmental pressure finally selected a microbiome adapted to changing N sources and availability.

It is important to mention that these results do not infer that expression or activity of these metabolisms will be necessarily increased in cultivated soil microbiomes, still, the highest abundance of these COGs could be reflected in a highest diversification and specialization. The presence of a highest copy number of strategic genes have already been linked to copiotrophic or oligotrophic life-styles [58].

Microbiomes of non cultivated soils showed a higher abundance of sequences related to Coenzyme A and acetyl-Coa metabolism than microbiomes of cultivated soils. It is known that acetyl-CoA is a fundamental building block and energy source [71]. High acetyl-CoA levels would indicate a "proliferative" or "fed" state, while low acetyl-CoA levels (and high CoA levels) would be indicative of a "quiescent" or "starved" state. Pantothenate kinase (PanK), the key enzyme in CoA synthesis, was also highest in non cultivated soil microbiomes [72]. In addition, it was stated that some oligotrophs preferentially use lipids as immediate and stored sources of carbon and energy in marine environments [58]. The observed abundance of Pank genes may ensure a correct CoA intracellular level in fasted moments, allowing proper lipid utilization. In addition, non cultivated metagenomic profiles showed higher abundance of sequences related to trehalose utilization. Trehalose is known to serve as energy source in many microorganisms [73]. Members of Actinobacteria genera Mycobacterium and family Frankiaceae are known to produce and/or utilize trehalose [74–76]. As mentioned above, our results showed higher abundance of Actinobacteria-related sequences in non cultivated soils. Moreover, sequences classified within family Frankiaceae were only present in these soils and the abundance of sequences assigned as Mycobacterium was higher than in cultivated soils (not shown). These observations are congruent with an oligotrophic strategy based on the use of storage components for carbon and energy sources.

The responses of the phylogenetic structure and metagenomic profile to agronomic land use were significantly correlated, suggesting some degree of correspondence between these different microbiome features. These results agree with previous observations done in soils from different biomes [9,77]. Moreover, Fierer et al. found similar a similar correlation between metagenomic and phylogenetic data for microbial communities of agricultural soils under different nitrogen gradients in experimental plots [10].

Correlation with soil properties

In addition, soil properties such as organic matter, phosphorus and nitrate levels explained most of the variability observed between cultivated and non cultivated soils. As mentioned, the highest amount of organic matter is found mostly in a recalcitrant form in these soils. Moreover, it was already established that nitrate accumulation exhibits a negative correlation with organic carbon availability [78]. In addition it has been proved that

organic matter source and quality played an important role in regulating the magnitude of carbon metabolism and could be as important as nutrient abundance in water environments [79]. Our results are congruent with these observations since non cultivated soils, with highest levels of organic carbon, presented metagenomic profiles with tendencies to oligotrophy and the best explanation for this scenario is the low lability of the recalcitrant forms of carbon and nitrogen.

Assessing the effects of different tillage systems on Pampas soil microbiomes

Our results also showed differences in the structure and composition of soil microbiomes between no-till and conventional tillage soils. Sequences related to phyla Gemmatimonadetes, candidate division TM7 and Acidobacteria, were highest in CT soils, while the abundances of Nitrospirae and candidate division WS3 were highest in NT soils. Tillage is the principal agent producing soil disturbance and subsequent soil structure modification [80,81]. The negative effects of CT in soil stabilization and macroaggregate losses were previously registered in Pampas soils [82]. It has also been shown that NT increases macroaggregate abundance and organic matter content [83].These increments are related with a more recalcitrant organic matter, with increased humic acid contents and nutrient retention [84–86]. The higher abundance of Nitrospirae in NT soils reinforces the idea of a community adapted to a better N mining in these environments. Moreover, we observed very low abundance of Nitrobacter related sequences in Balcarce soils and no significant difference was observed between CT and NT soils (not shown). Changes in ammonia oxidation due to a decrease in ammonia availability by humic substances have already been proposed in microcosm experiments [87]. If this is the case of Balcarce soils, higher humic acids in NT soils would be decreasing ammonia availability for oxidation into nitrite and therefore decreasing nitrate availability, compared to CT soils. The predicted highest stability of organic matter and abundance of macroaggregates in NT soils are probably generating more marked nitrite gradients than in CT soils. The highest abundance of Nitrospirae in NT may be reflecting a highest lineage diversity that would be better adapted to these gradients.

In addition, communities under NT showed higher abundance of sequences related to the order Syntrophobacterales (Deltaproteobacteria, Figure S4). These are known anaerobic and syntrophic organisms [88,89]. These characteristics may be an advantage in stable soils with higher number of highly humic macro aggregates since syntrophy is known to be important for community functioning in micro-environments with low nutrient levels [90]. On the other hand, CT soils presented higher relative abundance of Gammaproteobacteria related sequences. The order Xanthomonadales was the main responsible for these differences (Figure S4). This taxon has already been associated to CT practices [32].

Even though we could not find significant differences associated to tillage systems at the community structure level for metabolic profiles, COG categories related to intracellular trafficking and secretion, amino acid transport and metabolism and energy production and conversion were more abundant in CT microbiomes. These results suggest a tendency of CT microbiomes to a more copiotrophic life-style strategy than NT microbiomes.

Conclusion

Our results are consistent with the hypothesis that microbiomes exhibit different life- history and trophic strategies in Pampean soils under different land uses and tillage systems. Our data suggest that microbiomes of fertilized cultivated soils have more flexible metabolisms adapted to nutrient fluxes with tendencies to copiotrophy while microorganisms in non cultivated soils are better adapted to lowest external nutrient availability and homogenous environment. The lowest nutrient accessibility in non cultivated soils may be explained by the higher amount of humic substances, recalcitrant organic matter and the lack of fertilizer amendments. Moreover, NT soils, with most stable structure and highest macroaggregate abundance, presented microbiomes better adapted to recalcitrant environments; while CT microbiomes presented a higher tendency to copiotrophy.

This work is of major contribution to understand how historical changes in soil properties due to agronomical land use have altered the diversity and function of below-ground communities. The importance of high-throughput characterization for the reconstruction of pre- agricultural microbiomes is being reinforced nowadays [9]. Following this direction, our findings will be very useful in future restoration and monitoring programs of Argentine Pampas ecosystems.

Supporting Information

Figure S1 PCoA plots of Pampa production field soil microbiomes based on average Bray Curtis distance matrices. A) PCoA of cultivated and non cultivated soil microbiomes. Standard error ellipses show 95% confidence areas. B) PCoA biplot of soil properties that best explained variation in community structure. Correlations were calculated using BIOENV on average data of each sampled site (Mantel r = 0.6214, p≤0.05). Circles represent samples from "La Estrella", crosses represent "Criadero Klein" samples and triangles represent "La Negrita" samples.

Figure S2 Relative abundances of Cluster of Orthologous groups (COGs) in Pampa production field soil microbiomes. Bars represent ± 1 standard error. Only significant COGs are showed.

Figure S3 Comparison of tillage systems effects on the structure of metabolic profiles. PCoA plot based on Euclidean distance matrices. CA: conventional tillage; NT: no-tillage. Standard error ellipses show 95% confidence areas.

Figure S4 Relative abundances of reads assigned to orders within classes *Gammaproteobacteria* and *Deltaporteobacteria* in NT and CT soils.

Table S1 Soil chemical and physical properties.

Acknowledgments

We acknowledge Germán F. Domínguez and Guillermo A. Studdert from the "Manejo Sustentable del suelo" research group (from Facultad de Ciencias Agrarias, Universidad Nacional de Mar del Plata, Unidad Integrada Balcarce) for providing samples of the tillage systems experiment. We thank Gonzalo Berhongaray, Josefina de Paepe, and María Rosa Mendoza for his help during sampling in the Rolling Pampas. We also acknowledge Roxana Colombo and Marcelo Soria for helpful discussion and advice.

Author Contributions

Conceived and designed the experiments: BC NR RÁ AM MV. Performed the experiments: BC NR. Analyzed the data: BC. Contributed reagents/materials/analysis tools: BC NR RÁ AM MV. Wrote the paper: BC NR MV. Obtained permission for sampling: RÁ.

References

1. Foley JA, Ramankutty N, Brauman KA, Cassidy ES, Gerber JS, et al. (2011) Solutions for a cultivated planet. Nature 478: 337–342.
2. Tilman D, Balzer C, Hill J, Befort BL (2011) Global food demand and the sustainable intensification of agriculture. Proc Natl Acad Sci U S A 108: 20260–20264.
3. Tilman D, Cassman KG, Matson PA, Naylor R, Polasky S (2002) Agricultural sustainability and intensive production practices. Nature 418: 671–677.
4. Balmford A, Green RE, Scharlemann JPW (2005) Sparing land for nature: exploring the potential impact of changes in agricultural yield on the area needed for crop production. Global Change Biology 11: 1594–1605.
5. Power AG (2010) Ecosystem services and agriculture: tradeoffs and synergies. Philos Trans R Soc Lond B Biol Sci 365: 2959–2971.
6. Foley JA, Defries R, Asner GP, Barford C, Bonan G, et al. (2005) Global consequences of land use. Science 309: 570–574.
7. Bennett EM, Carpenter SR, Caraco NF (2001) Human Impact on Erodable Phosphorus and Eutrophication: A Global Perspective: Increasing accumulation of phosphorus in soil threatens rivers, lakes, and coastal oceans with eutrophication. BioScience 51: 227–234.
8. Scholes MC, Scholes RJ (2013) Dust Unto Dust. Science 342: 565–566.
9. Fierer N, Ladau J, Clemente JC, Leff JW, Owens SM, et al. (2013) Reconstructing the Microbial Diversity and Function of Pre-Agricultural Tallgrass Prairie Soils in the United States. Science 342: 621–624.
10. Fierer N, Lauber CL, Ramirez KS, Zaneveld J, Bradford MA, et al. (2012) Comparative metagenomic, phylogenetic and physiological analyses of soil microbial communities across nitrogen gradients. ISME J 6: 1007–1017.
11. Satorre EH, Slafer GA (1999) Wheat: Ecology and Physiology of Yield Determination: Taylor & Francis.
12. Hall AJ, Rebella CM, Ghersa CM, Culot JP (1992) Field crop systems of the Pampas. In: Pearson CJ, editor. Field Crop Ecosystems of the World. Amsterdam: Elsevier Science & Technology Books. pp. 413–450.
13. Alvarez R, Díaz RA, Barbero N, Santanatoglia OJ, Blotta L (1995) Soil organic carbon, microbial biomass and CO2-C production from three tillage systems. Soil and Tillage Research 33: 17–28.
14. Bernardos JN, Viglizzo EF, Jouvet V, Lértora FA, Pordomingo AJ, et al. (2001) The use of EPIC model to study the agroecological change during 93 years of farming transformation in the Argentine pampas. Agricultural Systems 69: 215–234.
15. Alvarez R (2001) Estimation of carbon losses by cultivation from soils of the Argentine Pampa using the Century Model. Soil Use and Management 17: 62–66.
16. Hevia GG, Buschiazzo DE, Hepper EN, Urioste AM, Antón EL (2003) Organic matter in size fractions of soils of the semiarid Argentina. Effects of climate, soil texture and management. Geoderma 116: 265–277.
17. Quiroga AR, Buschiazzo DE, Peinemann N (1996) Soil Organic Matter Particle Size Fractions in Soils of the Semiarid Argentinian Pampas. Soil Science 161.
18. Kassam A, Friedrich T, Shaxson F, Pretty J (2009) The spread of Conservation Agriculture: justification, sustainability and uptake. International Journal of Agricultural Sustainability 7: 292–320.
19. Alvarez R, Steinbach HS (2009) A review of the effects of tillage systems on some soil physical properties, water content, nitrate availability and crops yield in the Argentine Pampas. Soil and Tillage Research 104: 1–15.
20. Gomez E, Bisaro V, Conti M (2000) Potential C-source utilization patterns of bacterial communities as influenced by clearing and land use in a vertic soil of Argentina. Applied Soil Ecology 15: 273–281.
21. Aon MA, Cabello MN, Sarena DE, Colaneri AC, Franco MG, et al. (2001) I. Spatio-temporal patterns of soil microbial and enzymatic activities in an agricultural soil. Applied Soil Ecology 18: 239–254.
22. Agaras B, Wall LG, Valverde C (2012) Specific enumeration and analysis of the community structure of culturable pseudomonads in agricultural soils under no-till management in Argentina. Applied Soil Ecology 61: 305–319.
23. Nievas F, Bogino P, Nocelli N, Giordano W (2012) Genotypic analysis of isolated peanut-nodulating rhizobial strains reveals differences among populations obtained from soils with different cropping histories. Applied Soil Ecology 53: 74–82.
24. Zabaloy MC, Garland JL, Gómez MA (2008) An integrated approach to evaluate the impacts of the herbicides glyphosate, 2,4-D and metsulfuron-methyl on soil microbial communities in the Pampas region, Argentina. Applied Soil Ecology 40: 1–12.
25. Zabaloy MC, Gómez E, Garland JL, Gómez MA (2012) Assessment of microbial community function and structure in soil microcosms exposed to glyphosate. Applied Soil Ecology 61: 333–339.

26. Deng W, Xi D, Mao H, Wanapat M (2008) The use of molecular techniques based on ribosomal RNA and DNA for rumen microbial ecosystem studies: a review. Molecular Biology Reports 35: 265–274.

27. Pontes D, Lima-Bittencourt C, Chartone-Souza E, Amaral Nascimento A (2007) Molecular approaches: advantages and artifacts in assessing bacterial diversity. Journal of Industrial Microbiology & Biotechnology 34: 463–473.

28. Sogin ML, Morrison HG, Huber JA, Welch DM, Huse SM, et al. (2006) Microbial diversity in the deep sea and the underexplored "rare biosphere". Proceedings of the National Academy of Sciences 103: 12115–12120.

29. Fortunato CS, Herfort L, Zuber P, Baptista AM, Crump BC (2012) Spatial variability overwhelms seasonal patterns in bacterioplankton communities across a river to ocean gradient. ISME J 6: 554–563.

30. Delmont TO, Prestat E, Keegan KP, Faubladier M, Robe P, et al. (2012) Structure, fluctuation and magnitude of a natural grassland soil metagenome. ISME J 6: 1677–1687.

31. Figuerola ELM, Guerrero LD, Rosa SM, Simonetti L, Duval ME, et al. (2012) Bacterial Indicator of Agricultural Management for Soil under No-Till Crop Production. PLoS ONE 7: e51075.

32. Souza RC, Cantão ME, Vasconcelos ATR, Nogueira MA, Hungria M (2013) Soil metagenomics reveals differences under conventional and no-tillage with crop rotation or succession. Applied Soil Ecology 72: 49–61.

33. Lauber CL, Ramirez KS, Aanderud Z, Lennon J, Fierer N (2013) Temporal variability in soil microbial communities across land-use types. ISME J 7: 1641–1650.

34. Ramirez KS, Craine JM, Fierer N (2012) Consistent effects of nitrogen amendments on soil microbial communities and processes across biomes. Global Change Biology 18: 1918–1927.

35. Ramirez K, Lauber CL, Knight R, Bradford MF (2010) Consistent effects of nitrogen fertilization on soil bacterial communities in contrasting systems. Ecology: 3463–3470

36. Lavado RS (2008) La Región Pampeana: historia, características y uso de sus suelos. In: Alvarez R, editor. Materia Orgánica Valor agronómico y dinámica en suelos pampeanos. Buenos Aires: Editorial Facultad de Ingenieria. pp. 1–11.

37. Berhongaray G, Alvarez R, De Paepe J, Caride C, Cantet R (2013) Land use effects on soil carbon in the Argentine Pampas. Geoderma 192: 97–110.

38. Rascovan N, Carbonetto B, Revale S, Reinert M, Alvarez R, et al. (2013) The PAMPA datasets: a metagenomic survey of microbial communities in Argentinean pampean soils. Microbiome 1: 21.

39. Nelson DW, Sommers LE, Sparks DLE, Page ALE, Helmke PAE, et al. (1996) Total Carbon, Organic Carbon, and Organic Matter. Methods of Soil Analysis Part 3-Chemical Methods: Soil Science Society of America, American Society of Agronomy. pp. 961–1010.

40. Bremner JME, Sparks DLE, Page ALE, Helmke PAE, Loeppert RH (1996) Nitrogen-Total. Methods of Soil Analysis Part 3-Chemical Methods: Soil Science Society of America, American Society of Agronomy. pp. 1085–1121.

41. Kuo SE, Sparks DLE, Page ALE, Helmke PAE, H LR (1996) Phosphorus. Methods of Soil Analysis Part 3-Chemical Methods: Soil Science Society of America, American Society of Agronomy. pp. 869–919.

42. Rhoades JDE, Sparks DLE, Page ALE, Helmke PAE, H. LR (1996) Salinity: Electrical Conductivity and Total Dissolved Solids. Methods of Soil Analysis Part 3-Chemical Methods: Soil Science Society of America, American Society of Agronomy. pp. 417–435.

43. Gee GW, Bauder JWEKA (1986) Particle-size Analysis. Methods of Soil Analysis: Part 1—Physical and Mineralogical Methods: Soil Science Society of America, American Society of Agronomy. pp. 383–411.

44. Cole J, Wang Q, Cardenas E, Fish J, Chai B, et al. (2009) The Ribosomal Database Project: improved alignments and new tools for rRNA analysis. Nucleic acids research: 141–145.

45. Caporaso JG, Kuczynski J, Stombaugh J, Bittinger K, Bushman FD, et al. (2010) QIIME allows analysis of high-throughput community sequencing data. Nat Meth 7: 335–336.

46. Edgar RC (2010) Search and clustering orders of magnitude faster than BLAST. Bioinformatics 26: 2460–2461.

47. Caporaso JG, Bittinger K, Bushman FD, DeSantis TZ, Andersen GL, et al. (2010) PyNAST: a flexible tool for aligning sequences to a template alignment. Bioinformatics 26: 266–267.

48. Price MN, Dehal PS, Arkin AP (2009) FastTree: Computing Large Minimum Evolution Trees with Profiles instead of a Distance Matrix. Molecular Biology and Evolution 26: 1641–1650.

49. Lozupone C, Knight R (2005) UniFrac: a New Phylogenetic Method for Comparing Microbial Communities. Applied and Environmental Microbiology 71: 8228–8235.

50. DeSantis TZ, Hugenholtz P, Larsen N, Rojas M, Brodie EL, et al. (2006) Greengenes, a Chimera-Checked 16S rRNA Gene Database and Workbench Compatible with ARB. Applied and Environmental Microbiology 72: 5069–5072.

51. Wang Q, Garrity GM, Tiedje JM, Cole JR (2007) Naïve Bayesian Classifier for Rapid Assignment of rRNA Sequences into the New Bacterial Taxonomy. Applied and Environmental Microbiology 73: 5261–5267.

52. Kindt R, Coe R (2005) Tree Diversity Analysis: A Manual and Software for Common Statistical Methods for Ecological and Biodiversity Studies: World Agroforestry Centre.

53. Dixon P (2003) VEGAN, a package of R functions for community ecology. Journal of Vegetation Science 14: 927–930.

54. Meyer F, Paarmann D, D'Souza M, Olson R, Glass EM, et al. (2008) The metagenomics RAST server - a public resource for the automatic phylogenetic and functional analysis of metagenomes. BMC Bioinformatics 9: 386.

55. Fierer N, Bradford MA, Jackson RB (2007) Toward an ecological classification of soil bacteria. Ecology 88: 1354–1364.

56. Dion P, Nautiyal CS (2008) Microbiology of Extreme Soils: Springer.

57. Galantini J, Rosell R (2006) Long-term fertilization effects on soil organic matter quality and dynamics under different production systems in semiarid Pampean soils. Soil and Tillage Research 87: 72–79.

58. Lauro FM, McDougald D, Thomas T, Williams TJ, Egan S, et al. (2009) The genomic basis of trophic strategy in marine bacteria. Proceedings of the National Academy of Sciences 106: 15527–15533.

59. Nemergut DR, Townsend AR, Sattin SR, Freeman KR, Fierer N, et al. (2008) The effects of chronic nitrogen fertilization on alpine tundra soil microbial communities: implications for carbon and nitrogen cycling. Environmental Microbiology 10: 3093–3105.

60. Zhang H, Sekiguchi Y, Hanada S, Hugenholtz P, Kim H, et al. (2003) Gemmatimonas aurantiaca gen. nov., sp. nov., a Gram-negative, aerobic, polyphosphate-accumulating micro-organism, the first cultured representative of the new bacterial phylum Gemmatimonadetes phyl. nov. International Journal of Systematic and Evolutionary Microbiology 53: 1155–1163.

61. DeBruyn JM, Nixon LT, Fawaz MN, Johnson AM, Radosevich M (2011) Global Biogeography and Quantitative Seasonal Dynamics of Gemmatimonadetes in Soil. Applied and Environmental Microbiology 77: 6295–6300.

62. Maixner F, Noguera DR, Anneser B, Stoecker K, Wegl G, et al. (2006) Nitrite concentration influences the population structure of Nitrospira-like bacteria. Environmental Microbiology 8: 1487–1495.

63. Attard E, Poly F, Commeaux C, Laurent F, Terada A, et al. (2010) Shifts between Nitrospira- and Nitrobacter-like nitrite oxidizers underlie the response of soil potential nitrite oxidation to changes in tillage practices. Environmental Microbiology 12: 315–326.

64. Edwards AL, Reyes FE, Héroux A, Batey RT (2010) Structural basis for recognition of S-adenosylhomocysteine by riboswitches. RNA 16: 2144–2155.

65. Loenen WA (2006) S-adenosylmethionine: jack of all trades and master of everything? Biochem Soc Trans 34: 330–333.

66. Winkler WC, Breaker RR (2005) Regulation of bacterial gene expression by riboswitches. Annual Review of Microbiology 59: 487–517.

67. Nudler E, Mironov AS (2004) The riboswitch control of bacterial metabolism. Trends in biochemical sciences 29: 11–17.

68. Sañudo-Wilhelmy SA, Cutter LS, Durazo R, Smail EA, Gómez-Consarnau L, et al. (2012) Multiple B-vitamin depletion in large areas of the coastal ocean. Proceedings of the National Academy of Sciences.

69. Yan D (2007) Protection of the glutamate pool concentration in enteric bacteria. Proceedings of the National Academy of Sciences 104: 9475–9480.

70. Doucette CD, Schwab DJ, Wingreen NS, Rabinowitz JD (2011) α-ketoglutarate coordinates carbon and nitrogen utilization via enzyme I inhibition. Nat Chem Biol 7: 894–901.

71. Cai L, Tu BP (2011) On Acetyl-CoA as a Gauge of Cellular Metabolic State. Cold Spring Harbor Symposia on Quantitative Biology 76: 195–202.

72. Leonardi R, Rehg JE, Rock CO, Jackowski S (2010) Pantothenate Kinase 1 Is Required to Support the Metabolic Transition from the Fed to the Fasted State. PLoS ONE 5: e11011.

73. Elbein AD, Pan YT, Pastuszak I, Carroll D (2003) New insights on trehalose: a multifunctional molecule. Glycobiology 13: 17R–27R.

74. Barabote RD, Xie G, Leu DH, Normand P, Necsulea A, et al. (2009) Complete genome of the cellulolytic thermophile Acidothermus cellulolyticus 11B provides insights into its ecophysiological and evolutionary adaptations. Genome Research 19: 1033–1043.

75. Lopez MF, Fontaine MS, Torrey JG (1984) Levels of trehalose and glycogen in Frankia sp. HFPArI3 (Actinomycetales). Canadian Journal of Microbiology 30: 746–752.

76. Tropis M, Meniche X, Wolf A, Gebhardt H, Strelkov S, et al. (2005) The Crucial Role of Trehalose and Structurally Related Oligosaccharides in the Biosynthesis and Transfer of Mycolic Acids in Corynebacterineae. Journal of Biological Chemistry 280: 26573–26585.

77. Fierer N, Leff JW, Adams BJ, Nielsen UN, Bates ST, et al. (2012) Cross-biome metagenomic analyses of soil microbial communities and their functional attributes. Proceedings of the National Academy of Sciences 109: 21390–21395.

78. Taylor PG, Townsend AR (2010) Stoichiometric control of organic carbon-nitrate relationships from soils to the sea. Nature 464: 1178–1181.

79. Apple JK, del Giorgio PA (2007) Organic substrate quality as the link between bacterioplankton carbon demand and growth efficiency in a temperate salt-marsh estuary. ISME J 1: 729–742.

80. Bayer C, Mielniczuk J, Amado TJC, Martin-Neto L, Fernandes SV (2000) Organic matter storage in a sandy clay loam Acrisol affected by tillage and cropping systems in southern Brazil. Soil and Tillage Research 54: 101–109.

81. Langdale GW, West LT, Bruce RR, Miller WP, Thomas AW (1992) Restoration of eroded soil with conservation tillage. Soil Technology 5: 81–90.

82. Bongiovanni MD, Lobartini JC (2006) Particulate organic matter, carbohydrate, humic acid contents in soil macro- and microaggregates as affected by cultivation. Geoderma 136: 660–665.

83. Plaza-Bonilla D, Cantero-Martínez C, Viñas P, Álvaro-Fuentes J (2013) Soil aggregation and organic carbon protection in a no-tillage chronosequence under Mediterranean conditions. Geoderma 193–194: 76–82.

84. Jiao Y, Whalen JK, Hendershot WH (2006) No-tillage and manure applications increase aggregation and improve nutrient retention in a sandy-loam soil. Geoderma 134: 24–33.

85. Tivet F, de Moraes Sá JC, Lal R, Borszowskei PR, Briedis C, et al. (2013) Soil organic carbon fraction losses upon continuous plow-based tillage and its restoration by diverse biomass-C inputs under no-till in sub-tropical and tropical regions of Brazil. Geoderma 209–210: 214–225.

86. Slepetiene A, Slepetys J (2005) Status of humus in soil under various long-term tillage systems. Geoderma 127: 207–215.

87. Dong L, Córdova-Kreylos AL, Yang J, Yuan H, Scow KM (2009) Humic acids buffer the effects of urea on soil ammonia oxidizers and potential nitrification. Soil Biology and Biochemistry 41: 1612–1621.

88. Sieber JR, McInerney MJ, Gunsalus RP (2012) Genomic Insights into Syntrophy: The Paradigm for Anaerobic Metabolic Cooperation. Annual Review of Microbiology 66: 429–452.

89. McInerney MJ, Sieber JR, Gunsalus RP (2009) Syntrophy in anaerobic global carbon cycles. Current Opinion in Biotechnology 20: 623–632.

90. Kim HJ, Boedicker JQ, Choi JW, Ismagilov RF (2008) Defined spatial structure stabilizes a synthetic multispecies bacterial community. Proceedings of the National Academy of Sciences 105: 18188–18193.

Permissions

List of Contributors

Peter A. Lindsey
Panthera, New York, New York, United States of America
Mammal Research Institute, Department of Zoology and Entomology, University of Pretoria, Pretoria, Gauteng, South Africa

Vincent R. Nyirenda
Zambia Wildlife Authority, Chilanga, Lusaka, Zambia

Jonathan I. Barnes
Design & Development Services, Windhoek, Namibia

Matthew S. Becker
Department of Ecology, Montana State University, Bozeman, Montana, United States of America
Zambian Carnivore Programme, Mfuwe, Zambia

Rachel McRobb
South Luangwa Conservation Society, Mfuwe, Zambia

Craig J. Tambling
Centre for African Conservation Ecology, Department of Zoology, Nelson Mandela Metropolitan University, Port Elizabeth, South Africa

W. Andrew Taylor
Centre for Veterinary Wildlife Studies, Faculty of Veterinary Science, University of Pretoria, Pretoria, South Africa

Frederick G. Watson
Zambian Carnivore Programme, Mfuwe, Zambia
Division of Science and Environmental Policy, California State University Monterey Bay, Seaside, California, United States of America

Michael t'Sas-Rolfes
Cape Town, South Africa

Weston Anderson
Department of Geography and Environmental Engineering, The Johns Hopkins University, Baltimore, Maryland, United States of America
International Food Policy Research Institute, Washington, D.C., United States of America

Seth Guikema
Department of Geography and Environmental Engineering, The Johns Hopkins University, Baltimore, Maryland, United States of America

Ben Zaitchik
Department of Earth and Planetary Sciences, The Johns Hopkins University, Baltimore, Maryland, United States of America

William Pan
Nicholas School of Environment and Duke Global Health Institute, Duke University, Durham, North Carolina, United States of America

Miao Liu
State Key Laboratory of Forest and Soil Ecology, Institute of Applied Ecology, Chinese Academy of Sciences, Shenyang, China

Yanyan Xu
State Key Laboratory of Forest and Soil Ecology, Institute of Applied Ecology, Chinese Academy of Sciences, Shenyang, China
University of Chinese Academy of Sciences, Beijing, China

Yuanman Hu
State Key Laboratory of Forest and Soil Ecology, Institute of Applied Ecology, Chinese Academy of Sciences, Shenyang, China

Chunlin Li
State Key Laboratory of Forest and Soil Ecology, Institute of Applied Ecology, Chinese Academy of Sciences, Shenyang, China,
University of Chinese Academy of Sciences, Beijing, China

Fengyun Sun
State Key Laboratory of Forest and Soil Ecology, Institute of Applied Ecology, Chinese Academy of Sciences, Shenyang, China
University of Chinese Academy of Sciences, Beijing, China

Tan Chen
State Key Laboratory of Forest and Soil Ecology, Institute of Applied Ecology, Chinese Academy of Sciences, Shenyang, China
University of Chinese Academy of Sciences, Beijing, China

Francy Junio Gonçalves Lisboa
Soil Science Department, Agronomy Institute, Federal Rural University of Rio de Janeiro, Seropédica-RJ, Brazil

The James Hutton Institute, Craigiebuckler, Aberdeen, United Kingdom

Pedro R. Peres-Neto
Canada Research Chair in Spatial Modelling and Biodiversity; Université du Québec à Montréal, Département des sciences biologiques, Québec, Canada

Guilherme Montandon Chaer
Embrapa Agrobiologia, Seropédica-RJ, Brazil

Ederson da Conceição Jesus
Embrapa Agrobiologia, Seropédica-RJ, Brazil

Ruth Joy Mitchell
The James Hutton Institute, Craigiebuckler, Aberdeen, United Kingdom

Stephen James Chapman
The James Hutton Institute, Craigiebuckler, Aberdeen, United Kingdom

Ricardo Luis Louro Berbara
Soil Science Department, Agronomy Institute, Federal Rural University of Rio de Janeiro, Seropédica-RJ, Brazil

Katya E. Kovalenko
University of Windsor, Windsor, Ontario, Canada
Natural Resources Research Institute, University of Minnesota Duluth, Duluth, Minnesota, United States of America

Valerie J. Brady
Natural Resources Research Institute, University of Minnesota Duluth, Duluth, Minnesota, United States of America

Jan J. H. Ciborowski
University of Windsor, Windsor, Ontario, Canada

Sergey Ilyushkin
Colorado School of Mines, Golden, Colorado, United States of America

Lucinda B. Johnson
Natural Resources Research Institute, University of Minnesota Duluth, Duluth, Minnesota, United States of America

Misha Leong, Cla0ire Kremen and George K. Roderick
Department of Environmental Science Policy and Management, University of California, Berkeley, California, United States of America

Wei-Ta Fang
Graduate Institute of Environmental Education, National Taiwan Normal University, Taipei, Taiwan, ROC

Jui-Yu Chou
Department of Biology, National Changhua University of Education, Changhua, Taiwan, ROC

Shiau-Yun Lu
Department of Marine Environment and Engineering, National Sun Yat-sen University, Kaohsiung, Taiwan, ROC

Konstantina Zografou
Department of Biological Applications and Technologies, University of Ioannina, Ioannina, Greece

Vassiliki Kati
Department of Environmental and Natural Resources Management, University of Patras, Seferi, Agrinio, Greece

Andrea Grill
Department of Tropical Ecology and Animal Biodiversity, University of Vienna, Rennweg, Vienna, Austria

Department of Organismic Biology, University of Salzburg, Hellbrunnerstraße, Salzburg, Austria

Robert J. Wilson
Centre for Ecology and Conservation, University of Exeter Cornwall Campus, Penryn, United Kingdom, 6 P.O. Box 1220, Larissa, Greece

Elli Tzirkalli
Department of Biological Applications and Technologies, University of Ioannina, Ioannina, Greece

Lazaros N. Pamperis
6 P.O. Box 1220, Larissa, Greece

John M. Halley
Department of Biological Applications and Technologies, University of Ioannina, Ioannina, Greece

Dongmei He and Honghua Ruan
Faculty of Forest Resources and Environmental Science, and Key Laboratory of Forestry and Ecological Engineering of Jiangsu Province, Nanjing Forestry University, Nanjing, Jiangsu, China

Eric M. Gese
United States Department of Agriculture, Wildlife Services, National Wildlife Research Center, Department of Wildland Resources, Utah State University, Logan, Utah, United States of America

Craig M. Thompson
Department of Wildland Resources, Utah State University, Logan, Utah, United States of America

Vanessa M. Adams
Australian Research Council Centre of Excellence for Coral Reef Studies, James Cook University, Townsville, Queensland, Australia
Research Institute for the Environment and Livelihoods and Northern Australia National Environmental Research Program Hub, Charles Darwin University, Darwin, Northern Territory, Australia

Robert L. Pressey
Australian Research Council Centre of Excellence for Coral Reef Studies, James Cook University, Townsville, Queensland, Australia

Natalie Stoeckl
School of Business and Cairns Institute, James Cook University, Townsville, Queensland, Australia

Richard Schuster
Department of Forest and Conservation Sciences, University of British Columbia, Vancouver, British Columbia, Canada

Tara G. Martin
Department of Forest and Conservation Sciences, University of British Columbia, Vancouver, British Columbia, Canada
Ecosciences Precinct, CSIRO Ecosystem Sciences, Brisbane, Queensland, Australia

Peter Arcese1
Department of Forest and Conservation Sciences, University of British Columbia, Vancouver, British Columbia, Canada

Martin Patenaude-Monette
Groupe de recherche en écologie comportementale et animale, Département des sciences biologiques, Université du Québec à Montréal, Montréal, Québec, Canada

Marc Bélisle
Département de biologie, Université de Sherbrooke, Sherbrooke, Québec, Canada

Jean-François Giroux
Groupe de recherche en écologie comportementale et animale, Département des sciences biologiques, Université du Québec é Montréal, Montréal, Québec, Canada

Fazhu Zhao, Gaihe Yang, Xinhui Han, Yongzhong Feng and Guangxin Ren
College of Agronomy, Northwest A&F University, Yangling, Shaanxi, China; and The Research Center of Recycle Agricultural Engineering and Technology of Shaanxi

Province, Yangling, Shaanxi, China

Maria Julia de Lima Brossi
Cellular and Molecular Biology Laboratory, Center for Nuclear Energy in Agriculture, University of São Paulo, Piracicaba, SP, Brazil

Lucas William Mendes
Cellular and Molecular Biology Laboratory, Center for Nuclear Energy in Agriculture, University of São Paulo, Piracicaba, SP, Brazil

Mariana Gomes Germano

Brazilian Agricultural Research

Corporation, Embrapa Soybean, Londrina, PR, Brazi

Amanda Barbosa Lima
Cellular and Molecular Biology Laboratory, Center for Nuclear Energy in Agriculture, University of São Paulo, Piracicaba, SP, Brazil

Siu Mui Tsai
Cellular and Molecular Biology Laboratory, Center for Nuclear Energy in Agriculture, University of São Paulo, Piracicaba, SP, Brazil

Scott Devine
Warnell School of Forestry and Natural Resources, The University of Georgia, Athens, Georgia, United States of America

Daniel Markewitz
Warnell School of Forestry and Natural Resources, The University of Georgia, Athens, Georgia, United States of America

Paul Hendrix
Odum School of Ecology, The University of Georgia, Athens, Georgia, United States of America

David Coleman
Odum School of Ecology, The University of Georgia, Athens, Georgia, United States of America

Christoph D. D. Rupprecht
Environmental Futures Research Institute, Griffith University, Nathan, Queensland, Australia
Griffith School of Environment, Griffith University, Griffith University, Queensland, Australia

Jason A. Byrne
Environmental Futures Research Institute, Griffith University, Nathan, Queensland, Australia
Griffith School of Environment, Griffith University, Griffith University, Queensland, Australia

Erin Coulter Riordan and Philip W. Rundel
Department of Ecology and Evolutionary Biology, University of California Los Angeles, Los Angeles, California, United States of America

Jabi Zabala
Department of Zoology and Animal Cell Biology, Faculty of Science and Technology (UPV/EHU), Bilbao, Spain
Sebero Otxoa, 45, 5 B. 48480 Arrigorriaga, Biscay, Spain

Beatriz Dı´az
Department of Zoology and Animal Cell Biology, Faculty of Science and Technology (UPV/EHU), Bilbao, Spain
Department of Entomology, Aranzadi Science Society, San Sebastia´n, Guipu´ zcoa, Spain

Marta I. Saloña-Bordas
Department of Zoology and Animal Cell Biology, Faculty of Science and Technology (UPV/EHU), Bilbao, Spain

Toshikazu Kizuka
Center for Environmental Biology and Ecosystem Studies, National Institute for Environmental Studies, Tsukuba, Ibaraki, Japan

Munemitsu Akasaka
Center for Environmental Biology and Ecosystem Studies, National Institute for Environmental Studies, Tsukuba, Ibaraki, Japan
Faculty of Agriculture, Tokyo University of Agriculture and Technology, Fuchu, Tokyo, Japan

Taku Kadoya
Center for Environmental Biology and Ecosystem Studies, National Institute for Environmental Studies, Tsukuba, Ibaraki, Japan

Noriko Takamura
Center for Environmental Biology and Ecosystem Studies, National Institute for Environmental Studies, Tsukuba, Ibaraki, Japan

Ya-Lin Hu
State Key Laboratory of Forest and Soil Ecology, Institute of Applied Ecology, Chinese Academy of Sciences, Shenyang, People's Republic of China

Li-Le Hu
Chinese Research Academy of Environmental Sciences, Beijing, People's Republic of China

De-Hui Zeng
State Key Laboratory of Forest and Soil Ecology, Institute of Applied Ecology, Chinese Academy of Sciences, Shenyang, People's Republic of China

Belén Carbonetto
Instituto de Agrobiotecnología de Rosario (INDEAR), Predio CCT Rosario, Santa Fe, Argentina

Nicolás Rascovan
Instituto de Agrobiotecnología de Rosario (INDEAR), Predio CCT Rosario, Santa Fe, Argentina

Roberto Álvarez
Facultad de Agronomía, Universidad de Buenos Aires, Buenos Aires, Argentina

Alejandro Mentaberry
Departamento de Fisiología y Biología Molecular y Celular, Facultad de Ciencias Exactas y Naturales, Universidad de Buenos Aires, Buenos Aires, Argentina

Martin P. Vázquez
Instituto de Agrobiotecnología de Rosario (INDEAR), Predio CCT Rosario, Santa Fe, Argentina

Index